VOLUME NINETY ONE

SEMICONDUCTORS AND SEMIMETALS

Defects in Semiconductors

SERIES EDITORS

VOLUME NINETY ONE

SEMICONDUCTORS AND SEMIMETALS

Defects in Semiconductors

Edited by

LUCIA ROMANO
Department of Physics and Astronomy,
University of Catania,
Catania, Italy

VITTORIO PRIVITERA
National Research Council (CNR),
Institute of Microelectronics and Microsystems,
Catania, Italy

CHENNUPATI JAGADISH
Research School of Physics and Engineering,
Australian National University,
Canberra, Australia

ELSEVIER

AMSTERDAM • BOSTON • HEIDELBERG • LONDON
NEW YORK • OXFORD • PARIS • SAN DIEGO
SAN FRANCISCO • SINGAPORE • SYDNEY • TOKYO
Academic Press is an imprint of Elsevier

Academic Press is an imprint of Elsevier
225 Wyman Street, Waltham, MA 02451, USA
525 B Street, Suite 1800, San Diego, CA 92101-4495, USA
125 London Wall, London, EC2Y 5AS, UK
The Boulevard, Langford Lane, Kidlington, Oxford OX5 1GB, UK

First edition 2015

ISBN: 978-0-12-801935-1
ISSN: 0080-8784

For information on all Academic Press publications
visit our website at store.elsevier.com

CONTENTS

CONTRIBUTORS

Jodie E. Bradby
Department of Electronic Materials Engineering, Research School of Physics and Engineering, Australian National University, Canberra, Australian Capital Territory, Australia. (ch5)

Daniela Cavalcoli
Department of Physics and Astronomy, University of Bologna, Bologna, Italy. (ch7)

Anna Cavallini
Department of Physics and Astronomy, University of Bologna, Bologna, Italy. (ch7)

Beatrice Fraboni
Department of Physics and Astronomy, University of Bologna, Bologna, Italy. (ch7)

Bianca Haberl
Department of Electronic Materials Engineering, Research School of Physics and Engineering, Australian National University, Canberra, Australian Capital Territory, Australia, and Chemical and Engineering Materials Division, Oak Ridge National Laboratory, Oak Ridge, Tennessee, USA. (ch5)

Giuliana Impellizzeri
CNR-IMM MATIS, Catania, Italy. (ch3)

Naoya Iwamoto
University of Oslo, Physics Department, Center for Materials Science and Nanotechnology, Oslo, Norway. (ch10)

Mangalampalli S.R.N. Kiran
Department of Electronic Materials Engineering, Research School of Physics and Engineering, Australian National University, Canberra, Australian Capital Territory, Australia. (ch5)

Arne Nylandsted Larsen
Department of Physics and Astronomy/iNANO, Aarhus University, Aarhus, Denmark. (ch2)

Johan Lauwaert
Department Solid State Sciences, Ghent University, Gent, Belgium. (ch6)

Aaron G. Lind
Department of Materials Science and Engineering, University of Florida, Gainesville, Florida, USA. (ch4)

Thomas P. Martin
Department of Materials Science and Engineering, University of Florida, Gainesville, Florida, USA. (ch4)

Matthew D. McCluskey
Department of Physics & Astronomy, Washington State University, Pullman, Washington, USA. (ch8)

Abdelmadjid Mesli
Aix-Marseille Université, CNRS, Marseille Cedex, France. (ch2)

Enrico Napolitani
Dipartimento di Fisica e Astronomia, Università di Padova, Padova, and CNR-IMM MATIS, Catania, Italy. (ch3)

Peter Pichler
Technology Simulation, Fraunhofer Institute for Integrated Systems and Device Technology IISB, and University of Erlangen-Nuremberg, Erlangen, Germany. (ch1)

Michael A. Reshchikov
Department of Physics, Virginia Commonwealth University, Richmond, Virginia, USA. (ch9)

Nicholas G. Rudawski
Major Analytical Instrumentation Center, University of Florida, Gainesville, Florida, USA. (ch4)

Eddy Simoen
Department Solid State Sciences, Ghent University, Gent, Belgium. (ch6)

Bengt G. Svensson
University of Oslo, Physics Department, Center for Materials Science and Nanotechnology, Oslo, Norway. (ch10)

Henk Vrielinck
Department Solid State Sciences, Ghent University, Gent, Belgium. (ch6)

James S. Williams
Department of Electronic Materials Engineering, Research School of Physics and Engineering, Australian National University, Canberra, Australian Capital Territory, Australia. (ch5)

PREFACE

One sees qualities at a distance
and defects at close range
Victor Hugo

A crystal is a solid where the atoms form a periodic arrangement. Thermodynamically, a perfect crystal does not exist above 0 K, so point defects—places where the crystal's pattern is interrupted—will always be present in a crystal. Complexes can form between different kinds of point defects. The types and structures of these defects may have a profound effect on the properties of the materials.

Defects in semiconductors play a crucial role in determining the performance of electronic and photonic devices. Understanding the role of defects is crucial to explain several phenomena, from diffusion to gettering, or to draw theories on the materials' behavior, in response to electrical, optical, or mechanical fields.

Substitutional dopants form mobile pairs with the intrinsic point defects, i.e., vacancies and self-interstitials. During past decades, the majority of the defects and the mechanisms of their formation were elucidated with concurrent efforts in eliminating the unwanted defects. Models are now available to explain a variety of phenomena like dopant profile shapes; enhanced dopant diffusion; nonequilibrium effects caused by chemical reactions or irradiation damage; immobilization and reduced electrical activation of dopants via the formation of impurity phases, small clusters, and complexes with other impurities; and, finally, the pileup of dopants at interfaces and surfaces. The current state of knowledge about the actual diffusion mechanisms of dopants in silicon and germanium, processes - ion implantation and electron and proton irradiation - that perturb the intrinsic point defects, the formation of impurity phases, clusters, and complexes as well as associated effects on the intrinsic point defects, are presented in Chapters 1–3 of this book. Defects can play crucial role during phase transitions and contribute to develop new material phases; Chapter 4 reviews the origins of defects produced during the solid-phase regrowth of Si and the influence on resulting device performance. Nanoindentation (Chapter 5) can be used to study the deformation behavior of Si and Ge and their pressure-induced metastable phases, which can be of interest in the search, for example, for new semiconductor and superconducting behaviors.

Characterizations play a central role in a materials study, analytical techniques for the detection of electrically active defects in semiconductor materials, the operation principles, the strengths, and the weaknesses are outlined and illustrated in Chapter 6. Surface photovoltage spectroscopy (Chapter 7) allows the detection of electronic transitions (band-to-band, defect-band, and surface state-bands) on a huge range of semiconductors.

Silicon is the most studied and applied semiconductor, and even if there are still a lot of lacking answers about its physics, its research has been propellant to improve the know-how about semiconductor materials, their properties, and applications. The analytical techniques and the modeling developed for Si turned out to be very useful to characterize a variety of other materials in several fields of application. This book would like to cover the role of defects in various semiconductors that are widely used in industry and that can lead to future innovations. Therefore, we broaden our interest about germanium and some compound semiconductors such as ZnO, GaN, and SiC. The ZnO literature is vast and often contradictory. The purpose of Chapter 8 is to summarize reasonably well-established results on point defects in ZnO. The concentration of point defects in GaN is still relatively high. Point defects affect the performance of light-emitting devices and are also the main obstacle hindering the realization of high-power electronic devices. In Chapter 9, first-principles calculations are compared with the results from different experimental techniques in order to investigate the role of point defects in GaN. With the advancement in materials growth and increasing level of sophistication, point defects, dopants, impurities, as well as extended structural defects have evolved as crucial issues within the SiC community. Chapter 10 reviews recent progress in the understanding and control of the silicon and the carbon point defects, antisite defects, and hydrogen and transition metal impurities.

This book is aimed at researchers and students working on defects in semiconductors and book chapters were written by leading experts in the field. This book helps to define the field and prepares students for working in technologically important areas. It provides students with a solid foundation in both experimental methods and the theory of defects in semiconductors.

Lucia Romano
Vittorio Privitera
Chennupati Jagadish
Editors

CHAPTER ONE

Role of Defects in the Dopant Diffusion in Si

Peter Pichler[1]
Technology Simulation, Fraunhofer Institute for Integrated Systems and Device Technology IISB, Erlangen, Germany
University of Erlangen-Nuremberg, Erlangen, Germany
[1]Corresponding author: e-mail address: peter.pichler@iisb.fraunhofer.de

Contents

1. INTRODUCTION

In semiconductors, dopants reside predominantly on substitutional sites where they either provide free electrons (donors) or bind them (acceptors) to complete the valence-bond structure. The most successful concepts developed to describe dopant diffusion assume that the substitutional dopants form mobile pairs with the intrinsic point defects, i.e., vacancies and self-interstitials. These models allow to explain a variety of phenomena like different profile shapes observed for short and long diffusion times; the dependence of the profile form on the concentration of dopants; enhanced

Semiconductors and Semimetals, Volume 91
ISSN 0080-8784
http://dx.doi.org/10.1016/bs.semsem.2014.11.001

dopant diffusion below regions with high dopant concentration; non-equilibrium effects caused by chemical reactions like oxidation or nitridation at surfaces; immobilization and reduced electrical activation of dopants via the formation of impurity phases, small clusters and complexes with other impurities; and, finally, the pile-up of dopants at interfaces and surfaces. Due to the limited space, citation can be only exemplary. For a more extensive account of diffusion phenomena, the interested reader is referred to specific reviews in this field (Fahey et al., 1989; Pichler, 2004).

This chapter is structured as follows: In the first section, a methodology is explained which is commonly used in continuum simulation to describe the diffusion of dopants, intrinsic point defects, and other impurities as well as their interactions via coupled systems of continuity equations. In the following section, the diffusion of dopants via intrinsic point defects is discussed. This includes a review of the basic diffusion mechanisms, a derivation of the diffusion equations on the basis that dopant diffusion proceeds via a pair diffusion mechanism, and a discussion of the system behavior in terms of diffusion phenomena and diffusion profiles to be expected. The current state of knowledge about the actual diffusion mechanisms of dopants in silicon is summarized thereafter. In the subsequent section, processes are outlined that perturb the intrinsic point defects and lead to a variety of diffusion phenomena. Thereafter, the formation of impurity phases, clusters and complexes as well as associated effects on the intrinsic point defects are discussed. The chapter ends with an outline of interface segregation, a phenomenon that may lead to the loss of a substantial fraction of the dopants in a sample.

2. THE FRAMEWORK OF DIFFUSION–REACTION EQUATIONS

While pairing and dissolution reactions as well as migration of all kinds of point defects can be implemented directly in kinetic Monte Carlo approaches (see, e.g., Jaraiz, 2004), an indirect approach is required for continuum simulation. One such approach is to consider a number of point-like species, their diffusion, and possible reactions between them. Species in this sense refers to simple point defects like vacancies and self-interstitials as intrinsic point defects as well as dopant atoms on substitutional sites or other impurity atoms, but also to complexes between dopants and impurities with intrinsic point defect as well as clusters comprising dopants, intrinsic point defects, and other impurities. In the following, the framework of diffusion–reaction equations is briefly outlined. This framework is used in the subsequent sections to explain phenomena associated with the diffusion of dopants and typical forms of diffusion profiles. For a full account, the interested reader is referred to more extensive reviews in the field (e.g., Pichler, 2004, section 1.5).

Within the framework of diffusion–reaction equations, for each of the species considered, a continuity equation is solved. For the diffusion and reaction of species A, as an example, it would read

$$\frac{\partial C_A}{\partial t} = -\operatorname{div}(J_A) + R_A \tag{1}$$

with the flux J_A given for diffusion in an electrostatic field E by

$$J_A = -D_A \cdot \operatorname{grad} C_A - z_A \cdot \mu_A \cdot C_A \cdot E. \tag{2}$$

The terms t, C_A, R_A, D_A, and μ_A stand for time, concentration, a reaction term accounting for generation and loss due to quasi-chemical reactions, the diffusion coefficient and the mobility of the species, and div and grad are the divergence and gradient operators. The mobility is related to the diffusion coefficient by the Einstein relation $D_A/\mu_A = k \cdot T/q$ with k and q denoting Boltzmann's constant and elementary charge, respectively. In the tradition of early reviews in this field (e.g., Fair, 1981; Fichtner, 1983; Tsai, 1983; Willoughby, 1981), the charge state z_A has been defined here as the number of electrons associated (e.g., +1 for a singly negatively charged defects like ionized acceptors, −1 for a singly positively charged defect like an ionized donor, −2 for a doubly positively charged defect). It should be noted, though, that an association of the charge state with positive charges is likewise common (e.g., Fahey et al., 1989) and would manifest itself in a positive sign of the field term. While the definition of the charge state may not always be immediately apparently, it is easy to find it out from the equality (number of negative charges) or inequality (number of positive charges) of the signs of diffusion and field term. Written in terms of the electrostatic potential Ψ related to the electric field by $E = -\operatorname{grad} \Psi$, the diffusion flux (2) takes the familiar form

$$J_A = -D_A \cdot \operatorname{grad} C_A + z_A \cdot D_A \cdot C_A \cdot \operatorname{grad}\left(\frac{\Psi}{U_T}\right) \tag{3}$$

with the thermal voltage U_T introduced as abbreviation for $U_T = k \cdot T/q$.

The effects of the quasi-chemical reactions between the species considered are comprised in the reaction term R_A. In the following, to illustrate how quasi-chemical reactions between species can be taken into consideration within the framework of diffusion–reaction equations, let us consider reactions in the form

$$|\nu_A|A + |\nu_B|B \underset{k_{\leftarrow}}{\overset{k_{\rightarrow}}{\rightleftharpoons}} \nu_C C + \nu_D D \tag{4}$$

with the forward and backward reaction rates denoted by $k_{\rightarrow}$ and $k_{\leftarrow}$, respectively. The stoichiometric numbers ν denote how many of the respective species participate in the reaction. By definition, stoichiometric numbers appearing on the left-hand side are negative. Therefore, their absolute values were used in (4) for the sake of consistency. In chemistry, the concentrations of the species involved are usually given in the form of mole fractions. In crystals, it is more convenient to use site fractions $x = C/C^S$ defined as concentration C divided by the concentration of sites C^S for this defect in the lattice. For vacancies, as an example, C^S corresponds to the concentration of lattice sites C_{Si}. For bond-centered interstitial defects, as another example, the concentration of possible sites is twice that of lattice sites since there are four around each lattice atoms, which are shared among two neighboring atoms. Assuming ideally dilute concentrations so that the respective activity coefficients are unity, the site fractions of the defects are related to each other in equilibrium via the law of mass action

$$K = \prod_i x_i^{\nu_i} = \frac{x_C^{\nu_C} \cdot x_D^{\nu_D}}{x_A^{|\nu_A|} \cdot x_B^{|\nu_B|}} = \prod_i \theta_i \cdot \exp\left(-\frac{1}{k \cdot T} \cdot \sum_i \nu_i \cdot G_i^f\right) \tag{5}$$

with K denoting the equilibrium constant of the reaction. The θ_i stand for the—often neglected—numbers of geometrically equivalent and distinguishable configurations of defect i at a specific site, and G_i^f for the formation energy of the respective defect. When the result of a reaction is a single defect (e.g., C), the difference between its formation energy and the formation energies of the species from which it is formed can be seen as binding energy of the defect. Since "binding" corresponds to a lowering of the system energy upon formation of the defect and is associated with a positive value of the binding energy, it will be defined here as

$$G^B = G_A^f + G_B^f - G_C^f = -\sum_i \nu_i \cdot G_i^f. \tag{6}$$

However, it should be noted that there is no general consensus for the usage of the terms formation energy and binding energy in the literature.

It should be pointed out that only differences in formation energies are relevant for the right-hand side of (5). This leaves some freedom to define reference points for the formation energies. For dopants, it is customary to associate the reference state with a vanishing formation energy to the ionized, substitutional configuration. For electrons e^- and holes p^+, the formation energies are the Fermi level E_F and $-E_F$, respectively. Within the limits of Boltzmann statistics, the resulting contributions to the equilibrium constant of the reaction can be associated to the electron and hole concentrations n and p via

$$\frac{n}{n_i} = \exp\left(\frac{E_F - E_i}{k \cdot T}\right) \text{ and } \frac{p}{n_i} = \frac{n_i}{n} = \exp\left(\frac{E_i - E_F}{k \cdot T}\right) \tag{7}$$

with n_i and E_i corresponding to the charge carrier concentration and the Fermi level in intrinsically doped materials, respectively. For extrinsically doped semiconductors with dopant concentrations exceeding n_i, the Fermi level will move from E_i toward the valence or conduction band. Considering that the band edges contain the energy $-q \cdot \Psi$ of the electrons, the reference state can be selected in a way so that a vanishing electrostatic potential represents the intrinsic case. This allows to reformulate (7) in terms of the electrostatic potential

$$\frac{n}{n_i} = \exp\left(\frac{\Psi}{U_T}\right) \text{ and } \frac{p}{n_i} = \frac{n_i}{n} = \exp\left(-\frac{\Psi}{U_T}\right). \tag{8}$$

For self-interstitials and vacancies, G_I^f and G_V^f are usually associated with the energy needed to form one of them at a specific site under conservation of the number of host atoms in the system. The equilibrium concentration of such an intrinsic point defect X^i (X may stand for a vacancy V or a self-interstitial I) in charge state i follows formally from the reaction $0 \rightleftharpoons X^i + i\,p^+$ with the "0" symbolizing an undisturbed lattice as

$$C_{X^i}^{eq} = C_{X^i}^{S} \cdot \theta_{X^z} \cdot \exp\left(-\frac{G_{X^i}^f - i \cdot E_F}{k \cdot T}\right). \tag{9}$$

Considering that the relationship between the concentrations of a defect in particular charge states follows for any defect X (intrinsic point defects, impurities, and any complexes) formally from the reaction

$$X^j \rightleftharpoons X^i + \left(z_j - z_i\right) p^+ \tag{10}$$

in the form

$$\frac{C_{X^j}}{C_{X^i}} = \frac{C^S_{X^j}}{C^S_{X^i}} \cdot \frac{\theta_{X^j}}{\theta_{X^i}} \cdot \exp\left(-\frac{G^f_{X^j} - G^f_{X^i} - (z_j - z_i) \cdot E_F}{k \cdot T}\right), \tag{11}$$

one can deduce that the oversaturation of an intrinsic point defect

$$\frac{C_X}{C^{eq}_X} = \frac{\sum_i C_{X^i}}{\sum_i C^{eq}_{X^i}} = \frac{C_{X^i}}{C^{eq}_{X^i}} \tag{12}$$

is equal to the oversaturation of any of its charge states when the latter are in steady state. Often, it is advantageous to define a particular charge state r as a reference point for a certain defect and to use

$$\frac{C_{X^i}}{C_{X^r}} = \frac{C^S_{X^i}}{C^S_{X^r}} \cdot \frac{\theta_{X^i}}{\theta_{X^r}} \cdot \exp\left(-\frac{G^f_{X^i} - G^f_{X^r} - (z_i - z_r) \cdot E_F}{k \cdot T}\right) = \delta_{X^i} \cdot \left(\frac{n}{n_i}\right)^{z_i - z_r}, \tag{13}$$

with δ_i lumping all Fermi-level-independent factors. Formally, δ_{X^r} is unity. Within the validity range of Boltzmann statistics follows also that the relative concentration of the defect in a particular charge state is given by

$$\frac{C_{X^i}}{C_X} = \frac{C_{X^i}}{\sum_i C_{X^i}} = \frac{\delta_{X^i} \cdot \left(\frac{n}{n_i}\right)^{z_i}}{\sum_i \delta_{X^i} \cdot \left(\frac{n}{n_i}\right)^{z_i}}, \tag{14}$$

For each of the reactions r in the system, a reaction variable ζ_r can be defined that describes the extent to which the reaction has proceeded. When only one reaction like (4) has to be considered, the change in the concentrations of any of the species i involved in a closed, constant volume is given by

$$dC_i = \nu_i \cdot d\zeta, \tag{15}$$

and by

$$dC_i = \sum_r \nu_{i,r} \cdot d\zeta_r \tag{16}$$

when several parallel reactions have to be considered. The change of the reaction variables with time follows for the example of reaction (4) from the dynamic law of mass action in the form

$$\frac{d\zeta}{dt} = k_{\rightarrow} \cdot C_A^{|\nu_A|} \cdot C_B^{|\nu_B|} - k_{\leftarrow} \cdot C_C^{\nu_C} \cdot C_D^{\nu_D}. \tag{17}$$

Since R_A in (1) can be identified as the rate at which species A is generated by quasi-chemical reactions, it can be expressed from (16) as

$$R_A = \sum_r \nu_{A,r} \cdot \frac{d\zeta_r}{dt} \tag{18}$$

with the $d\zeta_r/dt$ given by (17).

For the important case of a binary quasi-chemical reaction involving species A, B, and C in the form

$$A + B \underset{k_{\leftarrow}}{\overset{k_{\rightarrow}}{\rightleftharpoons}} C, \tag{19}$$

the reaction terms result from the equations above in the form

$$R = R_A = R_B = -R_C = -k_{\rightarrow} \cdot C_A \cdot C_B + k_{\leftarrow} \cdot C_C. \tag{20}$$

It should be noted that the signs therein result from the stoichiometric numbers $\nu_A = \nu_B = -1$ and $\nu_C = 1$. They reflect that the concentrations of species A and B reduce when species A and B react to C and increase when species C dissolves to A and B, while for species C it is vice versa. The charm of binary reactions is that the forward reaction rate can be estimated from the theory of diffusion-limited reactions of Waite (1957). Given the diffusivities D_A and D_B of the reacting species, $k_{\rightarrow}$ becomes

$$k_{\rightarrow} = 4 \cdot \pi \cdot a_R \cdot (D_A + D_B). \tag{21}$$

The reaction radius a_R is expected to be on the order of few Angstrom when at least one of the species is electrically neutral and when no reaction barriers have to be considered. For reactions between charged species, an extension of Waite's theory is available from Debye (1942), and an extension to include diffusion barriers from Waite (1958).

For the general case of equations as (4), the theory of diffusion-limited reactions is no longer applicable. The forward reaction constant $k_{\rightarrow}$ could be assumed to be determined by the slowest reaction in the chain. However, no general theory exists for such a case. If only immobile species are included in the reaction equation, the physical meaning of $k_{\rightarrow}$ is completely lost since necessary reactions to bring them into a mobile state are not explicitly considered. On the other hand, with experimentally determined reaction constants, such equations can help to reduce the total number of equations to be solved and to reduce the respective computational efforts.

With the forward or backward reaction constant know from theory or experiment, the complementary reaction constant can be calculated from equilibrium considerations. In equilibrium, the time derivatives of all individual reaction variables (17) have to vanish. This allows to express the ratio of the forward and backward reaction constants as a ratio of concentrations in equilibrium or, via (5), in terms of the formation energies. Taking the binary reaction (19) as an example, the backward reaction constant can be expressed from equilibrium indicated by the superscript "eq" as

$$k_{\leftarrow} = k_{\rightarrow} \cdot \frac{C_{\mathrm{A}}^{\mathrm{eq}} \cdot C_{\mathrm{B}}^{\mathrm{eq}}}{C_{\mathrm{C}}^{\mathrm{eq}}}. \tag{22}$$

For self-interstitials and vacancies, equilibrium concentrations have a well-defined meaning. For impurities, a certain concentration in the volume considered has to be assumed.

While the diffusion of a particular defect species X — vacancies, self-interstitials, and pairs — could be accounted for with different continuity equations (1) for each of its charge states i, coupled by charging equations (10), it is customary to assume that the charge states come into steady state on a much shorter timescale than diffusion and reactions. This allows to reduce the number of equations by considering the total flux

$$J_{\mathrm{X}} = \sum_{\mathrm{i}} J_{\mathrm{X^i}} = \sum_{\mathrm{i}} -D_{\mathrm{X^i}} \cdot \mathrm{grad}\, C_{\mathrm{X^i}} + z_{\mathrm{i}} \cdot D_{\mathrm{X^i}} \cdot C_{\mathrm{X^i}} \cdot \mathrm{grad}\left(\frac{\Psi}{U_{\mathrm{T}}}\right) \tag{23}$$

instead with $D_{\mathrm{X^i}}$ standing for the diffusion coefficient of the defect X in charge state i. Using (11), any of the concentrations $C_{\mathrm{X^i}}$ can be expressed in terms of the total concentration C_{X} and the Fermi level. Within the validity range of Boltzmann statistics, this simplifies via (14) to

$$J_{\mathrm{X}} = -\sum\nolimits_{\mathrm{i}} \delta_{\mathrm{X^i}} \cdot D_{\mathrm{X^i}} \cdot \left(\frac{n}{n_{\mathrm{i}}}\right)^{z_{\mathrm{i}}} \cdot \mathrm{grad}\left(\frac{C_{\mathrm{X}}}{\sum\nolimits_{\mathrm{i}} \delta_{\mathrm{X^i}} \cdot \left(\frac{n}{n_{\mathrm{i}}}\right)^{z_{\mathrm{i}}}}\right). \tag{24}$$

Please note that the field term drops formally out in this form because of the application of the product rule for the calculation of the gradient of the concentration of charged defects after expressing it in terms of the electron concentration and the concentration of neutral defects via (13). Using the product rule again on the gradient in (24) would lead again to a flux in the form (3) with a Fermi-level-dependent diffusion coefficient and a Fermi-level-dependent charge state.

3. DIFFUSION OF SUBSTITUTIONAL DOPANTS VIA INTRINSIC POINT DEFECTS

Many impurities, notably the dopants, occupy predominantly substitutional sites in silicon. To exchange sites with neighboring atoms, a variety of direct mechanisms as well as mechanisms needing an interaction with intrinsic point defects were suggested in literature. These mechanisms are summarized in the first part of this section. Particularly concepts assuming pairs between impurities and intrinsic point defects to form, diffuse, and dissociate again have proven effective. After a derivation of such pair diffusion models on the basis of the methodology of diffusion–reaction equations, the behavior of the system with respect to the impurity profiles to be expected as well as prominent diffusion phenomena are explained.

3.1. Basic diffusion mechanisms

The most direct way of diffusion for substitutional dopants would be via some exchange of sites with neighboring atoms. The mechanisms suggested in early theoretical work were analyzed by Hu (1973a) and found to be highly unlikely because of energetic reasons. Later, a concerted exchange mechanism was suggested by Pandey (1986). While it was shown to lead for self-diffusion to similar activation energies as observed experimentally, convincing experimental evidence to support it has not been presented until then. With direct or indirect exchange with neighboring atoms being usually discarded, interactions with intrinsic point defects — vacancies and self-interstitials — are invoked to explain not only the formation of mobile dopant species but also nonequilibrium phenomena like enhanced and retarded diffusion of dopants during processes that involve chemical reactions at the surface of samples or transient diffusion phenomena after ion implantation.

Historically, the first concepts for the diffusion of impurities via intrinsic point defects were developed for diffusion phenomena in metals. As in the work of Steigman et al. (1939), particularly interaction with vacancies was usually taken into consideration to explain self-diffusion and impurity diffusion in such systems. A particular problem noted already by Steigman et al. was that the activation energy for impurity diffusion is generally smaller than the one for self-diffusion while equal values would have been expected for a simple vacancy mechanism. This discrepancy was explained first by Johnson (1939) who assumed that vacancy and impurity are energetically bound and may move as a unit. In this first pair diffusion model it was already described that the vacancy would move around the substitutional impurity and both would exchange sites occasionally. In the diamond lattice, the reorientation of the vacancy around the impurity requires the vacancy to pass a third coordination site. There, in order to perform a random walk as a pair, some binding of the vacancy to the impurity is required. While elaborate analyses are available to describe particularly correlation effects (Dunham and Wu, 1995; Hu, 1973b; Mehrer, 1971; Yoshida, 1971), pair diffusion is usually taken into consideration in a simplified framework. Within this approach suggested first by Yoshida et al. (1974), substitutional impurity atoms M_s react with vacancies V to form mobile pairs MV according to the quasi-chemical reaction

$$M_s + V \underset{k_{MV\leftarrow}}{\overset{k_{MV\rightarrow}}{\rightleftharpoons}} MV \tag{25}$$

with $k_{MV\rightarrow}$ and $k_{MV\leftarrow}$ standing for the respective forward and backward reaction constants.

A particular prediction of pair diffusion models is that a gradient in the vacancy concentration causes a flux of initially homogeneously distributed impurities in the same direction as the flux of the vacancies. This is a direct consequence of the formation of mobile pairs in proportion to the concentration of substitutional impurities and vacancies so that the gradient of the pairs replicates the gradient of the vacancies. In an alternative approach to pair diffusion, often called "non-Fickian" diffusion, correlation effects and pair binding were ignored. These models then predict that a flux of vacancies results in a flux of impurity atoms in the opposite direction (see, e.g., Aleksandrov et al., 1988; Kozlovskii et al., 1985; Maser, 1991). Experimentally, the best test case is antimony since this element is known to diffuse nearly exclusively via vacancies. Using protons and boron implantation into a homogeneous antimony background finally gave clear evidence of dopant diffusion in the same direction as the vacancy diffusion as predicted by the pair diffusion theory (Kozlovskii et al., 1984; Pichler et al., 1992).

For germanium, to explain why copper and nickel have high diffusivities and act as acceptors as well as recombination centers, van der Maesen and Brenkman (1955) suggested that both impurities may dissolve on substitutional and interstitial sites with the latter configuration being responsible for the fast diffusion. Their concept was extended by Frank and Turnbull (1956) who suggested that the conversion from the interstitial state M_i into the substitutional state M_s occurs via a reaction with a vacancy in the form

$$M_i + V \underset{k_{FT\leftarrow}}{\overset{k_{FT\rightarrow}}{\rightleftharpoons}} M_s \tag{26}$$

with $k_{FT\rightarrow}$ and $k_{FT\leftarrow}$ denoting the forward and backward reaction constants. This mechanism, usually referred to as Frank–Turnbull mechanism, is indispensable for modeling the diffusion of transition metals in silicon. For dopant diffusion in silicon, the reaction is an important part of indirect bulk recombination (see below).

The possible involvement of self-interstitials in dopant diffusion was indicated first in the work of Watkins (1965). After electron irradiation of aluminum-doped silicon, they found that aluminum atoms were introduced on interstitial sites with a similar rate as monovacancies. To explain why no self-interstitials were found, they postulated that the self-interstitials I generated by the electron irradiation were all trapped at substitutional aluminum atoms and ejected them to interstitial sites while restoring the silicon lattice. This process, written in the form

$$M_s + I \rightarrow M_i \tag{27}$$

is usually referred to as "Watkins replacement mechanism." Its significance for dopant diffusion was not immediately recognized, though. A similar mechanism was later proposed by Gösele et al. (1980) as an alternative to the Frank–Turnbull mechanism (26) to better explain the time dependence and characteristic U-shaped form of gold profiles in silicon. In this work, the authors suggested that interstitial gold atoms M_i may change to substitutional sites M_s by ejecting a silicon lattice atom to an interstitial site and occupying its original position. Termed "kick-out mechanism" by the authors, it can be written in the form

$$M_i \underset{k_{KO\leftarrow}}{\overset{k_{KO\rightarrow}}{\rightleftharpoons}} M_s + I \tag{28}$$

with $k_{KO\rightarrow}$ and $k_{KO\leftarrow}$ standing for the respective forward and backward reaction constants.

The dawn of dopant diffusion models involving self-interstitials came with the review article of Seeger and Chik (1968). The interstitialcy mechanism they proposed assumes that a self-interstitial next to an impurity atom displaces the impurity atom to an interstitial position and takes its place. The impurity on the interstitial site subsequently displaces another neighboring silicon atom to an interstitial site and, in this manner, has performed a diffusive jump. As in the case of vacancies, the self-interstitial needs to be bound to the impurity to explain why dopant diffusion has a smaller activation energy than self-diffusion. But for the interstitialcy mechanism, as already remarked by Hu (1973b), the binding potential does not have to be as far-ranging as for vacancies. To perform a random walk as a pair, the self-interstitial just needs to pass a second-nearest interstitial site while a vacancy has to pass a third coordination site. Within the methodology of pair diffusion theories, the formation of the bound self-interstitial-impurity pair MI is described by the quasi-chemical reaction

$$M_s + I \underset{k_{MI\leftarrow}}{\overset{k_{MI\rightarrow}}{\rightleftharpoons}} MI \tag{29}$$

with $k_{MI\rightarrow}$ and $k_{MI\leftarrow}$ denoting the respective forward and backward reaction constants.

When comparing the pair formation reaction (29) with the kick-out reaction in backward direction (28), it becomes apparent that the only difference between the two is that the migrating species is in the former case a bound pair of a substitutional impurity and a self-interstitial while it is an impurity interstitial in the latter case. Within the methodology of pair diffusion outlined in the next section, both are considered possible realizations of a migrating impurity-interstitial point defect and treated within exactly the same mathematical framework as impurity–vacancy pairs (25) and will not be distinguished further in the remainder of this chapter.

In addition to the reactions discussed above, the recombination of self-interstitials and vacancies in the bulk of semiconductors according to

$$I + V \underset{k_{B\leftarrow}}{\overset{k_{B\rightarrow}}{\rightleftharpoons}} 0 \tag{30}$$

needs to be taken into account with the "0" standing again for the undisturbed lattice, and $k_{B\rightarrow}$ and $k_{B\leftarrow}$ for the forward and backward reaction constants, respectively.

Particularly in highly doped regions, the concentration of intrinsic point defects may be exceeded by pairs of impurities with self-interstitials and vacancies. The reactions of such pairs with complementary intrinsic point defects like

$$\mathrm{MV} + \mathrm{I} \underset{k_{\mathrm{MV+I}\leftarrow}}{\overset{k_{\mathrm{MV+I}\rightarrow}}{\rightleftharpoons}} \mathrm{M_s} \tag{31}$$

and

$$\mathrm{MI} + \mathrm{V} \underset{k_{\mathrm{MI+V}\leftarrow}}{\overset{k_{\mathrm{MI+V}\rightarrow}}{\rightleftharpoons}} \mathrm{M_s} \tag{32}$$

then lead to the pairwise elimination of an interstitial and a vacancy and represent an efficient indirect recombination path (Loualiche et al., 1982; Mathiot and Pfister, 1984).

3.2. Pair diffusion models

Pair diffusion models as pioneered by Yoshida et al. (1974) assume that mobile pairs between impurities and vacancies or self-interstitials form according to the quasi-chemical reactions (25) and (29), respectively. Within the framework of diffusion–reaction equations discussed above, properties like diffusion coefficients, charge states, formation energies, and binding energies are attributed to these pairs.

Before discussing the diffusion of the pairs, let us discuss their concentration in a steady-state situation. For the example of an ionized donor $\mathrm{M_s^+}$ reacting with a neutral defect $\mathrm{X^0}$ (X may be either a vacancy V or a self-interstitial I) according to

$$\mathrm{M_s^+} + \mathrm{X^0} \underset{k_{\leftarrow}}{\overset{k_{\rightarrow}}{\rightleftharpoons}} \mathrm{MX^+}, \tag{33}$$

positively charged pairs $\mathrm{MX^+}$ will form. For all of the defects involved, the charge state is indicated as superscript symbols. Using (5), the concentration of the positively charged pairs follows as

$$C_{\mathrm{MX^+}} = C_{\mathrm{M_s^+}} \cdot C_{\mathrm{X^0}} \cdot \frac{1}{C_{\mathrm{Si}}} \cdot \frac{\theta_{\mathrm{MX^+}}}{\theta_{\mathrm{X^0}}} \cdot \exp\left(-\frac{G^{\mathrm{f}}_{\mathrm{MX^+}} - G^{\mathrm{f}}_{\mathrm{X^0}}}{k \cdot T}\right). \tag{34}$$

In this form, it was assumed that the concentrations of sites for defect X^0 and for the pair M_s^+ are equal. This is certainly true for vacancies and pairs with vacancies. Self-interstitials and self-interstitial-dopant pairs do not necessarily occupy the same sites. However, considering that a factor of two corresponds to an energy difference of 70 meV at 900 °C, such factors appear negligible in comparison to the uncertainties of determining formation energies from experiments or theory. Similarly, it was ignored that the concentrations of sites reduce when more and more dopants are introduced into the system. Finally, it should be noted that the ionized, substitutional configuration (with $\theta_{M_s^+} = 1$) was used as reference point for the formation energy of the dopants ($G^f_{M_s^+} = 0$). From the definition (6), $G^B_{MS^+} = G^f_{X^0} - G^f_{MS^+}$ follows as the binding energy of the pair. For pairs with acceptors, (34) would look the same except for the indicated charge states of the substitutional dopant and the pair which would then be negative. With the help of (9) and (12), the concentration of pairs can be rewritten in terms of the oversaturation C_X/C_X^{eq} of the intrinsic point defect in the form

$$C_{MX^+} = C_{M_s^+} \cdot \frac{C_X}{C_X^{eq}} \cdot \theta_{MX^+} \cdot \exp\left(-\frac{G^f_{MX^+}}{k \cdot T}\right) = \eta_{MX} \cdot C_{M_s^+} \cdot \frac{C_X}{C_X^{eq}} \tag{35}$$

with η_{MX} introduced as abbreviation for the temperature-dependent parameters.

It should be noted that the relationship between the concentrations of the pairs, the defects, and the dopants in (34) is independent of the Fermi level. This is because the reaction does not require any interaction with charge carriers. In extrinsically donor-doped material with its high electron density, one would expect the formation of neutral pairs and negatively charged pairs beside the neutral ones. In analogy to (34), the reactions $M_s^+ + X^- \leftrightarrow MX^0$ with singly negatively charged defects and $M_s^+ + X^= \leftrightarrow MX^-$ with doubly negatively charged defects would not involve any charge carriers and are described by Fermi-level-independent reaction constants. However, the concentrations of the charged defects X^- and $X^=$ will reflect that the Fermi level is closer to the conduction band than in intrinsically doped material and if we formulate the formation of a neutral charged pair in the form

$$M_s^+ + X^0 \underset{k_{\leftarrow}}{\overset{k_{\rightarrow}}{\rightleftharpoons}} MX^0 + p^+ \tag{36}$$

with p^+ representing a hole, we obtain from (5) the concentration of the neutral pairs as

$$C_{\mathrm{MX}^0} = C_{\mathrm{M}_s^+} \cdot C_{\mathrm{X}^0} \cdot \frac{1}{C_{\mathrm{Si}}} \cdot \frac{\theta_{\mathrm{MX}^0}}{\theta_{\mathrm{X}^0}} \cdot \exp\left(-\frac{G^{\mathrm{f}}_{\mathrm{MX}^0} - E_{\mathrm{F}} - G^{\mathrm{f}}_{\mathrm{X}^0}}{k \cdot T}\right). \tag{37}$$

It just remains to remark that an identical result would have been obtained if we had considered that the neutral pair results from a charging reaction with an electron n^- in the form $\mathrm{MX}^+ + n^- \leftrightarrow \mathrm{MX}^0$ or a hole in the form $\mathrm{MX}^+ \leftrightarrow \mathrm{MX}^0 + p^+$. Dividing the concentrations of the pairs in the neutral (37) and positive charge states (34) by each other, one can see that the ratio

$$\frac{C_{\mathrm{MX}^0}}{C_{\mathrm{MX}^+}} = \frac{\theta_{\mathrm{MX}^0}}{\theta_{\mathrm{MX}^+}} \cdot \exp\left(-\frac{G^{\mathrm{f}}_{\mathrm{MX}^0} - E_{\mathrm{F}} - G^{\mathrm{f}}_{\mathrm{MX}^+}}{k \cdot T}\right) = \delta_{\mathrm{MX}^0} \cdot \frac{n}{n_{\mathrm{i}}} \tag{38}$$

increases within the validity range of Boltzmann statistics linearly with the electron concentration. For donors, the positive charge state has been taken as reference state for the δ_{MX}, for acceptor pairs, it would be the negative charge state. Since C_{MX^+} does not depend on the Fermi level as long as $C_{\mathrm{M}_s^+}$ remains the same, we can conclude that the concentration of neutral pairs increases also linearly with the electron concentration. In analogy, one can conclude from

$$\frac{C_{\mathrm{MX}^-}}{C_{\mathrm{MX}^+}} = \frac{\theta_{\mathrm{MX}^-}}{\theta_{\mathrm{MX}^+}} \cdot \exp\left(-\frac{G^{\mathrm{f}}_{\mathrm{MX}^-} - 2 \cdot E_{\mathrm{F}} - G^{\mathrm{f}}_{\mathrm{MX}^+}}{k \cdot T}\right) = \delta_{\mathrm{MX}^-} \cdot \left(\frac{n}{n_{\mathrm{i}}}\right)^2, \tag{39}$$

that the concentration of negatively charged pairs increases quadratic with the electron concentration. For pairs with acceptors, one would have obtained a Fermi-level independent concentration of negatively charged pairs, a concentration of neutral pairs that increases linearly with the hole concentration, and a concentration of positively charged pairs that increases quadratic with the hole concentration. Evidently, the concentration of the substitutional atoms and the concentrations of the pairs have to sum up to the total concentration

$$C_{\mathrm{M}} = C_{\mathrm{M}_s} + \sum_{\mathrm{X,i}} C_{\mathrm{MX}^{\mathrm{i}}}, \tag{40}$$

from which the concentration of substitutional atoms can be calculated as a function of temperature, the total dopant concentration, the Fermi level, and the concentrations of the intrinsic point defects. C_{M_s} in (40) comprises in principle ionized as well as neutral substitutional dopants. However, the impurity states merge already at concentrations of some $10^{18}\ \mathrm{cm}^{-3}$ with the conduction or valance band (Altermatt et al., 2006a,b) so that a neglect of the neutral substitutional dopants is usually considered a good approximation.

To simulate the formation of pairs between dopants and intrinsic point defects dynamically, the pairing reactions (25) and (29) as well as the reactions of the pairs with complementary point defects via (31) and (32) have to be taken into account. As indicated above, it is generally assumed that steady state between the charge states of a defect is established on a much shorter timescale than those for diffusion and the reactions. Then, using the methodology of diffusion–reaction equations outlined above, the system of equations to be solved for one dopant species consists of the simple differential equation

$$\frac{\partial C_{M_s}}{\partial t} = R_{MI} + R_{MV} - R_{MV+I} - R_{MI+V} \tag{41}$$

for the substitutional dopant atoms, the continuity equations

$$\frac{\partial C_{MI}}{\partial t} = -\mathrm{div}(J_{MI}) - R_{MI} + R_{MI+V} \tag{42}$$

$$\frac{\partial C_{MV}}{\partial t} = -\mathrm{div}(J_{MV}) - R_{MV} + R_{MV+I} \tag{43}$$

for pairs, and the continuity equations

$$\frac{\partial C_I}{\partial t} = -\mathrm{div}(J_I) + R_{MI} + R_{MV+I} + R_{IV} \tag{44}$$

$$\frac{\partial C_V}{\partial t} = -\mathrm{div}(J_V) + R_{MV} + R_{MI+V} + R_{IV} \tag{45}$$

for the intrinsic point defects. Based on the number of equations, the system of coupled partial differential equations (41)–(45) is often referred to as five-stream model. With all reactions in the system being of the binary type (19), the reaction terms

$$R_{MI} = -k_{MI\rightarrow} \cdot C_{M_s} \cdot C_I + k_{MI\leftarrow} \cdot C_{MI}, \tag{46}$$

$$R_{MV} = -k_{MV\rightarrow} \cdot C_{M_s} \cdot C_V + k_{MV\leftarrow} \cdot C_{MV}, \tag{47}$$

$$R_{MV+I} = -k_{MV+I\rightarrow} \cdot C_{MV} \cdot C_I + k_{MV+I\leftarrow} \cdot C_{M_s}, \tag{48}$$

$$R_{\mathrm{MI+V}} = -k_{\mathrm{MI+V}\rightarrow} \cdot C_{\mathrm{MI}} \cdot C_{\mathrm{V}} + k_{\mathrm{MI+V}\leftarrow} \cdot C_{\mathrm{M_s}}, \tag{49}$$

$$R_{\mathrm{IV}} = -k_{\mathrm{IV}\rightarrow} \cdot (C_{\mathrm{I}} \cdot C_{\mathrm{V}} - C_{\mathrm{I}}^{\mathrm{eq}} \cdot C_{\mathrm{V}}^{\mathrm{eq}}) \tag{50}$$

have been defined as R in (20) so that R_{MI} and R_{MV} have to be accounted for with a negative sign in the continuity equations of the pairs while $R_{\mathrm{MV+I}}$ and $R_{\mathrm{MI+V}}$ have to be accounted for with a negative sing in the differential equation of the substitutional dopant atoms. The forward reaction constants $k_{\mathrm{XY}\rightarrow}$ lump the individual rates resulting from reactions of all charge states of defect X with all charge states of defect Y. The backward reaction constants can be obtained from (22). For the reaction of self-interstitials and vacancies, this leads to the familiar form (50). In the general case, both forward and backward reactions will be functions of the Fermi level and of temperature. The fluxes J_{X} are all of the form (24). For each additional dopant, three equations need to be added — one for the substitutional species and two for the pairs.

The system of equations sketched above needs to be completed by a relationship between the electric field and the concentrations of active defects. This relationship is obtained from the third of Maxwell's equations, which within the validity range of Boltzmann statistics can be written as

$$\mathrm{div}(\varepsilon \cdot \mathrm{grad}\,\Psi) = q \cdot \left(2 \cdot n_{\mathrm{i}} \cdot \sin h\left(\frac{\Psi}{U_{\mathrm{T}}}\right) + \sum_{\mathrm{X,i}} z_{\mathrm{i}} \cdot C_{\mathrm{X^i}} \right), \tag{51}$$

where ε denotes the permittivity. The summation comprises all charged defects. However, in practice, it suffices to include ionized donors and acceptors as well as charged complexes that limit the solubility of the dopants. The concentrations of ionized pairs of dopants and intrinsic point defects as well as the concentrations of ionized intrinsic point defects are usually considered to be too low in comparison to the majority carrier concentration. The importance of appropriate boundary conditions for the Poisson equation was emphasized in literature (Jung et al., 2004; Tsibizov et al., 2014). For some practical situations and assessing the system behavior below, it may suffice to assume local charge neutrality. This corresponds to neglecting the divergence of the electric displacement field and assuming that the term in parentheses on the right-hand side vanishes.

3.3. System behavior

Having formulated a system of partial differential equations that describe the diffusion of dopants by pair diffusion models within the framework of diffusion–reaction equations, the behavior of this system will be discussed in this section. To illustrate and explain the possible profile forms resulting for specific parameter combinations, a simplified pair diffusion model will be used. In this model, just one pairing reaction $M_s + X \rightleftharpoons MX$ between a substitutional impurity M_s with an intrinsic point defect X is considered. Without loss of generality we will use self-interstitials with the respective equations to represent the intrinsic point defects by replacing I by X. The system of equations then consists of the simple differential equation (41) for the substitutional dopant atoms and the continuity equations for the pair (42) and the intrinsic point defect (44). Of the reaction terms (46)–(50) only R_{MI} is nonzero.

Considering the continuity equation (42) for the pairs with the reaction term (44) and concentrating on the backward reaction, one can see from

$$\frac{\partial C_{MX}}{\partial t} = \ldots - k_{MX\leftarrow} \cdot C_{MX} \tag{52}$$

that the backward reaction constant $k_{MX\leftarrow}$ corresponds to the inverse time constant $\tau_{MX} = 1/k_{MX\leftarrow}$ of the dissolution of the pairs. An analysis of short process times, during which the pairing reaction has a significant influence on the profile form, has been pioneered by Cowern et al. (1990b). Following his analysis, the quantity $\lambda_{MX} = \sqrt{\tau_{MX} \cdot D_{MX}}$ can be seen as the mean projected path length of the pairs between their formation and their dissolution. Typical profile shapes as they would be expected for the diffusion of pairs with a constant diffusion coefficient from a source that maintains a constant surface concentration are shown in Fig. 1 as a function of the mean projected path length of the pairs λ_{MX} in relation to the macroscopic diffusion length of the dopants. As long as the diffusion length of the pairs is less than a tenth of the macroscopic dopant diffusion coefficient, the profile shape corresponds closely to the error function expected for the diffusion from a constant surface concentration with a constant diffusion coefficient. In case that the mean projected path length of the pairs approaches the macroscopic diffusion length, the profile shape becomes increasingly exponential. For the longest time, the substitutional concentration becomes smaller than the equilibrium surface concentration C_{MX}^{surf}. This is owed to the fact that the concentration of pairs was maintained at a constant value and dynamic equilibrium between the pairs and the substitutional concentration has not established yet. Since mean projected path lengths are on the order of few to 100 nm, such effects were observed particularly at low temperatures or for short process durations.

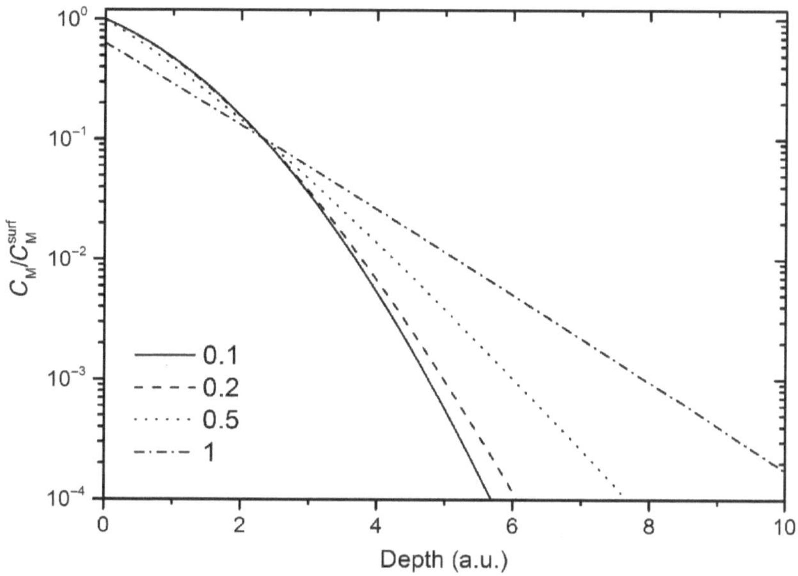

Figure 1 Depth profiles of a pair diffusion process with a constant pair diffusion coefficient, constant equilibrium surface concentration C_{MX}^{surf}, and rate limitation. The numbers of the legend refer to the ratio of the mean projected path length of the pairs and the macroscopic diffusion length of the dopants.

For sufficiently long times, the reaction terms (46)–(48) will come into a local equilibrium and the mean projected path length λ of the pairs will become much smaller than the macroscopic diffusion length. This allows to add the equations (41)–(43) to

$$\frac{\partial C_M}{\partial t} = \frac{\partial C_{M_s} + C_{MI} + C_{MV}}{\partial t} = -\mathrm{div}(J_{MI}) - \mathrm{div}(J_{MV}), \tag{53}$$

and to express the concentration of pairs in the fluxes J_{MI} and J_{MV} given by (24) in terms of the dopant concentration and the concentrations of the intrinsic point defects.

As emphasized by Cowern (1988), to calculate the concentration of pairs for arbitrary oversaturations of self-interstitials and vacancies, both the pairing reactions and the reactions of the pairs with the complementary point defects have to be taken into account. Using X to denote the defect with which the dopant forms the pair and $\overline{\mathrm{X}}$ to denote the complementary defect, the assumption of local equilibrium leads to

$$R_{MX} - R_{MX+\overline{X}} = -\left(k_{MX\rightarrow} \cdot C_X + k_{MX+\overline{X}\leftarrow}\right) \cdot C_{M_s} + \left(k_{MX\leftarrow} + k_{MX+\overline{X}\rightarrow} \cdot C_V\right) \cdot C_{MX} = 0, \tag{54}$$

from which the concentration of pairs can be expressed as

$$C_{MX} = \frac{k_{MX\rightarrow} \cdot C_X + k_{MX+\overline{X}\leftarrow}}{k_{MX\leftarrow} + k_{MX+\overline{X}\rightarrow} \cdot C_{\overline{X}}} \cdot C_{M_s}. \tag{55}$$

Since each of the reaction terms (46)–(49) have to vanish in equilibrium, one can rewrite (55) in the form

$$C_{MX} = \frac{k_{MX\rightarrow}}{k_{MX\leftarrow}} \cdot C_X^{eq} \cdot \frac{C_X/C_X^{eq} + \alpha_{MX}}{1 + \alpha_{MX} \cdot C_{\overline{X}}/C_{\overline{X}}^{eq}} \cdot C_{M_s}. \tag{56}$$

The partition parameter α_{MX} has been introduced as abbreviation for

$$\alpha_{MX} = \frac{k_{MX+\overline{X}\leftarrow}}{k_{MX\rightarrow} \cdot C_X^{eq}} \tag{57}$$

and describes to which extent the pair is formed via the pair-formation equation ($\alpha_{MX} = 0$) or from a substitutional dopant by creating also a complementary intrinsic point defect ($\alpha_{MX} = \infty$). The concentration of pairs in a nonequilibrium situation is governed by the second ratio on the right-hand side in (56) which is unity in equilibrium situations. In many cases, one of the intrinsic point defects will be in oversaturation while the complementary intrinsic point defect will be in undersaturation due to bulk recombination. To analyze this situation, Cowern (1988) considered two important limiting cases. In the first one, bulk recombination is so effective that the intrinsic point defects are in local equilibrium ($C_I \cdot C_V = C_I^{eq} \cdot C_V^{eq}$). The second ratio in (56) reduces then to

$$\frac{C_X/C_X^{eq} + \alpha_{MX}}{1 + \alpha_{MX} \cdot C_{\overline{X}}/C_{\overline{X}}^{eq}} = \frac{C_X}{C_X^{eq}}, \tag{58}$$

which would also have been obtained if only the pair reaction equation would have been considered. As a second limiting case, Cowern assumed that bulk recombination would be so ineffective that the complementary intrinsic point defect $\overline{X}$ maintains its equilibrium value. For such a case and $\alpha_{MX} > 0$, Cowern could show that the second ratio in (56)

$$\frac{C_X/C_X^{eq} + \alpha_{MX}}{1 + \alpha_{MX} \cdot C_{\overline{X}}/C_{\overline{X}}^{eq}} = \frac{C_X/C_X^{eq}}{1 + \alpha_{MX}} + \frac{\alpha_{MX}}{1 + \alpha_{MX}} \tag{59}$$

and with it the concentration of pairs remains smaller than predicted by (58). During irradiation or during the initial stage of postimplantation annealing, both intrinsic point defects are in strong oversaturation. If self-interstitials and vacancies are produced in equal numbers and have equal oversaturations, one obtains a constant value

$$\frac{C_X/C_X^{eq}+\alpha_{MX}}{1+\alpha_{MX}\cdot C_{\overline{X}}/C_{\overline{X}}^{eq}}=\frac{1}{\alpha_{MX}}\cdot\frac{C_{\overline{X}}^{eq}}{C_X^{eq}} \tag{60}$$

which depends only on the equilibrium concentrations of the intrinsic point defects and the partition parameter α_{MX}.

Pair diffusion models often concentrate on the pair formation reactions (46) and (47) and ignore the reactions (48) and (49) with the complementary point defects. Using again donors M_s^+ to illustrate the procedure, the term

$$\frac{C_{MX}}{\sum_i \delta_{MX^i}\cdot\left(\frac{n}{n_i}\right)^{z_i}}=\eta_{MX}\cdot C_{M_s^+}\cdot\frac{C_X}{C_X^{eq}}\cdot\frac{n}{n_i} \tag{61}$$

in the flux equations for the pairs (24) can be obtained from the pair concentrations (35), (38), and (39) derived for steady state of the pairing reaction. Inserting it into (24) gives the flux of pairs

$$J_{MX}=-\eta_{MX}\cdot\sum_i \delta_{MX^i}\cdot D_{MX^i}\cdot\left(\frac{n}{n_i}\right)^{z_i}\cdot\text{grad}\left(C_{M_s^+}\cdot\frac{C_X}{C_X^{eq}}\cdot\frac{n}{n_i}\right). \tag{62}$$

Before proceeding, it should be noted that the argument of the gradient operator in (62) contains the substitutional concentration $C_{M_s^+}$ rather than the total concentration C_M which would be needed to bring (53) into a form of a diffusion equation with an effective diffusion coefficient. In general, the substitutional concentration follows in local equilibrium indirectly from the requirement (40) that the substitutional concentration of dopants and the concentrations of the pairs sum up to the total dopant concentration. As long as the concentrations of the pairs are negligible in comparison to the total concentration, their concentrations will increase linearly with the oversaturation of the respective intrinsic point defect. Eventually, the concentrations of the pairs will lead to a reduction of the substitutional concentration. Any further increase of concentration of the respective point defect will lead to a sublinear increase of the concentration of the respective impurity pairs. Finally, in case of an extreme oversaturation of the intrinsic point defects, basically all dopants would be in pairs. Then, the diffusion of the dopants proceeds with the diffusion coefficient of the respective pairs and would not be able to increase further. For the diffusion of arsenic in silicon, there were suggestions (Heinrich et al., 1990; Novell and Law, 1992) that non-negligible concentrations of pairs of arsenic with intrinsic point defects could explain an apparent discrepancy in the experiments of Fahey et al. (1985). However, since then, the concept was not picked up again and it is generally assumed that the concentration of pairs is much smaller than the substitutional concentration for all technically relevant conditions. This allows to simply replace the concentration of substitutional donors $C_{\mathrm{M}_s^+}$ by the total concentration C_{M} in (62) which, inserted in (53), gives the familiar expression

$$\frac{\partial C_{\mathrm{M}}}{\partial t} = \sum_{\mathrm{X=I,V}} \operatorname{div}\left(D_{\mathrm{MX}} \cdot \operatorname{grad}\left(C_{\mathrm{M}} \cdot \frac{C_{\mathrm{X}}}{C_{\mathrm{X}}^{\mathrm{eq}}}\right) + D_{\mathrm{MX}} \cdot \frac{C_{\mathrm{X}}}{C_{\mathrm{X}}^{\mathrm{eq}}} \cdot \operatorname{grad}\left(\frac{\Psi}{U_{\mathrm{T}}}\right)\right) \tag{63}$$

with

$$D_{\mathrm{MX}} = \eta_{\mathrm{MX}} \cdot \sum_{\mathrm{i}} \delta_{\mathrm{MX^i}} \cdot D_{\mathrm{MX^i}} \cdot \left(\frac{n}{n_{\mathrm{i}}}\right)^{1+z_{\mathrm{i}}} \tag{64}$$

standing for the effective Fermi-level-dependent diffusion coefficient via pairs with the intrinsic point defect X under condition of thermal equilibrium for the intrinsic point defects. For acceptors, the sign of the field term in (63) would be negative and the exponent of the electron concentration in (64) would be $z_{\mathrm{i}} - 1$. As outlined above, the charge states of the pairs are expected to be singly negative ($z = 1$), neutral ($z = 0$), and singly positive ($z = -1$). Based on that, besides a Fermi-level-independent component, the diffusion coefficient D_{MX} has components which increase with n/n_{i} and $(n/n_{\mathrm{i}})^2$ for donors, and components which increase with p/n_{i} and $(p/n_{\mathrm{i}})^2$ for acceptors. For intrinsic concentrations and with the concentrations of the intrinsic point defects at equilibrium values, (63) reduces to a simple diffusion equation with a constant diffusion coefficient. This so-called intrinsic diffusion coefficient

$$D_{\mathrm{M}}^{\mathrm{i}} = \eta_{\mathrm{MI}} \cdot \sum_{\mathrm{i}} \delta_{\mathrm{MI^i}} \cdot D_{\mathrm{MI^i}} + \eta_{\mathrm{MV}} \cdot \sum_{\mathrm{i}} \delta_{\mathrm{MV^i}} \cdot D_{\mathrm{MV^i}} \tag{65}$$

lumps the contributions of all pairs of the dopant with self-interstitials and vacancies and is usually the most often investigated and reported parameter for dopants. As a sum of parameters that can all be assumed to have Arrhenius-like temperature behavior, $D_{\mathrm{M}}^{\mathrm{i}}$ cannot really be expected to have Arrhenius-like temperature dependence itself. However, experience shows that the temperature dependence of the intrinsic diffusion coefficient can be described with sufficient accuracy by an Arrhenius law. Depending on the dopant/substrate system, this may be motivated by the dominance of only one pair in only one charge state, by similar activation energies of similarly important pairs, by the limited temperature range, or by experimental uncertainties in general.

When discussing the profile form expected for extrinsically doped semiconductors, one has to note that the electrostatic potential will reflect the doping conditions via the Poisson equation (51) and constitutes via the field term in (63) an additional driving force for dopant redistribution. An analysis of this situation has been given by Smits (1958) and Lehovec and Slobodskoy (1961) by assuming charge neutrality. This allows to express the electrostatic potential in terms of the dopant concentrations. Inserting the electrostatic potential into (63) results for one dopant in the diffusion equation

$$\frac{\partial C_{\mathrm{M}}}{\partial t} = \mathrm{div}\left(f_{\mathrm{f}} \cdot \sum_{\mathrm{X=I,V}} D_{\mathrm{MX}} \cdot \mathrm{grad}\left(C_{\mathrm{M}} \cdot \frac{C_{\mathrm{X}}}{C_{\mathrm{X}}^{\mathrm{eq}}}\right)\right) \tag{66}$$

with the field-enhancement factor

$$f_{\mathrm{f}} = 1 + \frac{C_{\mathrm{M}}}{\sqrt{C_{\mathrm{M}}{}^2 + 4 \cdot n_{\mathrm{i}}{}^2}}. \tag{67}$$

Typical profiles resulting from the diffusion of dopants at high concentrations for point-defect concentrations at equilibrium can be found in Fig. 2. The shallowest profile has been obtained from the assumption of a constant diffusion coefficient and without considering field enhancement. For the next-deeper profile, field enhancement has been taken into account. The effects on the profile shape are limited, though, since f_{f} varies only between unity for small concentrations and two for large ones. Assuming that the effective diffusion coefficient varies with n/n_i and $(n/n_i)^2$ for donors results in increasingly steeper profiles. The latter, as an example, appears typical for arsenic diffusion at concentrations above $2 \times 10^{20}\ \mathrm{cm}^{-3}$ (Hoyt and Gibbons, 1986; Nylandsted Larsen et al., 1986).

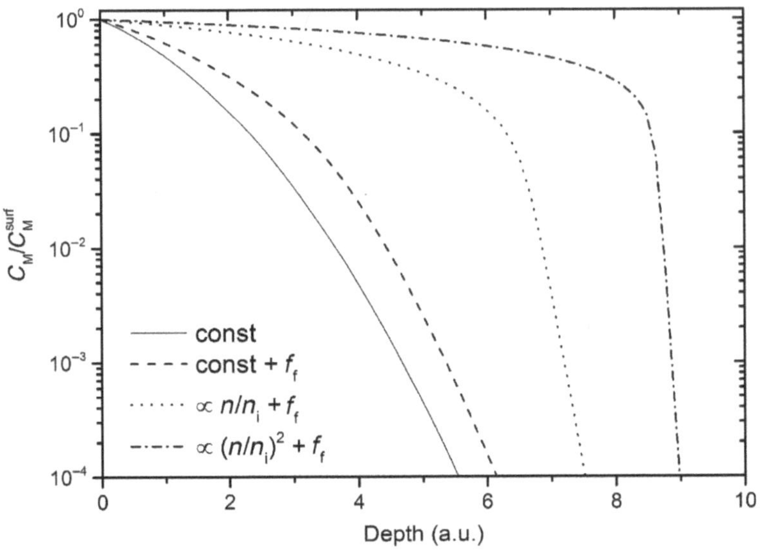

Figure 2 Depth profiles resulting from a diffusion process with a constant surface concentration C_M^{surf}. The legend indicates the assumptions for the effective diffusion coefficient: Constant or varying with n/n_i or $(n/n_i)^2$. f_f indicates that field enhancement has been taken into account.

In the derivation of the concentration of pairs (34), the assumption was made that the number of sites remains approximately constant whatever the concentration. This assumption breaks down at very high concentrations. In samples with a phosphorus background concentration exceeding 2×10^{20} cm^{-3}, a sharp increase of the diffusion of germanium, tin, arsenic, and antimony was found by Nylandsted Larsen et al. (1989, 1993). This phenomenon was interpreted by Mathiot and Pfister (1989) within the percolation theory of Stauffer (1979). Pair binding implies that the formation energy for an intrinsic point defect is lower in the vicinity of a dopant than in a pure silicon environment. When dopants are sufficiently close, their attractive potentials will overlap. Above a certain dopant concentration, a percolation cluster of interacting dopants will form. Within this cluster, the formation energy of the intrinsic point defects is decreased so that their concentration is expected to increase. In addition, the migration energy of the intrinsic point defects is reduced so that they diffuse faster. In the wake of the intrinsic point defects, also a rapid redistribution of the dopants would be expected. However, as argued, e.g., by Ramamoorthy and Pantelides (1996), the rapid diffusion should also lead to a rapid break-up of the percolation cluster via the formation of complexes. Later work by Bunea and

Dunham (1997) and Xie and Chen (1999) confirmed this and indicated that the sharp increase of the diffusivity observed should be a transient phenomenon.

As discussed further below, there are many technologically important processes during which the intrinsic point defects are out of equilibrium but rather homogeneously distributed within the areas containing dopants. This allows to ignore the gradients of the intrinsic point defects in (63) and the equation reduces to a diffusion equation with field term

$$\frac{\partial C_{\mathrm{M}}}{\partial t} = \mathrm{div}\left(D_{\mathrm{M}} \cdot \mathrm{grad}\left(C_{\mathrm{M}} \right) + D_{\mathrm{M}} \cdot \mathrm{grad}\left(\frac{\Psi}{U_{\mathrm{T}}} \right) \right) \tag{68}$$

with the diffusion coefficient

$$D_{\mathrm{M}} = D_{\mathrm{MI}} \cdot \frac{C_{\mathrm{I}}}{C_{\mathrm{I}}^{\mathrm{eq}}} + D_{\mathrm{MV}} \cdot \frac{C_{\mathrm{V}}}{C_{\mathrm{V}}^{\mathrm{eq}}} \tag{69}$$

which depends via D_{MI} and D_{MV} on the Fermi level and via $C_{\mathrm{I}}/C_{\mathrm{I}}^{\mathrm{eq}}$ and $C_{\mathrm{V}}/C_{\mathrm{V}}^{\mathrm{eq}}$ on the oversaturations of the intrinsic point defects. To characterize whether a dopant diffuses predominantly via self-interstitials or via vacancies, the fractional diffusivity f_{I} was introduced in literature as

$$f_{\mathrm{I}} = \frac{D_{\mathrm{MI}}}{D_{\mathrm{MI}} + D_{\mathrm{MV}}} \tag{70}$$

so that the diffusion coefficient under nonequilibrium conditions can be written also in the form

$$D_{\mathrm{M}} = D_{\mathrm{M}}^{\mathrm{eq}} \cdot \left(f_{\mathrm{I}} \cdot \frac{C_{\mathrm{I}}}{C_{\mathrm{I}}^{\mathrm{eq}}} + (1 - f_{\mathrm{I}}) \cdot \frac{C_{\mathrm{V}}}{C_{\mathrm{V}}^{\mathrm{eq}}} \right) \tag{71}$$

with $D_{\mathrm{M}}^{\mathrm{eq}} = D_{\mathrm{MI}} + D_{\mathrm{MV}}$ corresponding to the diffusion coefficient under equilibrium conditions which, for intrinsic doping conditions, corresponds to the intrinsic diffusion coefficient $D_{\mathrm{M}}^{\mathrm{i}}$. In silicon, boron was found to diffuse nearly entirely via self-interstitials and antimony nearly entirely via vacancies. Phosphorus appears to diffuse at low concentrations nearly entirely via self-interstitials while it diffuses predominantly via vacancies at high concentrations. Arsenic is a hybrid diffuser at all concentrations. In many situations, local equilibrium between the intrinsic point defects will be established on a much shorter timescale than the duration of the respective technological process. This allows to express the normalized impurity diffusion coefficient $D_{\mathrm{M}}/D_{\mathrm{M}}^{\mathrm{eq}}$ as a function of the oversaturation of the self-interstitials $C_{\mathrm{I}}/C_{\mathrm{I}}^{\mathrm{eq}}$. The result of the analysis is shown in Fig. 3 for selected values of the fractional diffusivities via self-interstitials f_I. Enhancement for the diffusion is observed for dopants diffusing via vacancies ($f_I < 0.5$) being in oversaturation ($C_I/C_I^{eq} < 1$) as well as for dopants diffusing via self-interstitials ($f_I > 0.5$) being in oversaturation ($C_I/C_I^{eq} > 1$). When dopants diffuse primarily via vacancies, retarded diffusion is observed for small oversaturations of self-interstitials. As soon as the oversaturation C_I/C_I^{eq} exceeds a value of $(1-f_I)/f_I$, the diffusion becomes enhanced again. For dopants diffusing primarily via self-interstitials, retarded diffusion is observed for $C_I/C_I^{eq} < 1$ as long as the oversaturation of self-interstitials is not below a value of $(1-f_I)/f_I$. For a fractional diffusivity via self-interstitials of $f_I = 0.5$, only enhanced diffusion will be observed in nonequilibrium situations. The smallest possible effective diffusion coefficient is $2 \cdot \sqrt{f_I \cdot (1-f_I)} \cdot D_M^{eq}$. This relationship was used in the early work of Gösele and Tan (1983) to estimate the fractional diffusivity of dopants via self-interstitials when their diffusion was found to be retarded. Alternative methods for a determination of f_I were suggested by Fahey et al. (1985) and Gossmann et al. (1997).

In some experimental set-ups, notably during postimplantation annealing, large gradients of the intrinsic point defects may occur. Such

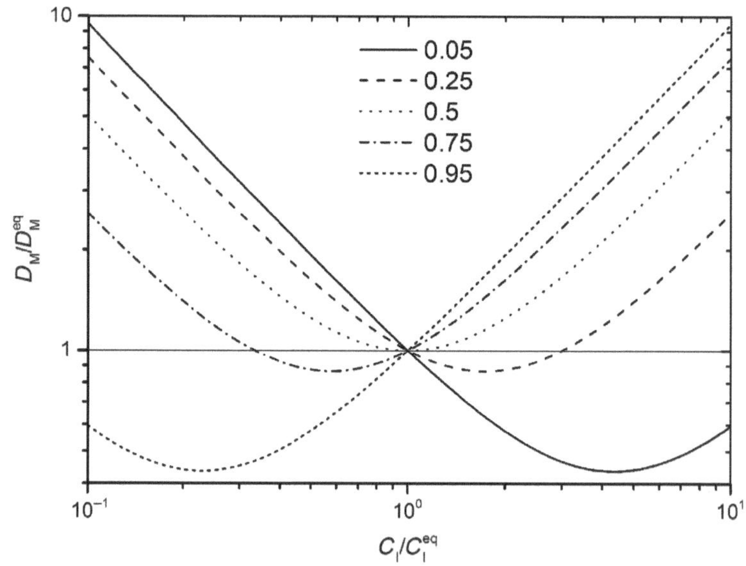

Figure 3 Normalized impurity diffusion coefficient when the intrinsic point defects are in a nonequilibrium situation ($C_I/C_I^{eq} = C_V^{eq}/C_V \neq 1$) for dopants with fractional diffusivities via self-interstitials f_I as given in the legend.

gradients, as discussed already above, lead to a transport of dopant atoms in the direction of the flux of the intrinsic point defects. This phenomenon was made responsible, e.g., by Rafferty et al. (1993) for a pile-up of dopants at the surface causing a reverse short-channel effect in MOSFETs. The concept was corroborated by Duffy et al. (2003) who demonstrated that the pile-up correlates with the depth of the end-of-range damage and, thus, with the gradient of the concentration of self-interstitials toward the surface.

While the influence of the intrinsic point defects on the diffusion of the dopants is apparent from the preceding discussions, it remains to point out that the diffusion of dopants may also influence the distribution of the intrinsic point defects. This influence results from the coupling terms R_{MI}, R_{MV}, $R_{\mathrm{MI+V}}$, and $R_{\mathrm{MV+I}}$ in the continuity equations of the intrinsic point defects (44) and (45). Simulations with our simplified system with a constant pair diffusion coefficient are shown in Fig. 4. The parameter varied is the ratio of the equilibrium transport capacities of the intrinsic point defects $D_{\mathrm{X}} \cdot C_{\mathrm{X}}^{\mathrm{eq}}$ and the pairs $D_{\mathrm{MX}} \cdot C_{\mathrm{MX}}^{\mathrm{surf}}$ where $C_{\mathrm{MX}}^{\mathrm{surf}}$ denotes the constant surface concentration of the pairs used as boundary condition in the simulation. It can be

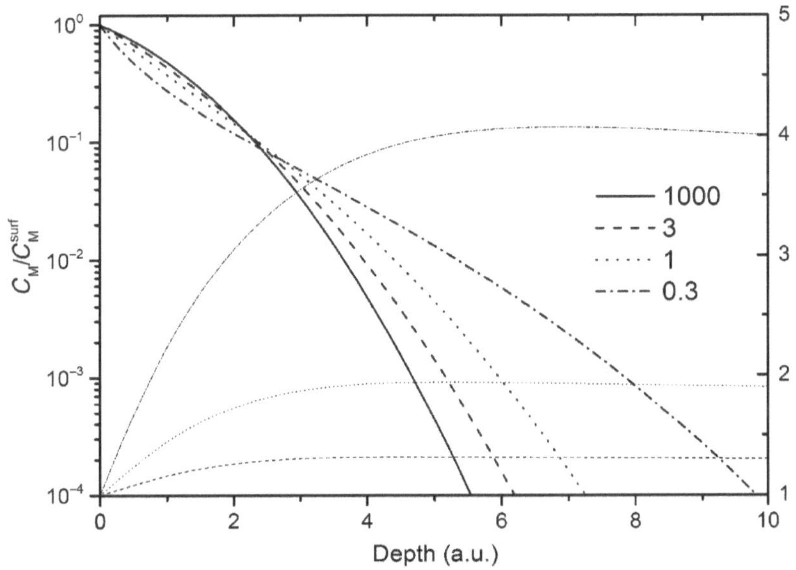

Figure 4 Depth profiles resulting from a diffusion process with a constant pair diffusion coefficient and a constant surface concentration $C_{\mathrm{M}}^{\mathrm{surf}}$. The corresponding oversaturation of intrinsic defects is drawn with thinner lines and shown on the right axis. The legend gives the ratio of the equilibrium transport capacities of intrinsic point defects and pairs.

seen that the intrinsic point defects maintain their equilibrium concentration as long as this ratio is much larger than unity. The profile shape of the dopants corresponds then again closely to the error function expected for the diffusion from a constant surface concentration with a constant diffusion coefficient. When the ratio of the equilibrium transport capacitances reduces and falls below unity, an oversaturation of intrinsic point defects forms in the bulk. The dopant profiles are then characterized by an inflection point and become increasingly deeper. This can be motivated by noting that the diffusion of pairs between dopants and intrinsic point defects bring both together into the bulk where they eventually dissociate. The dopants become substitutional while the intrinsic point defects are left over and increase their concentration in the bulk. The resulting oversaturation of intrinsic point defects in the bulk will be small if self-diffusion is effective enough to bring the left-over point defects rapidly back to the surface. The less effective self-diffusion is, the higher will be the oversaturation of the intrinsic point defects in the bulk. The parameters able to characterize the effectiveness of transport into the bulk and transport back to the surface are the aforementioned products of concentration and diffusion coefficient. The effect of dopant diffusion on the intrinsic point defects has first been recognized for phosphorus. It shows a prominent kink in high-concentration diffusion profiles (Duffy et al., 1968; Tsai, 1969) and was also found to accelerate the diffusion of other dopants separated spatially (Miller, 1960; Yeh, 1962). Similar effects were found for boron, although less pronounced (OrrArienzo et al., 1988).

The mathematical analysis of the influence of dopant diffusion on the intrinsic point defects has been pioneered by Schaake (1984) and Mathiot and Pfister (1985). Adding the continuity equations of the self-interstitials and vacancies in an appropriate way, one obtains

$$\frac{\partial C_{\mathrm{MI}}}{\partial t}+\frac{\partial C_{\mathrm{I}}}{\partial t}-\frac{\partial C_{\mathrm{MV}}}{\partial t}-\frac{\partial C_{\mathrm{V}}}{\partial t}=-\mathrm{div}(J_{\mathrm{MI}})-\mathrm{div}(J_{\mathrm{I}})+\mathrm{div}(J_{\mathrm{MV}})+\mathrm{div}(J_{\mathrm{V}}). \quad (72)$$

Since the intrinsic point defects diffuse much faster than the dopants, they will soon come into a steady state determined by the doping profile. This allows to assume that the sum of the time derivatives in (72) vanishes. Under certain constraining conditions such as independence of self-interstitials and vacancies or equilibrium between the two, the right-hand side can be integrated to obtain the point-defect concentrations as a function of space.

4. DOPANTS IN SILICON AND THEIR DIFFUSION MECHANISMS

In this section, the current state of knowledge about the diffusion of dopants in silicon is summarized. For a more complete account including parameters, the interested reader is referred to some reviews and research articles in this field (Jones, 1999; Pichler, 2004, chapter 5; Stolwijk and Bracht, 1998), and the references therein.

Of all the acceptor elements in silicon, boron has the highest solubility by far and one of the smallest diffusion coefficients. This makes it virtually the only choice for microelectronics. Still, its intrinsic diffusion coefficient is high, similar to that of phosphorus and by about one order of magnitude higher than those of arsenic and antimony. For boron, the enhanced diffusion in oxidizing ambient and retarded diffusion in nitriding ambient (see below) made clear that boron diffuses at least at low concentrations predominantly via self-interstitials. Boron interstitials were identified by electron paramagnetic resonance (EPR) measurements and correlated to deep level transient spectroscopy (DLTS) measurements (Harris et al., 1987). From these measurements it was concluded that the donor level of the boron interstitial is located above the acceptor level so that these levels are in negative-U ordering. Accordingly, the neutral charge state should not be stable. On the other hand, based particularly on the work of De Salvador et al. (2006, 2010), the current understanding of boron diffusion in silicon is that it proceeds predominantly via neutral pairs with self-interstitials. Under intrinsic conditions, the neutral pairs form by a reaction of substitutional boron atoms with neutral self-interstitials and the subsequent reaction with a hole. Under extrinsic p-doping conditions, the substitutional boron reacts predominantly with doubly positively charged self-interstitials, followed by the release of a hole. For the saddle points of the reactions, energies of 4.1 and 4.4 eV were determined, respectively. In addition to the neutral pair, a small contribution to diffusion comes from a negatively charged pair. It becomes significant only under counterdoping conditions. Negative and neutral pairs lead to a macroscopic diffusion coefficient that is independent of and linearly increasing with the hole concentration, respectively.

Aluminum as the acceptor element with the highest diffusion coefficient has some applications in power electronics since it allows to achieve similarly deep p-n junctions than with boron with a significantly smaller thermal budget. There is agreement that aluminum diffuses nearly entirely via self-interstitials. Because of its low solubility, the diffusion of aluminum alone will hardly occur under extrinsic conditions. Some diffusion studies indicated enhanced diffusion in an extrinsic boron background. The insufficient experimental basis does not allow to draw farther-reaching conclusions, though.

After having been used intensively at the beginning of the silicon semiconductor area, gallium has now a marginal role. This is caused in part also by its high diffusivity in silicon dioxide which makes it virtually impossible to use oxide as diffusion barrier. Several investigations made clear that gallium diffuses predominantly via self-interstitials. From the diffusion of gallium in an extrinsic boron background, gallium appears to diffuse under intrinsic conditions predominantly as negatively charged pair, i.e., with a macroscopic diffusion coefficient which is independent of the hole concentration. For extrinsic conditions, neutral and positively charged pairs contribute with the latter one dominating for very high hole concentrations. Diffusion via these pairs is reflected by a macroscopic diffusion coefficient with components that increase linearly and quadratic with the hole concentration. The experimental basis is limited, though.

Because of its high ionization energy, indium has been used for infrared detectors. Its segregation into oxides enables the fabrication of steep retrograde channel profiles. Indium appears to diffuse predominantly via complexes with self-interstitials. The solubility of indium is too low to cause extrinsic conditions during diffusion. Diffusion in an extrinsic boron background indicated enhanced diffusion. The insufficient experimental basis does not allow to draw farther-reaching conclusions, though.

Although being a Group-V element, nitrogen has been found to be an ineffective donor in silicon with a maximum concentration below $10^{16}\ \mathrm{cm}^{-3}$.

Of the acceptors, particularly because of the economic possibility of doping from the gas phase, phosphorus was the dopant of choice at the beginning of the silicon semiconductor area and still is for power semiconductors and solar cells. Because of its high solubility it is also used in microelectronics despite its rather high diffusion coefficient which exceeds those of arsenic and antimony at intrinsic conditions by roughly an order of magnitude. Its diffusion behavior was long a matter of dispute. Diffusion profiles after diffusion from the gas phase at high concentrations show prominent kink and tail features (Duffy et al., 1968; Tsai, 1969). In addition, phosphorus was found to accelerate the diffusion of other dopants separated spatially (Miller, 1960; Yeh, 1962), the so-called emitter-push effect in bipolar transistors. Based in part on the early identification of the phosphorus-vacancy pair, the so-called E-center, by a variety of experimental methods, models for the diffusion of phosphorus were based at first only on interactions with vacancies (e.g., Yoshida et al., 1974). Later, its enhanced diffusion in oxidizing ambient and retarded diffusion in nitriding ambient made clear that phosphorous diffuses at least at low concentrations predominantly via self-interstitials. From isoconcentration studies, this should be predominantly a neutral pair resulting in a macroscopic diffusion coefficient that increases linearly with the electron concentration. Complementary experiments made clear that the emitter-push effect is associated with an oversaturation of self-interstitials caused by the phosphorus. On that basis, many models were presented in literature based on the interaction of phosphorus with self-interstitials only. While it is undisputed now that phosphorus diffuses at low concentrations via self-interstitials, the arguments summarized by Mathiot and Pfister (1985) in favor of a vacancy mechanism appear still reasonable at high concentrations. In simulations, the reproduction of high-concentration phosphorus profiles remains a challenge and many models require assumptions like mobile clusters to reproduce the rather flat profiles. In the work of Mathiot and Pfister (1989), it was assumed that the diffusion at the highest concentrations is enhanced by the formation of a percolation cluster (see above).

Because of its high solubility and low diffusion coefficient, arsenic has become the most common donor element in microelectronics and nanoelectronics. While all other dopants diffuse at intrinsic concentrations nearly entirely via self-interstitials or vacancies, it became soon clear that arsenic diffuses to a substantial degree via both mechanisms. Based on a variety of experimental conditions including postimplantation annealing, Martinez-Limia et al. (2008) concluded that the fractional diffusivity of arsenic via self-interstitials increases from less than 0.05 at 800 °C to a maximum of 0.4 at around 1180 °C and decreases again for higher temperatures. Isoconcentration experiments indicated that arsenic diffuses predominantly via neutral pairs at intrinsic concentrations, complemented by a significant fraction of negatively charged pairs above a concentration of 2×10^{20} cm^{-3} (Hoyt and Gibbons, 1986; Nylandsted Larsen et al., 1986). The contributions of positive, neutral and negatively charged pairs lead to components of the macroscopic diffusion coefficient which are independent of the electron concentration or increase linearly and quadratic, respectively.

Antimony has an intrinsic diffusion coefficient similar to that of arsenic. Although its equilibrium solubility is smaller, significantly higher values

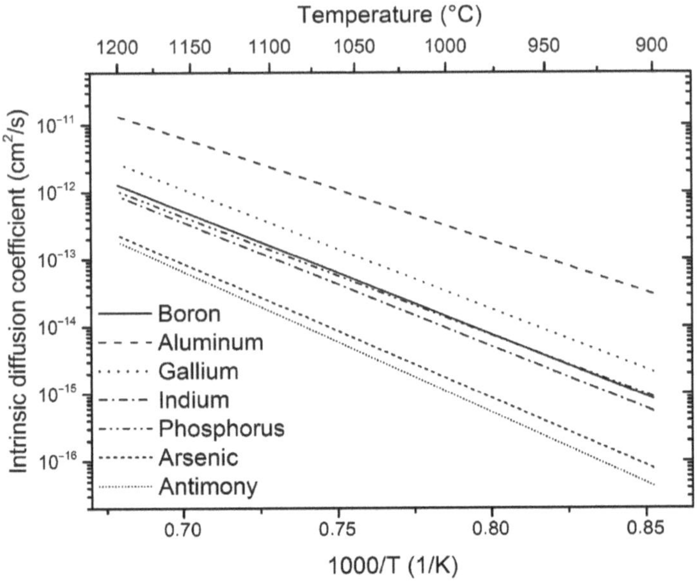

Figure 5 Intrinsic diffusion coefficients of the dopants in silicon reproduced from Pichler (2004).

demonstrated under nonequilibrium conditions in a strained environment make it a viable candidate for microelectronics again. Based on the general agreement that antimony diffuses nearly entirely via a vacancy mechanism, retarded rather than enhanced diffusion is expected during postimplantation annealing. The diffusion via vacancies made antimony also an element of choice for diffusion investigations as it allows to characterize an oversaturation of vacancies during high-temperature processing. Diffusion of antimony at extrinsic concentrations indicated that it diffuses via positive, neutral, and negatively charged pairs with vacancies. Accordingly, the macroscopic diffusion coefficient has components which are independent of the electron concentration or increase linearly and quadratic, respectively.

For an overview, the intrinsic diffusion coefficients of the dopants are reproduced in Fig. 5 from Pichler (2004) in the temperature range from 900 to 1200 °C.

5. NONEQUILIBRIUM PROCESSES

Technological processes often give rise to a perturbation of the concentrations of the intrinsic point defects. In this section, the most important phenomena will be briefly discussed. For an extensive review, the interested reader is referred particularly to section XIV of the review of Fahey et al. (1989).

During high-temperature processing in atmospheres containing oxygen or traces of water, the oxygen or water reacts with the silicon to silicon dioxide. In comparison to annealing in inert ambient, the diffusion of phosphorus (Masetti et al., 1973) and boron (Antoniadis et al., 1978) were found to be enhanced while the diffusion of antimony was found to be retarded (Mizuo and Higuchi, 1981). Supported by the growth and shrinkage of extended defects identified as agglomerates of self-interstitials, the current understanding of oxidation-enhanced and retarded diffusion is based on the explanation of Dobson (1971) and Hu (1974): During oxidation, not all of the silicon atoms from the consumed silicon layer are oxidized. Most of them will be incorporated into the growing silicon dioxide layer while a small fraction segregates into the silicon and increases there the concentration of self-interstitials. In simulation programs, the generation of self-interstitials is usually taken into consideration via the boundary condition for the self-interstitials. Previously, following the work of Lin et al. (1981), the oversaturation of self-interstitials was modeled as a function of the oxide growth rate dx_{ox}/dt in the form

$$C_I - C_I^{eq} = K \cdot \left(\frac{dx_{ox}}{dt}\right)^n \tag{73}$$

with exponents n between 0.2 and 0.3. Today, it is more common to assume in the tradition of Hu (1974) that a fraction Θ of the silicon volume consumed by the oxidation is injected as self-interstitials into the silicon volume. Formulated as boundary condition for the continuity equation of the self-interstitials, the flux $\langle J_I, \vec{n} \rangle$ perpendicular to the surface out of the silicon volume

$$\langle J_I, \vec{n} \rangle = k_I \cdot (C_I - C_I^{eq}) - \Theta \cdot C_{Si} \cdot \frac{dx_{ox}}{dt} \tag{74}$$

accounts for the recombination of self-interstitials at the surface and the injected self-interstitials. Often, the surface recombination rate k_I is assumed to depend on the oxidation rate to reproduce experimental findings (Law, 1991).

While the oxidation of (100)-oriented silicon samples is generally found to lead to an injection of self-interstitials, there was also evidence that oxidation of (111)-oriented silicon samples with high thermal budgets rather leads to an increase of the vacancy concentration (see, e.g., Tan and

Ginsberg, 1983). An explanation was suggested by Taniguchi et al. (1989) who considers in addition the segregation and diffusion of self-interstitials into the oxide.

In contrast to oxidation, nitridation of bare silicon surfaces in NH_3 leads to a retarded diffusion of boron and phosphorus as well as to an enhanced diffusion of antimony (e.g., Mizuo et al., 1983). While this was clear evidence for an oversaturation of vacancies and an undersaturation of self-interstitials, the mechanism behind was not clear. Whereas there is a clear correlation between oxide growth and self-interstitial oversaturation, nitride growth stops with a thickness of few nm while the effects on diffusion are observed on a much longer timescale. In addition, it was found that deposited nitride layers have a similar effect on dopant diffusion (Ahn et al., 1988). The experiment also indicated that stress in the nitride layer plays a direct role. However, the correlation with diffusion in silicon was unclear since the thick silicon was hardly strained by the thin nitride layer. An explanation was finally given by Cowern (2007) on the basis that the formation of a point defect in the bulk of a crystal requires the transfer of a host atom from or to the interface to the nitride layer. And there, the work done against the overlayer has to be taken into account for the calculation of the formation energy. Currently, it remains to be discussed which role such stress effects may have on the intrinsic point defects during oxidation.

Nitridation of oxide layers was found to lead to similar effects as oxidation (Mizuo et al., 1983). This is explained by an oxidation of the silicon by oxygen atoms liberated by the nitridation of the oxide.

Clear evidence for a strongly enhanced concentration of vacancies during the growth of cobalt and titanium silicides came from the observation of an enhanced diffusion of buried antimony layers by Honeycutt and Rozgonyi (1991). Previously, Hu (1987) reported the enhanced diffusion of buried antimony and boron layers during the annealing of silicon samples with a deposited tantalum silicide layer. While Hu concluded on a substantial vacancy component in the diffusion of boron, a clear understanding of such effects particularly in light of the complicated silicide reactions and possible strain effects according to the model of Cowern (2007) is still missing.

Closely associated with the generation of self-interstitials by oxidation is the injection of self-interstitials into the bulk by oxygen precipitation. With the oxidation of silicon being associated with a volume increase by a factor of more than two, the excess volume required for the precipitates has to be provided either via the injection of silicon atoms to interstitial positions or the consumption of vacancies. Similarly, the precipitation of phosphorus was found by Bourret and Schröter (1984) to be associated with an emission of self-interstitials in the bulk, which leads to an enhanced diffusion of the non-precipitated phosphorus atoms. Very high concentrations of boron were also found by Agarwal et al. (1999) to induce an increased concentration of self-interstitials. Associated with the formation of a silicon boride phase, the phenomenon is referred to as boron-enhanced diffusion.

The certainly largest source of nonequilibrium conditions is ion implantation. Due to the nonconservative nature of the implantation process, samples will experience a large oversaturation of self-interstitials during annealing. Readers interested in the phenomena associated are referred to "Ion implantation defects and shallow junctions in Si and Ge" by Enrico Napolitani and Giuliana Impellizzeri.

6. PRECIPITATES, CLUSTERS, AND COMPLEXES

Dopants being present in high concentrations, as they are desired particularly in contact regions, tend to become electrically inactive by the formation of dopant-rich precipitates and clusters. In this context, the term precipitate is used for volume defects comprising a large number of dopants while clusters will be used for point-like defects including only few dopant atoms. In addition, complexes with other dopants or other impurities may form. Within this section, basic knowledge from experiments and the implementation in simulation programs are briefly discussed.

6.1. Dopant phases and precipitates

Whenever the dopant concentration exceeds locally a certain limit known as solid solubility, an impurity-rich phase is expected to form beside the silicon phase according to the respective phase diagram (see, e.g., Massalski et al., 1990). Above the eutectic temperature of the system or the silicon-rich subsystem, the equilibrium phase is a liquid saturated with silicon. Below the eutectic temperature, the equilibrium phase is either a pure dopant phase (Al, Ga, In, Sb) or a silicide (SiB_6, SiP, SiAs). Dynamically, formation of the new phase can occur by spinodal decomposition or, more likely in silicon, by nucleation and growth of precipitates. However, being limited by kinetics, phase formation does not necessarily occur during a particular process even if the solid solubility is exceeded. And even if a dopant-rich phase forms, it is not sure whether it is the equilibrium phase.

Experimentally, phase formation was observed in a variety of systems. For boron, an SiB_6 layer was found to form during predeposition from the gas phase (see, e.g., Pignatel and Queirolo, 1980). This layer maintains the boron concentration at solid solubility during in-diffusion. During the predeposition of phosphorus from the gas phase, so many phosphorus atoms may segregate for particular process conditions from the growing phosphosilicate layer into the silicon that the solubility of phosphorus is exceeded and SiP precipitates form (Bourret and Schröter, 1984; Masetti et al., 1977). Nowadays, ion implantation allows to dope samples considerably above solid solubility. In such a system, the formation of SiAs precipitates was reported (Nobili et al., 1994). For antimony, precipitates having formed originally around the projected range of the implantation were found to dissolve and to form at the surface (Koffel et al., 2012).

Dopants in precipitates are usually electrically neutralized and immobilized. In addition, as mentioned above, dopant precipitation is often found to be associated with an injection of intrinsic point defects to accommodate the changes in molar volume. Although approaches have been reported in literature to model dopant precipitation (Brabec et al., 1989; Dunham, 1995), none of them is customarily used because of the computational overhead. Instead, simple models are commonly applied that consider diffusion only below solid solubility, model some first-order kinetics, or use cluster models as discussed below.

6.2. Dopant clusters

Already at concentrations below solid solubility, a variety of complexes may form that comprise dopant atoms and intrinsic point defects. A typical example is arsenic where the solid solubility is roughly one order of magnitude higher than the electron concentration (Nobili et al., 1994). The difference between the two is generally explained by the formation of clusters, i.e., small point-like structures that comprise only few dopant atoms and intrinsic point defects. In particular for arsenic, the clusters were thought to form from substitutional arsenic atoms, vacancies, and electrons according to

$$kV^l + m\mathrm{As}_s^+ + ne^- \underset{k_\leftarrow}{\overset{k_\rightarrow}{\rightleftharpoons}} (\mathrm{As}_m\mathrm{V}_k)^{n-m+k\cdot l} \tag{75}$$

with the charge states indicted as superscripts. Typically, the first clusters suggested comprised one vacancy and two to four arsenic atoms, and were electrically neutral. For such clusters, electrically active and unclustered arsenic atoms coincide and their concentration increases monotonically, albeit strongly sublinear, with the total concentration. To achieve a saturation of the electron concentration, Tsai et al. (1980) proposed a doubly positively charged cluster of three arsenic atoms. While such a cluster ensures a saturation of the unclustered concentration of dopants, no saturation of the electron concentration could result. Therefore, the authors had to assume in addition that the clusters become electrically neutral at room temperature. Direct evidence for the involvement of vacancies in such clusters came from extended X-ray-absorption fine-structure (EXAFS) measurements but also from reports that the deactivation of arsenic leads to an emission of self-interstitials (Rousseau et al., 1994).

While the discussion of arsenic clustering continued, boron clustering emerged as a new priority after it became clear that its solubility during postimplantation annealing showed a transient reduction with recovered equilibrium values only after long-time annealing (Cowern et al., 1990a). To explain such effects, clusters comprising boron atoms and self-interstitials were invoked. Supported by early *ab initio* simulations (Caturla et al., 1998), models soon assumed coexisting boron-interstitial clusters (BICs). With the notation B_mI_n referring to m boron atoms and n self-interstitials somehow arranged around m lattice sites, the growth of BICs is assumed to proceed either via the reaction with self-interstitials

$$B_mI_n + I \underset{k_{\leftarrow}}{\overset{k_{\rightarrow}}{\rightleftharpoons}} B_mI_{n+1} \tag{76}$$

or via the reaction with boron interstitials

$$B_mI_n + B_i \underset{k_{\leftarrow}}{\overset{k_{\rightarrow}}{\rightleftharpoons}} B_{m+1}I_{n+1}. \tag{77}$$

Development of such models was supported by numerous *ab initio* simulations. Comprising nowadays up to 10 boron atoms and self-interstitials, they were shown in many publications to be able to explain experiments in a wide range of experimental conditions. However, first indications that even much larger boron-interstitial complexes can form emerged already from the work of Cristiano et al. (2003). After dedicated atom probe measurements became available, it became clear that complexes with 5–50 boron atoms may form in the shape of platelets on (001) planes (Ngamo et al., 2009). Their influence on the transient deactivation and immobilization of boron is still unclear, though.

Inspired by the success of the BIC models but also due to the need to simulate transient activation phenomena in samples implanted with arsenic and phosphorus under conditions of an oversaturation of self-interstitials, the methodology of BICs was extended to arsenic and phosphorus and reactions with self-interstitials and vacancies (Harrison et al., 2006; Sahli et al., 2008) so that the respective models now cover phosphorus and arsenic coclustering for the full range of nonequilibrium conditions for the intrinsic point defects.

6.3. Ion pairs

In the classical sense, a pair of an acceptor ion and a donor ion on neighboring substitutional sites is referred to as ion pair. To characterize the stability of the pair, an ion pairing coefficient Ω is usually introduced in the form

$$\Omega = \frac{C_{\mathrm{P}}}{(C_{\mathrm{D}} - C_{\mathrm{P}}) \cdot (C_{\mathrm{A}} - C_{\mathrm{P}})} = \frac{4}{C_{\mathrm{Si}}} \cdot \exp\left(\frac{G^{\mathrm{B}}}{k \cdot T}\right) \tag{78}$$

with C_{P}, C_{D}, and C_{A} standing for the concentration of pairs and the total concentrations of donors and acceptors. G^{B} is the binding energy of the ion pair. Ion pairs are usually assumed to be electrically neutral and immobile. If one of the dopants, here for example the donor, is present in high concentration while the other is present in low concentration, the latter will have only a small effect on the diffusion of the former. On the other hand, in our example, the effective diffusion coefficient of the acceptors will be retarded to

$$D_{\mathrm{A}}^{\mathrm{eff}} = \frac{D_{\mathrm{A}}}{1 + \Omega \cdot C_{\mathrm{D}}}. \tag{79}$$

For high donor concentrations with $n/n_{\mathrm{i}} \gg 1$ and $p/n_{\mathrm{i}} \ll 1$, the diffusion coefficient D_{A} will reduce below its intrinsic value and saturate at the Fermi-level-independent terms of the diffusion coefficient in (64). However, with ion pairing, the effective acceptor diffusion coefficient will reduce indirectly proportional to the donor concentration.

In the work of Cowern (1989), ion pairing of boron with phosphorus, arsenic and antimony was described by an ion pairing coefficient of $\Omega = 0.17/n_{\mathrm{i}}$ with n_{i} being again the intrinsic charge carrier concentration. Interpreted in terms of a binding energy, this corresponds to 0.61 eV at 900 °C. Pairs of indium with donors were found to be energetically less favorable with an ion pairing coefficient of about $\Omega = 0.03/n_{\mathrm{i}}$ for pairs with phosphorus and arsenic, and about $\Omega = 0.015/n_i$ for pairs with antimony. At 900 °C, these values correspond to binding energies of about 0.44 and 0.37 eV, respectively.

Although usually assumed to be immobile, mobile ion pairs were suggested in some publications (Margesin et al., 1991; Solmi et al., 1995) to explain certain diffusion phenomena when donors and acceptors are present in similar and high concentrations.

7. INTERFACE SEGREGATION

A pile-up of donors at the silicon–silicon dioxide interface during oxidation has been reported early (Deal et al., 1965) and can be understood on the basis of the limited segregation of these elements into the silicon dioxide as a snow-plowing effect. However, it was reported soon that phosphorus (Johannessen et al., 1978; Schwarz et al., 1981) as well as arsenic and antimony (Sai-Halasz et al., 1985) pile-up even during inert annealing in a thin layer at the interface with a concentration that exceeds the bulk concentration by an order of magnitude. The current knowledge about such effects (see Dabrowski et al., 2002; Pei et al., 2008; Steen et al., 2008, and the references therein) is that the pile-up occurs within about one to few nm at the silicon side of the interface to silicon dioxide. There, they can be removed by a dip in HF or by an SC1/SC2 clean. Depending on the particular implant and annealing conditions, a substantial fraction of the implanted donors can be bound in the pile-up region. These dopants do not contribute significantly to conduction either because they are electrically inactive or because of their deleterious influence on mobility. Both the segregated doses of phosphorus and arsenic increase for low bulk concentrations linearly and for high bulk concentrations quadratic with the bulk concentration. While the quadratic increase has been explained also by the formation of donor pairs, one has to remark that the quadratic dependence may result already from the segregation itself. Assuming that the positive charge of a substitutional donor atom needs to be compensated in the reaction $D_s^+ + e^- \rightleftharpoons D_{seg}^0$ to an electrically neutral arsenic atom in the pile-up layer by an electron, the concentration of the latter will be proportional to the concentration of substitutional donors and to the electron concentration. For intrinsic donor concentrations, the electron concentration corresponds to the intrinsic charge carrier concentration, giving a linear relationship between segregated and substitutional atoms. For extrinsic donor concentrations, the electron concentration will increase with the donor concentration, resulting in a quadratic dependence.

Within continuum simulation programs, interface segregation is modeled via an approach introduced by Lau et al. (1989) and Orlowski (1989). In addition to the silicon and silicon dioxide phases, they introduced the interface as a third phase. The interface phase is characterized by a limited sheet concentration of sites N_M^S for dopants of which N_M can be assumed to be already occupied. Changes in the sheet concentration of dopants

$$\frac{\partial N_M}{\partial t} = J_t^{SiO_2} - J_e^{SiO_2} + J_t^{Si} - J_e^{Si} \tag{80}$$

result from the trapping and the emission of dopants from and to the silicon dioxide and silicon phases. While the emission of dopants

$$J_e = k_e \cdot N_M \tag{81}$$

can be assumed to be proportional to the sheet concentration of the dopants in the interface phase, trapping can only occur only as long as there are free sites in the interface phase. This is taken into account by modeling the trapping rate

$$J_t = k_t \cdot \left(N_M^S - N_M\right) \cdot C_M \tag{82}$$

as product of the sheet concentration of free sites $N_M^S - N_M$ and the concentration of dopants in the phase from which the dopant is trapped. In a modification proposed by Oh and Ward (1998), the emission rate (81) contains a similar term $C_M^{max} - C_M$ that should prevent that the donor concentration exceeds the "maximum soluble concentration of dopant in material." From the point of view of statistical thermodynamics, C_M^{max} should be rather the concentration of sites, i.e., the concentration of silicon atoms. And exactly what should be prevented is experimentally observed: During phosphorus predeposition from the gas phase, as mentioned above, so many phosphorus atoms segregate from the growing phosphosilicate glass into the silicon phase that the binary solubility phosphorus–silicon is exceeded there and SiP precipitates form (Bourret and Schröter, 1984; Masetti et al., 1977).

For boron, pile-up at the interface was observed too and made responsible for the reverse short-channel effect in MOSFETs (Rafferty et al., 1993) or suggested as an approach for the formation of shallow junctions (Wang et al., 2001). However, it was associated from the beginning with dopant transport caused by a large gradient of self-interstitials near the surface during postimplantation annealing. This was corroborated by the work of Duffy et al. (2003) who demonstrated that the effect is correlated with the depth of the end-of-range damage.

REFERENCES

Agarwal, A., Gossmann, H.J., Eaglesham, D.J., 1999. Boron-enhanced diffusion of boron: physical mechanisms. Appl. Phys. Lett. 74, 2331–2333.

Ahn, S.T., Kennel, H.W., Plummer, J.D., Tiller, W.A., 1988. Film stress-related vacancy supersaturation in silicon under low-pressure chemical vapor deposited silicon nitride films. J. Appl. Phys. 64, 4914–4919.

Aleksandrov, O.V., Kozlovskii, V.V., Popov, V.V., Samorukov, B.E., 1988. Diffusion of impurities from implanted silicon layers by rapid thermal annealing. Phys. Status Solidi A 110, K61–K65.

Altermatt, P.P., Schenk, A., Heiser, G., 2006a. A simulation model for the density of states and for incomplete ionization in crystalline silicon. I. Establishing the model in Si:P. J. Appl. Phys. 100, 113714.

Altermatt, P.P., Schenk, A., Schmithüsen, B., Heiser, G., 2006b. A simulation model for the density of states and for incomplete ionization in crystalline silicon. II. Investigation of Si: As and Si:B and usage in device simulation. J. Appl. Phys. 100, 113715.

Antoniadis, D.A., Gonzales, A.G., Dutton, R.W., 1978. Boron in near-intrinsic ‹100› and ‹111› silicon under inert and oxidizing ambients—diffusion and segregation. J. Electrochem. Soc. 125, 813–819.

Bourret, A., Schröter, W., 1984. HREM of SiP precipitates at the (111) silicon surface during phosphorus predeposition. Ultramicroscopy 14, 97–106.

Brabec, T., Schrems, M., Budil, M., Poetzl, H.W., Kuhnert, W., Pongratz, P., Stingeder, G., Grasserbauer, M., 1989. Modeling of antimony precipitation in silicon. J. Electrochem. Soc. 136, 1542–1545.

Bunea, M.M., Dunham, S.T., 1997. Lattice Monte Carlo simulations of vacancy-mediated diffusion and aggregation using ab-initio parameters. In: Diaz de la Rubia, T., Coffa, S., Stolk, P.A., Rafferty, C.S. (Eds.), Defects and Diffusion in Silicon Processing. Mat. Res. Soc. Symp. Proc, vol. 469. pp. 353–358.

Caturla, M.J., Johnson, M.D., Diaz de la Rubia, T., 1998. The fraction of substitutional boron in silicon during ion implantation and thermal annealing. Appl. Phys. Lett. 72, 2736–2738.

Cowern, N.E.B., 1988. General model for intrinsic dopant diffusion in silicon under non-equilibrium point-defect conditions. J. Appl. Phys. 64, 4484–4490.

Cowern, N.E.B., 1989. Ion pairing effects on substitutional impurity diffusion in silicon. Appl. Phys. Lett. 54, 703–705.

Cowern, N.E.B., 2007. Diffusion in a single crystal within a stressed environment. Phys. Rev. Lett. 99, 155903.

Cowern, N.E.B., Janssen, K.T.F., Jos, H.F.F., 1990a. Transient diffusion of ion-implanted B in Si: dose, time, and matrix dependence of atomic and electrical profiles. J. Appl. Phys. 68, 6191–6198.

Cowern, N.E.B., Janssen, K.T.F., van de Walle, G.F.A., Gravesteijn, D.J., 1990b. Impurity diffusion via an intermediate species: the B-Si system. Phys. Rev. Lett. 65, 2434–2437.

Cristiano, F., Hebras, X., Cherkashin, N., Claverie, A., Lerch, W., Paul, S., 2003. Clusters formation in ultralow-energy high-dose boron-implanted silicon. Appl. Phys. Lett. 83, 5407–5409.

Dabrowski, J., Müssig, H.J., Zavodinsky, V., Baierle, R., Caldas, M.J., 2002. Mechanism of dopant segregation to SiO_2/Si(001) interfaces. Phys. Rev. B 65, 245305.

Deal, B.E., Grove, A.S., Snow, E.H., Sah, C.T., 1965. Observation of impurity redistribution during thermal oxidation of silicon using the MOS structure. J. Electrochem. Soc. 112, 308–314.
Debye, P., 1942. Reaction rates in ionic solutions. Trans. Electrochem. Soc. 82, 265–272.
De Salvador, D., Napolitani, E., Mirabella, S., Bisognin, G., Impellizzeri, G., Carnera, A., Priolo, F., 2006. Atomistic mechanism of boron diffusion in silicon. Phys. Rev. Lett. 97, 255902.
De Salvador, D., Napolitani, E., Bisognin, G., Pesce, M., Carnera, A., Bruno, E., Impellizzeri, G., Mirabella, S., 2010. Boron diffusion in extrinsically doped crystalline silicon. Phys. Rev. B 81, 045209.
Dobson, P.S., 1971. The effect of oxidation on anomalous diffusion in silicon. Philos. Mag. 24, 567–576.
Duffy, M.C., Barson, F., Fairfield, J.M., Schwuttke, G.H., 1968. Effects of high phosphorus concentration on diffusion into silicon. J. Electrochem. Soc. 115, 84–88.
Duffy, R., Venezia, V.C., Heringa, A., Hüsken, T.W.T., Hopstaken, M.J.P., Cowern, N.E. B., Griffin, P.B., Wang, C.C., 2003. Boron uphill diffusion during ultrashallow junction formation. Appl. Phys. Lett. 82, 3647–3649.
Dunham, S.T., 1995. Modeling of the kinetics of dopant precipitation in silicon. J. Electrochem. Soc. 142, 2823–2828.
Dunham, S.T., Wu, C.D., 1995. Atomistic models of vacancy-mediated diffusion in silicon. J. Appl. Phys. 78 (4), 2362–2366.
Fahey, P., Barbuscia, G., Moslehi, M., Dutton, R.W., 1985. Kinetics of thermal nitridation processes in the study of dopant diffusion mechanisms in silicon. Appl. Phys. Lett. 46, 784–786.
Fahey, P.M., Griffin, P.B., Plummer, J.D., 1989. Point defects and dopant diffusion in silicon. Rev. Mod. Phys. 61, 289–384.
Fair, R.B., 1981. Concentration profiles of diffused dopants in silicon. In: Wang, F.F.Y. (Ed.), Impurity Doping Processes in Silicon. North-Holland, Amsterdam, pp. 315–442.
Fichtner, W., 1983. Process simulation. In: Sze, S.M. (Ed.), VLSI Technology. McGraw-Hill Internatioal Book Company, Auckland Bogotá Guatemala Hohannesburg Lisbon London Madrid Mexico new Delhi Panama Paris San Juan Sao Paulo Singapore Sidney Tokyo, pp. 385–444.
Frank, F.C., Turnbull, D., 1956. Mechanism of diffusion of copper in germanium. Phys. Rev. 104 (3), 617–618.
Gösele, U., Tan, T.Y., 1983. The nature of point defects and their influence on diffusion processes in silicon at high temperatures. In: Mahajan, S., Corbett, J.W. (Eds.), Defects in Semiconductors II. North-Holland, New York, NY, pp. 45–59.
Gösele, U., Frank, W., Seeger, A., 1980. Mechanism and kinetics of the diffusion of gold in silicon. Appl. Phys. 23, 361–368.
Gossmann, H.J., Haynes, T.E., Stolk, P.A., Jacobson, D.C., Gilmer, G.H., Poate, J.M., Luftman, H.S., Mogi, T.K., Thompson, M.O., 1997. The interstitial fraction of diffusivity of common dopants in Si. Appl. Phys. Lett. 71, 3862–3864.
Harris, R.D., Newton, J.L., Watkins, G.D., 1987. Negative-U defect: interstitial boron in silicon. Phys. Rev. B 36, 1094–1104.
Harrison, S.A., Edgar, T.F., Hwang, G.S., 2006. Interstitial-mediated arsenic clustering in ultrashallow junction formation. Electrochem. Solid-State Lett. 9, G354–G357.
Heinrich, M., Budil, M., Pötzl, H.W., 1990. Determination of diffusion parameters for arsenic. In: Wolford, D.J., Bernholc, J., Haller, E.E. (Eds.), Impurities, Defects and Diffusion in Semiconductors: Bulk and Layered Structures. Mat. Res. Soc. Symp. Proc, vol. 163. pp. 535–541.
Honeycutt, J.W., Rozgonyi, G.A., 1991. Enhanced diffusion of Sb-doped layers during Co and Ti reactions with Si. Appl. Phys. Lett. 58, 1302–1304.

Hoyt, J.L., Gibbons, J.F., 1986. Rapid thermal annealing of As in Si. In: Sedgwick, T.O., Seidel, T.E., Tsaur, B.Y. (Eds.), Rapid Thermal Processing. Mat. Res. Soc. Symp. Proc, vol. 52. pp. 15–22.
Hu, S.M., 1973a. Diffusion in silicon and germanium. In: Shaw, D. (Ed.), Atomic Diffusion in Semiconductors. Plenum Press, London, pp. 217–350.
Hu, S.M., 1973b. On interaction potential, correlation factor, vacancy mobility, and activation energy of impurity diffusion in diamond lattice. Phys. Status Solidi B 60, 595–603.
Hu, S.M., 1974. Formation of stacking faults and enhanced diffusion in the oxidation of silicon. J. Appl. Phys. 45, 1567–1573.
Hu, S.M., 1987. Point defect generation and enhanced diffusion in silicon due to tantalum silicide overlayers. Appl. Phys. Lett. 51, 308–310.
Jaraiz, M., 2004. Atomistic simulations in materials processing. In: Dabrowski, J., Weber, E.R. (Eds.), Predictive Simulation of Semiconductor Processing. Springer Series in Materials Science, vol. 72. pp. 73–109.
Johannessen, J.S., Spicer, W.E., Gibbons, J.F., Plummer, J.D., Taylor, N.J., 1978. Observation of phosphorus pile-up at the SiO_2-Si interface. J. Appl. Phys. 49, 4453–4458.
Johnson, R.P., 1939. A note on the hole theory of diffusion. Phys. Rev. 56, 814–818.
Jones, K., 1999. Dopants in silicon. In: Hull, R. (Ed.), Properties of Crystalline Silicon. EMIS Datareviews Series, vol. 20. INSPEC, London (Chapter 10).
Jung, M.Y.L., Gunawan, R., Braatz, R.D., Seebauer, E.G., 2004. Effect of near-surface band bending on dopant profiles in ion-implanted silicon. J. Appl. Phys. 95, 1134–1140.
Koffel, S., Pichler, P., Reading, M.A., Van den Berg, J., Kheyrandish, H., Hamm, S., Lerch, W., Pakfar, A., Tavernier, C., 2012. Precipitation of antimony implanted into silicon. ECS Trans. 41 (34), 9–17.
Kozlovskii, V.V., Lomasov, V.N., Gur'yanov, G.M., Kovarskii, A.P., 1984. Proton-stimulated diffusion of antimony in silicon. Sov. Phys. Semicond. 18 (5), 598–599.
Kozlovskii, V.V., Lomasov, V.N., Marushchak, N.V., 1985. Diffusion of a substitutional impurity in a crystal irradiated with ions. Sov. Phys. Tech. Phys. 30 (11), 1283–1285.
Lau, F., Mader, L., Mazure, C., Werner, C., Orlowski, M., 1989. A model for phosphorus segregation at the silicon-silicon dioxide interface. Appl. Phys. A 49, 671–675.
Law, M.E., 1991. Parameters for point-defect diffusion and recombination. IEEE Trans. Comput. Aided Des. 10, 1125–1131.
Lehovec, K., Slobodskoy, A., 1961. Diffusion of charged particles into a semiconductor under consideration of the built-in field. Solid State Electron. 3, 45–50.
Lin, A.M., Dutton, R.W., Antoniadis, D.A., Tiller, W.A., 1981. The growth of oxidation stacking faults and the point defect generation at Si-SiO interface during thermal oxidation of silicon. J. Electrochem. Soc. 128, 1121–1130.
Loualiche, S., Lucas, C., Baruch, P., Gailliard, J.P., Pfister, J.C., 1982. Theoretical model for radiation enhanced diffusion and redistribution of impurities. Phys. Status Solidi A 69, 663–676.
Margesin, B., Canteri, R., Solmi, S., Armigliato, A., Baruffaldi, F., 1991. Boron and antimony codiffusion in silicon. J. Mater. Res. 6, 2353–2361.
Martinez-Limia, A., Pichler, P., Steen, C., Paul, S., Lerch, W., 2008. Modeling and simulation of advanced annealing processes. Mater. Sci. Forum 573–574, 279–293.
Maser, K., 1991. Die Rolle der Überkreuz-Komponenten beim Dotandentransport im Festkörper. Exp. Tech. Phys. 39 (2), 169–180.
Masetti, G., Solmi, S., Soncini, G., 1973. On phosphorus diffusion in silicon under oxidizing atmospheres. Solid State Electron. 16, 1419–1421.
Masetti, G., Nobili, D., Solmi, S., 1977. Profiles of phosphorus predeposited in silicon and carrier concentration in equilibrium with SiP precipitates. In: Huff, H.R., Sirtl, E. (Eds.), Semiconductor Silicon 1977. Electrochem. Soc. Proc, vol. 77-2. pp. 648–657.

Massalski, T.B., Okamoto, H., Subramanian, P.R., Kacprzak, L. (Eds.), 1990. Binary Alloy Phase Diagrams, second ed. ASM International, Materials Park, OH.
Mathiot, D., Pfister, J.C., 1984. Dopant diffusion in silicon: a consistent view involving nonequilibrium defects. J. Appl. Phys. 55, 3518–3530.
Mathiot, D., Pfister, J.C., 1985. Evidences that P diffusion in Si is assisted mainly by vacancies. Appl. Phys. Lett. 47, 962–964.
Mathiot, D., Pfister, J.C., 1989. Dopant redistribution in heavily doped silicon: confirmation of the validity of the vacancy-percolation model. J. Appl. Phys. 66, 970–972.
Mehrer, H., 1971. Correlation factor for the diffusion of charged impurities in the diamond structure. Z. Naturforsch. 26a, 308–316.
Miller, L.E., 1960. Uniformity of junctions in diffused silicon devices. In: Gatos, H.C. (Ed.), Properties of Elemental and Compound Semiconductors. Interscience Publishers, New York, NY, pp. 303–322.
Mizuo, S., Higuchi, H., 1981. Retardation of Sb diffusion in Si during thermal oxidation. Jpn. J. Appl. Phys. 20, 739–744.
Mizuo, S., Kusaka, T., Shintani, A., Nanba, M., Higuchi, H., 1983. Effect of Si and SiO_2 thermal nitridation on impurity diffusion and oxidation induced stacking fault size in Si. J. Appl. Phys. 54, 3860–3866.
Ngamo, M., Duguay, S., Cristiano, F., Daoud-Ketata, K., Pareige, P., 2009. Atomic scale study of boron interstitial clusters in ion-implanted silicon. J. Appl. Phys. 105, 104904.
Nobili, D., Solmi, S., Parisini, A., Derdour, M., Armigliato, A., Moro, L., 1994. Precipitation, aggregation, and diffusion in heavily arsenic-doped silicon. Phys. Rev. B 49, 2477–2483.
Novell, P., Law, M.E., 1992. The effect of nondilute dopant-defect pair concentrations on arsenic diffusion. In: Proceedings of NUPAD IV. IEEE, New York, NY, pp. 41–44.
Nylandsted Larsen, A., Shiryaev, S.Y., Schwartz Sørensen, E., Tidemand-Petersson, P., 1986. Rapid thermal annealing of high concentration, arsenic implanted silicon single crystals. Appl. Phys. Lett. 48, 1805–1807.
Nylandsted Larsen, A., Andersen, P.E., Gaiduk, P., Kyllesbech Larsen, K., 1989. The effect of phosphorus background concentration on the diffusion of tin, arsenic and antimony in silicon. Mater. Sci. Eng. B 4, 107–112.
Nylandsted Larsen, A., Kyllesbech Larsen, K., Andersen, P.E., Svensson, B.G., 1993. Heavy doping effects in the diffusion of group IV and V impurities in silicon. J. Appl. Phys. 73, 691–698.
Oh, Y.S., Ward, D.E., 1998. A calibrated model for trapping of implanted dopants at material interface during thermal annealing. In: Technical Digest of the 1998 International Electron Devices Meeting (IEDM). IEEE, Piscataway, NJ, pp. 509–512.
Orlowski, M., 1989. New model for dopant redistribution at interfaces. Appl. Phys. Lett. 55, 1762–1764.
OrrArienzo, W.A., Glang, R., Lever, R.F., Lewis, R.K., Morehead, F.F., 1988. Boron diffusion in silicon at high concentrations. J. Appl. Phys. 63, 116–120.
Pandey, K.C., 1986. Diffusion without vacancies or interstitials: a new concerted exchange mechanism. Phys. Rev. Lett. 57 (18), 2287–2290.
Pei, L., Duscher, G., Steen, C., Pichler, P., Ryssel, H., Napolitani, E., De Salvador, D., Piro, A.M., Terrasi, A., Severac, F., Cristiano, F., Ravichandran, K., Gupta, N., Windl, W., 2008. Detailed arsenic concentration profiles at Si/SiO_2 interfaces. J. Appl. Phys. 104, 043507.
Pichler, P., 2004. Intrinsic Point Defects, Impurities, and Their Diffusion in Silicon. Springer-Verlag, Wien.
Pichler, P., Schork, R., Klauser, T., Ryssel, H., 1992. Direct experimental evidence for diffusion of dopants via pairs with intrinsic point defects. Appl. Phys. Lett. 60 (8), 953–955.

Pignatel, G.U., Queirolo, G., 1980. Further insight on boron diffusion in silicon obtained with Auger electron spectroscopy. Thin Solid Films 67, 233–238.

Rafferty, C.S., Vuong, H.H., Eshraghi, S.A., Giles, M.D., Pinto, M.R., Hillenius, S.J., 1993. Explanation of reverse short channel effect by defect gradients. In: Technical Digest of the 1993 International Electron Devices Meeting (IEDM). IEEE, Piscataway, NJ, pp. 311–314.

Ramamoorthy, M., Pantelides, S.T., 1996. Complex dynamical phenomena in heavily arsenic doped silicon. Phys. Rev. Lett. 76, 4753–4756.

Rousseau, P.M., Griffin, P.B., Plummer, J.D., 1994. Electrical deactivation of arsenic As a source of point defects. Appl. Phys. Lett. 65, 578–580.

Sahli, B., Vollenweider, K., Zographos, N., Zechner, C., 2008. Ab initio calculations of phosphorus and arsenic clustering parameters for the improvement of process simulation models. Mater. Sci. Eng. B 154–155, 193–197.

Sai-Halasz, G.A., Short, K.T., Williams, J.S., 1985. Antimony and arsenic segregation at Si-SiO_2 interfaces. IEEE Electron Device Lett. EDL-6, 285–287.

Schaake, H.F., 1984. The diffusion of phosphorus in silicon from high surface concentrations. J. Appl. Phys. 55, 1208–1211.

Schwarz, S.A., Barton, R.W., Ho, C.P., Helms, C.R., 1981. Studies of phosphorus pile-up at the Si-SiO_2 interface using Auger sputtering profiling. J. Electrochem. Soc. 128, 1101–1106.

Seeger, A., Chik, K.P., 1968. Diffusion mechanism and point defects in silicon and germanium. Phys. Status Solidi 29, 455–542.

Smits, F.M., 1958. Formation of junction structures by solid-state diffusion. Proc. IRE 46, 1049–1061.

Solmi, S., Valmorri, S., Canteri, R., 1995. Codiffusion of arsenic and boron implanted in silicon. J. Appl. Phys. 77, 2400–2406.

Stauffer, D., 1979. Scaling theory of percolation clusters. Phys. Rep. 54, 1–74.

Steen, C., Martinez-Limia, A., Pichler, P., Ryssel, H., Paul, S., Lerch, W., Pei, L., Duscher, G., Severac, F., Cristiano, F., Windl, W., 2008. Distribution and segregation of arsenic at the SiO_2/Si interface. J. Appl. Phys. 104, 023518.

Steigman, J., Shockley, W., Nix, F.C., 1939. The self-diffusion of copper. Phys. Rev. 56, 13–21.

Stolwijk, N.A., Bracht, H., 1998. Diffusion in silicon, germanium and their alloys. In: Beke, D.L. (Ed.), Diffusion in Semiconductors and Non-Metallic Solids. Springer, Berlin (Landolt-Börnstein, New Series III/33A, Chapter 2).

Tan, T.Y., Ginsberg, B.J., 1983. Observation of oxidation-enhanced and -retarded diffusion of antimony in silicon: the behavior of (111) wafers. In: Mahajan, S., Corbett, J.W. (Eds.), Defects in Semiconductors II. Mat. Res. Soc. Symp. Proc, vol. 14. pp. 141–145.

Taniguchi, K., Shibata, Y., Hamaguchi, C., 1989. Theoretical model for selfinterstitial generation at the Si/SiO_2 interface during thermal oxidation of silicon. J. Appl. Phys. 65, 2723–2727.

Tsai, J.C.C., 1969. Shallow phosphorus diffusion profiles in silicon. Proc. IEEE 57, 1499–1506.

Tsai, J.C.C., 1983. Diffusion. In: Sze, S.M. (Ed.), VLSI Technology. McGraw-Hill, New York, NY, pp. 169–218.

Tsai, M.Y., Morehead, F.F., Baglin, J.E.E., Michel, A.E., 1980. Shallow junctions by high-dose As implants in Si: experiments and modeling. J. Appl. Phys. 51, 3230–3235.

Tsibizov, A., Terterian, A., Zechner, C., 2014. Influence of Poisson equation boundary conditions and quantum corrections to carrier concentrations at material interfaces in TCAD process simulation. Phys. Status Solidi C 11, 101–104.

van der Maesen, F., Brenkman, J.A., 1955. On the behavior of rapidly diffusing acceptors in germanium. J. Electrochem. Soc. 102 (5), 229–234.

Waite, T.R., 1957. Theoretical treatment of the kinetics of diffusion-limited reactions. Phys. Rev. 107, 463–470.
Waite, T.R., 1958. General theory of bimolecular reaction rates in solids and liquids. J. Chem. Phys. 28, 103–106.
Wang, H.C.H., Wang, C.C., Chang, C.S., Wang, T., Griffin, P.B., Diaz, C.H., 2001. Interface induced uphill diffusion of boron: an effective approach for ultrashallow junction. IEEE Electron Device Lett. 22, 65–67.
Watkins, G.D., 1965. A review of EPR studies in irradiated silicon. In: Radiation Damage in Semiconductors. Dunod, Paris, pp. 97–113.
Willoughby, A.F.W., 1981. Double-diffusion processes in silicon. In: Wang, F.F.Y. (Ed.), Impurity Doping Processes in Silicon. North-Holland, Amsterdam, pp. 1–54.
Xie, J., Chen, S.P., 1999. Diffusion and clustering in heavily arsenic-doped silicon: discrepancies and explanation. Phys. Rev. Lett. 83, 1795–1798.
Yeh, T.H., 1962. Diffusion of phosphorus in silicon. Bull. Am. Phys. Soc. Ser. II 7, 332.
Yoshida, M., 1971. Diffusion of group V impurity in silicon. Jpn. J. Appl. Phys. 10 (6), 702–713.
Yoshida, M., Arai, E., Nakamura, H., Terunuma, Y., 1974. Excess vacancy generation mechanisms at phosphorus diffusion into silicon. J. Appl. Phys. 45 (4), 1498–1506.

CHAPTER TWO

Electron and Proton Irradiation of Silicon

Arne Nylandsted Larsen*[,1], Abdelmadjid Mesli[†]
*Department of Physics and Astronomy/iNANO, Aarhus University, Aarhus, Denmark
[†]Aix-Marseille Université, CNRS, Marseille Cedex, France
[1]Corresponding author: e-mail address: anl@phys.au.dk

Contents

1. INTRODUCTION

There are many motives for irradiating silicon with electrons and protons at MeV energies. Just to mention some of them, it could be to study the two native defects in silicon, the monovacancy and the self-interstitial, to study more complicated radiation induced defects such as the divacancy, trivacancy, di-interstitial, tri-interstitial, and impurity-vacancy and impurity-interstitial complexes, to study the penetration of charged particles into matter, or to make use of the created defects to control the charge carrier lifetimes in order to improve the switching characteristics of, e.g., silicon

Semiconductors and Semimetals, Volume 91
ISSN 0080-8784
http://dx.doi.org/10.1016/bs.semsem.2015.03.001

power devices. In this chapter, we will in particular be concerned with the different types of defects introduced in silicon during and after irradiation with protons and electrons at MeV energies, and we will limit our interest to defect complexes with less than three constituents as they are by far the most abundant in the case of light-particle irradiations.

In order to study the native defects and their complexes by experimental techniques such as, e.g., deep level transient spectroscopy (DLTS, see "Point Defects in GaN") concentrations of these defects of at least 10^{10} cm^{-3} are needed. The thermodynamic equilibrium concentrations of the native defects are vanishingly small at room temperature rendering experimental studies impossible, and, moreover, their migration is so fast that quenching from high temperature, where their concentrations are appreciable does not result in measurable concentrations unlike in the case of metals. Hence, they must be produced by nonequilibrium methods in which large quantities of Si atoms at lattice sites are removed from these sites, thus creating vacancies and self-interstitials. This can be done in a very controlled way by irradiation with light particles at MeV energies, and the very successful studies of the native defects and their complexes in silicon over the past more than 50 years, and their widespread use in the electronic industry, have only been made possible because of the feasibility of producing controlled amounts by irradiation of silicon with light particles at MeV energies.

The possibility of performing irradiations with the silicon target kept at cryogenic temperatures, where the created defects might be frozen-in, combined with *online* analysis facilities have made studies of the fate of the simple defects possible as the temperature is raised and they start to become mobile and interact with other defects or impurities to create large defects which will subsequently dissolve at certain temperatures influencing impurity diffusion, leakage currents, etc. In this chapter, we will follow some of these reactions.

There are quite a number of recent comprehensive reviews and books covering different aspects of the content of the present chapter from where we have been inspired and have received ideas: the very thorough book by Pichler (2004) on intrinsic point defects, impurities, and their diffusion in silicon, the review article by Auret and Deenapanray (2004) of deep level transient spectroscopy of defects in high-energy light-particle-irradiated Si, the review by Srour and Marshall (2003) of displacement damage effects in silicon devices, and finally but not least the many excellent reviews by Watkins (1986, 1991, 2000) on intrinsic defects in silicon.

2. IRRADIATION OF SILICON: GENERAL ISSUES

When an energetic particle such as the proton penetrates into a crystalline solid, like silicon, it will be slowed down by multiple collisions with silicon atoms and with bound or free electrons. For protons or heavier particles, these processes are termed nuclear stopping and electronic stopping, respectively, and for electrons they are normally called collision and radiative stopping, respectively. The nuclear/collision stopping process is responsible for the structural defects created during the slowing down of the energetic particle, and displacements of silicon atoms will occur as long as the particle can transfer energy in the collision process to a silicon atom that is larger than the displacement energy of the silicon atom, usually taken to be 15 eV. The maximum energy, T_{max}, which can be transferred in a head-on collision is in the nonrelativistic limit given by

$$T_{max} = \frac{4M_1M_2}{(M_1 + M_2)}E \tag{1}$$

where E is the energy of the incoming particle of mass M_1, and M_2 is the mass of the target atom. In the case of proton irradiation of silicon, this reduces to $T_{max} \approx 0.13 \times E$. Thus, the proton energy threshold for an atomic displacement in silicon is ~120 eV. In the case of electron irradiation, the relativistic expression should be used (Auret and Deenapanray, 2004):

$$T_{max} = \frac{2148}{Z}E^2 \tag{2}$$

where Z is the atomic number of the target atom, E is the energy in MeV, and T_{max} is in eV. In the case of silicon, this is reduced to $T_{max} \approx 153 \times E^2$, and the energy threshold for an atomic displacement in silicon with electrons is ~300 keV.

The relative importance of nuclear and electronic collisions depends on the energy of the incident particle and on its mass and charge. The nuclear and electronic stopping powers, defined as the mean energy loss per unit path length, for the proton in silicon are shown in Fig. 1. It appears that for all energies between 1 keV and 10 MeV the electronic stopping power is significantly larger than the nuclear stopping power, and even for a proton energy of 1 keV the stopping process in silicon will be accompanied by the creation of a large number of electron–hole pairs. To reach a regime where nuclear stopping is larger than the electronic, one must use heavier particles

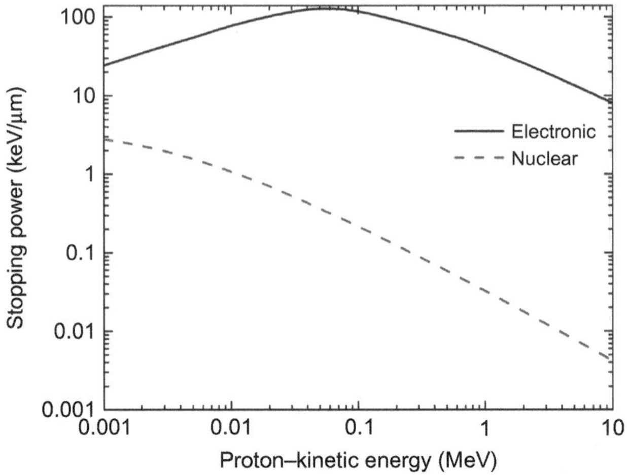

Figure 1 Stopping powers for protons in silicon as a function of energy (NIST, 2011).

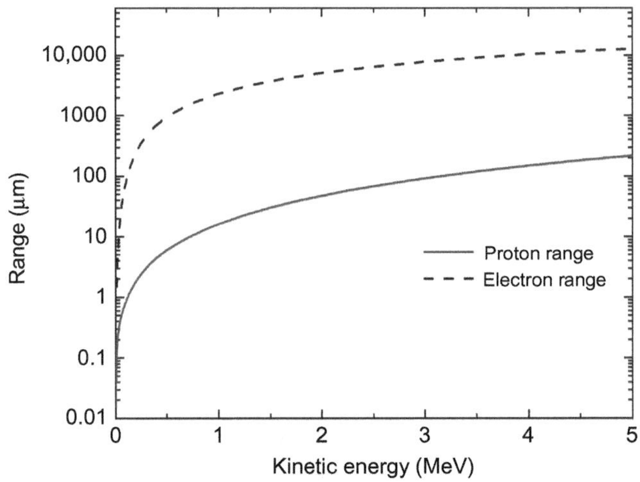

Figure 2 Proton and electron ranges in silicon as a function of energy. The proton ranges are the projected ranges and the electron ranges the continuous-slowing-down approximation (CSDA) ranges (NIST, 2011).

at low energy, e.g., α-particles at energies below ~2 keV. The proton range in silicon for a typical proton energy of 2 MeV is 48 μm as seen from Fig. 2, and for a typical silicon sample thickness of 200 μm, the proton energy must be larger than 5 MeV to penetrate through the sample.

The situation for irradiating silicon with electrons is quite different as seen from Fig. 2; for 2 MeV electrons, the range in silicon is about

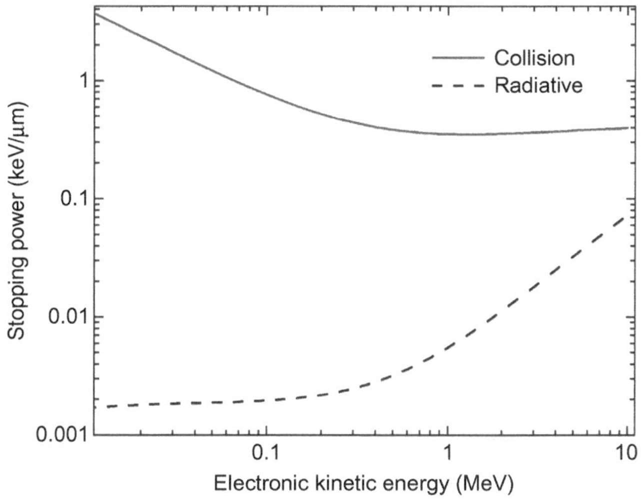

Figure 3 Stopping powers for electrons in silicon as a function of energy (NIST, 2011).

4 mm; thus, the electrons will penetrate through a typical silicon sample of thickness 200 μm. Moreover, the stopping power for electrons is almost independent of their energy between a few hundred keV to about 45 MeV, and certainly in the energy range below 10 MeV where most irradiations take place as seen in Fig. 3, resulting in a very uniform energy deposition as a function of depth. The overall result is that when irradiating with electrons of energies larger than about 300 keV and less than about 45 MeV the defect density will be uniform as a function of depth. This is not the case when irradiating with energetic protons in which case there will be an accumulation of defects around the end-of-range when the proton has lost most of its energy, and where consequently the importance of nuclear collision increases as shown in Fig. 1. This is nicely illustrated in the vacancy distribution shown in Fig. 4 from a SRIM simulation (Ziegler and Biersack, 2013) of an irradiation with 2 MeV protons.

3. THE ELEMENTARY ACT: FRENKEL PAIR

Before discussing the reactions of each constituent of the Frenkel pair, the vacancy and self-interstitial, leading to the formation of more complex defects, we describe below their individual electronic structures and diffusion mechanisms.

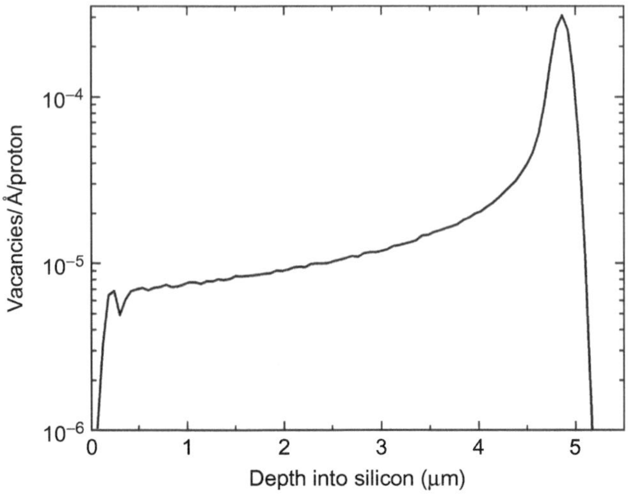

Figure 4 SRIM simulation of the number of vacancies/Å/proton as a function of depth in silicon produced during the slowing down of 2-MeV protons (Ziegler and Biersack, 2013).

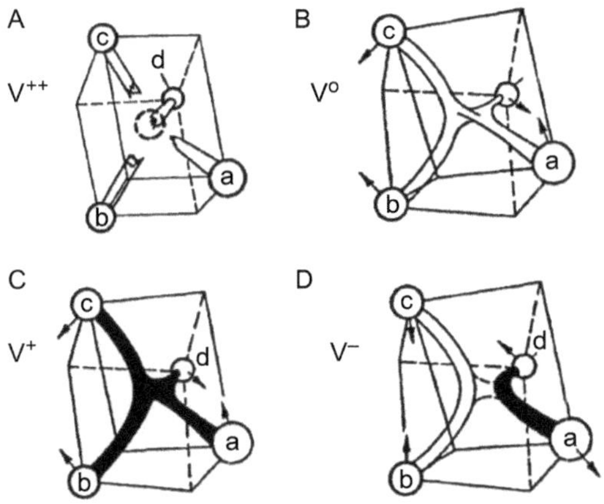

Figure 5 The four configurations of the silicon vacancy (Watkins, 1991). *Copyright Wiley-VCH Verlag GmbH&Co. KGaA. Reproduced with permission.*

3.1 The Vacancy

3.1.1 Electronic Structure and Electrical Level Positions

Figure 5 shows the four configurations of the vacancy, experimentally identified for the first time by electron paramagnetic resonance (EPR) (Watkins, 1964). The four structures differ by their respective charge states. Only V^+ and V^-, with an odd number of electrons, have been clearly identified. Complex considerations, which will not be detailed here, have shown that in the case of V^+, the unpaired electron is equally distributed over the four adjacent Si atoms, illustrated in bold in Fig. 5B. In the case of V^-, the unpaired electron is shared by a pair of atoms only. Both configurations and V^0, which was not observed by EPR, but essential to the existence of V^- and V^+, have the property to distort locally the lattice. Only V^{++} maintains its tetrahedral symmetry as shown in Fig. 5A. The distribution of the four charge states in the band gap is determined by energy constraints among which electrostatic requires that the binding of $n+1$ electron undergoes Coulomb repulsion from the n electrons already present, so that the expected binding energy distribution should obey the inequality,

$$E_{n+1} < E_n < E_{n-1} < E_{n-2} \tag{3}$$

where it is implicitly assumed that these values are incremented by a certain amount ΔE_n. However, electrostatic considerations, valid in vacuum, do not hold in a solid where elastic distortion induced by the creation of the vacancy affects the energy diagram by lowering all states leading to the so-called relaxed configuration illustrated in Fig. 6. The normal sequence described by Eq. (3), and shown in Fig. 6A, would not fundamentally be modified if the elastic forces were not coupled to the charge states and if they were weak. But this is not always the case and a strong coupling, itself depending on the charge states, may invert completely or partially this "normal" order. In the case of the vacancy, the elastic distortion is greater for the state V^+ than for V^{++} leading to the inversion shown in Fig. 6B. The

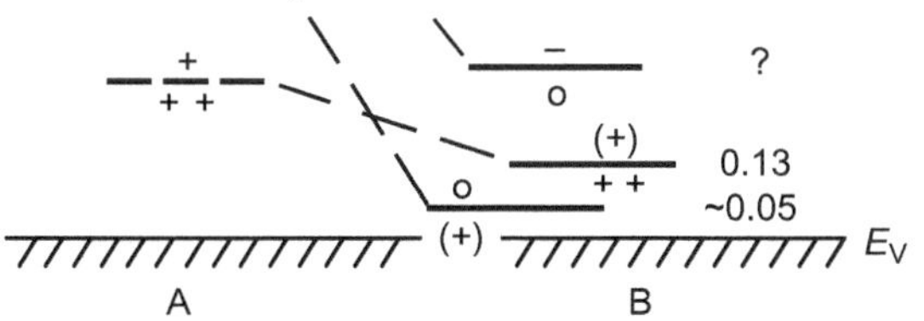

Figure 6 (A) Distribution of the four levels of the vacancy in unrelaxed configuration and (B) experimental result after Jahn–Teller relaxations revealing the negative-U property of the vacancy (Watkins, 1991). *Copyright Wiley-VCH Verlag GmbH&Co. KGaA. Reproduced with permission.*

mechanism behind this inversion, theoretically predicted (Baraff et al., 1979, 1980a,b) and experimentally observed (Newton et al., 1983; Watkins, 1964; Watkins and Troxell, 1980) in a semiconductor, is known as Jahn–Teller relaxation. Without going into this complex mechanism, responsible for the so-called negative-U, the following picture should be mentioned. Every charge state is represented by a harmonic oscillator vibrating around its equilibrium position. The distortion leads simply to a displacement of this equilibrium position. The one induced by the V^+ state is such that it compensates the Coulomb repulsion which the second hole should "feel" with respect to the first. The result is that the final state V^{++} is more stable, violating relation (3). The consequence is that the transition,

$$V^{++} \rightarrow V^0 \tag{4}$$

does not go through the intermediate unstable state V^+. The two holes are instead emitted simultaneously. Finally, it should be noticed that the only positions in the band gap that have been firmly established are those corresponding to the transitions $V^{+/0}$ and $V^{++/+}$. For the two other levels, although experimentally demonstrated and normally distributed in Si, their position can only be estimated to be greater than ~0.17 eV below the conduction band (Watkins, 1986). However, in irradiated $Si_{1-x}Ge_x$ two new levels appear which are not observed in pure Si. One of them appears at $E_C - 0.29$ eV for small x, and it has been argued that this is the $GeV^{=/-}$ level; the other appears at $E_V + 0.56$ eV for $x = 0.02$ and this signal has been tentatively assigned to the $GeV^{-/0}$ level. Thus, assuming that the presence of the Ge atom perturbs only slightly the electronic structure from that of the isolated vacancy it has been argued that these observed level positions must be very close to those expected for the acceptor states of the isolated vacancy in Si (Nylandsted Larsen et al., 2008).

3.1.2 Diffusion Mechanism

The diffusion of the vacancy has been one of the longest controversies in the field of defects in semiconductors, and it is still on-going and in good shape. As mentioned above, in the early 1960s, Watkins (1986) studied irradiation-induced vacancies in silicon by EPR and found a migration energy of the vacancy of 0.33 eV. It was soon after established that the vacancy-migration energy depends on the charge state of the vacancy, and values of 0.18 ± 0.02 eV for V^{2-}, 0.45 ± 0.04 eV for V^0, and 0.32 ± 0.02 eV for V^{2+} were determined from measurements at cryogenic temperatures (Watkins, 1986). These values were later on confirmed by several groups using different experimental techniques (Watkins, 1986). High temperature diffusion experiments carried out shortly after these first EPR studies, however, resulted in a vacancy-migration energy as high as 1.7 eV (Fairfield and Masters, 1967). Similar values were recently obtained by more direct measurements. Bracht et al. (2003) determined a value of 1.8 ± 0.5 eV from radiation enhanced self-diffusion experiments at temperatures between 780 and 872 °C using 2-MeV proton irradiations for various times and proton fluxes, and Ranki and Saarinen (2004) determined a value of 1.3 ± 0.4 eV from positron-annihilation experiments of thermal vacancies generated at high temperature in highly As- and P-doped Si. In a recent review of self-diffusion studies in silicon by means of isotopically enriched silicon, Bracht et al. (2014) found that the temperature dependence of self-diffusion is most consistently described with temperature-dependent vacancy formation and migration enthalpies, and they find the best description with a vacancy-migration energy varying from 0.5 eV at 700 °C to 0.8 eV at 1400 °C. We note that the low temperature value is not very different from the cryogenic value found by Watkins (1986). The temperature dependence of the vacancy properties could represent a defect that gets more extended or spread-out with increasing temperature. This concept was first proposed by Seeger and Chik (1968), but it could not be verified at that time due to experimental difficulties that limited self-diffusion studies to a narrow temperature range. Thus, there are strong indications of a temperature-dependent vacancy-migration energy with a small value for temperatures below ~700 °C, and a significant higher value of the order of 1 eV or more at high temperatures. This conclusion has been disputed by some researcher, and as late as in 2008 Watkins (2008) concluded "*there is only one vacancy, and nothing magical happens at elevated temperatures, contrary to what has often been suggested*" based on studies of self-diffusion in silicon by Shimizu et al. (2007) indicating a vacancy-migration energy of ~0.4 eV at high temperature. The value of 0.4 eV for high temperature diffusion of the vacancy was later on disputed by Bracht and Chroneos (2008).

Already 40 years ago, however, Van Vechten (1974) developed a theoretical model for the migration of the vacancy in silicon which was able to explain the experimental observation of the temperature-dependent migration energy. The model assumes the existence of two modes of migration or transition states available to the vacancy, and that one of the states dominates experiments at high temperature because it has a larger entropy while the other transition state dominates experiments at low temperature because it has a lower enthalpy and entropy. The state of the vacancy at lattice sites remains the same at high and low temperature; it is the transition state which differs. The high temperature value of the migration enthalpy is explained by assuming that in addition to the bond-breaking term, which is dominant at low temperature, there is a contribution from long-ranged forces similar to those appearing in simple metals. Thus, there is no need of assuming an extended or spread-out vacancy at elevated temperature as Seeger and Chik have proposed in their early model (Seeger and Chik, 1968). In fact, Van Vechten rejects the Seeger–Chik hypothesis of an extended high-temperature form of the vacancy from pure thermodynamic arguments.

Very recently, Voronkov and Falster (2012) discussed the high temperature diffusivity of the vacancy in silicon based on a hypothesis similar to that of Seeger–Chik with the vacancy in two structural forms one of which is fast and the other is slow.

The vacancy has been observed to migrate at temperatures down to 4.2 K under minority-carrier injection (Barnes and Samara, 1986). The mechanism for this enhanced migration is related to the electronic excitation as electron–hole pairs recombine at the positively charged vacancy combined with a strong electron-lattice coupling. This mechanism known as REDR (recombination enhanced defect reaction) will be discussed below.

3.2 The Self-Interstitial

The first experimental investigations of radiation induced self-interstitials in silicon date more than 50 years back (Watkins, 1991). Since then a huge number of experimental and theoretical investigations have been reported in the literature (Pichler, 2004). A major difference with the vacancy is that the self-interstitial has so far never been caught experimentally allowing an easy determination of its electronic structure, level positions and ordering, and diffusion mechanism. We may, however, cite recent attempts to identify the self-interstitial by Mukashev et al. (1998) and Gorelkinskii et al. (2009). These investigations are still leading to controversial conclusions. Therefore, turning to theory may help getting some glimpse on the various issues related to this elementary defect.

Several groups have modeled the self-interstitial in silicon, starting with the work by Bar-Yam and Joanopolous (1984) and Car et al. (1984) using *ab initio* pseudo-potential local density quantum mechanical calculations. More recently Jones et al. (2009) have reviewed theoretical investigations, and compared them to their own calculations based on density-functional, pseudo-potential calculations carried out using the AIMPRO code. This review shows clearly that a complete consensus is not reached on theoretical basis either. The present section restricts thus to recalling the main theoretical outcome by the authors cited above.

3.2.1 Electronic Structure and Electrical Level Positions

Theory seems to exclude for the self-interstitial a negatively charged configuration for any position of the Fermi level. It is so far either neutral (Si_i^{0}) or positively charged (Si_i^{+} or Si_i^{++}). This has some consequence on the so-called Watkins replacement mechanism in n-type silicon, which shall be discussed below. On the basis of his own experimental observations of Al_i^{++} and B_i^{0}, Watkins (1964) has proposed possible atomic configuration of the two charge states Si_i^{++} and Si_i^{0}, displayed in Fig. 7, which seem in match with the calculations carried out by Bar-Yam and Joanopolous (1984) and Car et al. (1984). These authors disagree however with Jones et al. (2009) on the exact positions of the positive charge states in the band gap leaving a doubt as to the possibility for the self-interstitial to exhibit or not the negative-U character. Bar-Yam and Joanopolous (1984) and Car et al. (1984) propose that a first donor state (0/+) is predicted to be at 0.8 eV below the conduction band, while a second donor (+/++) is located at 0.4 eV above the singly charged state. The self-interstitial displays therefore the negative-U character as a consequence of an inversion of the ordering of the states in the band gap. It follows thus that in the switch between (++) and (0) the charge state (+) is not stable. Based on the marker method, which is estimated to be accurate to within 0.1–0.2 eV, Jones et al. (2009) found the first donor state (0/+) at 0.23 or 0.43 eV below the conduction

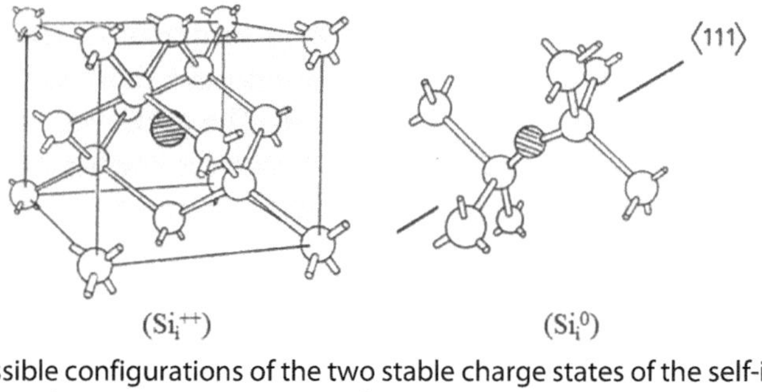

Figure 7 Possible configurations of the two stable charge states of the self-interstitial as proposed by Watkins (1964) on the basis of his replacement mechanism and the studies he carried on Al_i^{++} (left) and B_i^{0} (right). These are among the configurations predicted by Bar-Yam and Joanopolous (1984) and Car et al. (1984). *Copyright Wiley-VCH Verlag GmbH&Co. KGaA. Reproduced with permission.*

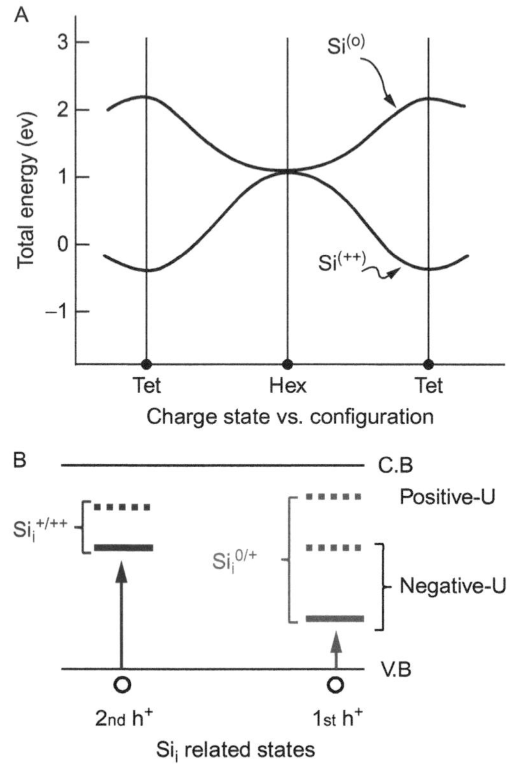

Figure 8 (A) Shows the stable configurations for relaxed and unrelaxed Si^0 and Si^{++} along the ⟨111⟩ direction and (B) represents the energy states as found by the theoretical groups cited in this review. As we can see negative-U ordering is not obvious for Jones et al. (2009), dashed lines, except in the case where the first donor is deep, while it seems to be evident for Bar-Yam and Joanopolous (1984) and Car et al. (1984).

band, depending on which marker is used for the calculation. The second donor state (+/++) was found at 0.26 eV below the conduction band. It is thus not obvious for Jones et al. (2009) whether the levels are distributed in negative-U order. The outcomes of the three theoretical approaches cited above are summarized in Fig. 8.

3.2.2 Diffusion Mechanism

The charge state dependent configuration displayed in Fig. 7, coupled to the high ionization rate during irradiation, naturally leads to the proposed athermal process for migration, known as Bourgoin–Corbett mechanism (Bourgoin and Corbett, 1978). Referring to Fig. 7A, the tetrahedral site is the most stable configuration for Si^{++}. The state (+) being unstable, capturing two electrons allows switching the charge state from Si^{++} to Si^{0}, the latter accommodating in a hexagonal configuration, more stable than the tetrahedral. An alternate cycling from tetrahedral Si^{++}, with the capture of holes, to hexagonal configuration of Si^{0} through electron capture, provides an effective athermal migration process leading to a long range migration even at 4.2 K as observed experimentally (Watkins, 1975a,b). Jones et al. (2009) have also demonstrated that such a mechanism is indeed possible. The migration energies for the thermal diffusion have been estimated theoretically by Jones et al. (2009). They found that the most mobile structure is the thermodynamically unstable positive charge state with an activation energy of 0.31 eV. The migration of the positive self-interstitial requires an activation energy of 0.53 eV, whereas the double positive charge state is the less mobile with an activation energy of 1.46 eV. In n-type silicon, interstitial impurities only appear at temperatures above 150 K (Watkins, 1986), indicating that the self-interstitial either starts to migrate at this temperature or is released from an immobile complex at this temperature. A migration temperature of 150 K is in agreement with the activation energy of migration of the neutral self-interstitial.

Some interesting results on the self-interstitial have been obtained with the grazing incidence diffuse X-ray scattering technique (Partyka et al., 2001) supporting the above discussed picture of the self-interstitial. The results from these measurements support the hypothesis that the motion of interstitial atoms at low temperatures is stimulated by electronic excitation. Moreover, thermally activated migration of the self-interstitial is found to start between 120 and 170 K, and the self-interstitials are observed to either recombine with vacancies or form immobile interstitial clusters.

4. BRANCHING REACTIONS: BASIC CONSIDERATIONS

The primary reaction is ignited by the energetic incident particle which, by kicking out a silicon atom from its natural site, leads to the formation of the Frenkel pair following the reaction:

$$0 \Longleftrightarrow V_{Si} + Si_i \qquad (5)$$

where 0 represents the perfect crystal and V_{Si} and Si_i are the vacancy and silicon self-interstitial, respectively. The reconstruction around the four broken bonds left by the vacancy leads to a distortion of the lattice resulting in two major consequences: (i) the normal sequence of the five different charge states of the vacancy, expected from pure electrostatic considerations, are violated as discussed above. Two of these states are inverted leading to the first observation of the negative-U property; (ii) a barrier is induced helping a significant fraction of the separated entities to survive the reverse reaction of annihilation. It is this fraction that is at the origin of all kind of stable defects formed following the complex hierarchy discussed below. An important and unavoidable ingredient, present while the reaction of Frenkel pair generation is taking place via reaction (5), is the ionization consisting of a huge amount of free electron–hole pairs. Reaction (5) leads thus to a situation out of equilibrium where the densities of both the vacancy and self-interstitial tend to decrease, either by surmounting the barrier of annihilation of by forming more stable defects. Silicon is by nature or by technological necessity not pure. It contains at least four major impurities: substitutional group V dopants (mostly phosphorus P_s), group-III dopants (mostly boron B_s), carbon C_s, and interstitial oxygen O_i. The two latter are considered to be the two main background species. It follows that some routes for complex formation are offered to the vacancy and self-interstitial. Among them, one is particularly appealing as it constitutes the first elegant demonstration of the so-called Watkins replacement mechanism (Watkins, 1964). This concept emerged from the experimental fact that self-interstitial has never been caught as a single entity, even when both the irradiation and observations are carried out at 4.2 K, while the vacancy could be isolated. It has been demonstrated that in p-type silicon irradiated with high energy electrons, the ratio $[V_{Si}]/[B_i] \approx 1$ (see Fig. 9 in Watkins, 1986). Such a one-to-one correspondence means that Si_i atoms, originating from the Frenkel pairs, have been totally converted to boron interstitial according to the simple reaction

$$Si_i \Longleftrightarrow B_i \tag{6}$$

This illustrates beautifully the first simplest form of Watkins replacement mechanism. In this case, the sequence of branching reaction (5) expressed above is in fact given by,

$$\begin{array}{l} 0 \Longleftrightarrow V_{Si} + Si_i \\ \qquad\qquad\quad \hookrightarrow B_i \end{array} \tag{7}$$

where in this special case, the free entities are the vacancy and interstitial boron. We have mentioned above that the Bourgoin–Corbett mechanism (Bourgoin and Corbett, 1978) gives a reasonable explanation for the conversion of self-interstitial into boron interstitial at low temperature. The situation regarding the behavior of the self-interstitial is however surprisingly different in n-type silicon stressing that the issue is still not clear as we shall see below.

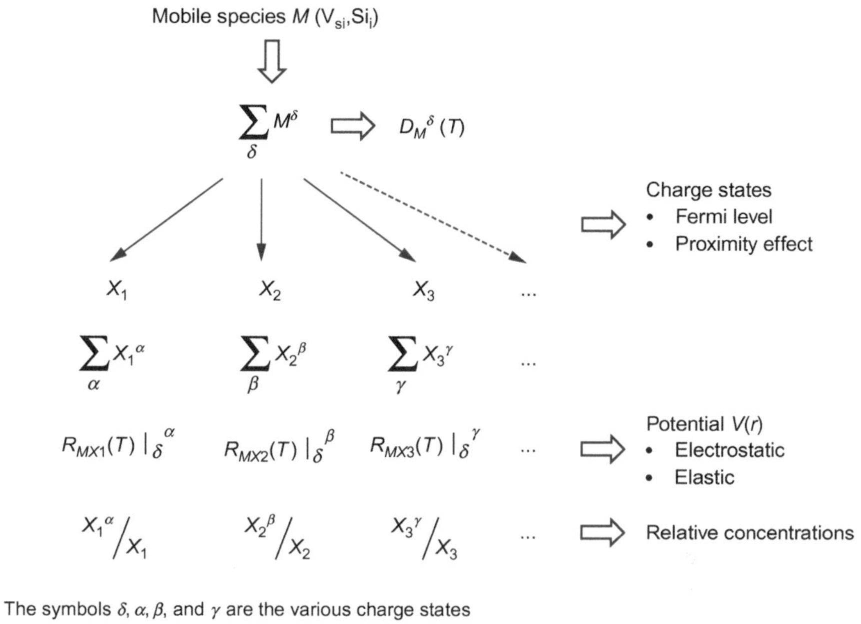

Figure 9 Illustration of the branching tree showing the routes towards the formation of complex defects. The main physical parameters determining the branching ratio are reported on the left.

Increasing the temperature will ignite other complex reactions in which silicon vacancy and interstitial boron will convert into more stable defects.

Let us denote in general cases by M either one of the mobile species (V_{Si}, Si_i, B_i, ...) and assume that interstitial oxygen O_i or any other partner in substitutional site remains immobile. Restricting to "bimolecular" type reactions, the branching reactions following the elementary acts (5), and possibly (6), can be written as,

$$M(\text{mobile species}) + X_j \overset{k_j,\, k_{-j}}{\Longleftrightarrow} MX_j \tag{8}$$

where X_j denote the various immobile species mentioned above. k_j and k_{-j} are the reaction rates representing, respectively, either the formation of the pair MX_j or its dissociation. The pair MM is also considered as one of the possible reactions taking place; it may represent the divacancy, or the di-interstitial, but also B_iB_s or C_iC_s which will as well be discussed below. More complex defects will also be treated further in this review, although they are outside the frame of reactions (8). It is worth mentioning that reaction (8) is rather compact. It represents in fact two or more separate master reactions, one for each moving species, namely V_{Si} and Si_i resulting from reaction (5), but also B_i or C_i resulting from reaction (7), which is the simplest form of Watkins replacement mechanism. The necessity for assigning to each mobile species its specific reactions is due to the totally different intrinsic properties of the reacting partners such as charge states, diffusion coefficient which might be charge state dependent, stability and affinity in binding with each other, etc. We may thus emphasize that once created, V_{Si} and Si_i, and possibly B_i or C_i. . ., will follow separate destinies. The major reason for this decoupling is related to the following points: (i) the way the lattice adapts around each primary species (the atomic arrangement is different for the vacancy and interstitials); (ii) the number, position and distribution of the various states for each species in relation to the way the lattice accommodates; and (iii) the strength of the carrier–phonon coupling around the species.

It is not difficult to imagine that reactions (8) may be tremendously hard tackling if some realistic simplifications are not put forward. The first one is to assume first order reactions limited by the diffusion of the mobile species $D_M(T)$. k_j will then obey the simple proportionality law,

$$k_j(T) \propto R_j(T) D_M(T) \tag{9}$$

where $R_j(T)$ defines the capture radius around species X_j. This parameter will be determined by the potential governing the interaction of M with X_j, which in turn will be determined by the respective charge states of the reactants. If the reactants are partially ionized, the rate $k_j(T)$ given by relation (9) may also split into two components, one for each charge state. The parameters on the right hand side of this relation will each be modulated by the Fermi function $f(x, T)$, where x stands for the position in the material. Although the species may be uniformly distributed, their charge population may be due to an external electric field which is not necessarily uniform. The population will thus depend on the relative positions of the Fermi level (E_F) and possibly the capture cross sections of the carriers. The Fermi level E_F is determined by the conductivity of the material and the temperature, both considered as intrinsic parameters. It will also depend on the ionization rate which depends on the characteristics of the incident particle (energy E, mass M, flux F), defined as extrinsic parameters. The same reasoning is valid for the rate $k_{-j}(T)$ controlling the reverse reaction, a rate which is mainly characterized by the binding energy of the pair and the attempt frequency for dissociation. This last parameter may in some circumstances be dependent on the free carriers available. Some examples will illustrate this aspect.

The second assumption is the absence of any gradient of the mobile species bringing $D_M(T)$ to depend on temperature only, removing Fick's law from any consideration. This is the case with high energy electron or proton irradiation where the Frenkel pairs and all cited extrinsic impurities are uniformly distributed over the observation area (a few μm). The potential $V(r)$ determining the capture radius $R_i(T)$ might be very simple if we restrict to electrostatic considerations but may become very complicated if charge screening and lattice distortion in close proximity to the reacting partners need to be taken into account. In the simplest case of purely electrostatic considerations, $R_i(T)$ varies between 50 Å (at 300 K) and 500 Å (at 30 K) for a coulomb potential ($\propto 1/r$), to about 5 Å for a polarization potential ($\propto 1/r^4$) governing the interaction between two partners, one being charged and the other neutral. In the case of short interacting potential, the binding requires that the partners come very close to each other. The local atomic distortions they induce will therefore play a dominant role making any simulation hazardous without a precise knowledge of the potential and local deformation. Therefore, relying on the experimental observations is the only way to tackle the problem. But it requires a perfect knowledge of the material and the experimental conditions.

The hierarchy established among all possible interactions is directly observed in the final proportions of MX_j, determining which pair is more likely to form for a given set of conditions (concentrations of the species, Fermi level determining the various charges states and their proportion, temperature, etc.). For instance, in the case of a random distribution of the interacting particles, the fact that the mobile species may be attracted by an immobile impurity through a Coulomb potential does not mean that this type of interaction has a prevailing priority. A binding between a charged mobile species and neutral immobile impurity may be more likely to form if the latter is found in much higher concentration than any other oppositely charged species. The diagram displayed in Fig. 9 summarizes the branching paths, which will be a guide when displaying the experimental observations recorded over more than 40 years of research.

Reactions (5)–(8) displayed above, and the diagram shown in Fig. 9, constitute qualitatively the basis of the so-called Watkins replacement mechanism and beyond. The formation of the pairs M_2, or MM according to reaction (8), and more generally M_n ($n > 2$), constitute particular cases. In silicon, it is the case of the divacancy V_2, trivacancy V_3, but also the di-interstitial $(Si_i)_2$ and tri-interstitial $(Si_i)_3$, etc. These complexes have been identified and some of them are considered below.

A last rather exotic effect, barely noticed in the literature (Kimerling et al., 1988), deserves to be mentioned; it is the "proximity charging effect." In the analysis given above, it was assumed that the interacting species keep their respective levels unchanged in the band gap no matter how close to each other they are. In reality, this is not the case, and when two oppositely charged species are close to each other, their energy states are disturbed by the Coulomb interaction, "pushing" them toward the conduction band for the donor and toward the valence band for the acceptor. This effect coupled with the position of the Fermi level affects the charge of the species under consideration. For a donor impurity M approaching a shallow acceptor X_j, the charged population will be given by,

$$f_{\mathrm{P}}(r) = \frac{[M^0]}{[M^+]} = \exp\{[E_{\mathrm{F}} - E_{\mathrm{T}}(r)]/kT\} \tag{10}$$

where

$$E_{\mathrm{T}}(r) = E_{\mathrm{T}}(r=0) + \frac{q_M q_X}{4\pi\varepsilon_0\varepsilon_r rkT} \tag{11}$$

E_{F} represents the Fermi level and $E_{\mathrm{T}}(r=0)$ the unperturbed energy level of the mobile donor defect. It turns then out that relation (6) will not only be modulated by the Fermi function $f(x,T)$ but also by the proximity function $f_{\mathrm{P}}(r)$. In the above equations, x is the macroscopic distance in the material and r is the microscopic separation between the reactants. As can be seen from this relation if one of the interaction species is neutral, or the distance r is infinite, the proximity effect vanishes. Finally, in Eq. (11), the approaching distance r is assumed to be large enough for the dielectric constant $\varepsilon_r(r)$ to keep its macroscopic definition.

At this stage, it is crucial to make a clear distinction between two situations having radically different consequences on the branching ratio. One situation deals with reactions (5)–(8), all taking place simultaneously. Its holds when the irradiation is performed at or above the critical temperature T_c at which the Frenkel constituents are both highly mobile. Considering that the self-interstitial is mobile already at 4.2 K, T_c is essentially determined by the mobility of the neutral charge state of the vacancy as shown in Fig. 5. Depending on the charge state of the vacancy, T_c ranges from 80 to 200 K (Watkins, 1986).

When the irradiation is performed above 200 K, reactions (5)–(7) are quasi-instantaneous, leaving reaction (8) acting alone but in a strong ionizing atmosphere. The material is thus rather intrinsic with the Fermi level located at midgap. In such a case, electrostatic considerations are rather minor, and the determining factor will be the concentrations of the partners. When the irradiation is stopped, all kind of complex defects are supposed to be formed and stabilized in the lattice. A few interesting exceptions have recently been discovered and will be discussed in this review. If on the other hand the irradiation is carried out at very low temperature, the Frenkel defects, V_{Si} and, indirectly Si_i, may be caught in an isolated form allowing tracing their destiny by increasing the temperature in a controlled way. Whether reactions (5) and (7) follow each other or take part simultaneously has an impact on the formation of important defects. Based on the above mentioned considerations, the pair MM can only be formed in the situation where M is neutral otherwise the repulsive Coulomb interaction dominates preventing any binding of M with itself. The only way of allowing the pair to form is thus to create it as a primary defect during the impact of the irradiating particle and the lattice. We shall show for instance how to avoid the creation of the divacancy in silicon.

Finally, it should be mentioned that all the above considerations are based on the implicit assumption that thermal equilibrium prevails. In other words, reaction (8) is supposed to take place under equilibrium (no ionization) after reaction (5) is stopped. The high degree of ionization prevailing in reaction (5) forbids indeed any equilibrium process. We briefly mentioned the Bourgoin–Corbett mechanism but we can also cite the REDR (Lang and Kimerling, 1974; Weeks et al., 1975). The latter mechanism can be activated even in the situation where reaction (6) follows reaction (5). A recent work on the well-known E-center has illustrated how a situation out of equilibrium may help disentangling an issue overlooked for over 40 years (Nylandsted Larsen et al., 2006).

4.1 Branching Reactions of the Vacancy

As mentioned above, if the irradiation takes place at cryogenic temperatures, the vacancy is immobile while the self-interstitial has, at least in p-type silicon, already transformed into boron interstitial via the Watkins replacement mechanism (Watkins, 1976; Weeks et al., 1975). This point will be discussed in more detail in the next section. We are thus left to follow the destiny of the vacancy upon annealing in its quest of binding to other species present in the lattice while undergoing a long migration process. The diagram displayed in Fig. 10 splits the reactions into two groups, one for each type of material, bearing in mind that at least one impurity, oxygen, is common

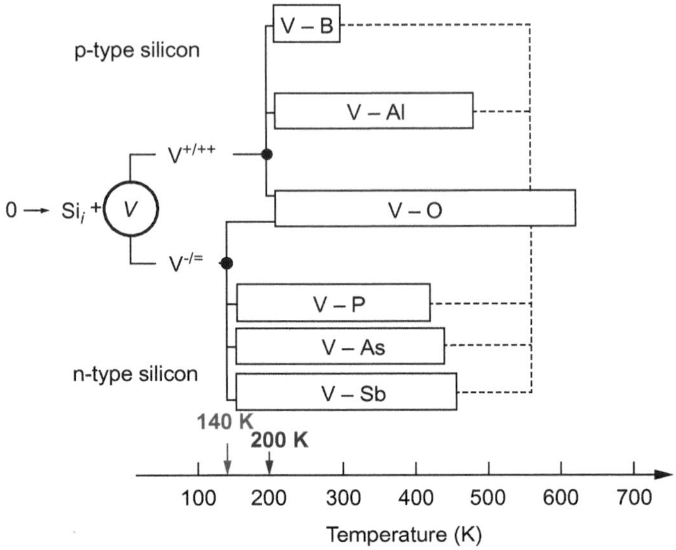

Figure 10 Branching reactions of the vacancy into various defects. The dots represent the origin of the two branches initiated, respectively, in p- and n-type silicon. The arrows mark these turning points. The temperature scale shows the thermal stability of the defects.

to both materials. We may also cite carbon, but so far this impurity has no or little affinity with the vacancy. Carbon will rather be a marker for self-interstitial as will be shown in the next section. The diagram shown in Fig. 10 is based on the assumption that the material is pure enough to include only the dopant and the two unavoidable background impurities, oxygen and carbon. Starting with p-type material, we may notice that group-III impurities are negatively charged. At low temperatures, the Fermi level is rather close to the valence band establishing the double positive charge character of the vacancy, V^{++}. Potentially, a Coulomb attraction exists then between the p-type dopant and the vacancy. Therefore, once V^{++} becomes mobile at around 200 K (see Fig. 11), its binding with boron becomes effective, forming the V-B pairs. For V–Al, which will not be discussed here, details can be found elsewhere (Kimerling, 1977; Watkins, 1967). V–B introduces a level in the band gap at 0.31 eV above the valence band (Kimerling, 1977). This pair forms even in materials doped to low concentrations ($[B] \sim 10^{13}$ cm^{-3}). At such low concentrations, two dopant atoms are in average far apart by about 5000 Å forcing thus the vacancy V^{++} to make tens of jumps before feeling the Coulomb attraction by the negatively charged dopant atoms. At low temperature, the interacting distance may be

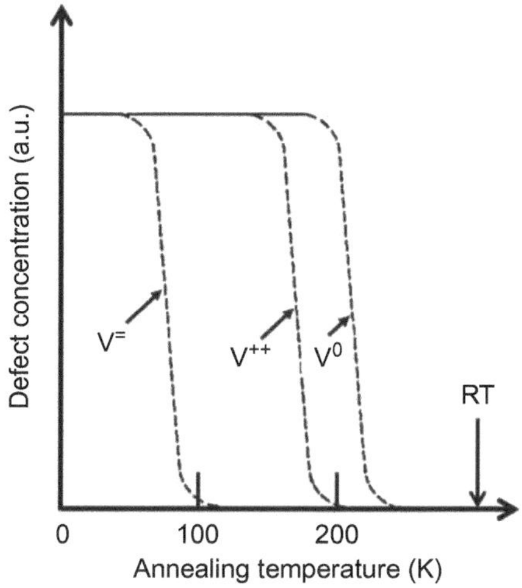

Figure 11 Schematic of the thermal stabilities of the three charge states of the vacancy (Watkins, 1991). *Copyright Wiley-VCH Verlag GmbH&Co. KGaA. Reproduced with permission.*

of the order of 500 Å as mentioned above. However, in spite of such a strong coulomb interaction, these pairs appear only in materials having low oxygen content. This illustrates the subtle interplay of the various ingredients determining which reaction prevails. Interstitial oxygen is neutral, forming a dimer with two adjacent silicon atoms. Thus, based on the sole electrostatic argument, the pair V–O is expected to be less likely to form than V–B. However, as mentioned above, electrostatic forces are not always the determining factors. The oxygen concentration in bulk grown silicon is seldom below 10^{16} cm^{-3}, a value always much larger than practical doping concentrations. Therefore, the formation of the pair V–O, also called the A-center, is more likely to form than the V–B pairs and, as shown in the diagram above, it is the most stable defect. The higher stability of V–O pair leads to an increase of its concentration after subsequent annealing as predicted in Fig. 10. Above 250 K vacancies released from V–B diffuse and may bind to other "free" oxygen atoms in the lattice. This process is illustrated in Fig. 10 by the dashed lines, where for the sake of simplicity we assume that the total amount of available vacancies, created immediately after irradiation, is conserved. If V–B has received much less attention than other complexes, it is because irradiations are very often performed at room temperature where the pair is not stable and thus not observed.

The atomic configurations of V–B and V–O, as extracted from EPR studies (Watkins, 1976; Watkins and Corbett, 1961) are represented in

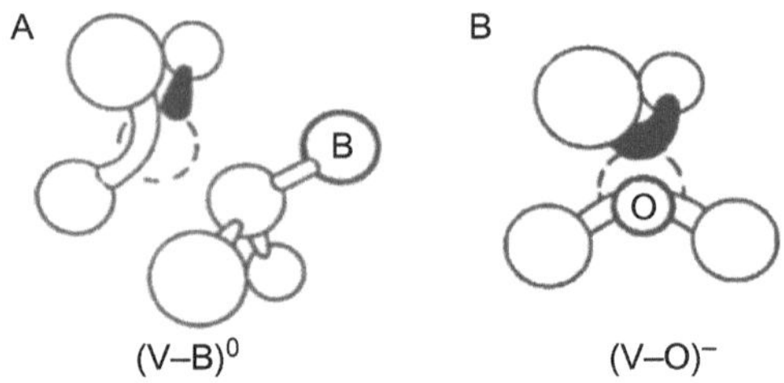

Figure 12 Atomic configurations of V–B and V–O extracted from EPR studies (Watkins, 1976, 1991; Watkins and Corbett, 1961). The dashed circle represents the vacancy and the bold bond represents the trapped unpaired electron. *Copyright Wiley-VCH Verlag GmbH&Co. KGaA. Reproduced with permission.*

Fig. 12, where the bold bonds indicate the localization of the trapped electron. For the V–B pair, boron sits in the next nearest neighbor to the vacancy, bringing an extra negative charge into the environment of the vacancy, and lowering its symmetry. Finally, the configuration of the A-center or V–O pair is represented in Fig. 12B. Hyperfine EPR spectra (Watkins and Corbett, 1961) reveal that the vacancy comes closer to an oxygen atom bridging two silicon atoms (Si–O–Si) nearby a dimer formed by two other silicon atoms as depicted in Fig. 12B. The associated energy of the degenerate level is located at 0.17 eV below the conduction band, appearing thus as a minority-carrier trap in p-type silicon.

In n-type silicon, the major impurities are the group V dopants, phosphorus, arsenic, antimony, and potentially bismuth although very rarely used, and of course oxygen as an unavoidable background contaminant. As mentioned above, carbon is also a natural contaminant, but with so far no observed affinity of bonding to a vacancy. In n-type silicon, at low temperature, the Fermi level is located in the upper part of the band gap close to the conduction band. The vacancy is then most likely double negatively charged $V^{=}$. Above 40 K, the dopant is positively charged leading thus to a coulomb interaction between the two species, $V^{=}$ and P^{+} for instance. The formation of the pair V–D, called also the E-center, where D represents the group V donors P, As, Sb, introduces an acceptor level in the range of 0.39–0.44 eV depending on the nature of the dopant (Evwaraye, 1977; Watkins and Corbett, 1964). This scheme was prevailing for over 40 years of research until recently when it was demonstrated that like many other point defects, the E-center is amphoteric: it introduces as well a donor level in the silicon band gap (Nylandsted Larsen et al., 2006).

In n-type materials, containing a fairly large amount of oxygen, both E- and A-centers can form simultaneously. The reason is that both defects are stable far above room temperature. According to the diagram shown in Fig. 10, both defects start to form when the vacancy becomes mobile, but contrary to p-type silicon the E-center seems to form no matter the amount of oxygen present in the lattice. Therefore, the branching ratio gives a priority to electrostatic interaction as will be demonstrated below when we discuss the donor level. The charge states and concentrations of the species under consideration are similar in both n- and p-type material, thus the key ingredient that determines the branching is the thermal stability. In brief, in both materials, the hierarchy is determined by the type of interacting potential, Coulombic for E-center versus polarization for the A-center, in which the thermal stability may favor one channel at the expense of the other as long as the corresponding concentrations are similar.

The E-center is technologically of prime importance and fundamentally relevant for several issues. Technologically, we may mention that diffusion of the important group-V dopants (P, As, Sb) takes place via this defect (Fahey et al., 1989), and electrical deactivation of donor atoms is often a result of the formation of this defect. Defect symmetry and Jahn–Teller aspects were for many years among the theoretical and experimental challenges (Elkin and Watkins, 1968; Watkins and Corbett, 1961).

It is well known that most vacancy related defects induce more than one charge state in the band gap. The E-center was among a few exceptions for over 40 years until a theoretical investigation postulated a second level associated to this defect, of donor type character located at about 0.2 eV above the valence band (Resende et al., 1999). This new donor state would be crucial to technology as a pertinent simulation of diffusion needs a mastering of the various state populations. Just after the theoretical predictions, a search for this state was undertaken. From a combination of counter doping of oxygen and carbon lean p-type silicon with phosphorus, electron irradiation at a temperature below the vacancy-migration temperature, optical excitation allowing a conversion of V^{2+} to V^{2-}, and Minority Carrier Transient Spectroscopy, it was possible to unambiguously identify the donor level of the E-center at E_V +0.27 eV (Nylandsted Larsen et al., 2006).

4.1.1 Annealing Under Nonequilibrium Conditions: The Case of E-Center

It is by far not unique that in semiconductors where the band gap is the seat of all dynamical processes, the thermal stability of many defects is charge state dependent. This is a logical consequence of what has been described above. If the electrostatics play a role in the formation of pairs such as the E-center, it is naturally expected that the same electrostatics also determine its stability. The E-center introduces both acceptor and donor states in the band gap, restricting, however, to n-type silicon, the only active state is the acceptor level which, depending on its position relative to the Fermi level, is either negative (Fermi level above) or neutral (Fermi level below). Assuming antimony as the doping atom, in the former case the configuration of the pair is $(V^{=}–Sb^{+})^{-}$, while in the latter case the configuration is $(V^{-}–Sb^{+})^{0}$. It is therefore straightforward to expect a stronger thermal stability for the negatively charge configuration, in which the Coulomb attraction between $V^{=}$ and P^{+} is stronger than in the case of the neutral configuration in which the Coulomb attraction involves two single charges. This effect was very early observed by Kimerling et al. (1971), Kimerling and Carnes (1971), and Evwaraye (1977). The former have described the role of the Fermi level in determining the fraction of the negative and neutral configuration. Figure 13 shows a 20-min isochronal annealing of the V–Sb pair in a Schottky barrier under, respectively, zero and reverse bias (Evwaraye,

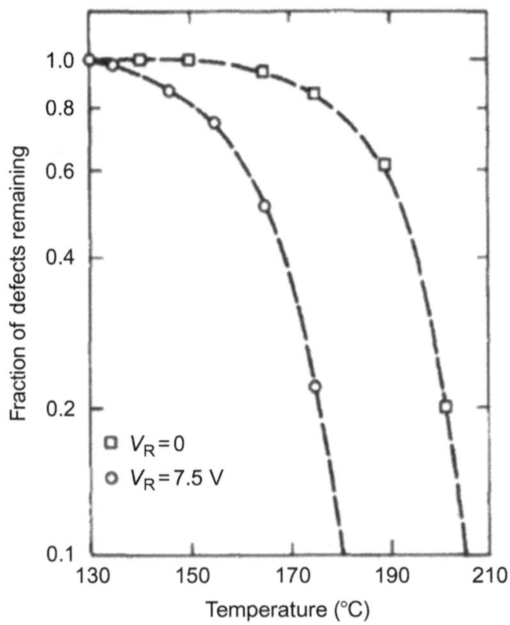

Figure 13 Isochronal anneal (20 min) of the Sb–V pair for the two annealing modes. *Reprinted with permission from Evwaraye (1977). Copyright 1977. AIP Publishing LLC.*

1977). Zero bias settles the E-center in the negatively charged configuration, $(V^{=}{-}Sb^{+})^{-}$, whereas the annealing under reverse bias removes one charge, settling the neutral configuration, $(V^{-}{-}Sb^{+})^{0}$. This experiment clearly demonstrates the enhanced annealing in reverse bias out of equilibrium conditions. Evwaraye (1977) could extract the activation energies for both state; namely 1.29 eV for $(V^{-}{-}Sb^{+})^{0}$ and 1.62 eV for $(V^{=}{-}Sb^{+})^{-}$. The difference is very close to the binding energy of an electron in the acceptor state responsible for the transition,

$$V^{=} \longrightarrow V^{-} + e^{-} \tag{12}$$

Several years later Barnes and Samara (1986) investigated the role of injecting both electrons and holes in the stability of the E-center, via its acceptor state. Annealing in an electron–hole pair atmosphere is another out of equilibrium situation. This process is followed by a nonnegligible rate of e–h pair recombination, releasing an amount of energy equivalent to the band gap. By forward biasing a pn junction to a level of 5 A/cm^2, Barnes and Samara could demonstrate that the annealing activation energy dropped from ~1.30 eV, extracted for the configuration $(V^{-}{-}Sb^{+})^{0}$, to 0.48 eV. The mechanism behind is well known as the REDR which has since been very well documented (Lang and Kimerling, 1974; Weeks et al., 1975). Figure 14 displays a simple illustration of the three annealing channels: zero bias, reverse bias, and forward bias, the latter leading to the REDR mechanism. The channel labeled (1) corresponds to annealing under reverse bias during which the defect is in neutral charge state. In case of the V–Sb above. the activation energy for annealing is $E_a(1) = 1.29$ eV. Channel labeled (2) holds under equilibrium (zero bias) where the defect is negatively charged

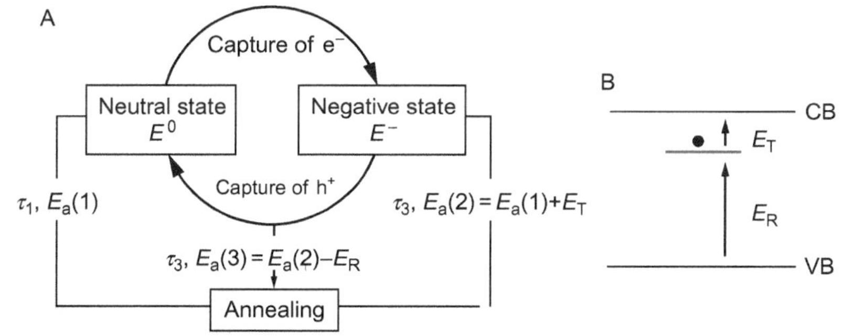

Figure 14 (A) Illustration of the three possible annealing channels as explained in the text, and (B) energy position of an occupied state with associated electron and hole transitions.

and the activation energy for annealing is higher, $E_a(2) = 1.62$ eV. Finally, if both electrons and holes are captured, the annealing is lowered even further to $E_a(2) - E_R$. It then follows a third channel labeled (3).

4.2 Branching Reactions of the Interstitial

The description given below on the branching reactions of the self-interstitial is largely inspired by a beautiful synthesis published many years ago by Kimerling et al. (1988) which, according to the present authors, has not received the attention it deserves. Therefore, the present contribution is a revived echo to the Kimerling et al.'s work. Figure 15 illustrates the branching reactions of the self-interstitial after its creation at cryogenic temperatures. As discussed above, the major difference between the vacancy and the self-interstitial is that the charge states of the latter were not possible to determine experimentally. Neither EPR nor DLTS have indeed been able beyond any doubt to identify this species. This is why, contrary to the vacancy, no charge states are indicated in Fig. 15. We are thus left to rely on what theory has told us. In p-type material, the self-interstitial converts to boron interstitial even at temperature as low as 4.2 K (Watkins, 1975b), while in n-type material it seems to be "waiting" somewhere into the lattice, in a nondetectable configuration, until the temperature is raised well above

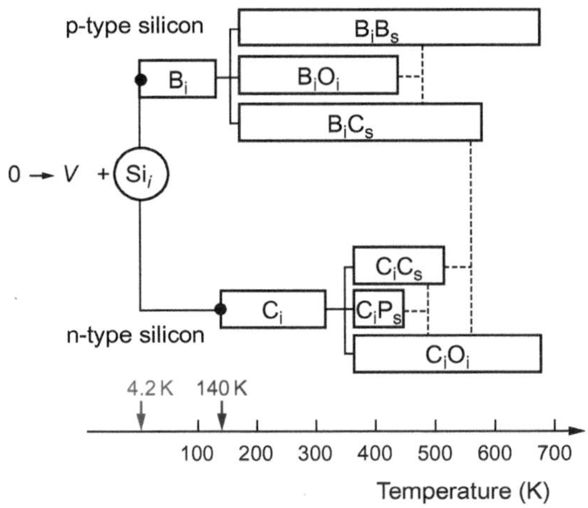

Figure 15 Branching reactions of the self-interstitial into various defects. The dots represent the two branches of the Watkins replacement mechanism taking place, respectively, in p- and n-type silicon. The arrows mark these turning points. The temperature scale shows the thermal stability of the defects. This diagram is inspired by Kimerling et al. (1988).

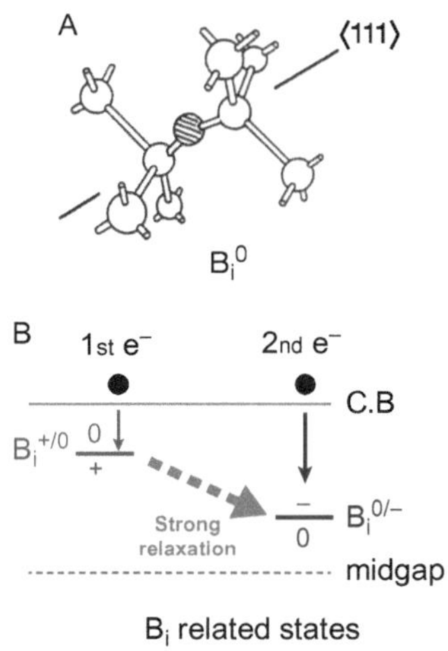

Figure 16 (A) The atomic configuration of interstitial boron (Watkins, 1975b, 1991); (B) shows the inverted distribution of its two coupled states following the negative-U character (Harris et al., 1982, 1987; Troxell and Watkins, 1980). *Copyright Wiley-VCH Verlag GmbH&Co. KGaA. Reproduced with permission.*

nitrogen temperature ($\approx$140 K) where it converts to carbon interstitial (Harris and Watkins, 1985).

EPR studies have allowed extracting the most probable configuration of interstitial boron (Watkins, 1975b). Figure 16A shows the atomic configuration of neutral boron, represented by hatched circle. It is displaced from its substitutional site but keeps bonding to the two nearest substitutional silicon atoms along the $\langle 111 \rangle$ direction. Figure 16B displays the two coupled levels associated with the interstitial boron which exhibits also the peculiar negative-U property similarly to the vacancy and self-interstitial. Here, the neutral state is unstable. Once the first electron is trapped at the donor level located at 0.13 eV below the conduction band, trapping of a second electron immediately follows on a deeper acceptor level located at 0.35 eV below the conduction band. This peculiar behavior is due to a large lattice relaxation following the capture of the first carrier (Harris et al., 1982, 1987; Troxell and Watkins, 1980). In a normal positive-U distribution, the acceptor state $B_i^{0/-}$ should be above the donor state $B_i^{+/0}$. In other words, the second electron is more tightly bound than the first, while we should expect the opposite in a normal distribution. Harris et al. (1982) could observe the unstable neutral state by combining infrared light and low temperatures to control charge population. At high temperatures, in p-type silicon containing carbon, a competition between boron and carbon in trapping self-interstitial occurs, as expected from the branching reactions shown in Fig. 15. The production of B_i and related defects quench as C_s becomes dominant. This has been technologically used to reduce the diffusion of implanted boron when ultra-shallow junctions are needed (Nishikawa et al., 1992).

Increasing further the temperature to about 220 K make interstitial boron to migrate forming complexes with the other existing background contaminants, namely oxygen (O_i) and carbon substitutional (C_s). If the material is highly doped the pair B_iB_s could also form and it is the most stable defect as shown in the branching scheme displayed in Fig. 15. As noted by the dashed lines in Fig. 15, the dissociation of B_iO_i occurring above 420 K liberates B_i which will bind either to B_s or C_s preserving mass conservation. Similar reasoning can be applied above 580 K where the pair B_iC_s dissociates liberating again B_i. So far, the atomic structures of these pairs have not been determined. However, their related levels in the band gap were identified. The pairs B_iB_s and B_iC_s introduce each a hole trap at 0.30 and 0.29 eV, respectively, above the valence band (Drevinsky and DeAngelis, 1985; Drevinsky et al., 1988) making their identification very tricky. Finally, the pair B_iO_i introduces an electron trap at 0.26 eV below the conduction band (Drevinsky and DeAngelis, 1985; Drevinsky et al., 1988).

In n-type material, the Watkins replacement mechanism does not involve the kick out of substitutional group-V dopant but rather substitutional carbon. We may thus expect that in n-type silicon, in the absence of boron and carbon, the self-interstitial would survive longer than in p-type silicon. The restriction of Watkins replacement mechanism to group-III impurities and carbon in silicon, excluding group-V dopants, may be a simple consequence of the fact that self-interstitial is either neutral or positively charged. In the latter case electrostatic repulsion forbids the interaction of $Si_i^{+/++}$ with positively charge group-V dopants. The conversion of self-interstitial to interstitial carbon is thus the only alternative constituting a second channel of interstitial defect reactions through carbon interstitial C_i, which is also a product of the Watkins replacement mechanism.

We mentioned above that in n-type silicon, Watkins replacement mechanism does not seem to occur until a temperature of the order of 140 K is reached while the same ingredients as in p-type silicon are present. Below this temperature, the self-interstitial seems thus to survive in a form which is either not active as neither EPR nor DLTS could detect any trace, or in an active form but beyond the capabilities of these techniques. Regarding this latter issue, a possible explanation of the lack of detection of self-interstitial in n-type silicon could be inherent to its possible negative-U character mentioned above. As illustrated in Fig. 8B, this peculiar property requires that only charge states populations with pairs of electrons or holes trapped at the defect leading, respectively to Si^{0} or Si^{++}, are likely to occur in significant amount. Thus, the absence of unpaired spins would explain why EPR was unable to observe any related signal. DLTS can, however, detect a change of population, even by pairs as it has been observed for the vacancy. Therefore, the only way of explaining why this technique was unable to observe any signal related to self-interstitial would be that its migration occurs at temperature lower than the DLTS peak corresponding to thermal emission of trapped pairs of electrons. In other words, the expected DLTS peak should be above 140 K, a temperature range where the self-interstitial converts irreversibly into carbon interstitial. A similar case has been observed for the Frenkel pair in germanium (Mesli et al., 2008).

The atomic configuration of interstitial carbon as observed by EPR (Song and Watkins, 1990; Watkins and Brower, 1976) is depicted in Fig. 17A. This defect introduces two coupled but normally distributed levels as shown in Fig. 17B. The acceptor state is located at 0.10 eV below the conduction band and the donor state is positioned at 0.27 eV above the valence

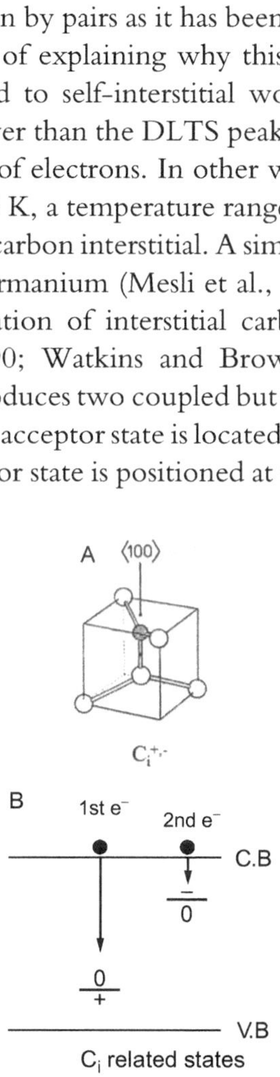

Figure 17 (A) The atomic configuration of interstitial carbon (Song and Watkins, 1990; Watkins, 1991; Watkins and Brower, 1976); (B) represents the normal distribution of its two coupled states (Asom et al., 1981). *Copyright Wiley-VCH Verlag GmbH&Co. KGaA. Reproduced with permission.*

band (Asom et al., 1981). Above room temperature, carbon interstitial starts migrating through the lattice and, following the branching hierarchy displayed in Fig. 15, it binds to group-V dopants, in the present case phosphorus, forming the pair C_iP_s, to oxygen forming the pair C_iO_i and, if the material contains a large concentration of substitutional carbon, it forms the pair C_iC_s in a similar way as B_iB_s in p-type silicon. The pair C_iO_i introduces a hole trap located at 0.36 eV above the valence band (Asom et al., 1981; Newman and Bean, 1971; Trombetta and Watkins, 1987). The pairs C_iO_i and B_iC_s are the two main defects which had to be avoided when studying the donor state of the E-center due to the fact that they are located in the same energy range (Nylandsted Larsen et al., 2006). This explains why a search for a material lean from oxygen and carbon was a prerequisite. All these defects are well documented in literature with a special mention to Watkins group which, without any doubt, came up with the major achievement.

Among all defects mentioned above, the C_iC_s pair deserves to be described in more detail. This defect is the preferred interstitial product when substitutional carbon is the main background impurity. During its migration above 140 K, C_i interacts with a C_s atom with a one-to-one conversion to C_iC_s (Leervad Pedersen et al., 1999). The atomic structure, illustrated in either of the two configurations displayed in Fig. 18, was identified via EPR studies mainly by Jellison (1982) and Song et al. (1988). As Fig. 18 suggests the stable configuration is charge state dependent. This property was

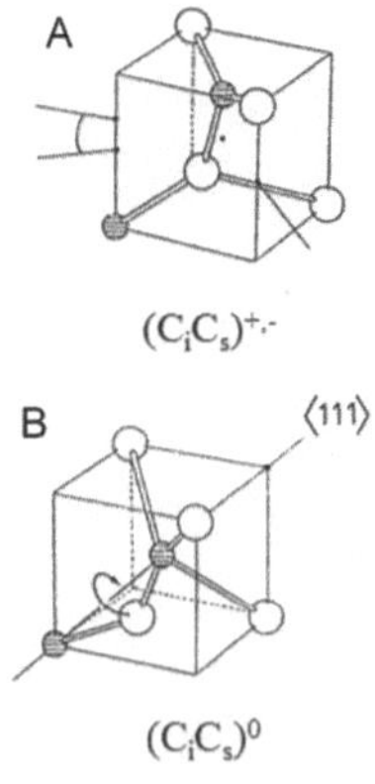

Figure 18 (A) Atomic structures for the positive and negative charge states of the pair C_iC_s, and (B) for the neutral state (Song and Watkins, 1990; Watkins, 1991). *Copyright Wiley-VCH Verlag GmbH&Co. KGaA. Reproduced with permission.*

not new when, in the eighties, Watkins group undertook the investigation of carbon-related defects. This period witnessed a wealth of cases of the so-called bistable electronics properties not only in silicon but also in various other semiconductors (a list of reviews is given in Song et al., 1988). In some situations, defects do not exhibit a frozen atomic configuration into the lattice. The permanent exchange between a defect and free carrier reservoirs (conduction or valence bands) may lead to a charge-controlled stability for each specific configuration. The pair C_iC_s may adopt two different configurations depending on its charge state as illustrated in Fig. 18A. Referring to Fig. 17A, the pair is formed with a C_s atom replacing one of the neighboring substitutional silicon atoms. The pair C_iC_s introduces two levels, an acceptor electron trap (−/0) at 0.17 eV below the conduction band (Jellison, 1982) and donor hole trap at 0.09 eV above the valence band (Song et al., 1988). The latter authors have demonstrated that each of these states has it metastable form. When the acceptor state is prepared in the neutral state, it changes its atomic configuration and the corresponding levels moves from 0.17 to 0.10 eV below the conduction band. The situation is similar for the donor-hole trap which moves from 0.09 eV above the valence band, when prepared in the positive state, to 0.05 eV when prepared in the neutral state. Figure 19 summarizes Song et al.'s investigation on the bistability of C_iC_s. The configuration labeled A correspond to the structure displayed in Fig. 18A, while configuration B represents the structure shown in Fig. 18B.

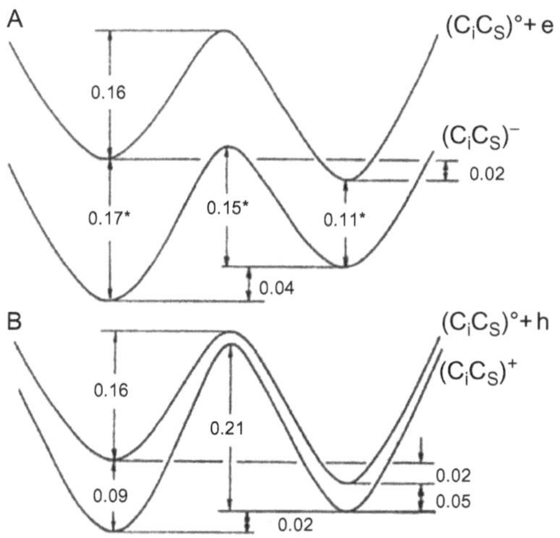

Figure 19 Configurational diagram of the pair C_iC_s showing the stable form for each charge state. See Song and Watkins (1990) for more details.

The branching reactions initiated by the self-interstitial and its main by-products B_i and C_i, resulting from Watkins replacement mechanism, show a wealth of reactions and complexities. Depending on the charge states involved, both long- and short-range migrations are expected to play a determining role in the interactions and the nature of emerging defects. The coupling to local phonons and the presence of free carriers led to the discovery of new phenomena such as bistability, which is a particular case of the more general mechanism of multistability. It has also led to highlight the origin of athermal processes and of the recombination enhanced defect reactions (Lang and Kimerling, 1974; Weeks et al., 1975). The fundamental understanding of the dynamics of most of the observed defects allowed also clarifying the concept of negative-U, now firmly established for several point defects in silicon such as the vacancy and boron interstitial, and to a lesser extent self-interstitial. It is of no doubt that the same studies carried out in more complex semiconductors will give rise to new phenomena.

5. COMPLEXES CONSISTING OF THE NATIVE DEFECTS

5.1 The Divacancy

The divacancy has been one of the most studied irradiation-induced defects in silicon. It was thoroughly addressed for the first time some 40 years ago by Watkins and Corbett (1965) using electron irradiations at low sample temperature and EPR. The model of the divacancy deduced from their studies consists of two vacancies resulting from adjacent atom sites as shown in Fig. 20. These authors were able to observe the divacancy in two different charge states, the singly positive and singly negative. However, they established that the divacancy can exist in four charge states—singly positive,

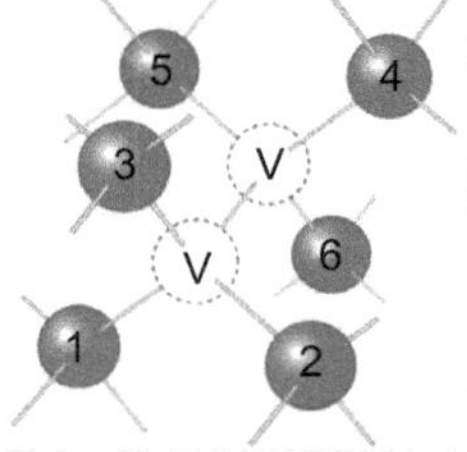

Figure 20 Schematic view of the divacancy. Vacancies are depicted by dashed circles (Iwata et al., 2008). *Reproduced with permission from the American Physical Society. Copyright 2008.*

neutral, singly negative, and doubly negative states—and gave estimates of the associated level positions in the band gap. The existence of the four different charge states have been demonstrated by numerous experiments during the last 40 years, and DLTS measurements have established a (0/+)-donor level at $\sim E_V + 0.20$ eV, a (−/0)-acceptor level at $\sim E_C - 0.42$ eV, and a (=/−)-acceptor level at $\sim E_C - 0.43$ eV (Mikelsen et al., 2005).

The divacancy can be formed either as a primary damage event or as a result of the interaction of two migrating vacancies. In the former case, Corbett and Watkins (1965) showed from the energy dependence of the production rate of the divacancy that the threshold for divacancy production is approximately a factor of two higher than that of the monovacancy resulting in a minimum electron energy for divacancy production of about 450 keV and a minimum proton energy of about 250 eV. In the case of the formation of the divacancy as a secondary defect, its concentration depends on the charge state of the migrating vacancy and thereby on the position of the Fermi level. Thus, in p-type silicon, in which the vacancy is positively charged, the probability of forming the divacancy as a secondary defect is smaller than in n-type silicon where the monovacancy can be both neutral and negatively charged. From electrostatic considerations and following the hierarchy tree outline in Fig. 9, vacancies need to be neutral to form divacancies.

In the early study by Watkins and Corbett (1965) it was found that the divacancy annealed at about 290 °C in CZ-silicon with high oxygen concentration but at about 350 °C in FZ-silicon with low oxygen concentration. They estimated the activation energy of diffusion to be ~ 1.3 eV and the binding energy to be ≥ 1.6 eV; thus, the dissociation energy must be higher than about 1.6 eV. As discussed by Evwaraye and Sun (1976) dissociation can still be a competing or dominant process with diffusion in systems where there are relatively few sinks. The activation energy of diffusion of the divacancy in the neutral charge state has recently been determined by Mikelsen et al. (2005) to be 1.30 ± 0.02 eV with a preexponential factor of $4.4 \pm 1.9 \times 10^8\ s^{-1}$, and in CZ-silicon the loss of V_2 and the formation of V_2O display a close one-to-one proportionality.

5.2 The Trivacancy

Until recently, the structure and electronic levels of the trivacancy were not identified without ambiguity although it is one of the most abundant defects after high energy electron irradiation with energies higher than about 5 MeV. Tentative models of EPR spectra measured about 40 years ago had, however, included the trivacancy (Lee and Corbett, 1974). Recently, Markevich and coworkers published a number of papers combining DLTS investigations and density-functional modeling calculations (Coutinho et al., 2012; Markevich et al., 2009), and they have been able to describe the structure of the trivacancy and to establish its energy-level structure in the band gap.

Markevich et al. (2009) and Coutinho et al. (2012) found that the trivacancy is a bistable center in the neutral charge state, and that the fourfold coordinated (FFC) configuration shown in Fig. 21D is the ground state, whereas V_3 is metastable by 0.26 eV in the (110)-planar configuration (the part of hexagonal ring (PHR) structure) shown in Fig. 21A. Immediately after electron irradiation, V_3 was only found in the PHR configuration, but after storage of the samples for a few weeks at room temperature or after a short annealing at 50–100 °C it transformed from the PHR configuration to the FFC configuration. It could be restored to the PHR configuration by application of a forward-bias injection at room temperature, and the transformation was found to be fully reversible. In the 2+, +, −, and 2- charge states, the PHR configuration was found to be the most stable. In the FFC configuration, the V_3 defect has only an acceptor level at $E_C - 0.075$ eV, while in the PHR configuration it has four levels, two acceptor levels at $E_C - 0.34$ and $E_C - 0.455$ eV and two donor levels at $E_V + 0.12$ and E_V

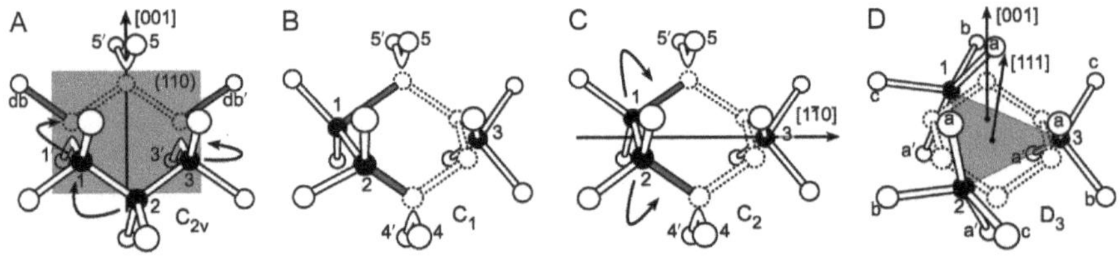

Figure 21 Atomic structure of stable and metastable V_3 complexes in silicon along with high symmetry crystallographic directions and planes. The PHR→FFC transformation mechanism is depicted by the arrows indicating the motion of three core Si atoms (1, 2, and 3) from substitutional sites in V_3[PHR] (A) to interstitial sites in V_3[FFC] (D). These are shown in black for a better perception of their placement with respect to the vacant sites (drawn with a dashed line). Intermediate structures along the mechanism are shown in (B) and (C). Broken bonds are shown as solid sticks (red sticks in the online version), reconstructed radical pairs Si_i–Si_i' are shown as "banana" bonds, and Si atoms that are slightly perturbed (or unperturbed) from their lattice sites are shown in white (Coutinho et al., 2012). *Reproduced with permission from the American Physical Society. Copyright 2008.*

+0.23 eV. Calculated energy-level positions from *ab initio* calculations are in very good agreement with these experimentally observed energy-level positions.

The transformation from the PHR configuration to the FFC configuration is found to proceed with an activation energy and preexponential factor of 1.16 ± 0.02 eV and 2.75×10^{13} s^{-1}, respectively, when the annealing is done with an applied reverse bias, and with 1.22 ± 0.02 eV and 1.6×10^{14} s^{-1}, respectively, when annealed without bias. Thus, the transformation takes place with similar rates and activation energies for both neutral and singly negatively charged V_3 defects. The V_3 defect anneals at around 250 °C with the activation energy 1.47 ± 0.04 eV and a reaction rate which is dependent on the oxygen concentration of the silicon samples indicating the involvement of oxygen in the elimination reaction.

5.3 The Di-Interstitial

The di-interstitial in silicon has been discussed for more than 40 years and has been the target of a large number of experimental and theoretical investigations over the years. It seems, however, fair to state that until now there is no unambiguous experimental identification of the di-interstitial but rather a number of circumstantial evidences pointing to an identification.

Which type of defect does the theoretical modeling of the di-interstitial suggest? The most recent calculations using first principles density methods employing Gaussian orbitals (Eberlein et al., 2001) or tight-binding molecular dynamics simulations (Cogoni et al., 2005; Kim et al., 1999; Posselt et al., 2005; Richie et al., 2004) point to a structure consisting of a center atom I_0 and dumbbell atoms $I_1 - I_2$ as shown in Fig. 22-K in which three atoms share the regular lattice site located at the origin of the axis, and the dumbbell is aligned parallel to the [110] direction. This configuration has a formation energy of 4.93 eV (Kim et al., 1999). According to Eberlein et al. (2001), there is a slightly modified configuration which has a comparable formation energy in which the dumbbell is added to a site near a bond center (Fig. 22-C). Both models have C_{1h} symmetry. Kim et al. (1999) demonstrated that four equivalent C_{1h} di-interstitial configurations, distinguished by the location of the center atom, can be constructed with three interstitials sharing one regular lattice site, and that thermal averaging motions among these four C_{1h} configurations with an energy barrier of 0.5 eV can result in a D_{2d} symmetry at room temperature. Thus, the low temperature symmetry is C_{1h} and the room temperature symmetry D_{2d}.

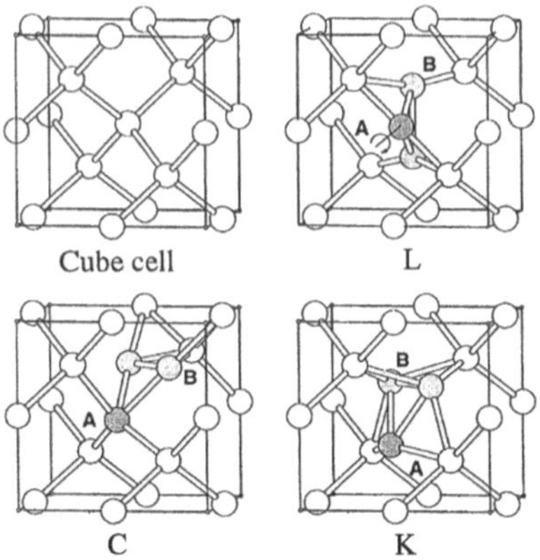

Figure 22 Models of I_2. *Reprinted from Eberlein et al. (2001), Copyright 2001, with permission from Elsevier.*

A donor level at E_V +0.1 eV and an acceptor level at $E_V+0.6$ eV were estimated for the C_{1h} di-interstitial by Kim et al. (1999); for the motionally average state, these two levels become almost degenerate and form a deep level at $E_V+0.4$ eV. Eberlein et al. (2001) found for the C_{1h} configuration, using the AIMPRO method, only one level in the band gap, a donor level at $E_V+0.53$ eV, and for the slightly modified configuration mentioned above also one donor level, now at $E_V+0.39$ eV.

There are several recent theoretical determinations of the diffusion energy of the neutral di-interstitial based on different methods. Eberlein et al. (2001) reports a preliminary value of 0.5 eV using the AIMPRO method, while Kim et al. (1999) using tight-binding molecular dynamics simulations find 0.7 ± 0.1 eV. Cogoni et al. (2005) using tight-binding molecular dynamics simulations coupled with the temperature-accelerated dynamics method find 0.89 eV and a diffusivity prefactor of 1.9×10^{-3} cm^2 s^{-1}. On the other hand, Posselt et al. (2005) using molecular dynamics simulations with the Stillinger–Weber potential find a somewhat lower value of 0.2 eV. Finally, Du et al. (2006) using tight-binding molecular dynamics and density-functional simulations find 0.3 eV. Thus, the theoretical diffusion energies scatter from 0.2 to 0.9 eV.

Let us now concentrate on the experimental information on the di-interstitial, and compare them as we go along with the theoretical predictions discussed above. The first mention of an experimental identification of the di-interstitial was by Lee et al. (1976) almost 40 years ago. They made a very comprehensive EPR study of a particular defect in silicon called the

P6 center induced by proton or neutron irradiation. The suggested microscopic structure of this defect was that of a positive charge state of two self-interstitials adjacent to a substitutional atom along a ⟨100⟩ axis (Fig. 22L). The defect was found to anneal at 127 °C with a migration energy estimated to be 0.6 ± 0.2 eV. An interesting feature was that only at low temperature (200 K) the defect symmetry was C_2 or C_{1h}, at high temperature (315 K) the defect symmetry became D_2 owing to a motional effect. From studies of the recovery kinetics of stress-induced alignments, this transition from C_2 to D_2 symmetry was found to be thermally activated with an activation energy of 0.6 ± 0.2 eV and a jump frequency of $\sim 10^8$ s^{-1}. A reanalysis of these EPR results was later done by Lee (1998), resulting in a slightly modified structure for the di-interstitial. In this structure, the dumbbell is rotated toward the (110) plane. Lee (1998) stated that the P6 center is stable at room temperature and anneals at 170 °C. This annealing temperature is supported by EPR investigations of the P6 defect by Pierreux and Stesmans (2005), showing that at 200 °C no traces of this defect remain.

The P6 defect observed by EPR resembles in many respects the di-interstitial defect as theoretically described by Kim et al. (1999) although the symmetry, as inferred by Eberlein et al. (2001), is not correct: the di-interstitial as described by Kim et al. (1999) possesses C_{1h} symmetry at low temperatures, whereas the P6 defect possesses C_2 symmetry at low temperature. It must be noted, however, that Lee (1998) ascribes the P6 defect to the positive charge state of the di-interstitial, I_2^+, whereas the di-interstitial structure with C_{1h} symmetry which Kim et al. (1999) finds is the di-interstitial in the neutral charge state, I_2^0, and different symmetry of the two charge states could undoubtly be expected; Kim et al. (1999) finds that the C_{2v} di-interstitial, which is metastable for the neutral charge state, becomes a stable defect when it is positively charged.

In support of the stability at room temperature are the results from the grazing incidence diffuse X-ray scattering investigations by Partyka et al. (2001) who find that the di-interstitial is immobile at room temperature. We must, however, be careful when we compare experimentally observed annealing behavior with theoretical estimates of the diffusion energy. As discussed above, the di-interstitial is theoretically expected to have a donor level in the lower part of the band gap not too far from mid-gap. If, e.g., 1-Ω cm p-type Si is used for the investigations then the di-interstitial will be positively charged for temperatures below ~500 K, and neutral for temperatures above. In n-type Si, the defect is always neutral. Therefore, a difference in anneal temperature for the positive and neutral charge states can easily be expected.

A number of different levels in the band gap observed using DLTS have been associated with the di-interstitial (Lefèvre, 1982; Meese, 1981) based in particular on the correlation to the annealing characteristics of the P6 defect. A general inconvenience of DLTS on irradiation-induced levels in the lower half of the band gap, where the di-interstitial associated levels are expected to be according to theory, is that there are several levels related to interstitial C and O, growing in and disappearing at room temperature, rendering an identification of a di-interstitial associated line very difficult unless care is taken to reduce the carbon and oxygen concentrations in the silicon samples. Nylandsted Larsen et al. (2013) have recently identified a level at E_V +0.357 eV in oxygen and carbon lean FZ p- and n-type silicon irradiated with 2-MeV protons which grows in at room temperature and anneals at ~425 K when positively charged and ~475 K when neutral. In p-type Si, the defect is only formed as a primary defect, whereas in n-type, it is also formed as a secondary defect. The behavior of this defect resembles in many respects the behavior of the P6 defect and was tentatively assigned to the same defect.

Very recently, Londos et al. (2013) have studied CZ-Si after fast neutron irradiation, and have in particular focused on the 533 cm^{-1} band which in many respects resemble the P6 defect. The band is observed to begin to decay at ~170 °C with an activation energy of 0.88 ± 0.03 eV.

5.4 The Tri-Interstitial

A large number of theoretical investigations have been devoted to the tri-interstitial, and some of them have recently been reviewed by Keunen and Stesmans (2009) in connection with their discussion of the atomic nature of the Si-B4 EPR center; we will come back to that shortly. One of the most recent theoretical investigations is by Carvalho et al. (2005) who did local density-functional calculations using the AIMPRO code. They investigated five structures for I_3, these structures are shown in Fig. 23. They find that the I_3–IV structure has the lowest energy; this is also found by Richie et al. (2004) using tight-binding molecular dynamics. Estreicher et al. (2001) using *ab initio* molecular dynamics simulations find that the I_3–III has the lowest energy; this structure, however, is demonstrated by Carvalho et al. (2005) to relax to the relatively more stable I_3-II. Cogoni et al. (2005), using temperature-accelerated tight-binding molecular dynamics simulations, also find the I_3-IV structure to have the

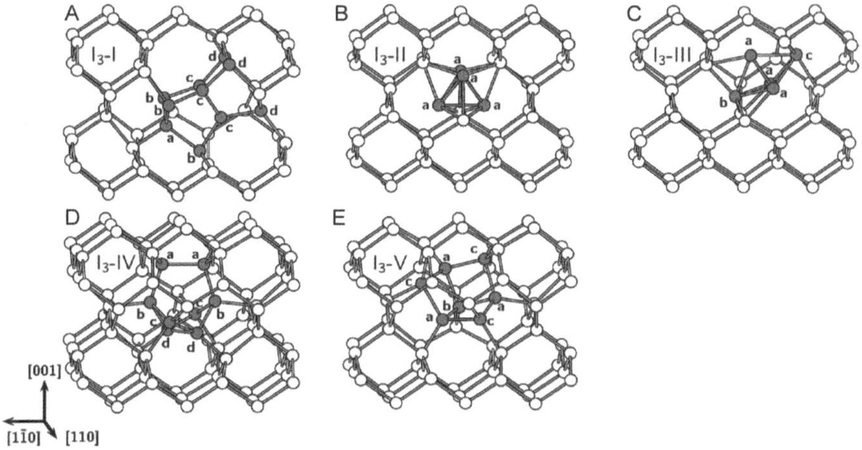

Figure 23 The I_3 structures. For clarity, Si atoms at the defect core are represented as gray spheres (Carvalho et al., 2005). *Reproduced with permission from the American Physical Society. Copyright 2008.*

lowest energy. Moreover, they estimated the migration energy and prefactor to be 1.71 eV and 0.21 $cm^2 s^{-1}$, respectively, thus, the defect starts to migrate at temperatures around ~200 °C. Among the five I_3 structures studied by Carvalho et al. (2005), only the I_3-I structure has a level in the band gap, a donor level at E_V +0.15 eV. This was confirmed recently by Leitão et al. (2011) who found the level to be at E_V +0.18 eV. Thus, only the I_3-I structure will be electrically active, will show photoluminescence activity, and is the only tri-interstitial that could be detected by EPR as this method can only detect defects with an unpaired spin.

Both the W photoluminescent (PL) center and the Si-B5 EPR center have been linked with the tri-interstitial defect. The W-center shows a zero-phonon emission line at 1018 meV; it starts to appear in the PL spectra at around room temperature and is maximized after annealing at ~300 °C and disappears at ~400 °C (Leitão et al., 2011). The symmetry of the W-center is trigonal, showing a unique phonon sideband at 565 cm^{-1}. The Si-B4 paramagnetic center has recently been reinvestigated by Keunen and Stesmans (2009). The characteristics of the Si-B4 center are, in the main, that its ingrowth starts at ~200 °C, it peaks at 250–300 °C and disappears at ~400 °C, has trigonal symmetry, and that it is an intrinsic defect and not related to vacancy clusters. The W and Si-B5 centers are in many respects similar and resemble the I3-I structure; this led Carvalho et al. (2005) to conclude that the calculated characteristics of the I_3-I support the assignment of this structure to the W-center, although this structure is a high energy, metastable I_3 defect. Keunen and Stesmans (2009) conclude similarly that the I_3-I is slightly preferable for the Si-B4 defect but express doubts about its high formation energy.

Observation of the tri-interstitial by DLTS has to our knowledge never been reported. A reason for that could be that the experimentally determined donor level of the di-vacancy is situated at E_V +0.18 eV, which is similar to the level position expected for the tri-interstitial. Moreover, the anneal temperatures are very similar. Thus, it will be difficult to distinguish between the DLTS spectra of the donor level of the di-vacancy and the tri-interstitial.

6. SUMMARY

Irradiation of silicon by light particles (electrons and protons) in the MeV energy range has been used for more than 50 years with great success to produce simple defects, formed from the elementary bricks: vacancy and self-interstitial. This way of producing defects revealed to be very powerful in controlling defect concentrations above equilibrium, allowing consistent conclusions to be drawn leading to valuable insights into their properties. Today, the accumulated knowledge on simple point defects in silicon comes to a great extent from such irradiations. In semiconductors, and particularly silicon, the band gap is the major seat of all kind of defect transformations, mainly governed by charge state populations and exchange with allowed bands. A number of properties could then be analyzed such as the mechanism of complex formation, their structure, stability, diffusivity, and electrical activity.

The knowledge of vacancy-type defects from the monovacancy to the trivacancy and impurity-vacancy complexes is well established although interesting and surprising new information still pop up such as the existence of the donor level of the E-center, and gaps in our knowledge are still existing; examples are the lack of direct information on the positions in the band gap of the acceptor levels of the monovacancy, and the temperature dependence of the vacancy-migration energy.

The interstitial type impurity defects are also well understood, and we have to move to one of the simplest defects, the monointerstitial, to find real gaps in our knowledge on simple irradiation-induced defects. Although this simple defect has been studied intensively for more than 50 years without significant breakthroughs, there are indication that a possible way to catch it could be by choosing irradiation conditions where the free carrier production from electronic stopping is minimized thereby switching off the Bourgoin–Corbett low temperature athermal diffusion mechanism.

This review aims at summarizing the most basic information which would hopefully enable understanding, or predicting, the properties of defects not only in other more complex systems, such as III–V dilute ternaries, III-antimonides, ZnO, and others but also in more elementary semiconductors down scaled to nanometer size. In these latter systems, the surface-to-volume ratio is very important and new interactions might be expected modifying and even enriching the defects properties mentioned above.

REFERENCES

Asom, M.T., Benton, J.L., Sauer, J.L., Kimerling, L.C., 1981. Interstitial defect reactions in silicon. Appl. Phys. Lett. 51, 256–258.

Auret, D.F., Deenapanray, P.N.K., 2004. Deep level transient spectroscopy of defects in high-energy light-particle irradiated Si. Crit. Rev. Solid State Mater. Sci. 29, 1–44.

Baraff, G.A., Kane, E.O., Schlüter, M., 1979. Silicon vacancy: a possible "Anderson Negative-*U*" system. Phys. Rev. Lett. 43, 956–959.

Baraff, G.A., Kane, E.O., Schlüter, M., 1980a. Simple parameterized model for Jahn-Teller systems: vacancy in *p*-type silicon. Phys. Rev. B21, 3563–3570.

Baraff, G.A., Kane, E.O., Schlüter, M., 1980b. Theory of the silicon vacancy: an Anderson negative-U system. Phys. Rev. B21, 5662–5686.

Barnes, C.E., Samara, G.A., 1986. Forward bias induced annealing of the E center in silicon. Appl. Phys. Lett. 48, 934–936.

Bar-Yam, Y., Joanopolous, J.D., 1984. Barrier to migration of the silicon self-interstitial. Phys. Rev. Lett. 52, 1129–1132.

Bourgoin, J.C., Corbett, J.W., 1978. Enhanced diffusion mechanisms. Radiat. Eff. 36, 157–188.

Bracht, H., Chroneos, A., 2008. The vacancy in silicon: a critical evaluation of experimental and theoretical results. J. Appl. Phys. 104, 0761081–0761083.

Bracht, H., Fage Pedersen, J., Zangenberg, N., Nylandsted Larsen, A., Haller, E.E., Lulli, G., Posselt, M., 2003. Radiation enhanced silicon self-diffusion and the silicon vacancy at high temperatures. Phys. Rev. Lett. 91, 2455021–2455024.

Bracht, H., Kube, R., Hüger, E., Schmidt, H., 2014. Properties of point defects in silicon: new results after a long-time debate. Solid State Phenom. 205–206, 151–156.

Car, R., Kelly, P.J., Oshiyama, A., Pantelides, S.T., 1984. Microscopic theory of atomic diffusion mechanisms in silicon. Phys. Rev. Lett. 52, 1814–1817.

Carvalho, A., Jones, R., Coutinho, J., Briddon, P.R., 2005. Density-functional study of small interstitial clusters in Si: comparison with experiments. Phys. Rev. B 72, 1552081–1552085.

Cogoni, M., Uberuaga, B.P., Voter, A.F., Colombo, L., 2005. Diffusion of small self-interstitial clusters in silicon: temperature-accelerated tight-binding molecular dynamics simulations. Phys. Rev. B 71, 1212031–1212034.

Corbett, J.W., Watkins, G.D., 1965. Production of di-vacancies and vacancies by electron irradiation of silicon. Phys. Rev. 138, A555–A560.

Coutinho, J., Markevich, V.P., Peaker, A.R., Hamilton, B., Lastovskii, S.B., Murin, L.I., Svensson, B.J., Rayson, M.J., Briddon, P.R., 2012. Electronic and dynamical properties of the silicon trivacancy. Phys. Rev. B 86, 1741011–17410113.

Drevinsky, P.J., DeAngelis, H.M., 1985. Interactions of interstitial silicon with Bs and Oi in electron-irradiated silicon. In: Kimerling, L.C., Parsey Jr., J.M. (Eds.), Proceeding of the Thirteenth International Conference on Defects in Semiconductors. The Metallurgical Society of AIME, Warrendale, PA, pp. 807–813.

Drevinsky, P.J., Caefer, C.E., Tobin, S.P., Mikkelson Jr., J.C., Kimerling, L.C., 1988. Influence of oxygen and boron on defect production in irradiated silicon. In: Stavola, M., Pearton, S.J., Davies, G. (Eds.), Defects in Electronic Materials vol. 104. Materials Research Society Symposium Proceedings, Pittsburgh, pp. 167–172.

Du, Y.A., Hennig, R.G., Wilkins, J.W., 2006. Diffusion mechanisms for silicon di-interstitials. Phys. Rev. B 73, 2452031–2452035.

Eberlein, T.A.G., Pinho, N., Jones, R., Coomer, B.J., Goss, J.P., Briddon, P.R., Öberg, S., 2001. Self-interstitial clusters in silicon. Physica B 308–310, 454–457.

Elkin, E.L., Watkins, G.D., 1968. Defects in irradiated silicon: electron paramagnetic resonance and electron-nuclear double resonance of the arsenic- and antimony-vacancy pairs. Phys. Rev. 174, 881–897.

Estreicher, S.K., Gharaibeh, M., Fedders, P.A., Ordejón, P., 2001. Unexpected dynamics for self-interstitial cluster in silicon. Phys. Rev. Lett. 86, 1247–1250.

Evwaraye, A.O., 1977. Electron-irradiation damage in antimony-doped silicon. J. Appl. Phys. 48, 734–738.

Evwaraye, A.O., Sun, E., 1976. Electron-irradiation-induced divacancy in lightly doped silicon. J. Appl. Phys. 47, 3776–3780.

Fahey, P.M., Griffin, P.B., Plummer, J.D., 1989. Point defects and dopant diffusion in silicon. Rev. Mod. Phys. 61, 289–384.

Fairfield, J.M., Masters, B.J., 1967. Self-diffusion in intrinsic and extrinsic silicon. J. Appl. Phys. 38, 3148–3154.

Gorelkinskii, Yu.V., Abdullin, Kh.A., Mukashev, B.N., Turmagambetov, T.S., 2009. Self-interstitials and related defects in irradiated silicon. Physica B 404, 4579–4582.

Harris, R.D., Watkins, G.D., 1985. Interstitial related defects in n-type. In: Kimerling, L.C., Parsey, J.M. (Eds.), Proceedings of 13th International Conference on Defects in Semiconductors, pp. 799–805.

Harris, R.D., Newton, J.L., Watkins, G.D., 1982. Negative-U properties for interstitial boron in silicon. Phys. Rev. Lett. 48, 1271–1274.

Harris, R.D., Newton, J.L., Watkins, G.D., 1987. Negative-U defect: interstitial boron in silicon. Phys. Rev. B36, 1094–1104.

Iwata, J., Shiraishi, K., Oshiyama, A., 2008. Large-scale density-functional calculations on silicon divacancies. Phys. Rev. B 77, 1152081–1152089.

Jellison Jr., G.E., 1982. Transient capacitance studies of an electron trap at $E_c–E_T = 0.105$ eV in phosphorus-doped silicon. J. Appl. Phys. 53, 5715–5719.

Jones, R., Carvalho, A., Goss, J.P., Briddon, P.R., 2009. The self-interstitial in silicon and germanium. Mater. Sci. Eng. B 159–160, 112–116.

Keunen, K., Stesmans, A., 2009. Atomic nature of the Si-B4 paramagnetic center assessed by multifrequency electron spin resonance. Phys. Rev. B 80, 1952071–19520713.

Kim, J., Kirchhoff, F., Aulbur, W.G., Wilkins, J.W., Khan, F.S., Kresse, G., 1999. Thermally activated reorientation of di-interstitial defects in silicon. Phys. Rev. Lett. 83, 1990–1993.

Kimerling, L.C., 1977. Defect states in electron-bombarded silicon: capacitance transient analyses. In: Inst. Conf. Ser. 31, London, pp. 221–230.

Kimerling, L.C., Carnes, C.P., 1971. Annealing of electron-irradiated n-type silicon: illumination and fluence dependence. J. Appl. Phys. 42, 3548–3552.

Kimerling, L.C., DeAngelis, H.M., Carnes, C.P., 1971. Annealing of electron-irradiated n-type silicon: I. donor concentration dependence. Phys. Rev. B3, 427–433.

Kimerling, L.C., Asom, M.T., Benton, J.L., 1988. Interstitial defect reactions in silicon. Mater. Sci. Forum 38–41, 141–150.
Lang, D.V., Kimerling, L.C., 1974. Observation of recombination-enhanced defect reactions in semiconductors. Phys. Rev. Lett. 33, 489–492.
Lee, Y.-.H., 1998. Silicon di-interstitial in ion-implanted silicon. Appl. Phys. Lett. 73, 1119–1121.
Lee, Y.-.H., Corbett, J.W., 1974. EPR study of defects in neutron-irradiated silicon: quenched-in alignment under <110>-uniaxial stress. Phys. Rev. B 9, 4351–4361.
Lee, Y.-.H., Gerasimenko, N.N., Corbett, J.W., 1976. EPR study of neutron-irradiated silicon: a positive charge state of the <100> split di-interstitial. Phys. Rev. B 14, 4506–4520.
Leervad Pedersen, T.P., Nylandsted Larsen, A., Mesli, A., 1999. Carbon-related defects in proton-irradiated, *n*-type epitaxial $Si_{1-x}Ge_x$. Appl. Phys. Lett. 75, 4085–4087.
Lefèvre, H., 1982. Annealing behavior of trap-centers in silicon containing A-swirl defects. Appl. Phys. 29, 105–111.
Leitão, J.P., Carvalho, A., Coutinho, J., Pereira, R.N., Santos, N.M., Ankiewicz, A.-O., Sobolev, N.A., Barroso, M., Lundsgaard Hansen, J., Nylandsted Larsen, A., Briddon, P.-R., 2011. Influence of Ge content on the optical properties of X and W centers in dilute Si-Ge alloys. Phys. Rev. B 84, 1652111–16521110.
Londos, C.A., Antonaras, G., Chroneos, A., 2013. Di-interstitial defect in silicon revisited. J. Appl. Phys. 114, 1935131–1935134.
Markevich, V.P., Peaker, A.R., Lastovskii, S.B., Murin, L.I., Coutinho, J., Torres, V.J.B., Briddon, P.R., Dobaczewski, L., Monakhov, E.V., Svensson, B.G., 2009. Trivacancy and trivacancy-oxygen complexes in silicon: experiments and ab initio modeling. Phys. Rev. B 80, 2352071–2352077.
Meese, J.M., 1981. A review of ntd-induced defects in silicon. In: Narayan, J., Tan, T.Y. (Eds.), Defects in Semiconductors. North-Holland, Amsterdam, pp. 225–240.
Mesli, A., Dobaczewski, L., Bonde Nielsen, K., Kolkovsky, Vl, Christian Petersen, M., Nylandsted Larsen, A., 2008. Low-temperature irradiation-induced defects in germanium: *in situ* analysis. Phys. Rev. B78, 1620216–1652021.
Mikelsen, M., Monakhov, E.V., Alfieri, G., Avset, B.S., Svensson, B.G., 2005. Kinetics of divacancy annealing and divacancy-oxygen formation in oxygen-enriched high-purity silicon. Phys. Rev. B 72, 1952071–1952076.
Mukashev, B.N., Abdullin, Kh.A., Gorelkinskii, Yu.V, 1998. Self-interstitials in silicon irradiated with light ions. Phys. Stat. Solidi A 168, 73–85.
Newman, R.C., Bean, A.R., 1971. Irradiation damage in carbon- doped silicon irradiated at low temperatures by 2 MeV electrons. Radiat. Eff. 8, 189–193.
Newton, J.L., Chatterjee, A.P., Harris, R.D., Watkins, G.D., 1983. Negative-U properties of the lattice vacancy in silicon. Physica 116B, 219–223.
Nishikawa, S., Tanaka, A., Yamali, T., 1992. Reduction of transient boron diffusion in preamorphized Si by carbon implantation. Appl. Phys. Lett. 60, 2270–2272.
NIST, 2011. Stopping-power and range tables for electrons, protons, and helium ions. In: Berger, M.J., Coursey, J.S., Zucker, M.A., Chang, J. (Eds.). http://www.nist.gov/pml/data/star/ Physical Measurement Laboratory, The National Institute of Standards and Technology (NIST).
Nylandsted Larsen, A., Mesli, A., Bonde Nielsen, K., Kortegaard Nielsen, H., Dobaczewski, L., Adey, J., Jones, R., Palmer, D.W., Briddon, P.R., Oberg, S., 2006. E center in silicon has a donor level in the band gap. Phys. Rev. Lett. 97, 1064021–1064024.
Nylandsted Larsen, A., Bro Hansen, A., Mesli, A., 2008. Irradiation-induced defects in SiGe. Mater. Sci. Eng. B 154–155, 85–89.

Nylandsted Larsen, A., Juul Pedersen, H., Christian Petersen, M., Privitera, V., Gurimskaya, Y., Mesli, A., 2013. In-growth of an electrically active defect in high-purity silicon after proton irradiation. J. Appl. Phys. 114, 2237061–2237066.

Partyka, P., Zhong, Y., Nordlund, K., Averback, R.S., Robinson, I.M., Ehrhart, P., 2001. Grazing incidence diffuse x-ray scattering investigation of the properties of irradiation-induced point defects in silicon. Phys. Rev. B 64, 2352071–2352078.

Pichler, P., 2004. Intrinsic Point Defects, Impurities, and Their Diffusion in Silicon. Springer-Verlag, Wien.

Pierreux, P., Stesmans, A., 2005. Electron spin resonance study of paramagnetic centers in neutron-irradiated heat-treated silicon. Phys. Rev. B 71, 1152041–1152048.

Posselt, M., Gao, F., Zwicker, D., 2005. Atomistic study of the migration of di- and tri-interstitials in silicon. Phys. Rev. B 71, 2452021–24520212.

Ranki, V., Saarinen, K., 2004. Formation of thermal vacancies in highly As and P doped Si. Phys. Rev. Lett. 93, 2555021–2555024.

Resende, A., Jones, R., Öberg, S., Briddon, P.R., 1999. Calculations of electrical levels of deep centers: application to Au-H and Ag-H defects in silicon. Phys. Rev. Lett. 82, 2111–2114.

Richie, D.A., Kim, J., Barr, S.A., Hazzard, K.R.A., Hennig, R., Wilkins, J.W., 2004. Complexity of small silicon self-interstitial defects. Phys. Rev. Lett. 92, 0455011–0455014.

Seeger, A., Chik, K.P., 1968. Diffusion Mechanisms and point defects in silicon and germanium. Phys. Stat. Solidi 29, 455–542.

Shimizu, Y., Uematsu, M., Itoh, K.M., 2007. Experimental evidence of the vacancy-mediated silicon self-diffusion in single-crystalline silicon. Phys. Rev. Lett. 98, 0959011–0959014.

Song, L.W., Watkins, G.D., 1990. EPR identification of the single-acceptor state of interstitial carbon in silicon. Phys. Rev. B42, 5759–5764.

Song, L.W., Zhan, X.D., Benson, B.W., Watkins, G.D., 1988. Bistable defect in silicon: the interstitial-carbon–substitutional-carbon pair. Phys. Rev. Lett. 60, 460–463.

Srour, J.R., Marshall, P.W., 2003. Review of displacement damage effects in silicon devices. IEEE Trans. Nucl. Sci. 50, 653–670.

Trombetta, J.M., Watkins, G.D., 1987. Identification of an interstitial carbon-interstitial oxygen complex in silicon. Appl. Phys. Lett. 51, 1103–1105.

Troxell, J.R., Watkins, G.D., 1980. Interstitial boron in silicon: a negative-U system. Phys. Rev. B22, 921–931.

Van Vechten, J.A., 1974. Enthalpy of vacancy migration in Si and Ge. Phys. Rev. B 10, 1482–1506.

Voronkov, V.V., Falster, R., 2012. The diffusivity of the vacancy in silicon: is it fast or slow? Mater. Sci. Semicond. Process. 15, 697–702.

Watkins, G.D., 1964. A review of EPR studies in irradiated silicon. Radiation Damage in Semiconductors. Paris-Royaumont, Dunod, Paris, pp. 97–113, 1965.

Watkins, G.D., 1967. Defects in irradiated silicon: electron paramagnetic resonance and electron-nuclear double resonance of the aluminum-vacancy pair. Phys. Rev. 155, 802–815.

Watkins, G.D., 1975a. Defects in irradiated silicon: EPR and electron-nuclear double resonance of interstitial boron. Phys. Rev. B12, 5824–5839.

Watkins, G.D., 1975b. EPR studies of the lattice vacancy and low temperature damage processes in silicon. In: Inst. Phys. Conf. Ser. 23, London, pp. 1–22.

Watkins, G.D., 1976. EPR of a trapped vacancy in boron-doped silicon. Phys. Rev. B13, 2511–2518.

Watkins, G.D., 1986. The lattice vacancy in silicon. In: Pantelides, S.T. (Ed.), Deep Centers in Semiconductors: A State-of-the-Art Approach. Gordon and Breach Science Publishers, New York, pp. 147–183.

Watkins, G.D., 1991. Intrinsic point defects in semiconductors. In: Schröter, W. (Ed.), Materials Science and Technology: A Comprehensive Treatment. Electronic Structure and Properties of Semiconductors, vol. 4. VCH Publishers Inc., New York, Weinheim, pp. 105–141.

Watkins, G.D., 2000. Intrinsic defects in silicon. Mater. Sci. Semicond. Process. 3, 227–235.

Watkins, G.D., 2008. The vacancy in silicon: identical diffusion properties at cryogenic and elevated temperatures. J. Appl. Phys. 103, 1061061–1061062.

Watkins, G.D., Brower, K.L., 1976. EPR observation of the isolated interstitial carbon atom in silicon. Phys. Rev. Lett. 36, 1329–1332.

Watkins, G.D., Corbett, J.W., 1961. Defects in irradiated silicon. I. Electron spin resonance of the Si-A center. Phys. Rev. 121, 1001–1014.

Watkins, G.D., Corbett, J.W., 1964. Defects in irradiated silicon: electron paramagnetic resonance and electron-nuclear double resonance of the Si-*E* center. Phys. Rev. 134, A1359–A1377.

Watkins, G.D., Corbett, J.W., 1965. Defects in irradiated silicon: electron paramagnetic resonance of the divacancy. Phys. Rev. 138, A543–A555.

Watkins, G.D., Troxell, J.R., 1980. Negative-U properties for point defects in silicon. Phys. Rev. Lett. 44, 593–596.

Weeks, J.D., Tully, J.C., Kimerling, L.C., 1975. Theory of recombination-enhanced defect reactions in semiconductors. Phys. Rev. B12, 3286–3292.

Ziegler, J.F., Biersack, J.P., 2013. The stopping and range of ions in matter—SRIM. http://www.srim.org/.

CHAPTER THREE

Ion Implantation Defects and Shallow Junctions in Si and Ge

Enrico Napolitani[*,†,1], **Giuliana Impellizzeri**[†]
*Dipartimento di Fisica e Astronomia, Università di Padova, Padova, Italy
†CNR-IMM MATIS, Catania, Italy
[1]Corresponding author: e-mail address: enrico.napolitani@unipd.it

Contents

1. INTRODUCTION

Over the past decades, information technology had an enormous and a continuously increasing impact in our daily life and society. This has been made possible by the aggressive scaling of the metal oxide semiconductor field effect transistors (MOSFETs), i.e., the building block of microelectronic (and, now, nanoelectronic) integrated circuits, following the well-known Moore's law (ITRS, 2013). The driving force for the continuous huge research efforts to maintain this trend over the years has been the performance and the economic outcomes deriving from device miniaturization.

One of the key issues in MOSFET scaling has been for decades, since the 1980s, the miniaturization of doped source/drain regions. The most common doping process consists in the localized introduction of the dopant in the substrate by the ion implantation process, followed by a thermal annealing in order to remove the ion implantation damage and electrically activate the dopant. Progresses in doping technology have led in the late 1990s to the fabrication of ultra-shallow junctions (USJs) by the

Semiconductors and Semimetals, Volume 91
ISSN 0080-8784
http://dx.doi.org/10.1016/bs.semsem.2015.01.001

introduction of ion implanter facilities working with energies below 1 keV with projected ranges of few nanometers and/or by complex processing recipes involving preamorphization, impurity co-implantation, and/or advanced annealing strategies with high temperatures and short times, such as rapid thermal annealings (RTAs) and "spike" anneals.

The major problem of the ion implantation doping originates from the point defects, i.e., self-interstitials (I) and vacancies (V), injected within the substrate as a result of the energetic ion collisions. Upon annealing, these defects undergo complex phenomena (Fahey et al., 1989; Pichler, 2004) such as migration, annihilation, formation of clusters, and extended defects. More importantly, these defects can interact with dopants driving detrimental effects for technology, such as excessive junction broadening and precipitation of dopants in electrically inactive clusters.

As it has been underlined earlier in the chapter "Role of Defects in the Dopant Diffusion in Si" by Pichler, point defects are responsible for the diffusion and clustering mechanisms of dopants through the reactions:

$$A_s(\text{immobile, active}) + X \Leftrightarrow AX(\text{mobile}) \tag{1}$$

$$AX(\text{mobile}) + A_s(\text{immobile, active}) \Leftrightarrow A_2X(\text{immobile, inactive}) \tag{2}$$

where A_s represents the dopant in substitutional position, where it is immobile and electrically active, and X indicates an I or V defect. The first reaction (1) is the responsible for dopant diffusion. The second reaction (2) represents the simplest one among the chain reactions leading to immobile and inactive dopant-defect clusters.

A typical recipe for the fabrication of an USJ in Si is shown in Fig. 1, where B is implanted at an energy of 500 eV in a preamorphized Si and then activated by a spike annealing at high temperature. The B concentration depth profiles measured by secondary ion mass spectrometry (SIMS) reported in the figure show a significant broadening, together with an immobile peak next to the surface, where B is in the form of electrically inactive clusters. Both phenomena are produced through reactions (1) and (2) by the I's emission from dissolving extended defects originating from the ion implantation damage.

The USJ formation processes are thus intimately linked with the evolution of ion implantation damage. Most of the progresses in doping technology achieved in the past decades are due to the understanding of the ion implantation damage and to the development of methods to control its evolution,

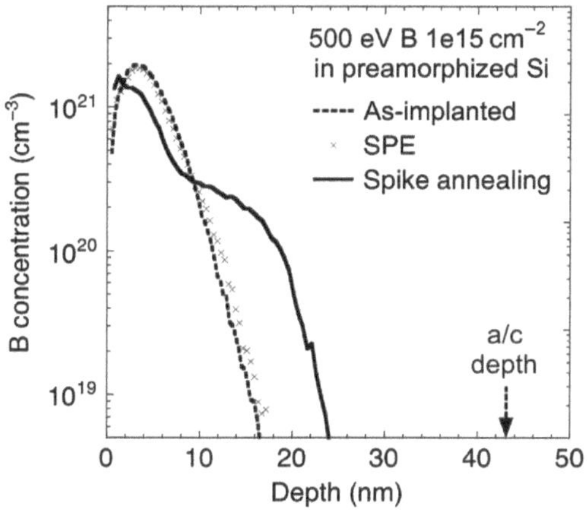

Figure 1 Concentration profiles of B implanted at 500 eV 10^{15} cm^{-2} in preamorphized Si before (dashed line) and after post-implantation annealing at 550 °C (×symbol) to induce the solid-phase epitaxy, or at 1050 °C with a spike process (continuous line) to remove the ion implantation damage. The post-implantation amorphous/crystalline (a/c) interface is indicated with an arrow.

through the so-called point defect engineering. Scope of this chapter is to review some important aspects of the generation and accumulation of damage during the ion implantation process, its evolution during the annealing, and its role in the fabrication of USJs in Si and Ge. Particular emphasis will be put on the point engineering strategies and alternative processing solutions.

Most of the subjects will be focused on silicon, where the understanding is more mature. Otherwise, several aspects on ion implantation damage and USJ formation in germanium will also be treated. Germanium, after decades of oblivion, has attracted increasing interest during the past decade as a promising high-mobility channel material for future ultra-scaled MOSFET devices (Claeys and Simoen, 2007; Pillarisetty, 2011). The implementation of Ge will require an update of the current understanding, in particular, concerning the effects of ion implantation doping in association with annealing on the structural and electrical properties of Ge. There are similarities, in terms of defect evolution, between Ge and Si, but also significant differences (Chroneos and Bracht, 2014; Mirabella et al., 2013; Simoen et al., 2012; Vanhellemont and Simoen, 2007). For example, in contrast to Si where both I and V defects play a role in the behavior of dopants, Ge is essentially dominated by vacancies. This latter aspect, considering the mainly interstitial nature of the implantation damage evolution, makes important differences as we will see later in this chapter.

2. GENERATION AND ACCUMULATION OF ION IMPLANTATION DAMAGE

The energy release of an ion along its track is usually described in the framework of the collision cascade theory, where part of the ion energy is transferred to the crystal through individual atomic displacements producing pairs of a self-interstitial and a vacancy defect (called I–V Frenkel pairs). The energy required to displace a single atom from its lattice site, E_{dis}, is assumed to be in the 12–15 eV range for both Si and Ge at relatively low energies (Thompson et al., 1977). Concerning the way by which the damage accumulates during the ion implantation, the simplest approach to describe it (the reader interested to go into deep on the matter is referred to Claeys and Simoen, 2009, Mayer and Nastasi, 2006, and Pelaz et al., 2004) is by the modified Kinchin–Pease formulation (Thompson et al., 1977), which can be rewritten in terms of defect fraction f_D as follows:

$$f_D = \frac{0.42\nu_e}{N}\frac{1}{E_{dis}^{eff}},$$

where ν_e is the energy density released with elastic nuclear collisions, E_{dis}^{eff} is the effective displacement energy, and N is the target atomic concentration. ν_e can be calculated through the collision cascade theory or by means of a Monte Carlo code (Ziegler, 2004), whereas f_D can be estimated experimentally by Rutherford Backscattering Spectrometry measurements in channeling conditions, or by strain measurements through an appropriate calibration procedure (Bai and Nicolet, 1991; Decoster and Vantomme, 2009; Napolitani et al., 2014). This allows to estimate E_{dis}^{eff}, as reported in Fig. 2, for implantation of ions of different masses both in silicon and germanium. E_{dis}^{eff} clearly results to deviate significantly from E_{dis}. In fact, E_{dis}^{eff} accounts for many complex mechanisms involved in damage accumulation besides the simple creation of I's and V's by atomic collisions (which is governed only by E_{dis}), such as their migration, clustering, recombination, as well as other effects such as energy (thermal) spikes. As a general trend, E_{dis}^{eff} increases by decreasing the implanted ion mass, as for lighter ions the collisional cascades are dilute; thus, I and V migration and recombination are favored causing $E_{dis}^{eff} > E_{dis}$. On the contrary for heavier ions, E_{dis}^{eff} decreases to values significantly lower than E_{dis} and even approaches the total heats of melting in Si and Ge (0.8 and 0.5 eV, respectively). This has been explained thanks to cooperative displacement mechanisms related to thermal spikes

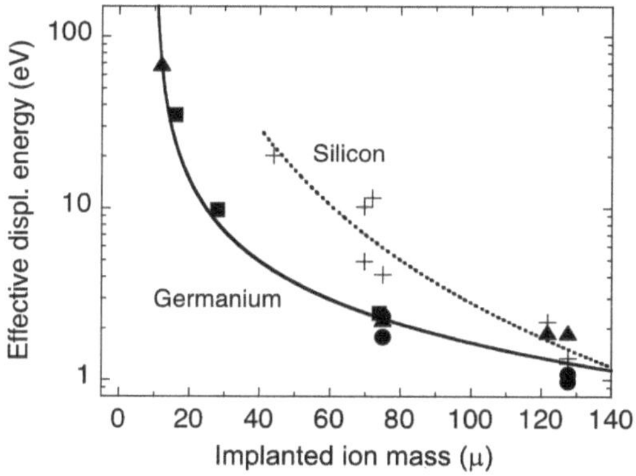

Figure 2 Effective displacement energy values for ion implantation in silicon (crosses, from Thompson et al., 1977) and germanium (squares from Napolitani et al., 2014, circles from Thompson et al., 1977, and triangles from Foti et al., 1977). Best fit of germanium data (continuous line) is taken from Napolitani et al. (2014) and best fit of silicon data (dotted line) is obtained with the curve $y = 3.7 \times 10^5\ x^{-2.6}$.

where local melting occurs, inducing enhanced defect generation in a region even larger than the calculated straggle cylinder (Foti et al., 1977; Napolitani et al., 2014; Thompson et al., 1977).

It is also clear from Fig. 2 that E_{dis}^{eff} in Si is higher than in Ge. This is indicative of the higher tendency of Ge to accumulate damage during ion implantation. The lower values of E_{dis}^{eff} reported for Ge in Fig. 2 suggest that the higher damaging in Ge cannot be ascribed only to the higher energy density released in elastic nuclear collisions (due to the higher atomic mass of Ge) but also to other effects such as a reduced mobility of defects in the collision cascades in Ge with respect to Si.

Beyond a certain ion implantation threshold, the damage fraction saturates (Decoster and Vantomme, 2009; Pelaz et al., 2004) and an amorphous layer forms. The underlying mechanisms are extremely complex and many different models have been proposed to describe them, as reviewed by Pelaz et al. (2004). However, one of the simplest ways to describe the onset of an amorphous phase, both in Si and Ge, is in the framework of the critical damage energy density (CDED) model (Claverie et al., 1988; Dennis and Hale, 1978; Hobler and Otto, 2003). This model assumes that the crystalline-amorphous is a first-order transition where the lattice spontaneously relaxes toward the amorphous phase, if its energy increases due to the nuclear collisions above a certain critical value E_{dc}. The fluence $D(z)$ needed to amorphize the crystal at a depth z results to be $D(z) = E_{dc}/\nu_e(z)$, where $\nu_e(z)$ is the energy density released with elastic nuclear collisions at a depth z. In Si,

$E_{dc} = 2$–5 eV/atom irrespective of the ion mass and energy at temperatures below 200 K (Hobler and Otto, 2003), which corresponds to 0.15–0.4 displacements per atom assuming $E_{dis} = 12$ eV. However, if the ion implantation is done at room temperature, and for ion masses lower than Si, E_{dc} tends to increase significantly by decreasing the ion mass (Claverie et al., 1989), similarly to what observed for E_{dis}^{eff} at subamorphizing fluencies (Fig. 2).

Also in Ge, amorphization kinetics can be described satisfactorily within the CDED model (Impellizzeri et al., 2010; Koffel et al., 2006, 2009a). A marked temperature dependence is found, similarly to what observed in Si, with a critical energy density of 5 eV/atom around room temperature and 1.5 eV/atom at liquid nitrogen temperature (Impellizzeri et al., 2010). The CDED approach is suitable for describing the crystalline-amorphous transition up to a Ge fluence of 5×10^{15} at cm^{-2}. Beyond that, a further phase transition occurs with the onset of a honeycomb surface morphology (Romano et al., 2010; Yates et al., 2012).

Upon annealing, the amorphous/crystalline (a/c) interface regrows through solid-phase epitaxy (SPE) and a band of defects is left close to the original position of the a/c interface. Readers interested in SPE in Si and related phenomena are referred to the chapter "Defective Solid-Phase Epitaxial Growth of Si" by Rudawski.

3. DAMAGE EVOLUTION AND DEFECT AGGLOMERATES

As underlined earlier in this book (see the chapter "Role of Defects in the Dopant Diffusion in Si" by Pichler), point defects have a central role in the mechanisms governing diffusion and clustering of dopants, in particular due to the link (proportionality) between point defect concentrations and dopant diffusivities (see Eq. 71 in the chapter "Role of Defects in the Dopant Diffusion in Si" by Pichler). Therefore, it is of crucial importance to discuss how the ion implantation damage evolves during the post-implantation annealing. A paradigmatic example is the transient-enhanced diffusion (TED) in Si where, during the thermal annealing, the excess of self-interstitials generated by the ion implantation provokes detrimental broadening and clustering effects on dopants (Cowern et al., 1990; Jain et al., 2002; Michel et al., 1987; Stolk et al., 1997). One of the earlier and clearer evidences of TED is reported in Fig. 3. A significant tail diffusion, well above the equilibrium B diffusivity, is observed lasting for less than an hour at 800 °C, whereas the peak above $\approx 3 \times 10^{18}$ cm^{-3} is nearly immobile due to B clustering in electrically inactive defects. Due to its technological

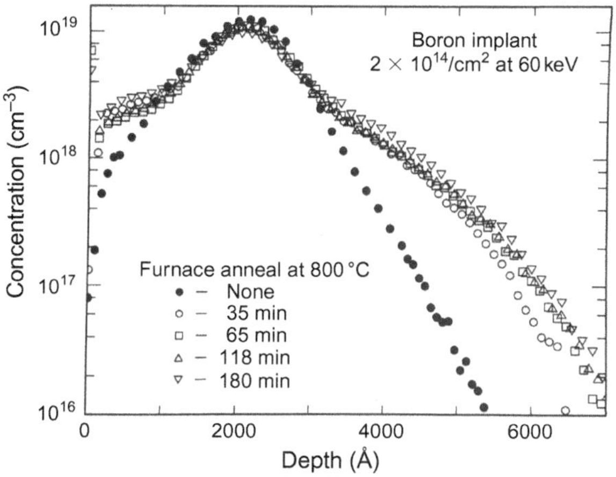

Figure 3 SIMS depth profiles of B implanted in Si after several annealing times at 800 °C, showing transient-enhanced diffusion (TED) in the profile tail and negligible diffusion in the peak due to B clustering. *Reprinted with permission from Michel et al. (1987). Copyright 1987, AIP Publishing LLC.*

importance for MOSFET scaling, TED of B and other dopants in Si was object of intense studies during the past decades, leading to a detailed and comprehensive understanding and modeling of the defect evolution mechanisms and of their role in dopant diffusion and clustering, as will be reviewed hereafter.

In spite of their extreme complexity, the phenomena occurring during the early stages of the damage annealing are satisfactorily described in a relatively large range of implant conditions with the so-called "+1" model (Giles, 1991). This model assumes that the Frenkel pairs rapidly recombine during the early stages of the annealing without significantly affecting the dopant behavior, and only the excess of interstitials that is generated as soon as the implanted ion becomes substitutional determines the subsequent evolution of the damage and most of the effects on dopants and impurities. Thus upon sample heating, as a net effect of the ion implantation the incident ion adds one interstitial to the lattice, without adding the corresponding vacancy. This view is confirmed by the absence of TED when annealing Sb-implanted Si, since Sb diffuses by the vacancy-mediated mechanism (Jain et al., 2002).

The "+1" model holds true for common implant conditions in Si, such as room temperature, medium/light ions (ions with masses close or below Si), moderate or high fluencies (i.e., higher than 10^{13} ions/cm^2), but there are several remarkable exceptions. For example, due to momentum transfer from primary ions, interstitials are generated slightly deeper in the substrate than vacancies. This causes a substrate rich in interstitials and a surface region rich in vacancies that can thus more easily escape I–V annihilation (Jain et al., 2002; Noda et al., 2000; Pelaz et al., 1998). As a result, a "+*n*" model, with $n > 1$, appears to be more appropriate for ions with relatively high mass and/or energy. *n* values greater than 1 can also be observed for very low implant fluences ($<10^{12}$ ions/cm^2), where the populations of I and V are so dilute that the probability for an I to interact with a dopant becomes greater than to interact with a vacancy (Pelaz et al., 1999). *n* values lower than 1 can instead be observed when implanting non-substitutional impurities, such as argon (Giles, 1993), fluorine (El Mubarek and Ashburn, 2003; Impellizzeri et al., 2004), helium (Raineri et al., 2000), or hydrogen. In the case of the latter three elements, a net enrichment in vacancies can even be observed, due to their tendency to form agglomerates with vacancies and thus stabilizing them.

The panorama depicted above changes considerably for amorphizing implants, where upon moderate annealing SPE sweeps away most of the damage, and only the non-amorphized region contributes to the excess of point defects in the lattice. In this case, since the as-implanted I profile is shifted toward the bulk compared to the vacancies, a net I concentration is always observed, inducing the formation of the so-called end-of-range (EOR) defects (Jones and Venables, 1991).

From the above, it is clear that annealing of ion-implanted silicon results, except in the case of extreme deviations from the "1" model, in an excess of interstitial defects. Depending on the implant fluence, energy, and thermal budget, I's can then evolve in agglomerates of different sizes and types, spanning from very small interstitial clusters (ICs) not visible by TEM (Cowern et al., 1999) to {311} rod-like extended defects (Eaglesham et al., 1994) and, at higher thermal budgets, to dislocation loops (DLs) (Cristiano et al., 2000; Stolk et al., 1997), as shown in Fig. 4.

After almost two decades of investigations and debates, a full comprehension of the evolution of the self-interstitial agglomerates was finally reached at the early 1900s. The I's, that after the early stages of the annealing are in the form of di-interstitials (Corbett et al., 1981), during prolonged annealings evolve by forming agglomerates which grow in size and reduce their density as a function of time, following an Ostwald ripening mechanism. According to this mechanism, I's are exchanged between the clusters by sustaining an I supersaturation, S_I, around each defect equal to $\exp(E_f/k_B T)$, where E_f is the formation energy of the cluster, i.e., the energy required to add an extra atom to it. An I concentration gradient, with a corresponding diffusion flux, is

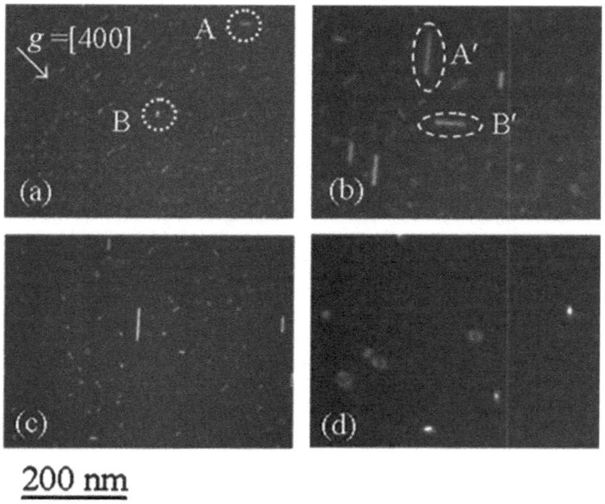

Figure 4 Plan-view TEM images of the extended defects formed after a 30 keV 10^{15} cm^{-2} Ge^{+} amorphizing implant and annealings at 800 °C for different times. After 15 s (a) only {311} rod-like defects are formed, after 120 s (b) the {311} defects have a reduced density and an increased size, after 300 s (c) a further reduction of the defect is observed, accompanied by the onset of dislocation loops (DLs), and after 2700 s (d) only few DLs are observed. *Reprinted with permission from Boninelli et al. (2006). Copyright 2006, AIP Publishing LLC.*

established between clusters having different E_f values. The driving force of the system is thus the reduction of the defect formation energy. As the formation energy typically decreases by decreasing the cluster size, a flux of I's is established from smaller clusters to bigger clusters, making the system to evolve by forming initially ICs, then {311} rod-like defects, and finally DLs. A gradient exists also toward the surface, which acts as sink for defects (Lamrani et al., 2004; Ortiz et al., 2004), making the Ostwald ripening nonconservative with the total amount of I's stored in defects decreasing with time up to a complete dissolution.

More importantly, the defects, during their evolution, sustain an average I supersaturation, S_I, which is the responsible for the TED of dopants. S_I is strictly related to the formation energies of the defects according to the Ostwald ripening theory. The formation energies of the different defects involved as a function of their size are reported in Fig. 5 (Cristiano et al., 2004). These are the key parameters describing the link between the defect evolution and TED, as highlighted in the right axis showing the corresponding I supersaturation, responsible for TED.

The thermal evolution of all types of extended defects therefore reflects the defect hierarchy in terms of formation energies. The excess of I's

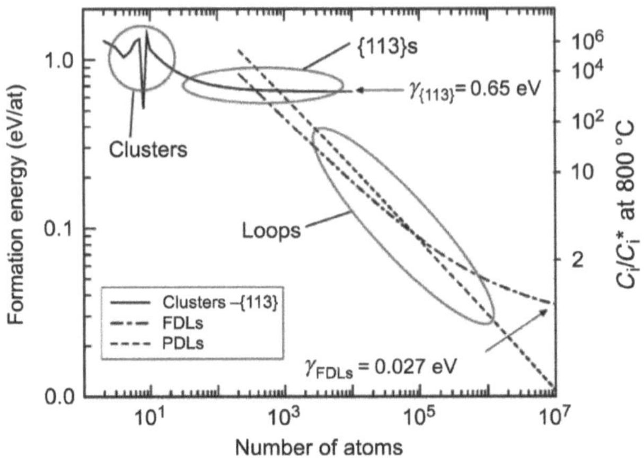

Figure 5 Formation energies (left axis) of the different Si self-interstitial agglomerates as a function of their size, and corresponding values of the self-interstitial supersaturation at 800 °C according to the Oswald ripening theory (right axis). *Reprinted from Cristiano et al. (2004). Copyright 2003, with permission from Elsevier.*

produced by the ion implantation evolves thus by forming initially ICs with relatively high formation energies (number of atoms $n < 10$ in Fig. 5) and then ICs evolve in {311} defects (Fig. 4A and B) with intermediate formation energies ($10 < n < 10^3$ in Fig. 5) growing in size until they form more stable DLs (Fig. 4C and D, and $n > 10^3$ in Fig. 5). Where the time and temperature dependences are regulated by the above formation energy hierarchy, by the I diffusivity, and by the I annihilation at the surface.

The formation energies of ICs in Fig. 5 (number of atoms <10) have been estimated experimentally by Cowern et al. (1999). As those clusters are below the TEM sensitivity, they measured S_I (Fig. 6) from the TED of epitaxially-grown B delta layer in Si-implanted Si and used inverse modeling to derive E_f as a function of the number of atoms per cluster. Interestingly, the energies of those ICs have two minima at $n = 4$ and 8, which have a strong impact on the defect evolution and on the corresponding S_I. This is quite different with respect to the monotonic decrease of the formation energy with increasing the defect size found for larger defects such as {311} and DLs.

Dopant clustering is another important aspect of TED linked with the ion implantation damage evolution, being responsible for the limited dopant activation. As underlined in the Chapter "Role of Defects in the Dopant Diffusion in Si" by Pichler, dopant clustering is governed by mechanisms conceptually similar to those governing the evolution of self-interstitial agglomerates. In this case, there are cluster–point defect reactions with

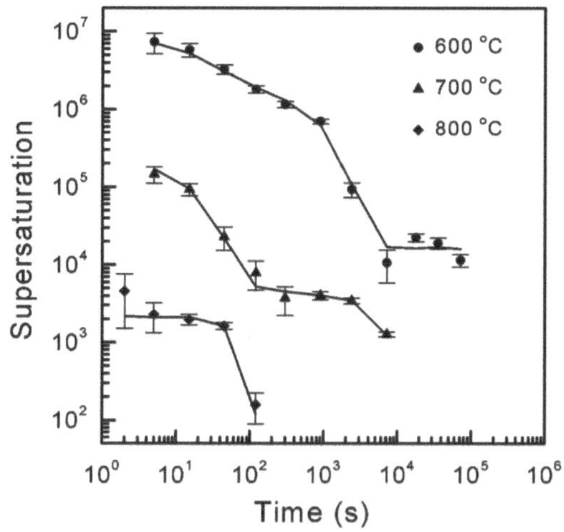

Figure 6 Interstitial supersaturation as a function of annealing temperature and time. Curves represent fits using an Ostwald ripening model. Nearly flat regions of the curves correspond to the dissolution of {311} defects while the faster decay at shorter times originates from the dissolution of small interstitial clusters. *Reprinted figure with permission from Cowern et al. (1999). Copyright 1999 by the American Physical Society.*

the further complication that clusters are composed by two species, namely the point defect and the dopant. The local thermal equilibrium is established both with the supersaturation of free I's and with the dopant atoms in a mobile configuration, the latter being influenced by the dopant concentration and by the I supersaturation and, thus, by the ion implantation damage evolution. Also in this case the key parameters are the formation energies of the different dopant clusters (Aboy et al., 2011). The experimental investigation of such parameters is challenging since the dopant-defect clusters are relatively small and hardly visible by TEM analyses. Boron-interstitial clusters (BICs) in Si have been directly observed only recently, both after B implantation at ultra-low energy (Cristiano et al., 2003) and after I injection by Si implantation in Si layers grown by molecular beam epitaxy (MBE) with B delta doping (Boninelli et al., 2007). They have been also observed by atom probe tomography (APT) just after B implantation (Cojocaru-Mirédin et al., 2009).

A method has been developed to experimentally access the energetics of the BIC formation and dissolution in Si (De Salvador et al., 2005; Mirabella et al., 2003), that is conceptually similar to the method described above (Fig. 6) to investigate the IC evolution. As shown in Fig. 7, the BICs are produced in a MBE-grown B box by Si implantation and annealing, and their dissolution is promoted during further prolonged annealing cycles.

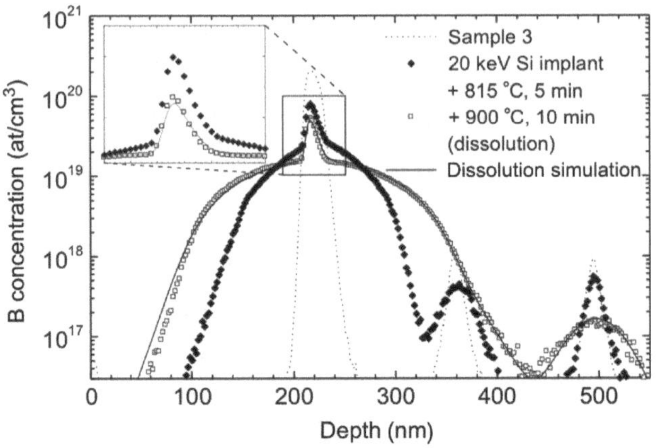

Figure 7 SIMS boron concentration profile of a Si sample grown with a B box (2×10^{20} B/cm^3) and an array of B delta layers before (dotted line) and after post-growth processing. Silicon implantation (20 keV, 10^{14} Si/cm^2) and thermal annealing induce a B clustering (diamonds). Further annealing at 900 °C induces B clusters dissolution (open squares), whose simulation is also displayed (continuous line). The inset is a magnification, on a linear vertical scale, of the rectangular region. *Reprinted figure with permission from De Salvador et al. (2005). Copyright 2005, AIP Publishing LLC.*

The concentrations of diffusing B and immobile clustered B are extracted from the chemical concentration profiles, whereas the I's supersaturation is measured from the TED observed in low-concentrated B delta-doped layers. An accurate analysis of the diffusion data through a rate diffusion model, and a comparison with the TEM investigations (Boninelli et al., 2007), revealed that B clusters dissolve through two distinct and coexisting paths with different energy barriers and rates. The faster one (3.6 eV) is predominant for small clusters formed at low B concentrations (10^{19} B/cm^3), while the slower one (4.8 eV) is characteristic of bigger clusters, detectable by TEM, formed above the solubility limit of boron (2×10^{20} B/cm^3). A quite comprehensive model describing the BIC formation and evolution has been recently published (Aboy et al., 2011), providing a large set of formation energies relative to B_nI_m clusters with n and m as large as eight atoms.

Compared to silicon, in germanium there are very few experimental evidences of ion implantation defect evolution. The main reason resides in the extremely low equilibrium transport capacity of I's in Ge compared to that of V's (Werner et al., 1985), making Ge basically dominated by vacancies. It is worth noting that the diffusion mechanisms in Ge are mediated by vacancies for all the dopants except for boron. This clearly influences also the ion implantation damage evolution, which has an inherent interstitial nature (see, for example, the discussion above on the "+1" model). As a consequence, the possibility to form EOR defects in Ge after SPE has been questioned for a long time and their presence has been revealed only recently (Koffel et al., 2009b) by TEM analyses. The EOR defects in Ge are in the form of small dots, much smaller and less stable than in Si. In addition, they are formed by I's, as can be deduced by the positive perpendicular strain they induce in the lattice (Bisognin et al., 2008; Koffel et al., 2009b).

The first experimental evidence of boron TED in Ge has been published 1 year later by Napolitani et al. (2010). The paper shows that the dissolution of the EOR defects, promoted by thermal annealing, induces small but clearly detectable TED in an epitaxially-grown B delta located nearby (see Fig. 8). The B diffusion as a function of the annealing temperature resulted to be quantitatively correlated with the evolution of positive strain associated with the EOR damage, measured by high-resolution X-ray diffraction (HRXRD) analysis. This represents a clear evidence that the B diffusion in Ge is mediated by I's and also allowed to determine an energy barrier for EOR dissolution of 2.1 ± 0.3 eV. Such barrier is significantly lower than the dissolution barrier of 3.8 eV found for the {311} defects in Si, even if homologous temperatures are considered, consistently with the significantly smaller size of the defects observed in Ge.

Concerning ion-implanted vacancies, there are evidences that Ge easily enriches itself in vacancies. Upon annealing, vacancies can survive up to

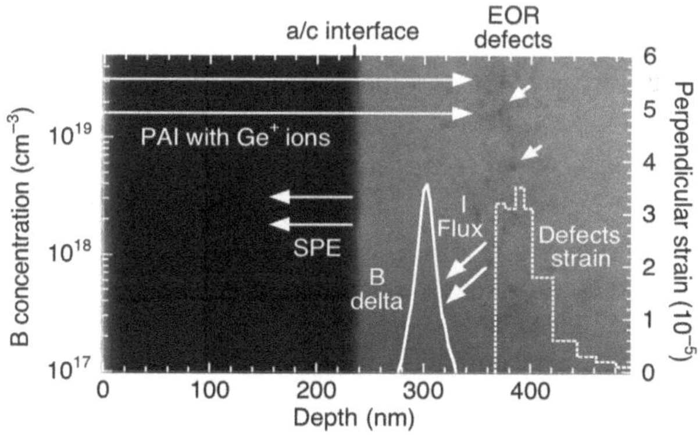

Figure 8 Ge sample grown by MBE with a B delta layer (continuous line) after preamorphization by Ge ion implantation and annealed at 300 °C in order to induce partial SPE and the formation of the EOR defects shown in the cross-TEM micrograph. HRXRD strain depth profile associated to the defects is also shown (dashed line). Upon further thermal annealings, the EOR dissolves releasing I's and promoting the boron TED, as discussed in the text. *Reprinted figure with permission from Napolitani et al. (2010). Copyright 2010, AIP Publishing LLC.*

400 °C in the form of divacancy complexes, as indicated by positron annihilation spectroscopy measurements (Slotte et al., 2008).

The comprehension of the defect behavior in Ge is still open. The ion-implanted defect evolution is overall not well understood. For example, based on evidences reported in literature of diffusion at room temperature of both I's (Bruno et al., 2009) and V's (Romano et al., 2010) defects, it is likely that upon annealing I–V annihilation and mechanisms such as the "+1" model observed in Si are active also in Ge, but there are no evidences yet. The difficulty in observing extended defects in Ge with respect to Si might suggest that the dominant interstitial nature of the damage evolution, due to I–V annihilation, is mitigated in Ge by the high equilibrium transport capacity of vacancies with respect to self-interstitials, thus preventing the evolution of damage into larger and more stable extended defects.

It is also not clear if the surface acts as sinks or repels I's. Experiments on the dissolution of the EOR damage would suggest a sink behavior (Boninelli et al., 2012), while a repelling behavior is claimed in order to explain the self-diffusion under proton irradiation (Bracht et al., 2009). In addition, the effect of different capping layers has to be understood in more detail in Ge due to their strong impact on the point defect annihilation and dopant (out-)diffusion and activation. Another intriguing aspect in Ge concerns the nature of the self-interstitial defects, which seems to be regulated by a completely new physics. Recent experimental and theoretical study by Cowern et al. (2013), focused on the migration length of mobile B in Ge upon different conditions (such as proton irradiation, TED, and equilibrium diffusion), revealed that I's in Ge can exist in two distinct forms: a simple one with low entropy and low energy, and a complex one with high entropy and high energy. The latter dominates diffusion processes at high temperatures and is proposed to have a structure similar to that of an amorphous pocket, a sort of extended point defect that has been named *i-morph*. The impact of this new class of defects on diffusion and defect dynamics, and on ion implantation damage accumulation and evolution still has to be explored.

4. ROLE OF DEFECTS IN SHALLOW JUNCTION FORMATION IN SI

The defect evolution and dopant activation phenomena in USJ processing revealed several significant differences with respect to deeper junctions. This is mostly due to the proximity of the surface, as well as to the reduced defect generation at ultra-low energies. As shown by Privitera et al. (1999), TED in Si is observed also for B implanted at energies as low as 250 eV, even if to a lower extent with respect to deeper implants, due to surface proximity, with no extended defects visible by TEM (Collart et al., 1998). The electrical activation of B implanted at ultra-low energies appeared to be a hard task, due to the extreme confinement of the dopant profile in a narrow region at extremely high concentrations (Privitera et al., 2000).

The TED for B ion energies below 1 keV has been investigated in detail by Napolitani et al. (1999) and Schroer et al. (2000). As shown in Fig. 9, the total tail shift is smaller than that induced by implantation at higher energies and the transient time is orders of magnitude shorter, as can be easily realized

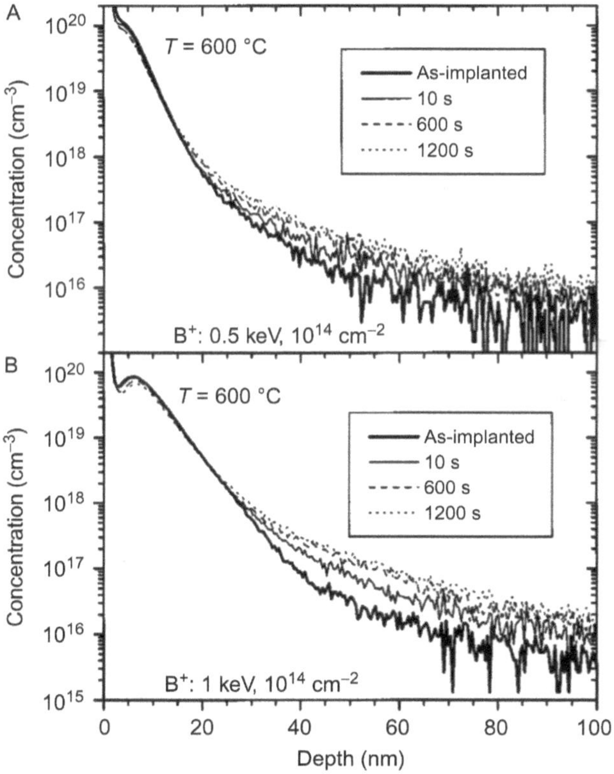

Figure 9 SIMS profiles of B implanted at 500 eV (A) and 1 keV (B) with a fluence of 1×10^{14} cm^{-2}, and after post-implantation annealings at 600 °C for several times, showing transient-enhanced diffusion already at low temperature and short times. The as-implanted profiles are also reported as a reference. *Reprinted figure with permission from Napolitani et al. (1999). Copyright 1999, AIP Publishing LLC.*

by comparison with Fig. 3. Figure 9 shows that two different processes take place: an ultrafast process, shorter than 10 s at 600 °C, and a fast one, decaying with a time constant of about 300 s at 600 °C and about 13 s at 750 °C, corresponding to an activation energy of 1.7 eV (Napolitani et al., 1999). The total tail shift due to the ultrafast contribution decreases with decreasing the implantation energy, while neither the magnitude of the total shift nor the time constant changes for the fast process. From these observations, it was concluded that two different sources are responsible for the TED in the ultra-low energy regime in the case of B: small/simple clusters composed of B and/or silicon atoms releasing interstitials are responsible for the fast process, whereas the so-called weakly bound excess interstitials (WBEIs) were proposed to be responsible for the ultrafast process (Napolitani et al., 1999).

The defects responsible for the two TED processes described above have been shown to be located at different depths (Schroer et al., 2000), by exploiting a method consisting in a controlled surface layer etching by repeated native oxide growth and removal. The defects responsible for the fast tail shift are located closer to the surface than the defects leading to the ultrafast tail shift. This has been interpreted as an effect due to a momentum transfer of the B ions to the displaced Si atoms, causing a variation of the ratio of B concentration to the excess I concentration with the depth. The shallower clusters have less amount of I's, which are more tightly bound to B-containing clusters, while the deeper ones are rich in I's, which can thus be in a WBEI configuration responsible for the ultrafast TED.

Implantation at ultra-low energies depicts thus a complex scenario where the extreme proximity of the surface, that acts as a sink for I's, inhibits the evolution of the damage up to the more large and stable defects, such as the {311}'s responsible for TED at higher implantation energies. In this contest, an important role is played by the very small clusters, similarly to the case of faster decay at shorter times of the TED of B promoted by Si self-implantation, as evidenced by Cowern et al. (1999) (Fig. 6). Such behavior has been satisfactorily modeled with a rate equation approach by Mannino et al. (2001).

It is worth noting that the activation energies for TED, even considering those related with the {311} dissolution, are generally much lower than the energy barriers for boron cluster dissolution. Consider, for example, the dissolution barrier of 4.8 eV found for the BICs investigated by De Salvador et al. (2005) and reported in Fig. 7. Since the differences in energy barriers are minimized at higher temperatures, it is clear that in order to dissolve the B clusters, and thus activate the boron, while minimizing its diffusion, it is necessary to consider activation annealings at high temperatures for short times. Several alternative annealing solutions have thus been explored during the past decades to meet this criterion. Flash lamp annealing (FLA) is probably one of the most successful, which has been pioneered for the activation of USJs in Si by Gebel et al. (2002). FLA consists in irradiating the Si wafer with an array of xenon flash lamps ignited by a high-voltage pulse, able to reduce the annealing time down to the ms range, with peak temperatures higher than 1300 °C. Given the promising results obtained by Gebel et al. (2002), this methodology has been subsequently developed in commercial flash-assisted apparatus (Lerch et al., 2005). Another promising annealing methodology that has been explored for USJ processing is the pulsed laser thermal annealing (LTA) in the melting regime (Fortunato et al., 2001). This procedure consists of irradiating the implanted sample with a pulsed laser beam such that the surface layer is melted for very short times (few tens of ns) to a depth governed by the laser energy density. The dopant redistributes across the melted layer during the liquid phase, with diffusivities of the order of 10^{-4} cm^2/s, and does not penetrate beyond the liquid/solid interface. Solidification of the melted layer results in the formation of high-quality crystalline material with a high fraction of electrically active dopant. Due to the extremely fast liquid-/solid-phase transition, metastable layers with concentrations of dopants even above the solid solubility have been demonstrated in Si (Monakhov and Svensson, 2005). Diffusion in liquid Si revealed some intriguing anomalies through the formation of a B pile-up close to the maximum melt depth, which have been recently explained thanks to the peculiar bonding character of group-IV semiconductors close to the melting temperature (Fisicaro et al., 2013).

Even if extremely effective in annealing the ion implantation damage, LTA does not leave a material free of defects (La Magna et al., 2007). However, being their nature mostly of vacancy type, this effect can have a beneficial role in annihilating I's and thus reducing the B diffusion (Mannino et al., 2005). Despite being highly promising for shallow junction formation, LTA still experiences implementation difficulties in technologies, due to integration issues.

Besides the above-mentioned advanced annealing strategies, intense efforts have been spent during the past decades in exploring complex implant recipes and methodologies with the objective of controlling the defect populations responsible for diffusion and activation issues. This branch of research and technology is commonly called "point defect engineering." A successful approach, as already mentioned, has been the introduction of a preamorphization step prior to shallow dopant implantation, followed by a thermal annealing to induce SPE and remove the EOR damage. Such a procedure is effective in avoiding both channeling during dopant implantation and also the overlap of the dopant with the ion implantation damage (Chao et al., 1996), with benefits on dopant activation and junction depth, even if TED is not completely removed. Other approaches involve, in conjunction with a preamorphization step, co-implantation with impurities such as C or F, which are able to introduce traps for I's.

Napolitani et al. (2001) demonstrated that C, if appropriately placed between the I source and the shallow B implant, can completely suppress the boron TED. They implanted B in a MBE-grown silicon sample, previously amorphized by silicon implantation, and subsequently annealed to induce SPE and electrically activate the dopant. They show that the backflow of silicon interstitials released by the EOR damage is completely trapped by a carbon-rich silicon layer interposed between the damage and the implanted boron. No appreciable TED is observed in the samples up to complete dissolution of the EOR damage, and complete electrical activation is obtained. The reactions between substitutional carbon and I's have been subsequently investigated and modeled in detail (Mirabella et al., 2002) by quantifying the effect of a layer rich in C grown by MBE, interposed between the near-surface self-interstitial source (produced by Si implantation) and a deeper B delta-doped layer. It has been shown that I trapping occurs through a C–I clustering reaction where on average two C atoms are needed to trap a self-interstitial. The resulting clusters are stable upon prolonged annealings.

Fluorine has been shown to be extremely effective in suppressing the boron TED in preamorphized Si. Impellizzeri et al. (2004, 2006) have demonstrated that F acts as a trap for interstitials causing a consistent reduction of the interstitial flux released by EOR defects during thermal treatments, with no direct chemical interaction between B and F. The mechanism for TED reduction is based on the incorporation in the crystalline Si after SPE of vacancy-type complexes, due to F, effective in sinking the I's. For this reason, the position of F with respect to B and the EOR defects is decisive for his effectiveness in suppressing TED and deactivation effects. However, F is strongly pushed by the moving amorphous/crystal interface, due to a segregation effect at the amorphous/crystalline interface that tends to accumulate F in the amorphous phase. The resulting F profile after SPE is significantly different from the implanted one (Impellizzeri et al., 2006; Mirabella et al., 2005). The redistribution of F during SPE has been the matter of intensive investigation, mainly by SIMS, and a full comprehensive model was elaborated to predict the final F redistribution by Mastromatteo et al. (2010). The model clearly puts in evidence that the F clustering is a crucial step in the F incorporation in Si, and it has to be taken into account to correctly reproduce the F profile during the whole SPE process. The F clustering during SPE of preamorphized Si has been investigated by TEM (Boninelli et al., 2008), showing after a partial SPE at 700 °C that cavities form both in amorphous Si (a-Si) and in crystal Si (c-Si), and by electrical measurements, showing a donor-like behavior of F complexes (Impellizzeri et al., 2007). More in detail, X-ray absorption spectroscopy (XAS) demonstrated that F is mostly incorporated in c-Si in the form of SiF_4 molecules after the SPE process (De Salvador et al., 2009). TEM and XAS results have been used to conclude that silicon cavities, observed by TEM, are filled by SiF_4 molecules. Direct evidence of F-rich clusters in correspondence of cavities came only recently thanks to APT 3D chemical analysis (Panciera et al., 2012), as reported in Fig. 10. It is clear that F has a more complex behavior with respect to C, but it has a very important advantage for shallow junction formation processing. In fact, in contrast to C–I interaction that produces very stable clusters, the interaction between F–V complexes (i.e., nanocavities filled with SiF_4) and I's tends to dissolve the complexes and make F able to diffuse away, thus eventually leaving a crystal free of F-related defects.

5. ROLE OF DEFECTS IN SHALLOW JUNCTION FORMATION IN GE

As underlined previously, in contrast to silicon where a leading role is played by I defects, germanium is dominated by V's (Chroneos and Bracht, 2014; Mirabella et al., 2013; Simoen et al., 2012; Vanhellemont and Simoen, 2007). The diffusion mechanisms are in fact mediated by vacancies for self-diffusion and all the dopants, except for boron (that shows an I-assisted diffusion). Furthermore, vacancies in Ge are present both in the neutral V^0 and in the doubly negatively charged $V^=$ states. Due to extrinsic effects on charged defects, V^0 prevails under p-type doping and $V^=$ prevails under

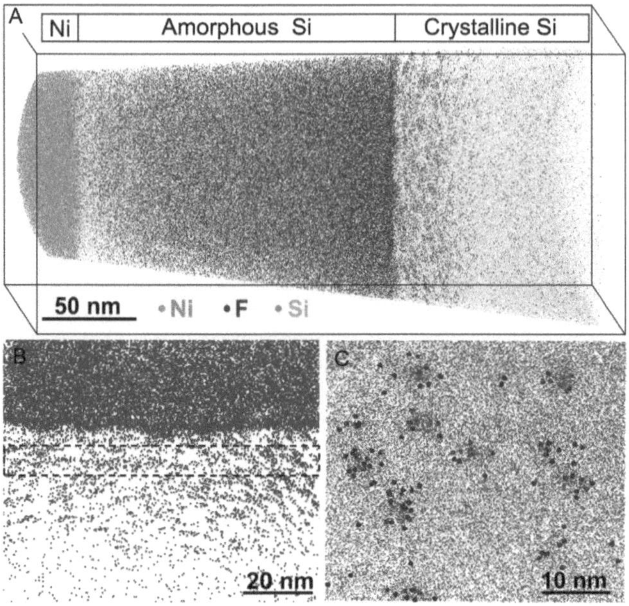

Figure 10 Atom probe tomography (APT) 3D reconstruction of a preamorphized Si sample implanted with F and annealed to induce a partial SPE: (A) the whole analyzed volume where F, Ni, Si, and F atoms are represented by black (blue in the Web version), mid-gray (green in the Web version), and light gray dots, respectively; (B) a vertical slice of the 3D volume evidencing F accumulation at a/c interface and F clusters in the c-Si; (C) an horizontal slice of the 3D volume taken in correspondence of the box in (B) showing Si (dots) and F (spheres) lateral distribution. F clusters are located in correspondence with regions denser in Si atoms, compatible with an artifact of the technique occurring in the presence of cavities. The surface has been covered with a Ni layer before the analysis. *Reprinted figure with permission from Panciera et al. (2012). Copyright 2012, AIP Publishing LLC.*

n-type doping, with a concentration that increases as the square of active donors concentration. This causes an extremely high diffusion coefficients for donor atoms at high concentrations which represent an issue for shallow junction technology. On the contrary, p-type dopants have relatively low equilibrium diffusivities (with B showing a much lower diffusivity than the other acceptors), with no extrinsic effects. Furthermore, donor-V pairing at high dopant concentrations promotes dopant clustering that limits the achievement of the high active donor levels required by future CMOS technology. The maximum equilibrium dopant solubilities (Fistul, 2004) are relatively high for acceptors, with value close to 5×10^{20} cm^{-3} for Al and

Ga, except for B which has a maximum solubility around 10^{18} cm^{-3}. Solubilities for donors are lower than for acceptors, with P and As having values of 2×10^{20} cm^{-3} and 10^{20} cm^{-3}, respectively, and Sb having a significant lower value of 10^{19} cm^{-3}.

Implantation of p-type dopant has been investigated in detail. Boron, due to its low solubility, upon implantation at subamorphizing fluences shows very low activation levels (Satta et al., 2005). However, Impellizzeri et al. (2009a) have shown that the presence of a high level of implantation-induced damage in the form of amorphous pockets enhances the electrical activation upon annealing. Implanting B in preamorphized Ge allows to reach extremely high values of active B after SPE: up to 6×10^{20} cm^{-3} (Mirabella et al., 2008). No significant diffusion of implanted B is observed in Ge, in agreement with the low equilibrium B diffusivity and low TED (Napolitani et al., 2010). Such high metastable levels of active B are indeed considerably stable (Impellizzeri et al., 2009a) and, even if subjected to an I flux due to EOR defects dissolution, B experiences only a partial and transient deactivation up to recovery of the high activation reached immediately after SPE (Panciera et al., 2010).

Ga implanted in Ge has been investigated by some authors (Heera et al., 2010; Hellings et al., 2011; Impellizzeri et al., 2009b; Satta et al., 2006a), with promising results. In particular, Impellizzeri et al. (2009b) implanted Ga at 77 K in order to avoid the formation of a surface honeycomb structure, which instead is observed if implanted at room temperature. After 1 h at 450 °C, the amorphous layer produced by the Ga^{+} implantation is completely regrown by SPE and a remarkable activation of about 7×10^{20} cm^{-3} is observed, with negligible diffusion. However, when increasing the temperature above 550 °C and up to 650 °C, a significant deactivation occurs well below the equilibrium solubility. This effect is quite surprising and it is explained as a thermally activated Ga clustering. Given that it occurs at concentrations below the equilibrium solubility, such clustering is presumably related with the ion implantation damage but the underlying microscopical mechanism is not clear yet.

The diffusion and electrical activation of Al^{+} implanted in Ge have been investigated in detail by Impellizzeri et al. (2012a). The Al implantation (with a fluence of 1×10^{15} cm^{-2} at 25 keV) induces the amorphization of a shallow surface layer of ~55 nm. After 1 h at 400 °C, the amorphous layer is completely regrown by SPE and a remarkable activation of 1×10^{20} cm^{-3} is obtained (Fig. 11). Such high activation level is maintained up to 600 °C with negligible diffusion, in agreement with the equilibrium diffusivity of

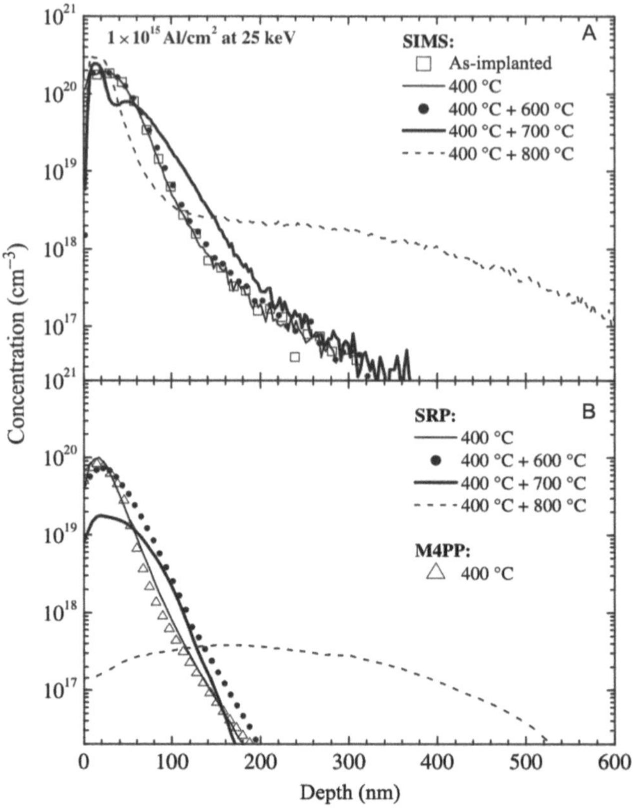

Figure 11 Chemical and carrier concentration profiles of Al implanted in Ge as measured, respectively, by secondary ion mass spectrometry (A) and by spreading resistance profiling or micro-four-point probe (B) just after implantation (open squares), after SPE at 400 °C for 1 h (continuous line), and after SPE plus thermal annealing at different temperatures. Negligible diffusion with activation as high as 10^{20} cm^{-3} is observed up to 600 °C. Deactivation and diffusion (toward the bulk and uphill) are observed at 700 °C and above. *Reprinted figure from Impellizzeri et al. (2012a). Copyright 2012, The Japan Society of Applied Physics.*

Al. It is worth to mention that EOR defects are observed at 400 °C which completely dissolve already at 500 °C in agreement with the dissolution temperature reported in literature (Napolitani et al., 2010). It is significant that the dissolution of the EOR defects and the related I flux does not induce any enhanced diffusion nor deactivation of Al, confirming that the interaction of Al with self-interstitials is not favored in agreement with a diffusion mechanism mediated by vacancies. For higher annealing temperatures (700 and 800 °C), Al shows a significant diffusion toward the bulk, compatible with the equilibrium diffusivity, but also an unexpected uphill diffusion next to the surface, where a significant deactivation is also observed. Such anomalous diffusion is explained in terms of the presence of dopant traps, placed next to the surface, able to make immobile and electrically inactive the dopant. Considering that the latter detrimental effects are present only at high temperatures, which are not technologically relevant, Al seems highly promising for the formation of shallow p^+/n junctions in Ge. Add-on benefits of implanted Al are that it has a lower projected range than B, and a lower tendency of Ga, due to its lower mass, to induce honeycomb surface morphology during implant.

The behavior of n-type implanted dopants is considerably worse with respect to p-type dopants, both in terms of diffusion and activation. The maximum activation levels reported so far for P or As (implanted and conventionally annealed) are about $5 \times 10^{19}\ cm^{-3}$, i.e., well below the maximum equilibrium solubilities reported above (Chui et al., 2005; Satta et al., 2006b; Vanhellemont and Simoen, 2012 and references therein). The reason for the limited activation is most probably due to the pairing between doubly charged vacancies and the dopant atoms. In fact, the resulting mobile negatively charged V-dopant complexes readily react with the positive substitutional donors due to the Coulomb attraction, causing an enhanced clustering (Chroneos and Bracht, 2014). This interpretation is supported by the observation of P clustering below the equilibrium solubility also in the case of n-type doping from a diffusion source, i.e., without any ion implantation damage (Brotzmann et al., 2008). Another possible effect that has been considered to explain the reduced donor activation is the acceptor-like character of the residual damage after ion implantation and the related counter-doping effect (Kim et al., 2011). Following this interpretation, some recipes have been successfully explored, in order to increase the activation of P through the reduction of damage, such as multiple implantation and multiple annealing of P (Kim et al., 2012), or Sb and P co-implantation (Kim et al., 2010) even if the interpretation of Sb and P co-implantation is not clear as also local strain compensation effects might come into play (Kim et al., 2011). The dissolution of the EOR damage has been claimed to have a beneficial effect, since the I flux released by the EOR defects reduces the V concentration and consequently the P clustering (Vanhellemont and Simoen, 2012).

Diffusion of ion-implanted n-type dopants has been simulated with diffusivities close to equilibrium, i.e., with no significant TED effects (Koffel et al., 2011), confirming that the ion implantation damage has no significant role in affecting the diffusion of implanted dopants in Ge. The diffusion is in any case high due to the high diffusivities in the extrinsic conditions usually employed for n+/p junctions.

Summarizing, no significant TED is observed for implanted dopants in Ge. P-type doping is not an issue, since the equilibrium diffusivities are low, and the activation levels at or above the equilibrium solubilities can be easily reached with a preamorphization step. N-type doping is puzzling due to the detrimental role of (equilibrium) $V^{=}$ in inducing an excessive junction broadening and dopant clustering. Given the above panorama, reducing diffusion and clustering of n-type dopants appears to be a hard task.

Point defect engineering strategies or alternative annealing solutions have been explored to improve the shallow junction fabrication. Impellizzeri et al. (2011) have explored F and As co-implantation evidencing a complex behavior. They observed that F implantation enriches the Ge matrix in V's (Sprouster et al., 2013), causing an enhanced diffusion of As within the layer amorphized by F and As implantation and subsequently regrown by SPE. In correspondence of the EOR region, F forms complexes with Ge interstitials that act as sinks for V's and induce an abrupt suppression of As diffusion. This effect might be beneficial to realize abrupt shallow n-type junctions. However, the As electrical activation under codoping with F resulted to be relatively poor (Impellizzeri et al., 2012b).

Pulsed LTA in the melting regime has instead successfully explored by several authors (Bruno et al., 2012; Heo et al., 2006) as an alternative method to improve n-type doping of Ge, similarly to what done in Si (as discussed previously in this chapter). In particular, Milazzo et al. (2014) have investigated the effect of LTA in As-implanted Ge. They observed box-like profiles after multiple shots, with no appreciable As loss from the surface and with activation levels as high as 10^{20} cm^{-3}, representing the highest As activation reported so far. LTA was also explored for Ge implanted with B by Impellizzeri et al. (2013) observing that B activation is limited by B out-diffusion and by clustering with O atoms diffusing from the surface native oxide toward the bulk.

REFERENCES

Aboy, M., Pelaz, L., Bruno, E., Mirabella, S., Boninelli, S., 2011. Kinetics of large B clusters in crystalline and preamorphized silicon. J. Appl. Phys. 110, 073524.

Bai, G., Nicolet, M., 1991. Defects production and annealing in self-implanted Si. J. Appl. Phys. 70, 649.

Bisognin, G., Vangelista, S., Bruno, E., 2008. High-resolution X-ray diffraction by end of range defects in self-amorphized Ge. Mater. Sci. Eng. B 154–155, 64–67.

Boninelli, S., Cherkashin, N., Claverie, A., Cristiano, F., 2006. Evidences of an intermediate rodlike defect during the transformation of {113} defects into dislocation loops. Appl. Phys. Lett. 89, 161904.
Boninelli, S., Mirabella, S., Bruno, E., Priolo, F., Cristiano, F., Claverie, A., De Salvador, D., Bisognin, G., Napolitani, E., 2007. Evolution of boron-interstitial clusters in crystalline Si studied by transmission electron microscopy. Appl. Phys. Lett. 91, 031905.
Boninelli, S., Impellizzeri, G., Mirabella, S., Priolo, F., Napolitani, E., Cherkashin, N., Cristiano, F., 2008. Formation and evolution of F nanobubbles in amorphous and crystalline Si. Appl. Phys. Lett. 93, 061906.
Boninelli, S., Impellizzeri, G., Alberti, A., Priolo, F., Cristiano, F., Spinella, C., 2012. Role of the Ge surface during the end of range dissolution. Appl. Phys. Lett. 101, 162103.
Bracht, H., Schneider, S., Klug, J.N., Liao, C.Y., Hansen, J.L., Haller, E.E., Larsen, A.N., Bougeard, D., Posselt, M., Wündisch, C., 2009. Interstitial-mediated diffusion in germanium under proton irradiation. Phys. Rev. Lett. 103, 31–34.
Brotzmann, S., Bracht, H., Simoen, E., Haller, E.E., Christensen, J.S., Werner, P., 2008. Diffusion and defect reactions between donors, C, and vacancies in Ge. I. Experimental results. Phys. Rev. B 77, 1–13.
Bruno, E., Mirabella, S., Scapellato, G.G., Impellizzeri, G., Terrasi, A., Priolo, F., Napolitani, E., De Salvador, D., Mastromatteo, M., Carnera, A., 2009. Mechanism of B diffusion in crystalline Ge under proton irradiation. Phys. Rev. B 80, 1–4.
Bruno, E., Scapellato, G.G., La Magna, A., Cuscunà, M., Napolitani, E., Boninelli, S., Priolo, F., Fortunato, G., Privitera, V., 2012. Anomalous transport of Sb in laser irradiated Ge. Appl. Phys. Lett. 101, 172110.
Chao, H.S., Griffin, P.B., Plummer, J.D., 1996. Influence of dislocation loops created by amorphizing implants on point defect and boron diffusion in silicon. Appl. Phys. Lett. 68, 3570.
Chroneos, A., Bracht, H., 2014. Diffusion of n-type dopants in germanium. Appl. Phys. Rev. 1, 011301.
Chui, C.O., Kulig, L., Moran, J., Tsai, W., Sarawat, K.C., 2005. Germanium n-type shallow junction activation dependences. Appl. Phys. Lett. 87, 091909.
Claeys, C., Simoen, E., 2007. Germanium-Based Technologies From Materials to Devices. Elsevier BV.
Claeys, C., Simoen, E., 2009. Extended Defects in Germanium Fundamental and Technological Aspects. Springer-Verlag, Berlin, Heidelberg.
Claverie, A., Vieu, C., Fauré, J., Beauvillain, J., 1988. Cross-sectional high-resolution electron microscopy investigation of argon-ion implantation-induced amorphization of silicon. J. Appl. Phys. 64, 4415.
Claverie, A., Roumili, A., Gessinn, N., Beauvillain, J., 1989. Kinetics of silicon amorphization by N+ implantation: dose rate and substrate temperature effects. Mater. Sci. Eng. B 4, 205–209.
Cojocaru-Mirédin, O., Mangelinck, D., Blavette, D., 2009. Nucleation of boron clusters in implanted silicon. J. Appl. Phys. 106, 113525.
Collart, E.J.H., Weemers, K., Gravesteijn, D.J., van Berkum, J.G.M., 1998. Characterization of low-energy (100 eV–10 keV) boron ion implantation. J. Vac. Sci. Technol. B 16, 280.
Corbett, J.W., Karins, J.P., Tan, T.Y., 1981. Ion-induced defects in semiconductors. Nucl. Instrum. Methods 182–183, 457–476.
Cowern, N.E.B., Janssen, K.T.F., Jos, H.F.F., 1990. Transient diffusion of ion-implanted B in Si: dose, time, and matrix dependence of atomic and electrical profiles. J. Appl. Phys. 68, 6191.
Cowern, N.E.B., Mannino, G., Stolk, P., Roozeboom, F., Huizing, H., van Berkum, J., Cristiano, F., Claverie, A., Jaraíz, M., 1999. Energetics of self-interstitial clusters in Si. Phys. Rev. Lett. 82, 4460–4463.

Cowern, N.E.B., Simdyankin, S., Ahn, C., Bennett, N.S., Goss, J.P., Hartmann, J.-M., Pakfar, A., Hamm, S., Valentin, J., Napolitani, E., De Salvador, D., Bruno, E., Mirabella, S., 2013. Extended point defects in crystalline materials: Ge and Si. Phys. Rev. Lett. 110, 155501.

Cristiano, F., Grisolia, J., Colombeau, B., Omri, M., de Mauduit, B., Claverie, A., Giles, L.F., Cowern, N.E.B., 2000. Formation energies and relative stability of perfect and faulted dislocation loops in silicon. J. Appl. Phys. 87, 8420.

Cristiano, F., Hebras, X., Cherkashin, N., Claverie, A., Lerch, W., Paul, S., 2003. Clusters formation in ultralow-energy high-dose boron-implanted silicon. Appl. Phys. Lett. 83, 5407.

Cristiano, F., Cherkashin, N., Hebras, X., Calvo, P., Lamrani, Y., Scheid, E., de Mauduit, B., Colombeau, B., Lerch, W., Paul, S., Claverie, A., 2004. Ion beam induced defects in crystalline silicon. Nucl. Inst. Methods Phys. Res. B 216, 46–56.

Decoster, S., Vantomme, A., 2009. Implantation-induced damage in Ge: strain and disorder profiles during defect accumulation and recovery. J. Phys. D. Appl. Phys. 42, 165404.

Dennis, J.R., Hale, E.B., 1978. Crystalline to amorphous transformation in ion-implanted silicon: a composite model. J. Appl. Phys. 49, 1119.

De Salvador, D., Napolitani, E., Bisognin, G., Carnera, A., Bruno, E., Mirabella, S., Impellizzeri, G., Priolo, F., 2005. Experimental evidences for two paths in the dissolution process of B clusters in crystalline Si. Appl. Phys. Lett. 87, 221902.

De Salvador, D., Bisognin, G., Napolitani, E., Mastromatteo, M., Baggio, N., Carnera, A., Boscherini, F., Impellizzeri, G., Mirabella, S., Boninelli, S., Priolo, F., Cristiano, F., 2009. Formation and incorporation of SiF[sub 4] molecules in F-implanted preamorphized Si. Appl. Phys. Lett. 95, 101908.

Eaglesham, D.J., Stolk, P.A., Gossmann, H.-J., Poate, J.M., 1994. Implantation and transient B diffusion in Si: the source of the interstitials. Appl. Phys. Lett. 65, 2305.

El Mubarek, H.A.W., Ashburn, P., 2003. Reduction of boron thermal diffusion in silicon by high energy fluorine implantation. Appl. Phys. Lett. 83, 4134.

Fahey, P.M., Griffin, P.B., Plummer, J.D., 1989. Point defects and dopant diffusion in silicon. Rev. Mod. Phys. 61, 289–384.

Fisicaro, G., Huet, K., Negru, R., Hackenberg, M., Pichler, P., Taleb, N., La Magna, A., 2013. Anomalous impurity segregation and local bonding fluctuation in l-Si. Phys. Rev. Lett. 110, 117801.

Fistul, V.I., 2004. Impurities in Semiconductors: Solubility, Migration, and Interactions. CRC Press LLC, Boca Raton.

Fortunato, G., Mariucci, L., Stanizzi, M., Privitera, V., Spinella, C., Coffa, S., Napolitani, E., 2001. Fabrication of ultra-shallow junctions with high electrical activation by excimer laser annealing. Mater. Sci. Semicond. Process. 4, 417–423.

Foti, G., Vitali, G., Davies, J.A., 1977. Molecular and atomic damage in germanium. Radiat. Eff. 32, 187–191.

Gebel, T., Voelskow, M., Skorupa, W., Mannino, G., Privitera, V., Priolo, F., Napolitani, E., Carnera, A., 2002. Flash lamp annealing with millisecond pulses for ultra-shallow boron profiles in silicon. Nucl. Inst. Methods Phys. Res. B 186, 287–291.

Giles, M.D., 1991. Transient phosphorus diffusion below the amorphization threshold. J. Electrochem. Soc. 138, 1160.

Giles, M.D., 1993. Transient phosphorus diffusion from silicon and argon implantation damage. Appl. Phys. Lett. 62, 1940.

Heera, V., Mücklich, A., Posselt, M., Voelskow, M., Wündisch, C., Schmidt, B., Skrotzki, R., Heinig, K.H., Herrmannsdörfer, T., Skorupa, W., 2010. Heavily Ga-doped germanium layers produced by ion implantation and flash lamp annealing: structure and electrical activation. J. Appl. Phys. 107, 053508.

Hellings, G., Rosseel, E., Simoen, E., Radisic, D., Petersen, D.H., Hansen, O., Nielsen, P.F., Zschätzsch, G., Nazir, A., Clarysse, T., Vandervorst, W., Hoffmann, T.Y., De Meyer, K., 2011. Ultra shallow arsenic junctions in germanium formed by millisecond laser annealing. Electrochem. Solid-State Lett. 14, H39.

Heo, S., Baek, S., Lee, D., Hasan, M., Jung, H., Lee, J., Hwang, H., 2006. Sub-15 nm n[sup +]/ p-germanium shallow junction formed by PH[sub 3] plasma doping and excimer laser annealing. Electrochem. Solid-State Lett. 9, G136.

Hobler, G., Otto, G., 2003. Status and open problems in modeling of as-implanted damage in silicon. Mater. Sci. Semicond. Process. 6, 1–14.

Impellizzeri, G., dos Santos, J.H.R., Mirabella, S., Priolo, F., Napolitani, E., Carnera, A., 2004. Role of fluorine in suppressing boron transient enhanced diffusion in preamorphized Si. Appl. Phys. Lett. 84, 1862.

Impellizzeri, G., Mirabella, S., Priolo, F., Napolitani, E., Carnera, A., 2006. Fluorine in preamorphized Si: Point defect engineering and control of dopant diffusion. J. Appl. Phys. 99, 103510.

Impellizzeri, G., Mirabella, S., Piro, A.M., Grimaldi, M.G., Priolo, F., Giannazzo, F., Raineri, V., Napolitani, E., Carnera, A., 2007. Fluorine counter doping effect in B-doped Si. Appl. Phys. Lett. 91, 132101.

Impellizzeri, G., Mirabella, S., Bruno, E., Piro, A.M., Grimaldi, M.G., 2009a. B activation and clustering in ion-implanted Ge. J. Appl. Phys. 105, 063533.

Impellizzeri, G., Mirabella, S., Irrera, A., Grimaldi, M.G., Napolitani, E., 2009b. Ga-implantation in Ge: electrical activation and clustering. J. Appl. Phys. 106, 013518.

Impellizzeri, G., Mirabella, S., Grimaldi, M.G., 2010. Ion implantation damage and crystalline-amorphous transition in Ge. Appl. Phys. A 103, 323–328.

Impellizzeri, G., Boninelli, S., Priolo, F., Napolitani, E., Spinella, C., Chroneos, A., Bracht, H., 2011. Fluorine effect on as diffusion in Ge. J. Appl. Phys. 109, 113527.

Impellizzeri, G., Napolitani, E., Boninelli, S., Privitera, V., Clarysse, T., Vandervorst, W., Priolo, F., 2012a. Aluminium implantation in germanium: uphill diffusion, electrical activation, and trapping. Appl. Phys. Express 5, 021301.

Impellizzeri, G., Napolitani, E., Boninelli, S., Sullivan, J.P., Roberts, J., Buckman, S.J., Ruffell, S., Priolo, F., Privitera, V., 2012b. Role of F on the electrical activation of As in Ge. ECS J. Solid State Sci. Technol. 1, Q44–Q46.

Impellizzeri, G., Napolitani, E., Boninelli, S., Fisicaro, G., Cuscunà, M., Milazzo, R., La Magna, A., Fortunato, G., Priolo, F., Privitera, V., 2013. B-doping in Ge by excimer laser annealing. J. Appl. Phys. 113, 113505.

ITRS—International Technology Roadmap for Semiconductors, 2013. http://www.itrs.net.

Jain, S.C., Schoenmaker, W., Lindsay, R., Stolk, P.A., Decoutere, S., Willander, M., Maes, H.E., 2002. Transient enhanced diffusion of boron in Si. J. Appl. Phys. 91, 8919.

Jones, K.S., Venables, D., 1991. The effect of implant energy, dose, and dynamic annealing on end-of-range damage in Ge^+-implanted silicon. J. Appl. Phys. 69, 2931.

Kim, J., Bedell, S.W., Maurer, S.L., Loesing, R., Sadana, D.K., 2010. Activation of implanted n-type dopants in Ge over the active concentration of 1×10[sup 20] cm [sup −3] using coimplantation of Sb and P. Electrochem. Solid-State Lett. 13, H12.

Kim, J., Bedell, S.W., Sadana, D.K., 2011. Improved germanium n^+/p junction diodes formed by coimplantation of antimony and phosphorus. Appl. Phys. Lett. 98, 082112.

Kim, J., Bedell, S.W., Sadana, D.K., 2012. Multiple implantation and multiple annealing of phosphorus doped germanium to achieve n-type activation near the theoretical limit. Appl. Phys. Lett. 101, 112107.

Koffel, S., Claverie, A., Benassayag, G., Scheiblin, P., 2006. Amorphization kinetics of germanium under ion implantation. Mater. Sci. Semicond. Process. 9, 664–667.

Koffel, S., Cherkashin, N., Houdellier, F., Hytch, M.J., Benassayag, G., Scheiblin, P., Claverie, A., 2009a. End of range defects in Ge. J. Appl. Phys. 105, 126110.
Koffel, S., Scheiblin, P., Claverie, A., Benassayag, G., 2009b. Amorphization kinetics of germanium during ion implantation. J. Appl. Phys. 105, 013528.
Koffel, S., Kaiser, R.J., Bauer, A.J., Amon, B., Pichler, P., Lorenz, J., Frey, L., Scheiblin, P., Mazzocchi, V., Barnes, J.-P., Claverie, A., 2011. Experiments and simulation of the diffusion and activation of the n-type dopants P, As, and Sb implanted into germanium. Microelectron. Eng. 88, 458–461.
La Magna, A., Privitera, V., Fortunato, G., Cuscunà, M., Svensson, B., Monakhov, E., Kuitunen, K., Slotte, J., Tuomisto, F., 2007. Vacancy generation in liquid phase epitaxy of Si. Phys. Rev. B 75, 235201.
Lamrani, Y., Cristiano, F., Colombeau, B., Scheid, E., Calvo, P., Schäfer, H., Claverie, A., 2004. Direct evidence of the recombination of silicon interstitial atoms at the silicon surface. Nucl. Inst. Methods Phys. Res. B 216, 281–285.
Lerch, W., Paul, S., Niess, J., McCoy, S., Selinger, T., Gelpey, J., Cristiano, F., Severac, F., Gavelle, M., Boninelli, S., Pichler, P., Bolze, D., 2005. Advanced activation of ultra-shallow junctions using flash-assisted RTP. Mater. Sci. Eng. B 124–125, 24–31.
Mannino, G., Whelan, S., Schroer, E., Privitera, V., Leveque, P., Svensson, B.G., Napolitani, E., 2001. An investigation on the modeling of transient enhanced diffusion of ultralow energy implanted boron in silicon. J. Appl. Phys. 89, 5381.
Mannino, G., Privitera, V., La Magna, A., Rimini, E., Napolitani, E., Fortunato, G., Mariucci, L., 2005. Depth distribution of B implanted in Si after excimer laser irradiation. Appl. Phys. Lett. 86, 051909.
Mastromatteo, M., De Salvador, D., Napolitani, E., Panciera, F., Bisognin, G., Carnera, A., Impellizzeri, G., Mirabella, S., Priolo, F., 2010. Fluorine redistribution and incorporation during solid phase epitaxy of preamorphized Si. Phys. Rev. B 82, 1–12.
Mayer, J., Nastasi, M., 2006. Ion Implantation and Synthesis of Materials. Springer-Verlag, Berlin.
Michel, A.E., Rausch, W., Ronsheim, P.A., Kastl, R.H., 1987. Rapid annealing and the anomalous diffusion of ion implanted boron into silicon. Appl. Phys. Lett. 50, 416.
Milazzo, R., Napolitani, E., Impellizzeri, G., Fisicaro, G., Boninelli, S., Cuscunà, M., De Salvador, D., Mastromatteo, M., Italia, M., La Magna, A., Fortunato, G., Priolo, F., Privitera, V., Carnera, A., 2014. N-type doping of Ge by As implantation and excimer laser annealing. J. Appl. Phys. 115, 053501.
Mirabella, S., Coati, A., Salvador, D., Napolitani, E., Mattoni, A., Bisognin, G., Berti, M., Carnera, A., Drigo, A., Scalese, S., Pulvirenti, S., Terrasi, A., Priolo, F., 2002. Interaction between self-interstitials and substitutional C in silicon: interstitial trapping and C clustering mechanism. Phys. Rev. B 65, 045209.
Mirabella, S., Bruno, E., Priolo, F., De Salvador, D., Napolitani, E., Drigo, A.V., Carnera, A., 2003. Dissolution kinetics of boron-interstitial clusters in silicon. Appl. Phys. Lett. 83, 680.
Mirabella, S., Impellizzeri, G., Bruno, E., Romano, L., Grimaldi, M.G., Priolo, F., Napolitani, E., Carnera, A., 2005. Fluorine segregation and incorporation during solid-phase epitaxy of Si. Appl. Phys. Lett. 86, 121905.
Mirabella, S., Impellizzeri, G., Piro, A.M., Bruno, E., Grimaldi, M.G., 2008. Activation and carrier mobility in high fluence B implanted germanium. Appl. Phys. Lett. 92, 251909.
Mirabella, S., De Salvador, D., Napolitani, E., Bruno, E., Priolo, F., 2013. Mechanisms of boron diffusion in silicon and germanium. J. Appl. Phys. 113, 031101.
Monakhov, E., Svensson, B., 2005. Enhanced boron diffusion in excimer laser preannealed Si. Appl. Phys. Lett. 86, 151902.

Napolitani, E., Carnera, A., Schroer, E., Privitera, V., Priolo, F., Moffatt, S., 1999. Microscopical aspects of boron diffusion in ultralow energy implanted silicon. Appl. Phys. Lett. 75, 1869.

Napolitani, E., Coati, A., De Salvador, D., Carnera, A., Mirabella, S., Scalese, S., Priolo, F., 2001. Complete suppression of the transient enhanced diffusion of B implanted in preamorphized Si by interstitial trapping in a spatially separated C-rich layer. Appl. Phys. Lett. 79, 4145.

Napolitani, E., Bisognin, G., Bruno, E., Mastromatteo, M., Scapellato, G.G., Boninelli, S., De Salvador, D., Mirabella, S., Spinella, C., Carnera, A., Priolo, F., 2010. Transient enhanced diffusion of B mediated by self-interstitials in preamorphized Ge. Appl. Phys. Lett. 96, 201906.

Napolitani, E., Bruno, E., Bisognin, G., Mastromatteo, M., De Salvador, D., Scapellato, G.G., Boninelli, S., Priolo, F., Privitera, V., Carnera, A., 2014. Role of ion mass on damage accumulation during ion implantation in Ge. Phys. Status Solidi 211, 118–121.

Noda, T., Odanaka, S., Umimoto, H., 2000. Effects of end-of-range dislocation loops on transient enhanced diffusion of indium implanted in silicon. J. Appl. Phys. 88, 4980.

Ortiz, C.J., Pichler, P., Fühner, T., Cristiano, F., Colombeau, B., Cowern, N.E.B., Claverie, A., 2004. A physically based model for the spatial and temporal evolution of self-interstitial agglomerates in ion-implanted silicon. J. Appl. Phys. 96, 4866.

Panciera, F., Fazzini, P.F., Collet, M., Boucher, J., Bedel, E., Cristiano, F., 2010. End-of-range defects in germanium and their role in boron deactivation. Appl. Phys. Lett. 97, 012105.

Panciera, F., Hoummada, K., Mastromatteo, M., De Salvador, D., Napolitani, E., Boninelli, S., Impellizzeri, G., Priolo, F., Carnera, A., Mangelinck, D., 2012. Investigation of fluorine three-dimensional redistribution during solid-phase-epitaxial–regrowth of amorphous Si. Appl. Phys. Lett. 101, 103113.

Pelaz, L., Gilmer, G.H., Jaraiz, M., Herner, S.B., Gossmann, H.-J., Eaglesham, D.J., Hobler, G., Rafferty, C.S., Barbolla, J., 1998. Modeling of the ion mass effect on transient enhanced diffusion: deviation from the "+1" model. Appl. Phys. Lett. 73, 1421.

Pelaz, L., Gilmer, G.H., Venezia, V.C., Gossmann, H.-J., Jaraiz, M., Barbolla, J., 1999. Modeling of the effects of dose, dose rate, and implant temperature on transient enhanced diffusion. Appl. Phys. Lett. 74, 2017.

Pelaz, L., Marqués, L., Barbolla, J., 2004. Ion-beam-induced amorphization and recrystallization in silicon. J. Appl. Phys. 96, 5947.

Pichler, P., 2004. Intrinsic Point Defects, Impurities, and Their Diffusion in Silicon. Springer-Verlag, Wien.

Pillarisetty, R., 2011. Academic and industry research progress in germanium nanodevices. Nature 479, 324–328.

Privitera, V., Napolitani, E., Priolo, F., Moffatt, S., La Magna, A., Mannino, G., Carnera, A., Picariello, A., 1999. Atomic transport properties and electrical activation of ultra-low energy implanted boron in crystalline silicon. Mater. Sci. Semicond. Process. 2, 35–44.

Privitera, V., Schroer, E., Priolo, F., Napolitani, E., Carnera, A., 2000. Electrical behavior of ultra-low energy implanted boron in silicon. J. Appl. Phys. 88, 1299.

Raineri, V., Saggio, M., Rimini, E., 2000. Voids in silicon by He implantation: from basic to applications. J. Mater. Res. 15, 1449–1477.

Romano, L., Impellizzeri, G., Tomasello, M.V., Giannazzo, F., Spinella, C., Grimaldi, M.G., 2010. Nanostructuring in Ge by self-ion implantation. J. Appl. Phys. 107, 084314.

Satta, A., Simoen, E., Clarysse, T., Janssens, T., Benedetti, A., De Jaeger, B., Meuris, M., Vandervorst, W., 2005. Diffusion, activation, and recrystallization of boron implanted in preamorphized and crystalline germanium. Appl. Phys. Lett. 87, 172109.

Satta, A., Simoen, E., Janssens, T., Clarysse, T., De Jaeger, B., Benedetti, A., Hoflijk, I., Brijs, B., Meuris, M., Vandervorst, W., 2006a. Shallow junction ion implantation in Ge and associated defect control. J. Electrochem. Soc. 153, G229.
Satta, A., Janssens, T., Clarysse, T., Simoen, E., Meuris, M., Benedetti, A., Hoflijk, I., De Jaeger, B., Demeurisse, C., Vandervorst, W., 2006b. P implantation doping of Ge: diffusion, activation, and recrystallization. J. Vac. Sci. Technol. B 24, 494.
Schroer, E., Privitera, V., Priolo, F., Napolitani, E., Carnera, A., 2000. Transient-enhanced diffusion of boron implanted at ultralow energies in silicon: localization of the source. Appl. Phys. Lett. 76, 3058.
Simoen, E., Mitard, J., Hellings, G., Eneman, G., De Jaeger, B., Witters, L., Vincent, B., Loo, R., Delabie, A., Sioncke, S., Caymax, M., Claeys, C., 2012. Challenges and opportunities in Ge pMOSFETs. Mater. Sci. Semicond. Process. 15, 588–600.
Slotte, J., Rummukainen, M., Tuomisto, F., Markevich, V.P., Peaker, A., Jeynes, C., Gwilliam, R.M., 2008. Evolution of vacancy-related defects upon annealing of ion-implanted germanium. Phys. Rev. B 78, 085202.
Sprouster, D.J., Campbell, C., Buckman, S.J., Impellizzeri, G., Napolitani, E., Ruffell, S., Sullivan, J.P., 2013. Defect complexes in fluorine-implanted germanium. J. Phys. D. Appl. Phys. 46, 505310.
Stolk, P.A., Gossmann, H.-J., Eaglesham, D.J., Jacobson, D.C., Rafferty, C.S., Gilmer, G.H., Jaraíz, M., Poate, J.M., Luftman, H.S., Haynes, T.E., 1997. Physical mechanisms of transient enhanced dopant diffusion in ion-implanted silicon. J. Appl. Phys. 81, 6031.
Thompson, D.A., Walker, R.S., Davies, J.A., 1977. Evidence for spike-effects in low-energy heavy-ion bombardment of Si and Ge. Radiat. Eff. 32, 135.
Vanhellemont, J., Simoen, E., 2007. Brother silicon, sister germanium. J. Electrochem. Soc. 154, H572–H583.
Vanhellemont, J., Simoen, E., 2012. On the diffusion and activation of n-type dopants in Ge. Mater. Sci. Semicond. Process. 15, 642–655.
Werner, M., Mehrer, H., Hochheimer, H., 1985. Effect of hydrostatic pressure, temperature, and doping on self-diffusion in germanium. Phys. Rev. B 32, 3930–3937.
Yates, B.R., Darby, B.L., Elliman, R.G., Jones, K.S., 2012. Role of nucleation sites on the formation of nanoporous Ge. Appl. Phys. Lett. 101, 131907.
Ziegler, J.F., 2004. SRIM-2003. Nucl. Inst. Methods Phys. Res. B 219–220, 1027.

CHAPTER FOUR

Defective Solid-Phase Epitaxial Growth of Si

Nicholas G. Rudawski*,[1], Aaron G. Lind†, Thomas P. Martin†
*Major Analytical Instrumentation Center, University of Florida, Gainesville, Florida, USA
†Department of Materials Science and Engineering, University of Florida, Gainesville, Florida, USA
[1]Corresponding author: e-mail address: ngr@ufl.edu

Contents

Semiconductors and Semimetals, Volume 91
ISSN 0080-8784
http://dx.doi.org/10.1016/bs.semsem.2014.11.005

1. INTRODUCTION

The solid-phase epitaxial growth (SPEG) process of Si refers to the epitaxial crystallization of an amorphous (α) layer of Si in direct contact with a single-crystal Si substrate (Csepregi et al., 1996). The process is shown schematically in Fig. 1. Typically, the process starts with a single-crystal substrate of specific crystallographic orientation as shown in Fig. 1A. Next, a layer extending from the surface of the substrate into the bulk is rendered amorphous via appropriate ion-implantation conditions (Christel et al., 1981; Elliman et al., 1987; Gibbons, 1972; Mayer et al., 1968; Morehead and Crowder, 1970; Morehead et al., 1972; Washburn et al., 1983; Ziegler, 2004) as shown in Fig. 1B. Subsequently, the substrate is annealed at an elevated temperature (T), which causes epitaxial growth to occur at the interface between the amorphous and crystalline regions (growth interface) as shown in Fig. 1C; the growth interface moves upward toward the wafer surface as the α-Si layer is consumed. Finally, when all the amorphous material is consumed, the crystallized region has the same crystallographic orientation as the underlying substrate as shown in Fig. 1D. This process is interchangeably referred to as solid-phase epitaxial regrowth, solid-phase epitaxial recrystallization, solid-phase epitaxy, and solid-phase epitaxial crystallization. Additionally, while SPEG was previously reported for cases where the α-Si layer was produced via deposition or laser annealing

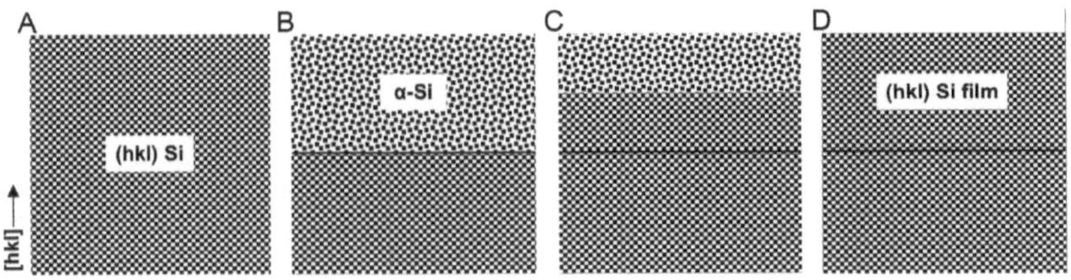

Figure 1 The macroscopic SPEG process: (A) a virgin (hkl) Si wafer, (B) amorphization of the near-surface region of the wafer via ion implantation, (C) partial crystallization of the α-Si layer via annealing to effect epitaxial growth at the amorphous/crystalline interface, and (D) completed crystallization of the entire α-Si layer resulting in a single-crystal film with same orientation at the substrate.

(Bonse et al., 2004; Costache et al., 2004; Izawa et al., 2007; Jia et al., 2004; Kiani et al., 2009, 2010), SPEG where ion implantation is used to produce the α-Si layer is of the most technological relevance and will be addressed here exclusively.

The SPEG process is routinely used in the manufacturing of modern Si-based integrated circuit devices for two main reasons: (1) the ability to accurately and finely control profiles of dopants introduced into a wafer via ion implantation and (2) the ability to enhance electrical activation of said dopants during subsequent annealing to effect SPEG. Both of these considerations are critical in resulting device performance.

In terms of dopant profile control, consider a Si substrate with a surface amorphous layer prepared via Si^+-implantation (so-called "preamorphization"). If a dopant species, such as B^+, As^+, or P^+, is ion-implanted within the preamorphized layer, the lack of order in the amorphous layer will prevent excessive broadening of the dopant concentration profile due to ion channeling along well-defined crystallographic channels (Myers and Wilson, 1980; Myers et al., 1979; Nishi et al., 1978; Robinson and Oen, 1963). This ability to prevent profile broadening is critical in resulting device performance (Kim et al., 2002; Larson and Snyder, 2006; Seong-Dong et al., 2000; Sze and Gibbons, 1966; Taur et al., 1998).

Subsequently, to activate the implanted dopant species, it is necessary to anneal the material at an elevated temperature. In the case where the dopant atoms were implanted into a preamorphized layer, SPEG will occur as a result of annealing. As the growth interface reaches the implanted dopant atoms, the dopants incorporate substitutionally due to the epitaxial nature of the process, and become electrically active. In fact, active dopant concentrations in excess of equilibrium solubility can be achieved via SPEG (Bousetta et al., 1991; Ozturk et al., 1988; Tsaur and Anderson, 1983; Vonborany and Kogler, 1993). In contrast, if activation of implanted dopant atoms is attempted in crystalline material, substitutional incorporation of the dopant can only occur if vacancies are present, and the dopants and vacancies have enough mobility to combine during annealing. In general, this method is far less effective in terms of the active dopant concentrations that can be achieved (Privitera et al., 1999).

Ideally, the SPEG process is perfect in that the initial α-Si layer crystallizes into a perfect single crystal after completion of SPEG. However, depending on the characteristics of the SPEG process, it is typical for defects to be introduced into the crystallized layer. Furthermore, the presence of these defects in Si-based devices is often detrimental to performance (Ashburn et al., 1977; Bull et al., 1979; Finetti et al., 1979; Sadana et al., 1977). Thus, this work reviews the understanding of defects associated with the SPEG process in ion-implanted Si. The different defects associated with the imperfect SPEG process will be reviewed along with the current understanding of how the defects are related to microscopic and macroscopic characteristics of the SPEG process.

1.1. Atomistic considerations

The atomistic nature of the SPEG process remains a topic of debate. However, it is widely accepted that SPEG is the result of some type of reaction occurring at the growth interface. Several models were previously advanced to explain the nature of the reaction at the growth interface including diffusion (Narayan and Holland, 1982) and bond rearrangement (Csepregi et al., 1978). However, it is worth pointing out that the SPEG process is very similar in many respects to liquid-phase epitaxial growth (Peteves and Abbaschian, 1991; Volmer and Marder, 1931; Wilson, 1900) due to the lack of long-range order in both amorphous and liquid phases.

To that end, for an atomically flat growth interface, the SPEG process can be modeled as consisting of (1) nucleating an island with critical size such that it is thermodynamically favorable, (2) propagating the edges of the island in the plane of the growth interface, and (3) nucleating a new island on top of the original one, once the original island reaches sufficiently large size (Rudawski and Jones, 2009). This process is shown at an atomic level in Fig. 2 for the case of SPEG occurring on a (001) wafer as viewed along an in-plane $\langle 110 \rangle$ direction.

It should be noted that the schematics shown in Fig. 2 assume that the atoms on the amorphous side of the growth interface add to the crystalline side of the interface in pairs, rather than one at a time. This assumption is based, in part, on the diamond cubic (DC) structure being equivalent to a face-centered cubic lattice with a 2-atom basis where the two atoms are identical. Thus, a pair of atoms in the DC structure can be treated analogously to a single atom for a face-centered cubic structure. Additionally, this same treatment is also used to explain the nature of dislocations and stacking faults in the DC structure (Hirth and Lothe, 1968) and can therefore be used to explain the generation of defects during SPEG. Of course, it should be noted that the validity of using atomic pairs to model SPEG is still a matter of debate.

A
B
C
D

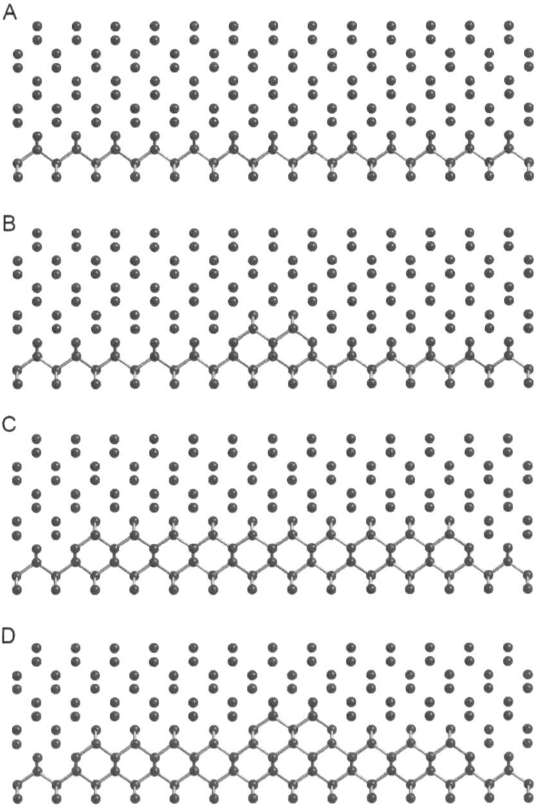

Figure 2 The atomistic nature of the SPEG process for (001) orientation as viewed along an in-plane ⟨110⟩ direction: (A) an atomically flat growth interface, (B) nucleation of an island of critical size, (C) propagation of island ledges in the plane of the growth interface, and (D) nucleation of another island of critical size on the original island (atoms in the crystalline state are shown as bonded).

On an atomistic level, there are several ways that defects can be introduced into the crystallizing layer during SPEG. In the case of an isolated growth interface, defects can form if atomic pairs do not add correctly to the growth interface during island nucleation or during ledge migration. However, defects can also be introduced during SPEG if two independent growth interfaces meet or if two portions of the same growth interface meet; if alignment of the two interfaces is not perfect upon meeting, defects can also form. These atomistic considerations of defect generation during SPEG will be returned to when discussing the individual types of defects.

1.2. Macroscopic considerations

The macroscopic kinetics of the SPEG process have been well studied within the context of α-Si layers with finite thickness (along the wafer normal direction) and infinite lateral (in-plane) extent; such α-Si layers are termed "blanket" layers and thus SPEG only occurs along the direction corresponding to the wafer normal, making the process one-dimensional. Extensive prior work established that the velocity of the growth interface is highly sensitive to the annealing temperature (Olson and Roth, 1988; Roth et al., 1990) used to effect SPEG, the crystallographic orientation (Csepregi et al., 1978) of the substrate, the presence of impurities (Csepregi et al., 1977; Kennedy et al., 1977; Lietoila et al., 1982; Morarka et al., 2010; Suni et al., 1982; Williams and Elliman, 1983; Williams and Short, 1983) within the α-Si layer, and the application of external stress (Aziz et al., 1991; Barvosa-Carter and Aziz, 2001; Barvosa-Carter et al., 1998, 2004; Chaki, 1991, 1994; Lu et al., 1989; Morarka et al., 2011; Nygren et al., 1985; Phan et al., 2001; Rudawski and Jones, 2009; Rudawski et al., 2007, 2008b,c,d, 2009a; Sage et al., 2000; Sklenard et al., 2013). The exact details of the effect each of these considerations has on the macroscopic nature of the SPEG process will not be described in detail here. That being said, microscopic (rather than macroscopic) considerations of the SPEG process are more critical when explaining the generation of defects during crystallization. However, within the context of each of the SPEG-related defects that will be subsequently discussed, the influence (if any) of each of these factors on the generation of each type of defect will be discussed.

2. ROLE OF INITIAL GROWTH INTERFACE MORPHOLOGY

Consider the case of a blanket amorphous layer formed on the surface of a single-crystal Si wafer via ion implantation. It is well known that the morphology and roughness of the resulting amorphous/crystalline (initial growth) interface are heavily dependent on the ion-implantation conditions. Specifically, as the ion energy used to effect amorphization is increased (Ziegler et al., 2010) or the ion mass decreased (Sadana et al., 1984), the roughness of the growth interface in the as-implanted state will increase and the amorphous to crystalline transition near the interface will be less abrupt. In terms of how this influences the quality of the SPEG process, the rougher the initial growth interface, the greater the defect density introduced into the crystallized layer (Glowinski et al., 1976; Jones et al., 1988; Sadana et al., 1984; Sands et al., 1985).

2.1. General observations of "hairpin" dislocations

Figure 3 shows on-axis bright-field cross-sectional transmission electron microscopy (BF-XTEM) images of a (001) Si wafer Si^+-implanted at 20 keV to a dose of 1.0×10^{15} cm^{-2}. In the as-implanted state, shown in Fig. 3A, an α-Si layer ~50 nm in thickness is evident with a flat initial growth interface. After annealing at $T = 600$ °C for 1.0 h the α-Si layer completely crystallized with no defects present in the crystallized layer as shown in Fig. 3B. For comparison, Fig. 4 shows an on-axis BF-XTEM images of a (001) Si wafer Si^+-implanted at 50, 100, and 200 keV to doses of 1.0×10^{15}, 1.0×10^{15}, and 3.0×10^{15} cm^{-2}, respectively. In the as-implanted state, shown in Fig. 4A, an α-Si layer ~350 nm in thickness is evident with a rough initial growth interface. After annealing at $T = 600$ °C for 1.0 h, the α-Si layer completely crystallized, but dislocations that extend from near the initial growth interface through the crystallized layer and terminating at the wafer surface are evident as shown in Fig. 4B. As previously described, these dislocations are perfect in nature with

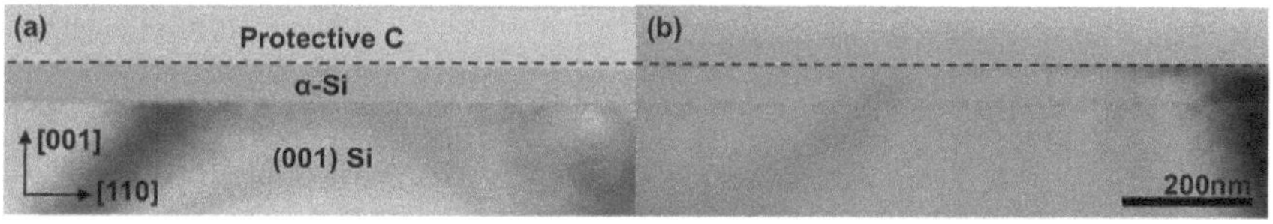

Figure 3 On-axis BF-XTEM images of (001) Si subjected to Si^+-implantation at 20 keV to a dose of 1.0×10^{15} cm^{-2}: (A) as-implanted and (B) after annealing at 600 °C for 1.0 h showing a crystallized layer free of defects (wafer surface indicated by the dotted line).

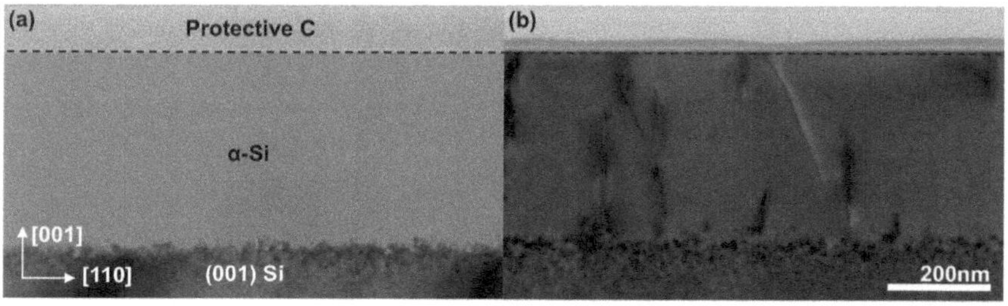

Figure 4 On-axis BF-XTEM images (001) Si subjected to Si^+-implantation at 50, 100, and 200 keV to doses of 1.0×10^{15}, 1.0×10^{15}, and 3.0×10^{15} cm^{-2}, respectively: (A) as-implanted and (B) after annealing at 600 °C for 1.0 h showing crystallized layer containing "hairpin" dislocations (wafer surface indicated by the dotted line).

Burgers vectors $b=(a/2)\langle 110\rangle$ and $t=\langle 110\rangle$ line directions and are commonly referred to as "hairpin" dislocations (Jones et al., 1988); the angle between b and t is 60°, and thus the dislocations have both shear and edge components, which is typical of perfect dislocations in Si (Hirth and Lothe, 1968). Typically, each dislocation has two segments that thread up to the surface from the initial growth interface; this often gives the dislocation a V-shape analogous to that of an open hairpin. Figure 5 shows a filtered high-resolution cross-sectional transmission electron microscopy (HR-XTEM) image of the core of a hairpin dislocation near the location of the initial growth interface as viewed along a segment with line direction $[1\bar{1}0]$; an extra $[\bar{1}\,\bar{1}1]$ lattice fringe near the dislocation core is indicated by the white arrow, which gives a Burgers vector for the dislocation of $b=(a/4)\langle 112\rangle$. This Burgers vector corresponds to the edge component of a 60° dislocation as viewed along the line direction, which confirms the expected nature of the defect. It should be noted that in Figs. 3B and 4B, a band of defects just beyond the initial position of the growth interface, known as end-of-range (EOR) defects, is present (Jones et al., 1988); whenever an α-Si layer formed via ion implantation is crystallized, EOR damage will always be present due to the super-saturation of interstitials just beyond the initial growth interface (Ziegler et al., 2010). However, as the EOR damage does not interact with the growth interface, EOR damage is not believed to have any influence on the SPEG process.

While it is evident even at the relatively large scale of Figs. 3A and 4A that the initial growth interface is rougher in the latter case, in actuality, roughness of the initial growth interface alone cannot explain the introduction of hairpin dislocations into the crystallized layer during SPEG. In

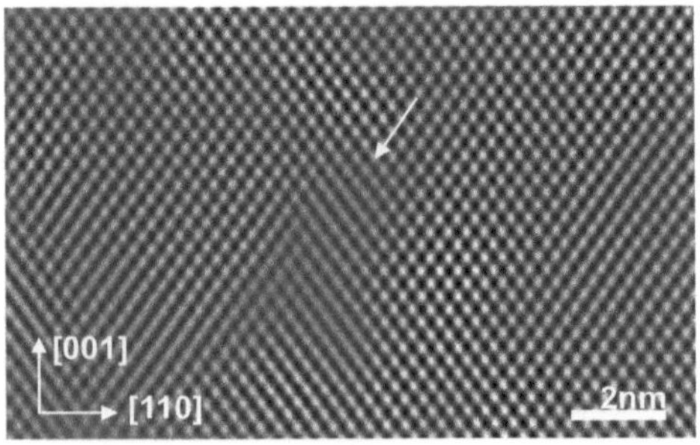

Figure 5 Fourier-filtered HR-XTEM image of the core of a "hairpin" dislocation near the initial growth interface in (001) Si subjected to Si^+-implantation at 50, 100, and 200 keV to doses of 1.0×10^{15}, 1.0×10^{15}, and 3.0×10^{15} cm^{-2}, respectively, after annealing at 600 °C for 1.0 h. An arrow indicates the presence of an extra $(\bar{1}\,\bar{1}1)$ lattice fringe.

principle, even if the growth interface is rough, the SPEG process should still not be defective as the "substrate" used for SPEG is still a single crystal without extended defects. Thus, to introduce hairpin defects into the crystallized layer during SPEG, there must be some disruption in the registry of the growth interface. Figure 6 shows high-magnification BF-XTEM images of the effect of ion-implantation conditions on growth interface morphology in the as-implanted state. For a (001) Si wafer Si^+-implanted at 20 keV to a dose of 1.0×10^{15} cm^{-2}, an abrupt transition from amorphous to crystalline is evident with peak to valley roughness ~10 nm, as shown in Fig. 6A. However, for a (001) Si wafer Si^+-implanted at 50, 100, and 200 keV to doses of 1.0×10^{15}, 1.0×10^{15}, and 3.0×10^{15} cm^{-2}, respectively, shown in Fig. 6B, the interface is much rougher with peak to valley roughness of ~30 nm; additionally, it is also evident that the transition from amorphous to crystalline is not perfectly abrupt. In fact, several small crystallites surrounded by α-Si are evident near the growth interface. Thus, upon annealing, the crystallites will also serve as "substrates" for the SPEG process. When the growth interface from a crystallite meets the growth interface from the underlying continuous single crystal, a defect will form if the registry between the two is not perfect (Sands et al., 1985).

The striking influence of starting growth interface roughness and morphology is also evident when viewed using bright-field plan-view transmission electron microscopy (BF-PTEM) with g_{220} as shown in Fig. 7. In the case of a (001) Si wafer Si^+-implanted at 20 keV to a dose of 1.0×10^{15} cm^{-2} (smooth initial growth interface with no near-interface crystallites) and annealed at $T = 600$ °C for 1.0 h, no hairpin dislocations are evident, as shown in Fig. 7A. However, in the case of a (001) Si wafer Si^+-implanted at 50, 100, and 200 keV to doses of 1.0×10^{15}, 1.0×10^{15}, and 3.0×10^{15} cm^{-2}, respectively (rough initial growth interface with many

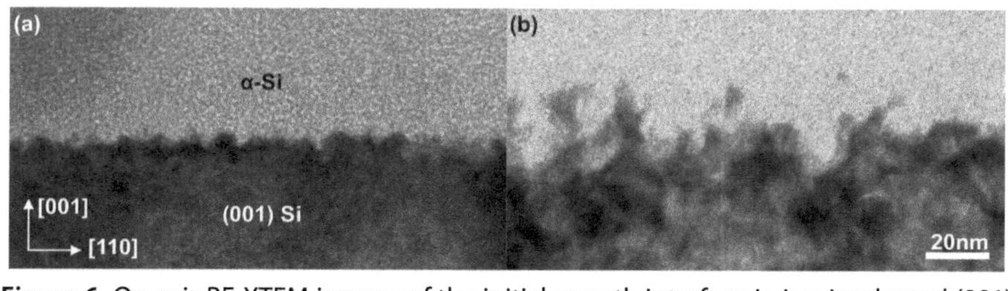

Figure 6 On-axis BF-XTEM images of the initial growth interface in ion-implanted (001) Si subjected to Si^+-implantation at (A) 20 keV to a dose of 1.0×10^{15} cm^{-2} and (B) 50, 100, and 200 keV to doses of 1.0×10^{15}, 1.0×10^{15}, and 3.0×10^{15} cm^{-2}, respectively (wafer surface indicated by the dotted line).

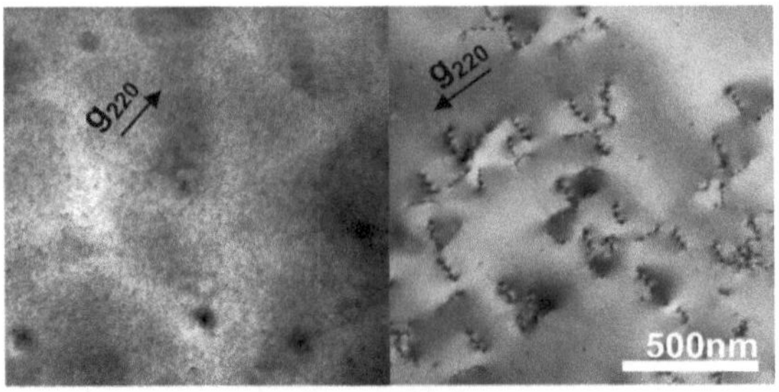

Figure 7 BF-PTEM images taken with g_{220} of (001) Si subjected to Si^{+}-implantation after annealing at 600 °C for 1.0 h: (A) Si^{+}-implantation at 20 keV to a dose of 1.0×10^{15} cm^{-2} and (B) Si^{+}-implantation at 50, 100, and 200 keV to doses of 1.0×10^{15}, 1.0×10^{15}, and 3.0×10^{15} cm^{-2}, respectively.

near-interface crystallites) and annealed at $T=600$ °C for 1.0 h, a high planar density of hairpin dislocations is clearly evident as shown in Fig. 7B.

2.2. Very-low temperature annealing

Interestingly, some reported that preforming "very-low" temperature annealing (VLTA) at $T<500$ °C for several hours can cause dissolution of the small crystallites in the α-Si layer just above the initial growth interface (Myers et al., 1985). The small diameter of the crystallites leads to a very high surface energy to volumetric free energy ratio, which theoretically makes it thermodynamically favorable for the crystallites to dissolve rather than grow upon annealing provided annealing is performed at a temperature low enough that SPEG is effectively suppressed. When the specimen is then annealed at temperatures high enough to effect SPEG, the resulting crystallized layer has fewer hairpin defects compared to the same specimen without the prior VLTA. The reported work on the use of VLTA to reduce hairpin defects notwithstanding, additional independent confirmation of its effectiveness is somewhat lacking.

2.3. Stress-induced enhancement to hairpin dislocation formation

Work by Rudawski et al. revealed that the application of uniaxial stress in the plane of the growth interface during SPEG did result in a higher density of hairpin dislocations compared to the case where the SPEG process was not stressed (Rudawski et al., 2007). It is still unclear why applied stress favored the nucleation of hairpin dislocations, but it was advanced that it may have been due to the applied stress increasing the misalignment between the growth interface from the substrate and the isolated crystallites rather than the ability of applied stress to alter the SPEG kinetics. Specifically, the crystallites are surrounded by an α-Si matrix, and α-Si was previously reported to exhibit viscoelastic behavior (Witvrouw and Spaepen, 1993). It is therefore plausible that the orientation of the isolated crystallites was changed by the application of applied stress as a viscoelastic matrix surrounded the crystallites. It should also be noted that the relationship between applied stress during SPEG and the resulting planar density of hairpin dislocations was quantified using cross-sectional, rather than plan-view, transmission electron microscopy imaging (Rudawski et al., 2007). As cross-sectional imaging for purposes of quantifying the planar density of defects is not ideal, the relationship between applied stress during SPEG and resulting hairpin dislocation density is somewhat questionable.

3. "BURIED" AMORPHOUS LAYERS

Ideally, when ion implantation is used to effect amorphization of the surface region of a Si wafer, the α-Si layer is continuous up to the surface of the wafer. However, depending on the ion-implantation conditions, it is possible to generate an α-Si layer that is not continuous to the wafer surface (Ziegler et al., 2010); the resulting α-Si layer is bounded on the top and bottom by single-crystal regions with nominally identical crystallographic orientation and typically referred to as being "buried." Upon annealing to effect SPEG, there are effectively two independent growth interfaces from which SPEG occurs. Defects can form when the two growth interfaces meet if the two interfaces are out of registry; these are often referred to as "clamshell" defects (Jones et al., 1988; Masters et al., 1970; Sadana et al., 1977, 1982).

Figure 8A shows an on-axis BF-XTEM image of a (001) Si wafer Si^+-implanted at 60 keV to a dose of 1.0×10^{15} cm^{-2} in the as-implanted state; an α-Si layer is ~120 nm-thick is evident. However, the α-Si layer does not extend all the way to the wafer surface, but terminates at a ~10 nm-thick layer of single-crystal Si that has nominally the same orientation as the substrate. Figure 8B shows a BF-XTEM image taken using g_{004} of the specimen in Fig. 8A after annealing at $T = 600$ °C for 1.0 h. A distinct line is evident at the meeting point of the two growth interfaces (indicated by arrows), which suggests the two growth interfaces were out of registry at the point of

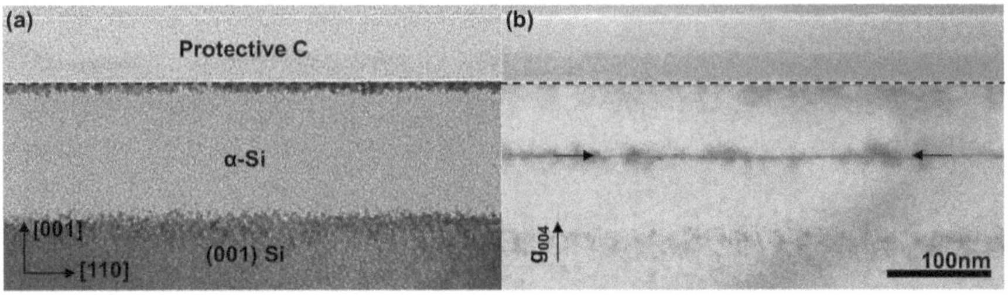

Figure 8 BF-XTEM images taken with g_{004} of (001) Si subjected to Si^+-implantation at 60 keV to a dose of 1.0×10^{15} cm^{-2}: (A) as-implanted and (B) after annealing at 600 °C for 1.0 h showing the presence of a clamshell defect (wafer surface indicated by the dotted line).

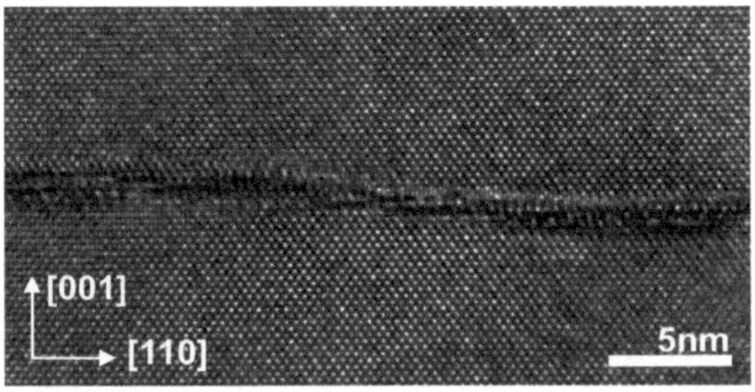

Figure 9 HR-XTEM image of the clamshell defect occurring in (001) Si subjected to Si^+-implantation at 60 keV to a dose of 1.0×10^{15} cm^{-2} after annealing at 600 °C for 1.0 h.

meeting; hence, a clamshell defect was generated. Figure 9 shows a HR-XTEM image of the clamshell defect shown in the annealed specimen presented in Fig. 8B; this disruption in registry at the meeting of the two growth interfaces is evident. Figure 10 shows the same annealed specimen presented in Fig. 8B imaged using weak-beam dark-field plan-view transmission electron microscopy (WBDF-PTEM) with g_{220}; the small bright irregularly shaped loops are the clamshell defects resulting from the meeting of the two growth interfaces; the lack of any ribbon contrast within the loops suggests the loops are perfect, rather than faulted in nature, which is consistent with other reports (Jones et al., 1988).

While it is evident from Fig. 10 that the clamshell defects are loop-like morphologically and that the habit plane of the defects must be (001) from Fig. 8B, the nature of the associated Burgers vectors remains unclear. It was originally reported that the Burgers vectors were also contained in the habit plane (Jones et al., 1988). However, as shown in Fig. 8B, the clamshell

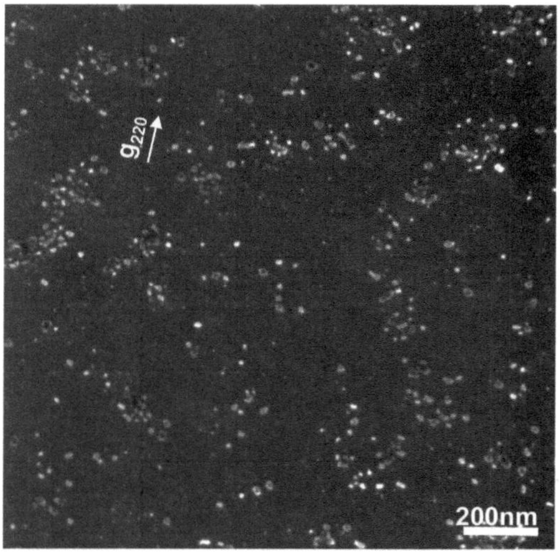

Figure 10 WBDF-PTEM image taken using g_{220} of clamshell defects occurring in (001) Si subjected to Si^{+}-implantation at 60 keV to a dose of 1.0×10^{15} cm^{-2} after annealing at 600 °C for 1.0 h.

defects are still in contrast when imaged using g_{004}. This suggests the Burgers vectors for the defects actually are not entirely contained in the habit plane as per the $g \cdot b = 0$ invisibility criterion and thus must have some out of plane component.

3.1. Clamshell defects for other wafer orientations

It should also be noted that while the case of clamshell defects occurring during SPEG for (001) wafers with a buried α-Si layer was discussed here, clamshell defects were also reported similarly for (111) wafers (Sadana et al., 1982). Similarly to the (001) case, the habit plane of the clamshell defects was also the same as the wafer orientation. However, the defects were faulted with Burgers vectors $(a/3)[111]$, which is a type of Frank partial dislocation. The nature of clamshell defects for other wafer orientations remains unknown; however, based on the differences in the defects for (001) and (111) wafers, it is likely that substrate orientation is highly influential.

4. SUBSTRATE ORIENTATION EFFECTS

As was previously shown, for SPEG on (001) wafers with blanket α-Si layers extending all the way to the wafer surface and free of impurities, the

SPEG process is by and large free of defects. The only exception to this is hairpin dislocations (which are perfect) that are introduced if small, isolated crystallites are generated in the α-Si layer just outside the initial growth interface, which typically results from high-energy ion implantation (Jones et al., 1988); no stacking faults or other extended defects are otherwise observed (Csepregi et al., 1976). In the case of SPEG on (011) wafers with blanket α-Si layers of a similar nature, the resulting growth is again free of stacking faults (Rudawski et al., 2009a). However, in the case of SPEG on (111) wafers, the resulting growth is typically highly defective (Csepregi et al., 1976; Martin-Bragado, 2012; Martin-Bragado and Sklenard, 2012; Rechtin et al., 1978).

4.1. Stacking sequence variation with substrate orientation

The dependence of the defectiveness of SPEG on substrate orientation can be understood by considering the differences in the stacking sequence of pairs of atoms during island nucleation between orientations as shown in Fig. 11. For the case of the (001) plane, the pairs add in an "ABABAB" type fashion, with the stacking sequence not close-packed; this makes any disruption of stacking sequence during SPEG on the (001) plane unfavorably energetically. For similar reasons, in the case of SPEG on (011) wafers, growth is also free of stacking faults. However, in the case of (111) orientation, the pairs add in an "ABCABCABC" configuration with the stacking close-packed. Thus, if the stacking sequence is changed to "ABCAB$\underset{\uparrow}{\overset{\downarrow}{A}}$BC" an isolated stacking fault is created (indicated by arrows), but the stacking sequence is still close-packed. Thus, the formation of a stacking fault costs very little energetically.

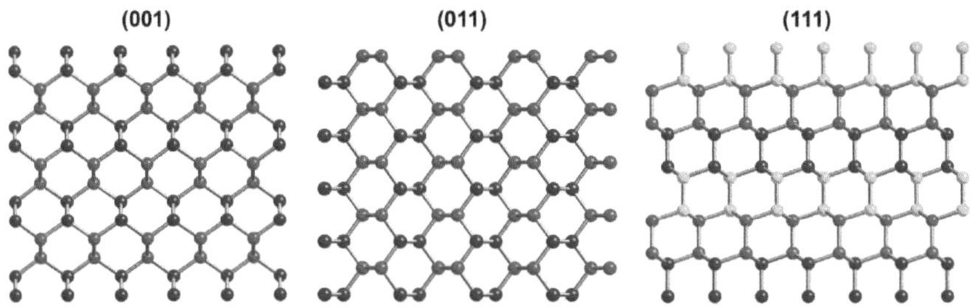

Figure 11 Atomistic schematics of the stacking sequence of atomic pairs for (001), (011), and (111) orientations as viewed along an in-plane ⟨110⟩ direction; changes in the stacking sequence are indicated by changes in color of the bilayer.

If there is no ordering to the disruption of the stacking sequence, then the layer will simply exhibit random stacking faults. However, it is common for the change in stacking sequence to exhibit ordering. This can result in domains of different Si polytypes as well as twins. In the case of polytypes, perhaps the most common are the 2H (or wurtzite) and 9R varieties. Consider if the stacking sequence is changed to "ABABABAB"; this will result in a four-layer domain of the 2H (or wurtzite) polytype of Si. Similarly, if the stacking sequence is changed to "ABCBCABAB," this will produce one layer of the 9R polytype. In addition to the 2H and 9R polytypes, several others have been identified experimentally, but the 2H and 9R are most typically observed (Fontcuberta i Morral et al., 2007; Gu et al., 2010; Liu and Wang, 2009; Lopez et al., 2009, 2011; McLarnan, 1981; Miyamoto and Hirata, 1978; Raffy et al., 2002; Wen et al., 2008). Of course, a disruption in the stacking sequence does not necessarily produce a random stacking fault or a different polytype. Consider the case of an "ABCAB$\underset{\uparrow}{\overset{\downarrow}{\text{C}}}$BACBA" stacking sequence. In this case, the stacking sequence is reflected across the indicated C plane of atoms, which results in the formation of a twin. Additionally, if the stacking sequence is "reflected" back to the original sequence after a short distance, such as in "ABCAB$\underset{\uparrow}{\overset{\downarrow}{\text{C}}}$BACBACB$\underset{\uparrow}{\overset{\downarrow}{A}}$BCABC," a "micro-twin" forms.

4.2. Observations of SPEG on (111) wafers

Figure 12 shows BF-XTEM images taken with g_{220} of a (111) Si wafer Si^+-implanted at 50, 100, and 200 keV to doses of 1.0×10^{15}, 1.0×10^{15}, and $3.0 \times 10^{15}\ cm^{-2}$, respectively. In the as-implanted case, shown in Fig. 12A, a continuous α-Si layer extends from the surface to ~350 nm into

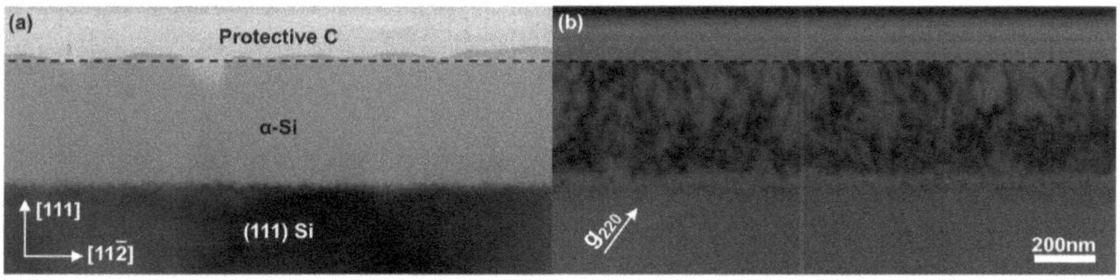

Figure 12 BF-XTEM images taken with g_{220} of (111) Si subjected to Si^+-implantation at 50, 100, and 200 keV to doses of 1.0×10^{15}, 1.0×10^{15}, and $3.0 \times 10^{15}\ cm^{-2}$, respectively: (A) as-implanted and (B) after annealing at 600 °C for 4.0 h (wafer surface indicated by the dotted line).

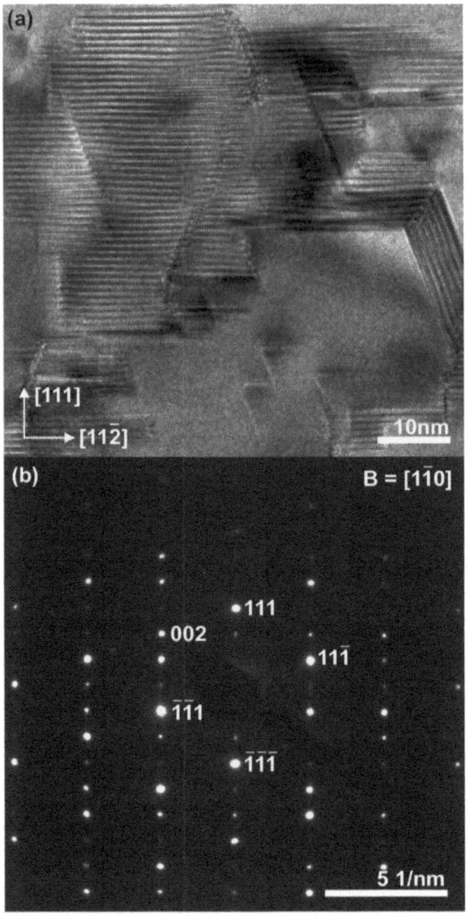

Figure 13 (A) HR-XTEM image of (111) Si subjected to Si^+-implantation at 50, 100, and 200 keV to doses of 1.0×10^{15}, 1.0×10^{15}, and 3.0×10^{15} cm^{-2}, respectively, after annealing at 600 °C for 4.0 h and (B) SAED pattern taken from the region shown in (A).

the wafer. After annealing at $T = 600$ °C for 4.0 h, shown in Fig. 12B, the SPEG process has completed, but the crystallized layer is highly defective. In fact, the quality of the SPEG process for this wafer orientation appears to have two different regimes (Csepregi et al., 1976). The initial ~80 nm of growth appears to be mostly free of defects with the rest of the growth appearing highly defective. Figure 13A shows a HR-XTEM image of the defective region of the specimen shown in Fig. 12B. Several isolated stacking faults are observed as well as twins and domains of 9R Si polytype. Interestingly, these defects appear to nucleate off both the (111) plane (wafer normal) as well as the $(11\overline{1})$ plane with normal direction ~72° away from the wafer normal; presumably, this also occurs for the other two {111} planes, but cannot be readily detected using a $B=[1\overline{1}0]$ specimen normal. The additional formation of stacking faults, twins, and 9R polytype domains using the three {111} planes oriented away from the wafer normal as "substrates" suggests that macroscopic faceting of the growth interface on the off-axis {111} planes is occurring or ledge migration is defective. In the former case, it was previously demonstrated that even through SPEG for (111) wafers is highly defective, no such macroscopic faceting of the growth interface is known to occur (Csepregi et al., 1976). This suggests the ledge migration process can become defective, but it is not yet clear as to if this is the case or as to why ledge migration would be defective for (111) orientation but not (001) or (011).

Selected area electron diffraction (SAED) was performed on the region of the specimen shown in Fig. 13A as presented in Fig. 13B. The spots due to the DC phase of Si are clearly evident and labeled accordingly. However, a slightly less intense array of spots with spacing one-third that of g_{111} for the DC phase and running parallel to g_{111} is also evident. This spacing of spots is equivalent to g_{0003} for the 9R polytype while the arrangement of spots indicates a view direction along the $B=[11\overline{2}0]$ zone axis (Miyamoto and Hirata, 1978). Additionally, a very dim array of spots with the same g_{0003} spacing of the 9R polytype but running parallel to $g_{11\overline{1}}$ is also evident, confirming the presence of the 9R polytype nucleating off the $(11\overline{1})$ plane with the same $B=[11\overline{2}0]$ zone axis orientation. In both cases, the orientation relationship between the DC and 9R phases is $\langle 1\overline{1}0\rangle_{DC}/\langle 11\overline{2}0\rangle_{9R}$ and $\{111\}_{DC}/\{0001\}_{9R}$. Lastly, it should also be noted that twinning on both the (111) and $(11\overline{1})$ planes, while evident in the HR-XTEM image shown in Fig. 13A, cannot be directly delineated from the 9R polytype using SAED along the $B=[1\overline{1}0]$ zone axis of the DC phase as there will be overlap of all spots due to twinning with some of spots from the 9R polytype domains.

5. STRESSED SPEG RESULTING FROM IMPURITIES

While morphological and crystallographic considerations are most obviously influential on the quality of the SPEG process, the presence of impurity atoms in sufficient concentration can also have profound effects due to the stress generated in the crystallizing layer resulting from incorporation during SPEG. This effect is most strikingly exhibited in the case of impurity elements that can assume the same DC crystal structure of Si, such as C or Ge.

Consider the case of such an impurity atom that is ion-implanted into an α-Si layer; the impurity concentration profile will be variable with a peak concentration located near the projected range of the ion (Ziegler et al., 2010). If it is then assumed that during the SPEG process that all the impurity atoms incorporate substitutionally due to the epitaxial nature of the process as well as the identical nature of the crystal structure of Si and impurity element, the crystallizing layer will be under a state of in-plane, biaxial stress. This arises due to changes in the lattice constant of the Si as greater amounts of the impurity species are incorporated in accordance with a Vegard's law approximation (Vegard, 1921). Thus, the magnitude of the in-plane biaxial stress will increase with increasing impurity concentration and reaches a maximum at the projected range of the impurity profile.

In the case of C, the lattice constant of diamond is smaller than that of Si, which places the crystallized layer into a state of in-plane biaxial tension. In comparison, the Ge lattice constant is larger than that of Si, so the crystallized layer will be placed into a state of in-plane biaxial tension. In either case, stacking faults and/or dislocations are introduced into the crystallized layer as a mechanism to relieve the stress due to the lattice mismatch (Paine et al., 1991), similarly to the case of growing a stressed film on a substrate via vapor-phase epitaxial growth (Bruinsma and Zangwill, 1987; Eaglesham and Cerullo, 1990; Frank and Vandermerwe, 1949; Venables, 2000; Volmer and Weber, 1926).

5.1. Impurity generating tensile stress (C)

Figure 14 shows HR-XTEM images of a (001) Si wafer after C^+-implantation at 10 keV to a dose of 1.0×10^{16} cm^{-2}. In the as-implanted state, shown in Fig. 14A, a continuous α-Si layer extending ~50 nm from the wafer surface is evident. However, after annealing at $T=800$ °C for 1.0 h, shown in Fig. 14B, approximately 20 nm of defect-free SPEG occurred, but subsequently, the resulting growth contains a large quantity of Shockley partial stacking faults of the type $b=(a/6)\langle 112\rangle$ (Hirth and Lothe, 1968); the depth of the stacking faults relative to the wafer surface is near the predicted peak C concentration (Ziegler et al., 2010). Additionally, it appears that individual grains of Si are nucleating

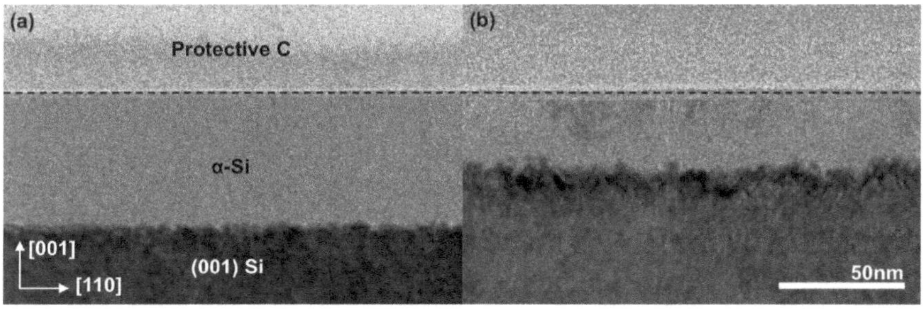

Figure 14 HR-XTEM images of (001) Si subjected to C^+-implantation at 10 keV to a dose of 1.0×10^{16} cm^{-2}: (A) as-implanted with superimposed C profile and (B) after annealing at 800 °C for 0.5 h showing the generation of stacking faults at the growth interface near the projected range of the C^+ profile (surface indicated by the dotted line).

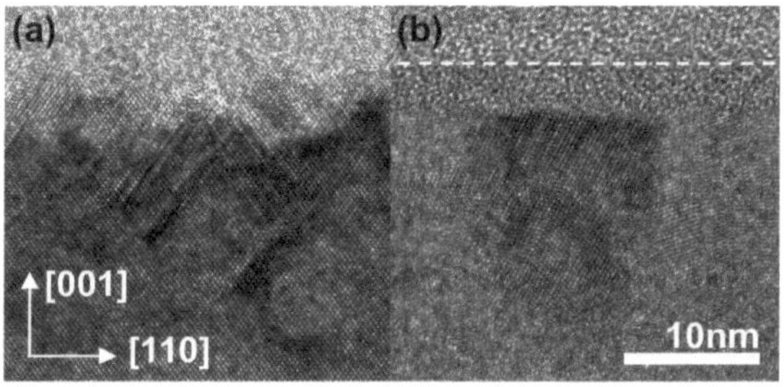

Figure 15 HR-XTEM images of (001) Si subjected to C^+-implantation at 10 keV to a dose of 1.0×10^{16} cm^{-2} after annealing at 800 °C for 0.5 h showing: (A) stacking faults generated at the resulting growth interface and (B) nucleation of Si crystallites near the wafer surface (wafer surface indicated by the dotted line).

from the wafer surface within the α-Si layer. Figure 15A and B shows high-magnification HR-XTEM images of the stacking faults at the resulting growth interface and grains nucleating at the wafer surface, respectively.

Based on simulations (Ziegler et al., 2010), the peak C concentration for C^+-implantation at 10 keV into Si to a dose of 1.0×10^{16} cm^{-2} is $\sim 2.0 \times 10^{21}$ cm^{-3}, which corresponds to a C mole fraction of $X_C \sim 0.04$. Using a Vegard's law approximation (Vegard, 1921), the lattice constant at the peak C concentration will be $a = 0.04a_C + 0.96a_{Si} = 0.536$ nm where $a_C = 0.357$ and $a_{Si} = 0.543$ nm are the lattice constants for diamond and Si, respectively. The in-plane biaxial strain is thus $\varepsilon_{11} = a_{Si}/a - 1 = 0.013$. The in-plane biaxial stress associated with this level of strain will be given by $\sigma_{11} = \varepsilon_{11}/(S_{11} + S_{12})$, where S_{11} and S_{12} are stiffness coefficients. Using the reported elastic properties of Si (Wortman and Evans, 1965), this gives

σ_{11} ~2.3 GPa. Thus, since the SPEG process is highly stressed, it is reasonable to conclude that the stacking faults were introduced during SPEG as a stress-relief mechanism.

The observation of individual Si grains nucleating on the wafer surface indicates that the SPEG process is competing with random nucleation and growth (RNG) to crystallize the α-Si layer. At $T = 800$ °C, the SPEG process for a 50 nm-thick α-Si layer free from impurities should be completed in <1 s (Olson and Roth, 1988). However, it is clear from Fig. 14 that the SPEG process is not complete, since ~30 nm of α-Si still remain. This is due to the presence of C, which is known to strongly retard macroscopic SPEG kinetics at high concentrations (Kennedy et al., 1977). In fact, the C concentration is high enough that SPEG is not just retarded but also completely halted. Thus, once growth interface motion stops, further annealing allows crystallization to occur via RNG, which is a much slower process than SPEG for crystallizing an α-Si layer. Thus, with further annealing, the remaining α-Si layer will be polycrystalline in nature, rather than single crystal as was previously reported (Rudawski et al., 2009c; Strane et al., 1996).

5.2. Impurity generating compressive stress (Ge)

Similarly to the case of high dose C^+-implantation as previously discussed, defective SPEG, primarily in the form of stacking faults near the peak impurity concentration, will also occur in cases of high dose Ge^+-implantation (Cristiano et al., 1999; Elliman and Wong, 1996; Paine et al., 1990, 1991). However, while the X_C necessary to produce defective SPEG is ~0.008, the Ge mole fraction necessary to produce defective SPEG is X_{Ge} ~0.16, which is ~20 times greater. In the case of pure Ge, the lattice constant is $a_{Ge} = 0.565$, which is ~4% greater than a_{Si}, whereas a_C is ~34% smaller than a_{Si}. Thus, assuming a Vegard's law approximation, X_{Ge} would have to be ~16 times X_C necessary to produce the same magnitude of lattice mismatch, which agrees well with the reported ~20-fold difference in threshold mole fractions necessary to produce defective SPEG for C^+- and Ge^+-implanted Si.

Careful work by Paine et al. (1991) also showed using PTEM that in addition to stacking faults, perfect edge, and 60° dislocations also form in the crystallized layer for high dose Ge^+-implantation as part of the stress-relief process; in fact this is somewhat expected as these perfect dislocations are more efficient at relieving the lattice mismatch than Shockley partial dislocations (Hirth and Lothe, 1968). Presumably, perfect dislocations also form in the crystallized layer for the case of high dose C^+-implantation, though this has not yet been experimentally verified. In part, this is due to the difficulty in preparing transmission electron microscopy specimens in such a manner than the perfect dislocations can be observed without being overwhelmed by the heavy presence of the stacking faults.

Interestingly, unlike the case of C^+-implantation, growth interface motion is never slowed or halted in the case of Ge^+-implantation, since Ge actually enhances macroscopic SPEG kinetics (Haynes et al., 1995; Kringhoj and Elliman, 1994; Lee et al., 1993). Thus, the SPEG process never competes with RNG to crystallize the α-Si layer, though the misfit-induced stacking faults will extend all the way to the wafer surface once SPEG completes (Elliman and Wong, 1996).

6. EXTERNALLY STRESSED SPEG

It was previously shown that the state of in-plane biaxial stress induced in the crystallizing layer from the substitutional incorporation of impurity atoms of sufficient concentration results in defective SPEG. Interestingly, it is also possible to apply an external uniform stress and influence the formation of defects during SPEG (Barvosa-Carter et al., 2004; Morarka et al., 2011; Phan et al., 2001; Rudawski et al., 2007, 2008b,c,d, 2009a; Sage et al., 2000; Sklenard et al., 2013). Consider a (001) Si wafer with a continuous α-Si layer extending to the wafer surface with a uniform in-plane uniaxial stress applied along [110] (σ_{11}). If this system is now annealed to effect SPEG, the growth interface will exhibit morphological instability as well as the generation of defects depending on the sign of the applied stress. Specifically, if $\sigma_{11} = 0$ (stress free) or $\sigma_{11} > 0$ (tensile), the growth interface will remain flat during SPEG and no defects will be generated, while if $\sigma_{11} < 0$ (compressive), the growth interface will roughen during SPEG and defects will be generated at the growth interface (Barvosa-Carter et al., 1998, 2004; Morarka et al., 2011; Phan et al., 2001; Rudawski et al., 2007, 2008b,c,d; Sage et al., 2000; Sklenard et al., 2013).

6.1. Growth interface (in)stability and defectiveness

Figure 16 presents on-axis BF-XTEM images showing the resulting growth interface morphology of a (001) Si wafer Si^+-implanted at 50, 100, and 200 keV to doses of 1.0×10^{15}, 1.0×10^{15}, and 3.0×10^{15} cm^{-2}, respectively, after annealing at $T = 500$ °C for 7.0 h for two different regimes of σ_{11} applied along the [110] direction. In the as-implanted case, shown in

Figure 16 On-axis BF-XTEM images of (001) Si subjected to Si^{+}-implantation at 50, 100, and 200 keV to doses of 1.0×10^{15}, 1.0×10^{15}, and 3.0×10^{15} cm^{-2}, respectively: (A) as-implanted, (B) after annealing at 500 °C for 7.0 h with $\sigma_{11}\geq0$ showing a stable resulting growth interface and (C) after annealing at 500 °C for 7.0 h with $\sigma_{11}<0$ showing a rough resulting growth interface.

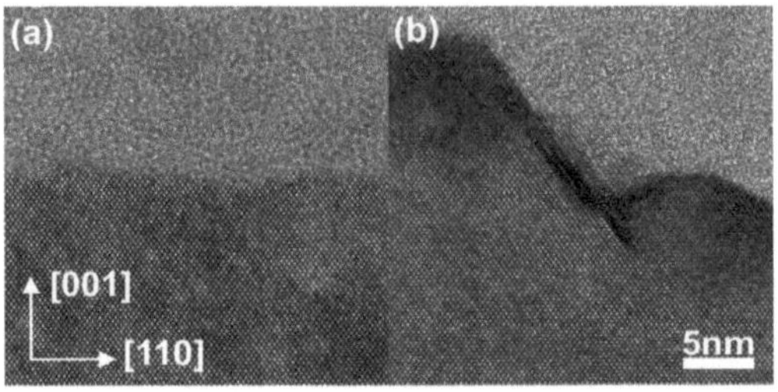

Figure 17 HR-XTEM images of (001) Si subjected to Si^{+}-implantation at 50, 100, and 200 keV to doses of 1.0×10^{15}, 1.0×10^{15}, and 3.0×10^{15} cm^{-2}, respectively, after annealing at 500 °C for 7.0 h (A) with $\sigma_{11}\geq0$ showing a growth interface free of defects and (B) with $\sigma_{11}<0$ showing a stacking fault at the growth interface.

Fig. 16A, initial growth interface is quite rough with peak to valley roughness of ~40 nm. In Fig. 16B, the growth interface after annealing is typical of the case of $\sigma_{11}\geq0$, while in Fig. 16C, the growth interface is typical for $\sigma_{11}<0$ In the case of $\sigma_{11}\geq0$, the growth interface appears nominally flat without significant perturbations, but in the case of $\sigma_{11}<0$, substantial roughing of the growth interface is evident. In fact, at higher magnifications, the atomic-level structure of the resulting growth interface is also different between the two cases. Figure 17 shows HR-XTEM images of the growth interfaces presented in Fig. 16B and C. In the case where annealing was performed with $\sigma_{11}\geq0$, shown in Fig. 17A, no disorder in the crystal lattice at the growth interface is evident. However, for the case where annealing was performed with $\sigma_{11}<0$, shown in Fig. 17B, the generation of stacking faults on $\{111\}$ facets at the growth interface is evident.

6.2. Origins and driving forces

For a growth interface to exhibit morphological instability, peaks in the interface must advance faster than troughs in the interface. Conversely, if the growth interface advances faster at troughs than at peaks, the interface will be stable against perturbations. It was first advanced that the stress-induced morphological instability of the growth interface was due to variation in the localized stress state changing the localized SPEG rate (Barvosa-Carter et al., 1998). In principle, for an interface with peaks and trough, variation in the localized stress state should exist due to differences in the elastic properties of amorphous and crystalline Si (Witvrouw and Spaepen, 1993). However, later work showed that variations in the localized stress state, even in the presence of substantial perturbations, were negligible (Morarka et al., 2011). It was thus advanced that a synergistic effect between the applied stress and the localized growth interface curvature was responsible for controlling morphology, though the origin of this is not yet understood.

It should also be noted that in the case where stress is applied externally in a uniform manner, the crystallizing film is in the exact same stress state as the underlying substrate; this is an important difference in comparison to the case where the crystallizing film is stressed due to the presence of impurity atoms. In the latter case, the substrate is effectively stress free while the film is crystallized in a stressed state; it has been extensively documented that highly stressed films grown on stress-free substrates via other methods (vapor-phase epitaxy, etc.) will exhibit roughening during growth as a result of growth mode transitions (Bruinsma and Zangwill, 1987; Eaglesham and Cerullo, 1990; Frank and Vandermerwe, 1949; Venables, 2000; Volmer and Weber, 1926) when the films are in a state of tension or compression. However, the fact that the substrate and film experience the same stress state when a uniform external stress is applied implies that roughening during stressed SPEG cannot be explained simply as a growth mode transition, particularly considering morphological instability is only observed in the presence of in-plane compression, and not tension.

The generation of stacking faults at the growth interface for the case where SPEG was performed with $\sigma_{11} < 0$, as shown in Fig. 17B, is also interesting. In cases where the SPEG process was stressed due to the constraint of the underlying substrate (as in the case of the incorporation of a high concentration of impurity atoms), the generation of defects serves as a mechanism to relax the stress in the film, but in the case where the film is under the same stress state as the substrate, no driving force for such relaxation exists. However, as is evident from Fig. 17B, there tends to be some localized faceting of the growth interface along $\{111\}$ planes. If SPEG occurs on these facets, it will likely be defective owing to the inherently defective nature of SPEG on (111) Si wafers as was previously discussed in detail (Csepregi et al.,

1976; Jones et al., 1988). It may thus be the case that the application of $\sigma_{11} < 0$ during SPEG causes the $\{111\}$ faceting and alters the macroscopic SPEG kinetics such that the facets may serve as "substrates" for SPEG to occur, but this is still unclear.

7. INTERACTION OF THE GROWTH INTERFACE WITH A SiO_x REGION

To this point, the SPEG process has only been considered for the case of a blanket α-Si layer that is infinite along the two in-plane dimensions of the wafer and finite along the wafer normal direction. Consider now the case of SPEG where one of the in-plane dimensions is now "semi-infinite" with one side of the initial growth interface terminating at a SiO_x layer running orthogonal to the growth interface, such as in the cases of an SiO_x-filled isolation trench or a trench with a surface oxidation layer. The case where both sides of the initial growth interface terminate at a SiO_x layer will also be considered.

7.1. Growth interface termination on one side

Figure 18A shows an on-axis BF-XTEM image of a (001) Si wafer Si^+-implanted at 40 keV to a dose of 1.0×10^{16} cm^{-2} in the vicinity of a SiO_x-filled trench in the as-implanted state. It is evident from Fig. 18A that the initial growth interface terminates on one side at the SiO_x-filled trench.

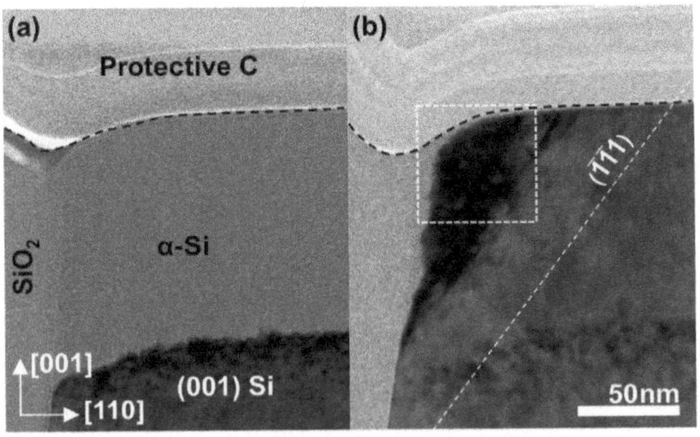

Figure 18 On-axis BF-XTEM images of a (001) Si wafer Si^+-implanted at 40 keV to a dose of 1.0×10^{16} cm^{-2} in the vicinity of a SiO_x-filled trench: (A) as-implanted showing termination of the initial growth interface at the SiO_x-filled trench and (B) annealed at 600 °C for 1.0 h showing a triangular region of defective Si bounded by the $(\bar{1}\bar{1}1)$ plane (wafer surface indicated by the dotted line).

Figure 18B shows the same specimen after annealing at $T=600$ °C for 1.0 h; a triangular region of defective Si starting from near the point of termination of the initial growth interface with the SiO_x layer and extending up to the wafer surface is clearly evident. Interestingly, one side of the triangular defective region appears to be bounded by a $\{111\}$ plane.

In fact, it was previously shown that the SPEG process in similar structures on (001) wafers results in faceting of the growth interface along the same $\{111\}$ plane with the growth interface pinned at the initial point of contact with the SiO_x-filled trench; this results in a triangular region of α-Si bound by the SiO_x-filled trench, the wafer surface, and the $\{111\}$ plane (Burbure et al., 2007). It should also be noted that similar faceting was observed for SPEG in the vicinity of SiO_x-filled trenches for wafer orientations other than (001) (Saenger et al., 2007a). For the SPEG process to continue, it must use the available $\{111\}$ plane as a "substrate" as this is more energetically favorable than nucleation and growth off of the SiO_x-filled trench or the wafer surface (Olson and Roth, 1988). As was previously described, SPEG on (111) wafers with blanket α-Si layers is highly prone to produce random stacking faults, polytype domains, and twins (Csepregi et al., 1976; Jones et al., 1988). Figure 19A shows a HR-XTEM image of the triangular defective region shown in Fig. 18B. The region consists primarily of 9R polytype domains and a few random stacking faults using the $(\bar{1}\,\bar{1}1)$ plane as a template; a few 9R polytype domains also formed using the (111) plane as a template. Figure 19B shows a SAED pattern taken from the region shown in Fig. 19A. The regular array of bright spots can be attributed the DC Si phase. However, a regular array of faint spots with spacing one-third that of g_{111} of the DC phase running parallel g_{111} is evident. Similarly to what was previously reported for the case of SPEG in (111) wafers with a blanket α-Si layer, this spacing of spots is equivalent to g_{0003} for the 9R polytype (Miyamoto and Hirata, 1978) while the arrangement of spots indicates a view direction along the $B=[11\bar{2}0]$ zone axis; the orientation relationship between the DC and 9R phases is correspondingly equivalent.

7.2. Growth interface termination on two sides

Previously, the case of a semi-infinite growth interface bounded on one side by a SiO_x region was considered. If the case is now considered where the growth interface is bounded on both sides by SiO_x regions, the same faceting of the growth interface along a $\{111\}$ plane followed by defective SPEG will also occur on the other bounded side. If the separation between the two sides

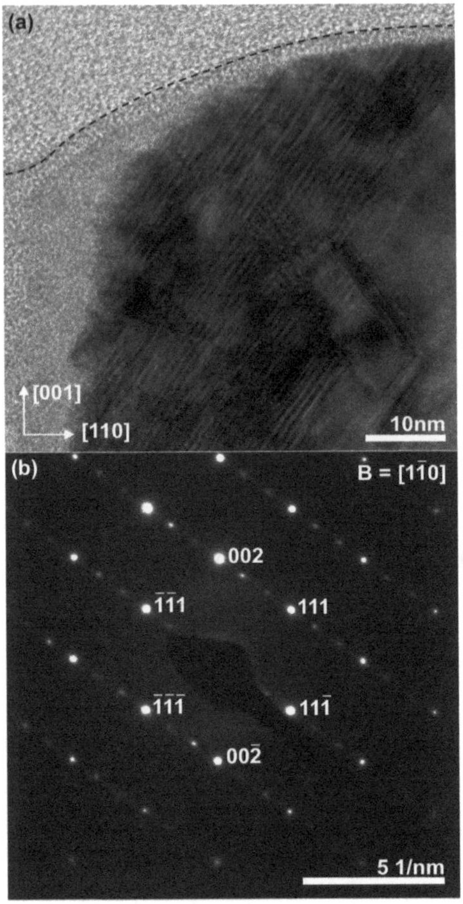

Figure 19 (A) HR-XTEM image of the defective region of (001) Si wafer Si^+-implanted at 40 keV to a dose of 1.0×10^{16} cm^{-2} in the vicinity of a SiO_x-filled trench after annealing at 600 °C for 1.0 h (wafer surface indicated by the dotted line) and (B) SAED pattern taken from region shown in (A).

is sufficiently large, there will be limited interaction between the SPEG processes occurring near the SiO_2 regions. However, when the two sides are brought very close together, in effect forming a "fin," the SPEG process is altered significantly. Furthermore, consideration of the SPEG process in fin structures is of great technological importance due to the current use of fin-type devices in current Si-based integrated circuits (Hisamoto et al., 2000). Figure 20 presents HR-XTEM images showing the SPEG process in a ~15 nm-wide fin structure As^+ and P^+-implanted at 25 and 8 keV

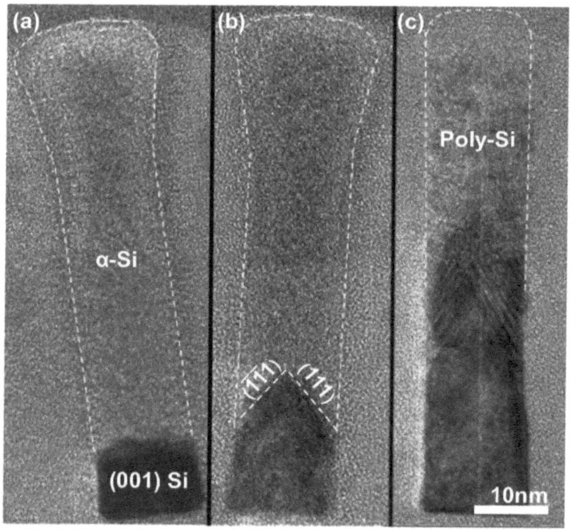

Figure 20 HR-XTEM images showing the SPEG process in a ~15 nm-wide fin structure As^+ and P^+-implanted at 25 and 8 keV to a doses of 3.0×10^{15} and 3.0×10^{15} cm^{-2}, respectively: (A) as-implanted, (B) after annealing at 600 °C for 1 min and (C) after rapid thermal annealing at 1050 °C (outline of fins indicated by the dotted line). *Adapted from Duffy et al. (2007). Copyright © 2007 by the American Institute of Physics; reprinted and modified by permission of the American Institute of Physics.*

to a doses of 3.0×10^{15} and 3.0×10^{15} cm^{-2}, respectively (Duffy et al., 2007). In the as-implanted case, shown in Fig. 20A, the top ~50 nm of the fin was rendered amorphous with an abrupt initial growth interface. After annealing at $T = 600$ °C for 1 min, shown in Fig. 20B, the growth interface faceted along $(\bar{1}\bar{1}1)$ and (111) planes with the majority of the top portion of the fin remains amorphous. However, for the case of a fin subjected to rapid thermal annealing at $T = 1050$ °C, presented in Fig. 20C, the entirety of the fin crystallized, but the top ~20 nm is now polycrystalline; the interface between the polycrystalline and the underlying single-crystal region also appears to be faceted along the same $(\bar{1}\bar{1}1)$ and (111) planes. The faceting of the growth interface along {111} planes suggests the initial growth interface was pinned due to the presence of a side oxidation layer (although the presence of such a layer was not readily observable in the HR-XTEM images shown in Fig. 20); this produces a growth interface with a sharp point with the top of the fin still amorphous. With further annealing, some SPEG occurs along the two {111} facets, but this process competes with RNG, which results in the top portion of the fin being polycrystalline.

7.3. SiO_x layer extent and growth interface pinning

From the presented work, it appears that the presence of any oxidation layer, no matter how miniscule in nature, may be enough to pin the growth interface and produce defective SPEG. This was partly confirmed by previous work where SPEG with a semi-infinite growth interface on a (001) wafer bordered on one side by an SiO_x-filled trench was studied (Burbure et al., 2007). Specifically, it was shown that the size of the triangular defective region generated during SPEG could be noticeably reduced if the SiO_x was etched from the trench and the structure annealed immediately to limit oxidation of the exposed trench sidewall.

8. DEFECTS OCCURRING AT MASK EDGES

The SPEG process has overwhelmingly been studied only for the idealized case of an initial growth interface that is blanket in nature and does not interact with anything. Unfortunately, such an idealized case is far from that typically encountered in typical Si-based device fabrication. Far more typical is the scenario where a masking layer, such as SiN_x or SiO_x, is deposited on the wafer surface and then openings in the masking layer are produced via etching. This typically produces openings that are rectilinear in nature and in the case of a (001) wafer, aligned with the in-plane $\langle 110 \rangle$ directions; this by far the most common configuration, but other orientation/direction combinations were also investigated previously (Saenger et al., 2005; Yang et al., 2006). If the wafer is now subjected to appropriate ion-implantation conditions, the surface of the exposed portions of the wafer will be rendered amorphous. However, the resulting α-Si layer is now finite in extent in one or both of the in-plane wafer directions (depending on the masking configuration) with the end points of the initial growth interface along the finite dimension terminating at the wafer surface near the edge of the masking. This produces an initial growth interface with variable crystallographic orientation near the mask edge. Depending on several factors, defects can form in the crystallized layer during SPEG near the mask edge (Cerva, 1991, 1997; Cerva and Kusters, 1989; Horiuchi et al., 1989; Saenger et al., 2007b).

8.1. Typical mask-edge defect structure

Figure 21A shows an on-axis BF-XTEM image of an α-Si layer with an initial growth interface with variable crystallographic orientation that

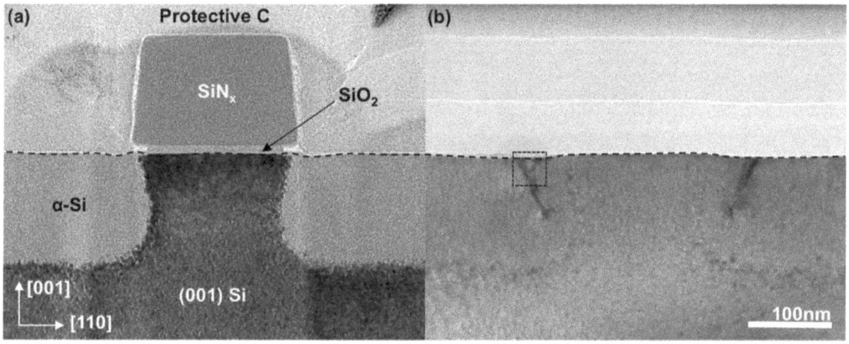

Figure 21 On-axis BF-XTEM images of (001) Si patterned with 120 nm-thick SiO_x/SiN_x lines running along $[1\bar{1}0]$ subjected to Si^+-implantation at 20 and 60 keV to doses of 1.0×10^{15} and 1.0×10^{15} cm^{-2}, respectively: (A) as-implanted and (B) annealed at 550 °C for 1.0 h (patterning removed prior to annealing) showing the generation defects in the crystallized layer near the mask edge (wafer surface indicated by the dotted line).

terminates at the wafer surface. In this case, ~10 nm of SiO_x and ~110 nm of SiN_x masking (~120 nm total thickness) were deposited on a (001) wafer. Subsequently, ~500 nm-wide line-shaped openings aligned with the in-plane $\langle 110 \rangle$ directions were etched into the masking layer to expose the underlying Si. Subsequently, Si^+-implantation at 20 and 60 keV to doses of 1.0×10^{15} and 1.0×10^{15} cm^{-2} was performed to amorphize the exposed regions of the wafer to a depth of ~120 nm. The variable orientation and surface termination near the mask edge of the initial growth interface are both clearly evident. Figure 21B shows an on-axis BF-XTEM image of the same structure after the SiO_x/SiN_x masking was removed via chemical etching and the sample subsequently annealed at T= 550 °C for 1.0 h; defects near the mask edge in the crystallized layer extending to the wafer surface are evident.

HR-XTEM imaging of the near-surface portion of the mask-edge defect indicated in Fig. 21B was performed to further study the nature of the defect as shown in Fig. 22. Disruptions to the stacking sequence along (111) and $(\bar{1}\bar{1}1)$ planes are evident, with a triangular defect-free region of material compassed by these two planes and the wafer surface. In the case of the defect-free region of material (region 2), the computed fast Fourier transform (FFT) shows the typical pattern of defect-free DC Si as viewed along the $B = [1\bar{1}0]$ zone axis with 111-type spots indicated. In the case of the stacking faults forming on the (111) plane (region 1), the stacking faults are random; this is confirmed by the FFT taken from this region, which

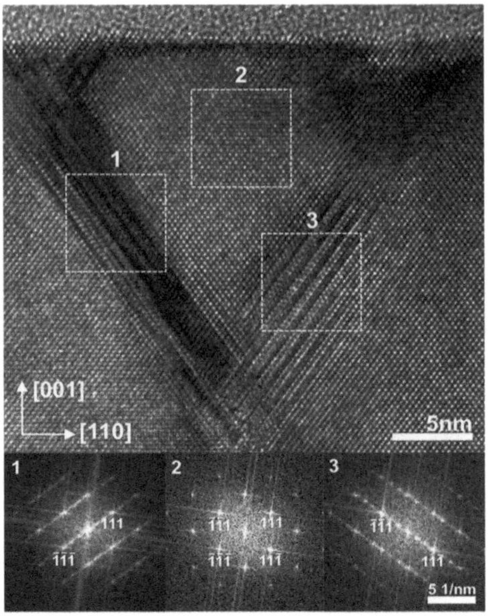

Figure 22 HR-XTEM image of the near-surface region of a mask-edge defect in (001) Si patterned with 120 nm-thick SiO_x/SiN_x lines running along $[1\bar{1}0]$ subjected to Si^+-implantation at 20 and 60 keV to doses of 1.0×10^{15} and 1.0×10^{15} cm^{-2}, respectively, and annealed at 550 °C for 1.0 h (patterning removed prior to annealing) showing different types of stacking faults. FFTs 1–3 were taken from the regions in the HR-XTEM image indicated by the boxes with the same numbers.

shows streaking of the spots along g_{111}. In comparison, for the stacking faults on the $(\bar{1}\,\bar{1}1)$ plane (region 3), the stacking faults are ordered in nature, leading to the formation of a domain of 9R polytype as viewed along a $\langle 11\bar{2}0\rangle$ zone axis. This is confirmed by the FFT taken from this region, which shows the presence of additional spots with spacing one-third $g_{\bar{1}\bar{1}1}$ running parallel to the same vector, which is g_{0003} for the 9R polytype (Miyamoto and Hirata, 1978). These results are similar in nature to SPEG on (111) wafers with a blanket α-Si layer and SPEG where the initial growth interface is pinned at the interface with an SiO_x layer.

To further analyze the mask-edge defects shown in Fig. 22, BF-PTEM imaging was performed with three different two-beam conditions as shown in Fig. 23. In the case of the image taken with g_{040}, shown in Fig. 23A, distinct lines of mask-edge defects are evident. However, when the defects are imaged with $g_{\bar{2}20}$, much of the line of mask-edge defects is no longer visible, though portions do remain in contrast, as shown in Fig. 23B. Lastly, when the defects are imaged with g_{220}, shown in Fig. 23C, the entire line of

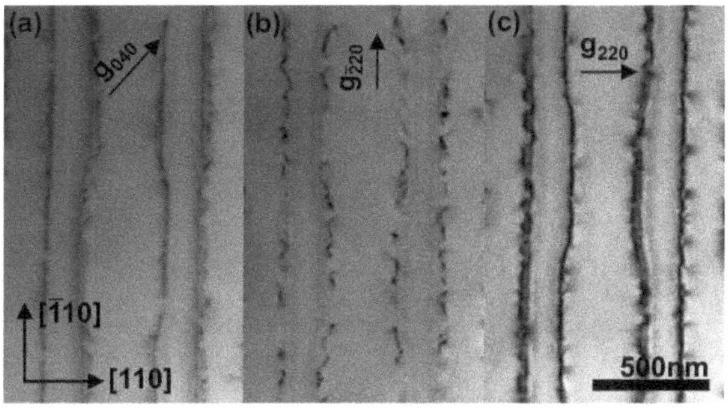

Figure 23 BF-PTEM images of mask-edge defects in (001) Si patterned with 120 nm-thick SiO_x/SiN_x lines running along $[1\bar{1}0]$ subjected to Si^+-implantation at 20 and 60 keV to doses of 1.0×10^{15} and 1.0×10^{15} cm^{-2}, respectively, and annealed at 550 °C for 1.0 h (patterning removed prior to annealing) taken with (A) g_{400}, (B) $g_{\bar{2}20}$, and (C) g_{220}.

mask-edge defects shows very strong contrast. As shown in the HR-XTEM image presented in Fig. 22, the mask-edge defects are stacking faults formed on the (111) and $(\bar{1}\bar{1}1)$ planes. Furthermore, as the generation of the stacking faults did not require the addition or removal of material from the crystal lattice, the Burgers vectors must be of the form $b = (a/6)\langle 112\rangle$, which are Shockley partial dislocations (Hirth and Lothe, 1968). For the (111) plane, the three possible Shockley partials are $b = (a/6)[11\bar{2}]$, $(a/6)[1\bar{2}1]$, and $(a/6)[\bar{2}11]$ while for the $(\bar{1}\bar{1}1)$ plane, the three Shockley partials are $b = (a/6)[112]$, $(a/6)[1\bar{2}\bar{1}]$, and $(a/6)[\bar{2}\bar{1}1]$.

If it is assumed that the distribution of Shockley partials is uniform, in the case of imaging using g_{040}, it is expected that all Shockley partials should be in strong contrast as $|g \cdot b| \geq 2/3$ in this case. However, for imaging with $g_{\bar{2}20}$, most of the line of mask-edge defects is out of contrast, which is unexpected if the distribution of Shockley partials is evenly distributed since only the $b = (a/6)[11\bar{2}]$ and $(a/6)[112]$ partials are expected to be out of contrast and $|g \cdot b| = 1$ for the other four partials. Lastly for the case of imaging with g_{220} where the line of mask-edge defects is very distinct, all Shockley partials are expected to be in contrast, but only the $(a/6)[11\bar{2}]$ and $(a/6)[112]$ partials are expected to show strong contrast as $|g \cdot b| = 1/3$ for the other four partials. These results suggest the distribution of Shockley partials is nonuniform with the $b = (a/6)[11\bar{2}]$ and $(a/6)[112]$ partials favored over the other four possibilities. It is worth noting that for these two Burgers vectors, the dot product with the elongated direction of the patterning, $[1\bar{1}0]$, is zero, but it is unclear otherwise why these two partial dislocations would be favored.

8.2. Simultaneous presence of stacking faults and perfect dislocations

The question also arises as to why some of the mask-edge defects are still in contrast when imaged with $g_{2\bar{2}0}$, as shown Fig. 23B. This can be explained by considering other reports of mask-edge defects in similar structures. Specifically, others reported mask-edge defects to consist of perfect 60° dislocations (Cerva, 1991, 1997; Cerva and Kusters, 1989; Shin et al., 2001b), which were not shown to be present in the mask-edge defects presented in Fig. 22. However, if the nature of the mask-edge defects varies spatially as one moves up and down the mask edge, this explains the lack of detection of these two defects in the specific region examined. Moreover, upon closer inspection, Fig. 23B shows contrast from the remaining portion of the mask-edge defects that is consistent with perfect dislocations, which suggests both stacking faults and perfect dislocations constitute the line of mask-edge defects.

8.3. Growth interface templating and impingement

The observation that the line of mask-edge defects consists of stacking faults suggests the growth interface templates along the two {111} planes, while the presence of perfect dislocations indicates two parts of the growth interface are impinging during SPEG (analogously to hairpin and clamshell defects). In fact, the templating of the growth interface and growth interface self-impingement were both shown to occur during the SPEG process in patterned material (Morarka et al., 2009). In terms of why this templating and self-impingement occurs, this is partly due to the differences in the velocities of the portions of the growth interface with [001] and [110] normal directions (Csepregi et al., 1978), but is also due to the growth interface being curved under the mask edge and being pinned at the wafer surface near the mask edge due to the presence of an oxidation layer. Regarding self-impingement, SPEG along [110] is slower than SPEG along [001] (Csepregi et al., 1978); however, SPEG along [110] is still sufficiently fast such that the portion of the growth interface with [110] normal impinges on the portion of the growth interface with [001] normal; this explains the origin of self-impingement and the generation of perfect dislocations under the mask edge, which is analogous to hairpin and clamshell defect formation. However, templating of the growth interface along the {111} planes is not as straightforward to explain. Partially, this is due to the growth interface with [110] normal being pinned at the surface, which is analogous to the scenario presented in Fig. 18 except for the difference in interface normal direction. However, this only explains templating along the {111} plane with normal tilted *away* from the mask edge. For the {111} plane with normal tilted *toward* the mask edge, simulations indicated the templating to be due to local variation in the growth interface velocity owing to the variable crystallographic orientation of the growth interface (Csepregi et al., 1978) and the growth interface being curved (Morarka et al., 2008, 2009). However, the exact origin of this is still not entirely clear.

8.4. Proximity of initial growth interface to wafer surface

Previous work established that when the starting growth interface is closer to the wafer surface, mask-edge defects are more difficult to form during SPEG (Rudawski et al., 2009b); the closer the initial growth interface is to the wafer surface, the portion of the growth interface with [001] normal effectively crystallizes the entire α-Si layer before the portion of the interface with [110] normal can impinge. Templating of the growth interface was also lessened when the initial growth interface was closer to the wafer surface, but the origin of this, similar to the origin of templating in general, is unclear.

8.5. Stress effects

As previously noted for the structure investigated here, the SiO_x/SiN_x patterning was removed prior to annealing to effect SPEG. This was done because such patterning is known to induce significant stresses in the substrate (Chidambaram et al., 2006; Hu, 1978, 1979, 1991) and stress is known to influence macroscopic SPEG kinetics (Aziz et al., 1991; Chaki, 1991, 1994; Nygren et al., 1985; Rudawski et al., 2008c,d). However, it is important to consider the role that applied stress may have on the SPEG process in patterned material, since the presence of stresses of substantial magnitude is ubiquitous in current Si-based device fabrication (Hu, 1991). Interestingly, extensive prior work showed that the application of different forms of mechanical stress during SPEG could alter the morphology of the evolving growth interface thus favoring or impeding mask-edge defect formation (Olson et al., 2006; Rudawski et al., 2006, 2008a, 2009b; Shin et al., 2001a,b,c). Specifically, when uniaxial tension was applied along the in-plane [110] direction during SPEG, the impingement of the two portions of the growth interface was retarded, which resulted in defect-free growth (Olson et al., 2006; Rudawski et al., 2006, 2008a, 2009b). In contrast, when uniaxial compression was applied along the in-plane [110] direction during

SPEG, the impingement of the two portions of the growth interface was enhanced, which resulted in defective growth (Olson et al., 2006; Rudawski et al., 2006, 2008a, 2009b).

8.6. Impurity effects

The data shown here used material free of impurity atoms to avoid complications, since impurities are known to profoundly alter the SPEG kinetics in wafers with blanket α-Si layers (Csepregi et al., 1977; Kennedy et al., 1977; Lietoila et al., 1982; Suni et al., 1982; Williams and Elliman, 1983; Williams and Short, 1983). However, it is important to consider the role of impurities (particularly electrically active ones) on mask-edge defect formation during SPEG, since impurity implantation is a critical part of Si-based device fabrication. In fact, previous work did investigate the role of an electrically active impurity concentration profile introduced into a masked Si wafer via ion implantation on the SPEG process (Rudawski et al., 2008a). The impurity concentration along the portion of the growth interface with [110] normal will then be variable according to the ion-implantation conditions, with peak concentration occurring at the projected range of the ion (Ziegler et al., 2010). This will result in the growth interface with [110] normal at the projected range of the ions moving faster than the surrounding portions during SPEG, as per the reported enhancement to SPEG kinetics in the presence of electrically active impurities (Csepregi et al., 1977; Lietoila et al., 1982; Suni et al., 1982; Williams and Elliman, 1983; Williams and Short, 1983). Obviously, this enhancement to the velocity of the growth interface with [110] normal favors impingement with the portion of the growth interface with [001] normal, and thus favors mask-edge defect formation.

9. SUMMARY

Understanding the origins of defects produced during the SPEG process of Si is of great interest due to the importance of SPEG in Si-based device fabrication and the influence SPEG-related defects can have on resulting device performance. Ideally, the SPEG process is perfect in that the crystallized layer (or region) is a perfect single crystal. However, it was shown here that for many conditions that the SPEG process is often not perfect. Specifically, initial growth interface roughness, interaction of independent growth interfaces, and substrate orientation are influential in the quality of the SPEG process. The addition of impurities of sufficient concentration also leads to defective SPEG due to the stresses generated in the crystallizing layer resulting from substitution of the impurities into the Si lattice. Interestingly, it was also shown that application of external stress during SPEG could also introduce defects into the crystallizing layer. Interaction of the growth interface with a SiO_x region also can result in defective SPEG via pinning of the growth interface at the point of contact resulting in templating of the growth interface along crystallographic orientations favorable for defective SPEG. Lastly, crystallization of layers in masked wafers often leads to the introduction of defects near the edge of the masking due to self-impingement of the growth interface as well as templating of the growth interface along crystallographic orientations favorable for defective SPEG. Furthermore, the surface proximity of the initial growth interface, the presence of stress, and the presence of impurities can all influence the generation of mask-edge defects during SPEG.

ACKNOWLEDGMENTS

The Major Analytical Instrumentation Center at the University of Florida is acknowledged for use of the focused ion beam and transmission electron microscope facilities used to produce the vast majority of the data presented in this work. R. G. Elliman (Australian National University) is also acknowledged for thoughtful discussions and for providing the ion-implantation services used to produce the specimens for studying SPEG in (111) wafers.

REFERENCES

Ashburn, P., Bull, C., Nicholas, K.H., Booker, G.R., 1977. Effects of dislocations in silicon transistors with implanted bases. Solid State Electron. 20, 731–740.

Aziz, M.J., Sabin, P.C., Lu, G.Q., 1991. The activation strain tensor—nonhydrostatic stress effects on crystal growth kinetics. Phys. Rev. B 44, 9812–9816.

Barvosa-Carter, W., Aziz, M.J., 2001. Time-resolved measurements of stress effects on solid-phase epitaxy of intrinsic and doped Si. Appl. Phys. Lett. 79, 356–358.

Barvosa-Carter, W., Aziz, M.J., Gray, L.J., Kaplan, T., 1998. Kinetically driven growth instability in stressed solids. Phys. Rev. Lett. 81, 1445–1448.

Barvosa-Carter, W., Aziz, M.J., Phan, A.V., Kaplan, T., Gray, L.J., 2004. Interfacial roughening during solid phase epitaxy: interaction of dopant, stress, and anisotropy effects. J. Appl. Phys. 96, 5462–5468.

Bonse, J., Brzezinka, K.W., Meixner, A.J., 2004. Modifying single-crystalline silicon by femtosecond laser pulses: an analysis by micro Raman spectroscopy, scanning laser microscopy and atomic force microscopy. Appl. Surf. Sci. 221, 215–230.

Bousetta, A., Vandenberg, J.A., Valizadeh, R., Armour, D.G., Zalm, P.C., 1991. Ultra low-energy (100–2000 eV) boron implantation into crystalline and silicon-preamorphized silicon. Nucl. Instrum. Methods Phys. Res. B 55, 565–568.

Bruinsma, R., Zangwill, A., 1987. Morphological transitions in solid epitaxial overlayers. Europhys. Lett. 4, 729–735.

Bull, C., Ashburn, P., Booker, G.R., Nicholas, K.H., 1979. Effects of dislocations in silicon transistors with implanted emitters. Solid State Electron. 22, 95–104.

Burbure, N., Rudawski, N.G., Jones, K.S., 2007. Effect of oxide on trench edge defect formation in ion-implanted silicon. Electrochem. Solid-State Lett. 10, H184–H185.
Cerva, H., 1991. High-resolution electron microscopy of diamond hexagonal silicon in low-pressure chemical vapor-deposited polycrystalline silicon. J. Mater. Res. 6, 2324–2336.
Cerva, H., 1997. Defects below mask edges in silicon induced by amorphizing implantations. Diffus. Defect Data A 148, 103–121.
Cerva, H., Kusters, K.H., 1989. Defect formation in silicon at mask edge during crystallization of an amorphous implantation layer. J. Appl. Phys. 66, 4723–4728.
Chaki, T.K., 1991. Hydrostatic pressure-enhanced solid-phase epitaxy. Phil. Mag. Lett. 63, 303–307.
Chaki, T.K., 1994. Solid-phase epitaxy—effects of irradiation, dopant, and pressure. Phys. Status Solidi A 142, 153–166.
Chidambaram, P.R., Bowen, C., Chakravarthi, S., Machala, C., Wise, R., 2006. Fundamentals of silicon material properties for successful exploitation of strain engineering in modern CMOS manufacturing. IEEE Trans. Electr. Dev. 53, 944–964.
Christel, L.A., Gibbons, J.F., Sigmon, T.W., 1981. Displacement criterion for amorphization of silicon during ion-implantation. J. Appl. Phys. 52, 7143–7146.
Costache, F., Kouteva-Arguirova, S., Reif, J., 2004. Sub-damage-threshold femtosecond laser ablation from crystalline Si: surface nanostructures and phase transformation. Appl. Phys. A 79, 1429–1432.
Cristiano, F., Nejim, A., Suprun-Belevich, Y., Claverie, A., Hemment, P.L.F., 1999. Formation of extended defects and strain relaxation in ion beam synthesised SiGe alloys. Nucl. Instrum. Methods Phys. Res. B 147, 35–42.
Csepregi, L., Mayer, J.W., Sigmon, T.W., 1976. Regrowth behavior of ion-implanted amorphous layers on (111) silicon Appl. Phys. Lett. 29, 92–93.
Csepregi, L., Kennedy, E.F., Gallagher, T.J., Mayer, J.W., Sigmon, T.W., 1977. Reordering of amorphous layers of Si implanted with P-31, As-75, and B-11 ions. J. Appl. Phys. 48, 4234–4240.
Csepregi, L., Kennedy, E.F., Mayer, J.W., Sigmon, T.W., 1978. Substrate-orientation dependence of epitaxial regrowth rate from Si-implanted amorphous Si. J. Appl. Phys. 49, 3906–3911.
Csepregi, L., Gyulai, J., Lau, S.S., 1996. The early history of solid phase epitaxial growth. Mater. Chem. Phys. 46, 178–180.
Duffy, R., Van Dal, M.J.H., Pawlak, B.J., Kaiser, M., Weemaes, R.G.R., Degroote, B., et al., 2007. Solid phase epitaxy versus random nucleation and growth in sub-20 nm wide fin field-effect transistors. Appl. Phys. Lett. 90, 241912.
Eaglesham, D.J., Cerullo, M., 1990. Dislocation-free stranski-krastanow growth of Ge on Si (100). Phys. Rev. Lett. 64, 1943–1946.
Elliman, R.G., Wong, W.C., 1996. Kinetic roughening and smoothing of the crystalline, Äìamorphous interface during solid phase epitaxial crystallization of GeSi alloy layers. Appl. Phys. Lett. 69, 2677–2679.
Elliman, R.G., Williams, J.S., Brown, W.L., Leiberich, A., Maher, D.M., Knoell, R.V., 1987. Ion-beam-induced crystallization and amorphization of silicon. Nucl. Instrum. Methods Phys. Res. B 19–20, 435–442.
Finetti, M., Galloni, R., Mazzone, A.M., 1979. Influence of impurities and crystalline defects on electron-mobility in heavily doped silicon. J. Appl. Phys. 50, 1381–1385.
Fontcuberta i Morral, A., Arbiol, J., Prades, J.D., Cirera, A., Morante, J.R., 2007. Synthesis of silicon nanowires with wurtzite crystalline structure by using standard chemical vapor deposition. Adv. Mater. 19, 1347–1351.
Frank, F.C., Vandermerwe, J.H., 1949. One-dimensional dislocations. I. static theory. Proc. R. Soc. Lond. A 198, 205–216.

Gibbons, J.F., 1972. Ion implantation in semiconductors. II. Damage production and annealing. Proc. Inst. Electr. Eng. 60, 1062–1096.
Glowinski, L.D., Tu, K.N., Ho, P.S., 1976. Direct observations of defects in implanted and post-annealed silicon wafers. Appl. Phys. Lett. 28, 312–313.
Gu, L., Yu, Y., Sigle, W., Usami, N., Tsukimoto, S., Maier, J., et al., 2010. Direct bandgap measurements in a three-dimensionally macroporous silicon 9R polytype using monochromated transmission electron microscope. Appl. Phys. Lett. 97, 213102.
Haynes, T.E., Antonell, M.J., Lee, C.A., Jones, K.S., 1995. Composition dependence of solid-phase epitaxy in silicon-germanium alloys—experiment and theory. Phys. Rev. B 51, 7762–7771.
Hirth, J.P., Lothe, J., 1968. Theory of Dislocations. McGraw-Hill, New York, NY.
Hisamoto, D., Lee, W.C., Kedzierski, J., Takeuchi, H., Asano, K., Kuo, C., et al., 2000. FinFET—a self-aligned double-gate MOSFET scalable to 20 nm. IEEE Trans. Electr. Dev. 47, 2320–2325.
Horiuchi, M., Tamura, M., Aoki, S., 1989. Three-dimensional solid-phase epitaxial regrowth from As^{+}-implanted Si. J. Appl. Phys. 65, 2238–2242.
Hu, S.M., 1978. Film-edge-induced stress in silicon substrates. Appl. Phys. Lett. 32, 5–7.
Hu, S.M., 1979. Film-edge-induced stress in substrates. J. Appl. Phys. 50, 4661–4666.
Hu, S.M., 1991. Stress-related problems in silicon technology. J. Appl. Phys. 70, R53–R80.
Izawa, Y., Setsuhara, Y., Hashida, M., Sasaki, R., Nagai, H., Yoshida, M., 2007. Ultrathin amorphous Si layer formation by femtosecond laser pulse irradiation. Appl. Phys. Lett. 90, 044107.
Jia, J., Li, M., Thompson, C.V., 2004. Amorphization of silicon by femtosecond laser pulses. Appl. Phys. Lett. 84, 3205–3207.
Jones, K.S., Prussin, S., Weber, E.R., 1988. A systematic analysis of defects in ion-implanted silicon. Appl. Phys. A 45, 1–34.
Kennedy, E.F., Csepregi, L., Mayer, J.W., Sigmon, T.W., 1977. Influence of O^{16}, C^{12}, N^{14}, and noble gases on crystallization of amorphous Si layers. J. Appl. Phys. 48, 4241–4246.
Kiani, A., Venkatakrishnan, K., Tan, B., 2009. Micro/nano scale amorphization of silicon by femtosecond laser irradiation. Optics Exp. 17, 16518–16526.
Kiani, A., Venkatakrishnan, K., Tan, B., 2010. Direct laser writing of amorphous silicon on Si-substrate induced by high repetition femtosecond pulses. J. Appl. Phys. 108, 074907.
Kim, S.D., Park, C.M., Woo, J.C.S., 2002. Advanced source/drain engineering for box-shaped ultrashallow junction formation using laser annealing and pre-amorphization implantation in sub-100-nm SOICMOS. IEEE Trans. Electr. Dev. 49, 1748–1754.
Kringhoj, P., Elliman, R.G., 1994. Solid-phase epitaxial crystallization of strain-relaxed $Si_{1-x}Ge_x$ alloy layers. Phys. Rev. Lett. 73, 858–861.
Larson, J.M., Snyder, J.P., 2006. Overview and status of metal S/D Schottky-barrier MOSFET technology. IEEE Trans. Electr. Dev. 53, 1048–1058.
Lee, C., Haynes, T.E., Jones, K.S., 1993. Kinetics of solid-phase epitaxial regrrowth in amorphized $Si_{0.88}Ge_{0.12}$ measured by time-resolved reflectivity. Appl. Phys. Lett. 62, 501–503.
Lietoila, A., Wakita, A., Sigmon, T.W., Gibbons, J.F., 1982. Epitaxial regrowth of intrinsic, P^{31}-doped and compensated ($P^{31}+B^{11}$ doped) amorphous Si. J. Appl. Phys. 53, 4399–4405.
Liu, X., Wang, D., 2009. Kinetically-induced hexagonality in chemically grown silicon nanowires. Nano Res. 2, 575–582.
Lopez, F.J., Hemesath, E.R., Lauhon, L.J., 2009. Ordered stacking fault arrays in silicon nanowires. Nano Lett. 9, 2774–2779.

Lopez, F.J., Givan, U., Connell, J.G., Lauhon, L.J., 2011. Silicon nanowire polytypes: indentification by Raman spectroscopy, generation mechanism, and misfit strain in heterostructures. ACS Nano 5, 8958–8966.

Lu, G.Q., Nygren, E., Aziz, M.J., Turnbull, D., White, C.W., 1989. Interferometric measurement of the pressure-enhanced crystallization rate of amorphous Si. Appl. Phys. Lett. 54, 2583–2585.

Martin-Bragado, I., 2012. Importance of twin defect formation created by solid-phase epitaxial growth: an atomistic study. Scr. Mater. 66, 186–189.

Martin-Bragado, I., Sklenard, B., 2012. Understanding Si (111) solid phase epitaxial regrowth using Monte Carlo modeling: bi-modal growth, defect formation, and interface topology. J. Appl. Phys. 112, 024327.

Masters, B.J., Fairfield, J.M., Crowder, B.L., 1970. Radiochemical determination of damage profiles in silicon. Radiat. Eff. 6, 57–61.

Mayer, J.W., Eriksson, L., Picraux, S.T., Davies, J.A., 1968. Ion implantation of silicon and germanium at room temperature. Analysis by means of 1.0 MeV helium ion scattering. Can. J. Phys. 46, 663–673.

McLarnan, T.J., 1981. The numbers of polytypes in close-packings and related structures. Z. Kristallogr. 155, 269–291.

Miyamoto, Y., Hirata, M., 1978. Polytypism and amorphousness in silicon whiskers. J. Phys. Soc. Jpn. 44, 181–190.

Morarka, S., Rudawski, N.G., Law, M.E., 2008. Level set modeling of the orientation dependence of solid phase epitaxial regrowth. J. Vac. Sci. Technol. B 26, 357–361.

Morarka, S., Rudawski, N.G., Law, M.E., Jones, K.S., Elliman, R.G., 2009. Modeling two-dimensional solid-phase epitaxial regrowth using level set methods. J. Appl. Phys. 105, 053701.

Morarka, S., Rudawski, N.G., Law, M.E., Jones, K.S., Elliman, R.G., 2010. Effect of n- and p-type dopants on patterned amorphous regrowth. J. Vac. Sci. Technol. B 28, C1F1–C1F5.

Morarka, S., Jin, S., Rudawski, N.G., Jones, K.S., Law, M.E., Elliman, R.G., 2011. Interface stability in stressed solid-phase epitaxial growth. J. Vac. Sci. Technol. B 29, 041210.

Morehead, F.F., Crowder, B.L., 1970. A model for the formation of amorphous Si by ion bombardment. Radiat. Eff. 6, 27–32.

Morehead, F.F., Crowder, B.L., Title, R.S., 1972. Formation of amorphous silicon by ion bombardment as a function of ion, temperature, and dose. J. Appl. Phys. 43, 1112–1118.

Myers, D.R., Wilson, R.G., 1980. Alignment effects on implantation profiles in silicon. Radiat. Eff. Defects Solids. 47, 91–94.

Myers, D.R., Wilson, R.G., Comas, J., 1979. Considerations of ion channeling for semiconductor microstructure fabrication. J. Vac. Sci. Technol. 16, 1893–1896.

Myers, E., Rozgonyi, G.A., Sadana, D.K., Maszara, W., Wortman, J.J., Narayan, J., 1985. Ge^+ preamorphization of Si: effects of dose and very low temperature thermal treatment on extended defect formation during subsequent SPE. In: MRS Proceedings, Vol. 52. Materials Research Society, Warrendale, PA.

Narayan, J., Holland, O.W., 1982. Solid-phase epitaxial growth in ion-implanted silicon. Phys. Status Solidi A 73, 225–236.

Nishi, H., Inada, T., Sakurai, T., Kaneda, T., Hisatsugu, T., Furuya, T., 1978. Uniform doping of channeled-ion implantation. J. Appl. Phys. 49, 608–613.

Nygren, E., Aziz, M.J., Turnbull, D., Poate, J.M., Jacobson, D.C., Hull, R., 1985. Effect of pressure on the solid-phase epitaxial regrowth rate of Si. Appl. Phys. Lett. 47, 232–233.

Olson, G.L., Roth, J.A., 1988. Kinetics of solid phase crystallization in amorphous silicon. Mater. Sci. Rep. 3, 1–77.

Olson, C.R., Kuryliw, E., Jones, B.E., Jones, K.S., 2006. Effect of stress on the evolution of mask-edge defects in ion-implanted silicon. J. Vac. Sci. Technol. B 24, 446–449.

Ozturk, M.C., Wortman, J.J., Osburn, C.M., Ajmera, A., Rozgonyi, G.A., Frey, E., et al., 1988. Optimization of the germanium preamorphization conditions for shallow-junction formation. IEEE Trans. Electr. Dev. 35, 659–668.

Paine, D.C., Howard, D.J., Stoffel, N.G., Horton, J.A., 1990. The growth of strained $Si_{1-x}Ge_x$ alloys on (001) silicon using solid-phase epitaxy. J. Mater. Res. 5, 1023–1031.

Paine, D.C., Howard, D.J., Stoffel, N.G., 1991. Strain relief in compositionally graded $Si_{1-x}Ge_x$ formed by high-dose ion-implantation. J. Electron. Mater. 20, 735–746.

Peteves, S.D., Abbaschian, R., 1991. Growth kinetics of solid–liquid Ga interfaces. 2. Theoretical. Metall. Trans. A 22, 1271–1286.

Phan, A.V., Kaplan, T., Gray, L.J., Adalsteinsson, D., Sethian, J.A., Barvosa-Carter, W., et al., 2001. Modelling a growth instability in a stressed solid. Modell. Simul. Mater. Sci. Eng. 9, 309–325.

Privitera, V., Napolitani, E., Priolo, F., Moffatt, S., La Magna, A., Mannino, G., et al., 1999. Atomic transport properties and electrical activation of ultra-low energy implanted boron in crystalline silicon. Mater. Sci. Semicond. Process. 2, 35–44.

Raffy, C., Furthmuller, J., Bechstedt, F., 2002. Properties of hexagonal polytypes of group-IV elements from first-principles calculations. Phys. Rev. B 66, 075201.

Rechtin, M.D., Pronko, P.P., Foti, G., Csepregi, L., Kennedy, E.F., Mayer, J.W., 1978. Electron microscopy study of defect structures in recrystallized amorphous layers of self-ion-irradiated (111) silicon. Philos. Mag. A 37, 605–620.

Robinson, M.T., Oen, O.S., 1963. The channeling of energetic atoms in crystal lattices. Appl. Phys. Lett. 2, 30–32.

Roth, J.A., Olson, G.L., Jacobson, D.C., Poate, J.M., 1990. Kinetics of solid-phase epitaxy in thick amorphous Si layers formed by MeV ion-implantation. Appl. Phys. Lett. 57, 1340–1342.

Rudawski, N.G., Jones, K.S., 2009. Atomistic considerations of stressed epitaxial growth from the solid phase. Scr. Mater. 61, 327–330.

Rudawski, N.G., Siebein, K.N., Jones, K.S., 2006. Effect of uniaxial stress on solid phase epitaxy in patterned Si wafers. Appl. Phys. Lett. 89, 082107.

Rudawski, N.G., Jones, K.S., Gwilliam, R., 2007. Solid phase epitaxy in uniaxially stressed (001) Si. Appl. Phys. Lett. 91, 172103.

Rudawski, N.G., Jones, K.S., Elliman, R.G., 2008a. Influence of As on the formation of mask-edge defects during stressed solid phase epitaxy in patterned Si wafers. J. Vac. Sci. Technol. B 26, 435–438.

Rudawski, N.G., Jones, K.S., Gwilliam, R., 2008b. Dopant-stress synergy in Si solid-phase epitaxy. Appl. Phys. Lett. 92, 232110.

Rudawski, N.G., Jones, K.S., Gwilliam, R., 2008c. Kinetics and morphological instabilities of stressed solid–solid phase transformations. Phys. Rev. Lett. 100, 165501.

Rudawski, N.G., Jones, K.S., Gwilliam, R., 2008d. Stressed solid-phase epitaxial growth of ion-implanted amorphous silicon. Mater. Sci. Eng. R-Rep. 61, 40–58.

Rudawski, N.G., Jones, K.S., Gwilliam, R., 2009a. Stressed solid-phase epitaxial growth of (011) Si. J. Mater. Res. 24, 305–309.

Rudawski, N.G., Jones, K.S., Morarka, S., Law, M.E., Elliman, R.G., 2009b. Stressed multidirectional solid-phase epitaxial growth of Si. J. Appl. Phys. 105, 081101.

Rudawski, N.G., Whidden, L.R., Craciun, V., Jones, K.S., 2009c. Amorphization and solid-phase epitaxial growth of C-cluster ion-implanted Si. J. Electron. Mater. 38, 1926–1930.

Sadana, D.K., Fletcher, J., Booker, G.R., 1977. Structural and electrical profiles for double damage layers in ion-implanted silicon. Electron. Lett. 13, 632–633.

Sadana, D.K., Washburn, J., Booker, G.R., 1982. Recrystallization of buried amorphous layers and associated electrical effects in P^+-implanted Si. Philos. Mag. B 46, 611–633.

Sadana, D.K., Maszara, W., Wortmann, J.J., Rozgonyi, G.A., Chu, W.K., 1984. Germanum implantation into silicon—an alternate pre-amorphization rapid thermal annealing procedure for shallow junction formation. J. Electrochem. Soc. 131, 943–945.
Saenger, K.L., de Souza, J.P., Fogel, K.E., Ott, J.A., Reznicek, A., Sung, C.Y., et al., 2005. Amorphization/templated recrystallization method for changing the orientation of single-crystal silicon: an alternative approach to hybrid orientation substrates. Appl. Phys. Lett. 87, 221911.
Saenger, K.L., de Souza, J.P., Fogel, K.E., Ott, J.A., Sung, C.Y., Sadana, D.K., et al., 2007a. A study of trench-edge defect formation in (001) and (011) silicon recrystallized by solid phase epitaxy. J. Appl. Phys. 101, 024908.
Saenger, K.L., Fogel, K.E., Ott, J.A., Sung, C.Y., Sadana, D.K., Yin, H., 2007b. Mask-edge defects in hybrid orientation direct-Si-bonded substrates recrystallized by solid phase epitaxy after patterned amorphization. J. Appl. Phys. 101, 084912.
Sage, J.F., Barvosa-Carter, W., Aziz, M.J., 2000. Morphological instability of growth fronts due to stress-induced mobility variations. Appl. Phys. Lett. 77, 516–518.
Sands, T., Washburn, J., Myers, E., Sadana, D.K., 1985. On the origins of structural defects in BF_2^+-implanted and rapid-thermally-annealed silicon—conditions for defect-free regrowth. Nucl. Instrum. Methods Phys. Res. B 7–8, 337–341.
Seong-Dong, K., Cheol-Min, P., Woo, J.C.S., 2000. Advanced model and analysis for series resistance in sub-100 nm CMOS including poly depletion and overlap doping gradient effect. In: 2000 International Electron Devices Meeting Technical Digest, pp. 723–726.
Shin, Y.G., Lee, J.Y., Park, M.H., Kang, H.K., 2001a. The effect of stress on solid-phase epitaxial regrowth in As^+-implanted two-dimensional amorphized Si. Jap. J. Appl. Phys. 40, 6192–6197.
Shin, Y.G., Lee, J.Y., Park, M.H., Kang, H.K., 2001b. Effects of stress on solid-phase epitaxial regrowth and corner defect generation in As^+-implanted, two-dimensional amorphized Si. J. Cryst. Growth 231, 107–114.
Shin, Y.G., Lee, J.Y., Park, M.H., Kang, H.K., 2001c. Effects of stress on the microstructure of the corner defect in As^+-implanted, two-dimensional amorphized Si. J. Cryst. Growth 233, 673–680.
Sklenard, B., Barbe, J.-C., Batude, P., Rivallin, P., Tavernier, C., Cristoloveanu, S., et al., 2013. An atomistic investigation of the impact of in-plane uniaxial stress during solid phase epitaxial regrowth. Appl. Phys. Lett. 102, 151907.
Strane, J.W., Lee, S.R., Stein, H.J., Picraux, S.T., Watanabe, J.K., Mayer, J.W., 1996. Carbon incorporation into Si at high concentrations by ion implantation and solid phase epitaxy. J. Appl. Phys. 79, 637–646.
Suni, I., Goltz, G., Grimaldi, M.G., Nicolet, M.A., Lau, S.S., 1982. Compensating impurity effect on epitaxial regrowth rate of amorphized Si. Appl. Phys. Lett. 40, 269–271.
Sze, S.M., Gibbons, G., 1966. Avalanche breakdown voltages of abrupt and linearly graded p-n junctions in Ge, Si, GaAs, and GaP. Appl. Phys. Lett. 8, 111–113.
Taur, Y., Wann, C.H., Frank, D.J., 1998. 25 nm CMOS design considerations. In: 1998 International Electron Devices Meeting Technical Digest, pp. 789–792.
Tsaur, B.Y., Anderson, C.H., 1983. Dual ion-implantation technique for formation of shallow p^+-n junctions in silicon. J. Appl. Phys. 54, 6336–6339.
Vegard, L., 1921. The constitution of the mixed crystals and the filling of space of the atoms. Z. Phys. 5, 17–26.
Venables, J., 2000. Introduction to surface and thin film processes. Cambridge University Press, Cambridge, UK.
Volmer, M., Marder, M., 1931. The theory of the linear crystalisation rate of under cooled melts and under cooled firm modifications. Z. Phys. Chem. Abt. A 154, 97–112.
Volmer, M., Weber, A., 1926. Germ-formation in oversaturated figures. Z. Phys. Chem. Stoechiom. Verwandtschafts. 119, 277–301.

Vonborany, J., Kogler, R., 1993. The influence of preamorphization on the properties of shallow p^+-n junctions in silicon radiation detectors. Nucl. Instrum. Methods Phys. Res. A 326, 42–46.
Washburn, J., Murty, C.S., Sadana, D., Byrne, P., Gronsky, R., Cheung, N., et al., 1983. The crystalline to amorphous transformation in silicon. Nucl. Instrum. Methods Phys. Res. 209, 345–350.
Wen, B., Zhao, J., Bucknum, M.J., Yao, P., Li, T., 2008. First-principles studies of diamond polytypes. Diam. Relat. Mater. 17, 356–364.
Williams, J.S., Elliman, R.G., 1983. Role electronic processes in epitaxial recrystallization of amorphous semiconductors. Phys. Rev. Lett. 51, 1069–1072.
Williams, J.S., Short, K.T., 1983. The influence of electronic processes on the epitaxial regrowth rate in ion-implanted silicon. Nucl. Instrum. Methods Phys. Res. 209, 767–771.
Wilson, H.A., 1900. On the velocity of solidification and viscosity of super-cooled liquids. Philos. Mag. 50, 238–250.
Witvrouw, A., Spaepen, F., 1993. Viscosity and elastic constants of amorphous Si and Ge. J. Appl. Phys. 74, 7154–7161.
Wortman, J.J., Evans, R.A., 1965. Young's modulus, shear modulus, and Poisson's ratio in silicon and germanium. J. Appl. Phys. 36, 153–156.
Yang, M., Chan, V.W.C., Chan, K.K., Shi, L., Fried, D.M., Stathis, J.H., et al., 2006. Hybrid-orientation technology (HOT): opportunities and challenges. IEEE Trans. Electr. Dev. 53, 965–978.
Ziegler, J.F., 2004. SRIM-2003. Nucl. Instrum. Methods Phys. Res. B 219, 1027–1036.
Ziegler, J.F., Ziegler, M.D., Biersack, J.P., 2010. SRIM—the stopping and range of ions in matter (2010). Nucl. Instrum. Methods Phys. Res. B 268, 1818–1823.

CHAPTER FIVE

Nanoindentation of Silicon and Germanium

Mangalampalli S.R.N. Kiran*, Bianca Haberl*,†, Jodie E. Bradby*, James S. Williams*,1

*Department of Electronic Materials Engineering, Research School of Physics and Engineering, Australian National University, Canberra, Australian Capital Territory, Australia
†Chemical and Engineering Materials Division, Oak Ridge National Laboratory, Oak Ridge, Tennessee, USA
1Corresponding author: e-mail address: Jim.Williams@anu.edu.au

Contents

1. INTRODUCTION AND BACKGROUND

This chapter focuses on the deformation behavior of Si and Ge under nanoindentation loading. In Si, deformation by indentation (and the measured hardness of the material) at room temperature is almost entirely

Semiconductors and Semimetals, Volume 91
ISSN 0080-8784
http://dx.doi.org/10.1016/bs.semsem.2014.12.002

determined by a phase transformation from the diamond cubic (dc-Si) phase or an amorphous Si film to a denser metallic (β-Sn) phase at a pressure of around 11 GPa (Clarke et al., 1988; Gerk and Tabor, 1978; Gilman, 1993; Gridneva et al., 1972; Haberl et al., 2012). On unloading, the β-Sn phase does not transform back to dc-Si but to the metastable r8 and bc8 Si phases (Bradby et al., 2000a, 2001; Domnich and Gogotsi, 2002; Kailer et al., 1999). The r8 phase is predicted to be a narrow bandgap semiconductor (Malone et al., 2008a,b) and bc8 is experimentally found to be a semimetal (Besson et al., 1987). Hence, r8 might be expected to exhibit improved absorption across the solar spectrum (Malone et al., 2008a) and bc8 (in nanocrystalline form) should to give rise to "multiple exciton generation" and thus to efficient next-generation photovoltaics (Wippermann et al., 2013). In dc-Ge, the deformation mode at room temperature depends on the specific indentation conditions, with both plastic deformation by propagation of lattice defects and phase transformation possible (Bradby et al., 2002; Jang et al., 2005a; Kailer et al., 1999; Oliver et al., 2009a). For amorphous Si and Ge films (a-Si and a-Ge), phase transformations can also occur but there is significant plastic deformation of the amorphous phase under some conditions (Follstaedt et al., 2004; Haberl et al., 2004, 2006; Oliver et al., 2008, 2009b).

In reviewing the indentation behavior of Si and Ge materials, it is first instructive to give a brief overview of the behavior under loading in opposed anvil high-pressure devices, most commonly a diamond anvil cell (DAC). Figure 1 shows a schematic diagram (on the left) of the pressure-induced Si phases that have been observed under such DAC pressure. When pressure is applied to dc-Si in a DAC, it undergoes a transformation at around 11 GPa to the metallic β-Sn phase (Jamieson, 1963), which is about 22% more dense than dc-Si (Mujica et al., 2003). Further loading causes a series of transformations to increasingly dense metallic phases up to and beyond 80 GPa (Mujica et al., 2003). On pressure release, the reverse transformation sequence through the metallic phases occurs until (β-Sn)-Si is reached. Thereafter, (β-Sn)-Si does not directly transform to dc-Si but through the rhomohedral r8-Si phase (Piltz et al., 1995) to a body-centered cubic bc8-Si phase which is metastable at ambient pressure (Wentorf and Kasper, 1963). Whereas the kinetic barriers to phase transformation between the metallic phases are low and their pressure-induced transitions are reversible (Malone and Cohen, 2012a; Needs and Mujica, 1995), there is a considerable kinetic barrier to the formation of dc-Si from (β-Sn)-Si, and hence, alternate metastable r8 and bc8 phases are favored. Indeed, when metastable

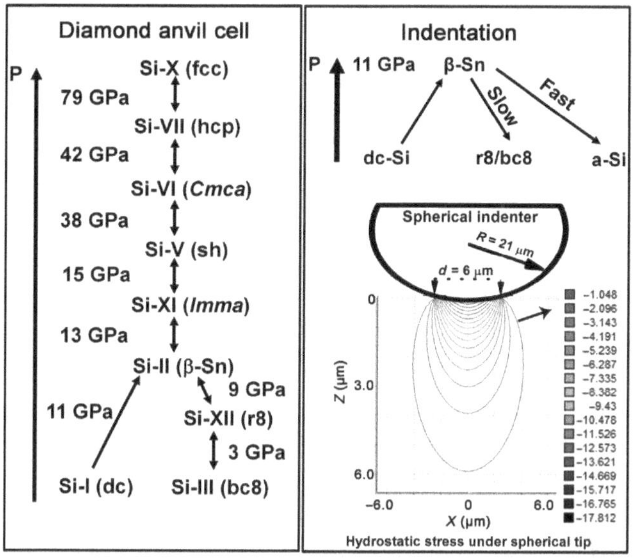

Figure 1 Possible phase transformation pathways in dc-Si under pressure with both the diamond anvil cell and nanoindentation. The right bottom figure (Wong, 2013) shows the hydrostatic stress distribution in Si under an ~21-μm radius spherical microindenter at a load of 450 mN. Stresses are higher directly under the contact and decrease with depth.

bc8-Si is annealed, it transforms to the hexagonal-diamond (hd) phase at 200 °C (Wentorf and Kasper, 1963), which in turn transforms to the stable dc-Si phase at 750 °C (Brazhkin et al., 1992). The r8 phase instead transforms directly to dc-Si at 255 °C without any intermediate phases (Haberl et al., 2015).

The notation of the high-pressure Si phases in Fig. 1 needs some comment. Historically, the roman numerals for Si phases were designated in the sequence of their discovery. Since this labeling can often lead to confusion (particularly since the corresponding Ge notation is not consistent with the Si notation) in the remainder of this chapter, we use the other commonly employed notation based on the crystallographic structure referred to in brackets on the left side of Fig. 1.

The right-hand portion of Fig. 1 summarizes the phase transformation behavior under indentation that was briefly referred to above. Once the (β-Sn)-Si phase forms, it is not possible via point loading to induce further transformations to more dense phases since the soft, unconfined (β-Sn)-Si phase prohibits the required higher pressures from being achieved. In contrast, the setup of the DAC always provides the confinement that allows increasing pressure to be applied and hence the further nucleation of more dense metallic phases. For indentation, as we will show in subsequent sections, the unloading rate can lead to transformation from (β-Sn)-Si to either a mixed r8/bc8-Si phase under sufficiently slow unloading or an a-Si end phase, when the unloading rate is too fast to allow nucleation of r8/bc8 crystalline phases. Hence, there is a significant difference between the end phases observed from DAC pressurization and point contact loading under indentation. In the lower panel on the right in Fig. 1, we show a typical plot of the hydrostatic pressure under a large spherical indenter tip (~21 μm radius) at a load of 450 mN using the Elastica code (Wong, 2013). In this case, the hydrostatic pressure under the indenter tip exceeds the β-Sn transformation pressure out to a diameter of about 6 μm, indicating that the transformed zone can be quite large under indentation.

Using a-Si as the starting material in DAC pressurization yields a more complex picture. Some authors (Garg et al., 2011; Pandey et al., 2011) have observed crystallization into metallic phases followed by the formation of r8-Si upon decompression, while others have reported a pressure-induced transformation to a metallic high-density a-Si (HDA) phase (rather than the crystalline (β-Sn)-Si phase) which was found to transform back to the low-density form of a-Si on pressure release (Daisenberger et al., 2007; McMillan et al., 2005; Shimomura et al., 1974). Moreover, other studies have found that only sufficiently high pressures (higher than in the dc-Si case) yield crystalline metallic phases, followed by the (partial) nucleation of bc8-Si (Imai et al., 1996; Minomura, 1981) or full reamorphization (Haberl et al., 2013) upon decompression. The study by Haberl et al. (2013) examined the literature in light of a dependence of the phase transformation behavior on the exact nature of the starting a-Si used (i.e., morphology and impurity content). They concluded that pure, voidless a-Si would follow the same transformations as dc-Si, but that impurities and porosity prevent this to varying degrees. This has not, however, been fully explored experimentally as yet.

For Ge under the application of pressure by both DAC and indentation loading, the literature suggests that the situation is more complicated than for Si. Although essentially the same phases as observed for dc-Si upon compression form also from dc-Ge during DAC loading (albeit at higher pressures) (Mujica et al., 2003), there is considerable disagreement in the literature on the metastable Ge phases that result from (β-Sn)-Ge on pressure release and the conditions under which they are formed. Variously, bc8-Ge, r8-Ge, hexagonal diamond (hd-Ge), simple tetragonal (st12-Ge), and even dc-Ge and a-Ge have been reported as end phases (Brazhkin et al., 1995; Hanfland and Syassen, 1990; Lyapin et al., 1996; Nelmes et al., 1993; Wosylus et al., 2008). Nelmes et al. (1993) explained these different phase observations in terms of unloading rate in a DAC: slow unloading was reported to result in predominantly the kinetically stable st12-Ge phase, whereas fast unloading led to bc8-Ge which anneals at room temperature to hd-Ge. Similarly, Lyapin et al. (1996) and Brazhkin et al. (1995) attributed the different phase formation in a DAC to a temperature dependence: (slow) decompression at elevated temperatures results in dc-Ge, at room temperature in st12-Ge, at dry ice temperatures in bc8-Ge, and at even lower temperatures in a-Ge. However, a recent DAC study (Haberl et al., 2014) to clarify the influence of shear stress and unloading rate on the transformation from (β-Sn)-Ge has shown that the hydrostatic conditions enabled by a gas pressure medium produce the r8 phase (followed by bc8-Ge) on depressurization at room temperature regardless of the rate, whereas with no pressure medium (i.e. large pressure gradients and high shear) st12-Ge is observed to be the predominant end phase. This result suggests that high-pressure gradients and shear stress during unloading favor st12-Ge, whereas more hydrostatic conditions result in the r8-Ge and bc8-Ge phases (Haberl et al., 2014). The authors also suggest that such control over the formation of the metastable phases is necessary for their eventual exploitation. In particular, st12-Ge and hd-Ge are expected to be of interest due to their kinetic stability at room temperature (Brazhkin et al., 1992) and the predicted properties as superconductor and direct bandgap material, respectively (Malone and Cohen, 2012b).

With a-Ge as starting material, DAC pressurization has also been reported to induce (β-Sn)-Ge and a similar range of crystalline phases on pressure release but the situation is more complex. Some authors observe the formation of the unstable bc8-Ge phase followed by the kinetically stable hd-Ge structure (Imai et al., 1996; Minomura, 1981), while others report (like for a-Si) a metallic high-density form of a-Ge under pressure rather than (β-Sn)-Ge (Barkalov et al., 2010; Cicco et al., 2008; Coppari et al., 2009; Principi et al., 2005). On unloading, the reappearance of a low-density a-Ge phase was observed in some cases (Cicco et al., 2008; Principi et al., 2005). Indeed, Coppari et al. (2009) have suggested that voids observed in the type of a-Ge used in their experiments may account for crystalline/amorphous differences observed in their experiments. This may suggest that the differences between the transformation behavior of a-Ge and dc-Ge reported in the literature may relate, just as for a-Si, to morphology and impurity content within the form of a-Ge used for the various DAC experiments.

For dc-Ge under indentation, the different modes of deformation (plastic flow versus phase transformation) referred to above complicate the situation even further. However, even when metastable phases are observed on unloading, there has been disagreement in the literature on the possible Ge end phases and the phase evolution during unloading, as we illustrate below. Nevertheless, some recent studies have, to some extent, unraveled this complex behavior in Ge and we also briefly review these studies in this chapter.

In Section 2, the behavior of crystalline Si is reviewed, with a focus on the phase evolution and microstructure. In Section 3, the indentation results for amorphous and deposited Si films are presented. In both these sections, the sparse data available on indentation above room temperature as well as the annealing behavior of the residual metastable phases are briefly presented. In Section 4, we review *in situ* electrical measurements for Si under indentation, which has assisted in following the semiconductor-to-metallic transition on loading and further transformations to insulating phases on unloading. Also in this section, *ex situ* electrical measurements on the semiconducting properties of the metastable r8/bc8 end phases are reviewed. The indentation behavior of both crystalline and amorphous Ge is reviewed in Section 5. Finally, a brief summary is given in Section 6.

2. CRYSTALLINE Si

2.1 Observations from nanoindentation load/unload curves

Although it has been shown in several earlier studies that point contact loading (mainly with Vickers indenters under rapid loading) could induce phase transformations in Si (Clarke et al., 1988; Gridneva et al., 1972), it was only after the development of instrumented nanoindentation techniques that detailed analyses of indentation-induced phase transformations began (Jang et al., 2005b; Pharr, 1992; Weppelmann et al., 1993). These studies used Berkovich and spherical indenters under controlled loading and unloading conditions and were able to quantify several important parameters of the material directly from indentation curves, such as material stiffness and elastic modulus, the onset of plastic deformation and hardness. As a result, it was found that plasticity and hardness in Si could be related to the onset of a phase transformation (Weppelmann et al., 1993). For example, the departure from elastic behavior during indentation loading, and hence the onset of plasticity, was controlled by a phase transformation to (β-Sn)-Si under the tip. As a result, the hardness of Si at room temperature was found to correlate very well with the measured transformation pressure of 11–12 GPa (Haberl et al., 2012; Weppelmann et al., 1993). In addition, discontinuities observed in nanoindentation load–unload curves (applied force/pressure as a function of tip penetration) could provide valuable and reliable *in situ* information on the evolution and nature of indentation-induced deformation by phase transformation, particularly when the information was supplemented with *ex situ* Raman and electron microscopy data, as we show below.

Figure 2 shows typical nanoindentation curves of load versus penetration depth for Si under fast and slow unloading conditions using a Berkovich indenter tip. During loading, the onset of plasticity (related to a phase change to (β-Sn)-Si under the indenter tip) can often be determined from a small change in slope of the loading curve. This is most noticeable for spherical indentation (Bradby et al., 2000b, 2001; Williams et al., 1999) where the stress field under the indenter tip is more uniform than for pointed indenters such as Berkovich, and can be identified by differentiating the loading curve (see inset in Fig. 2). This discontinuity is referred to as a pop-in. During

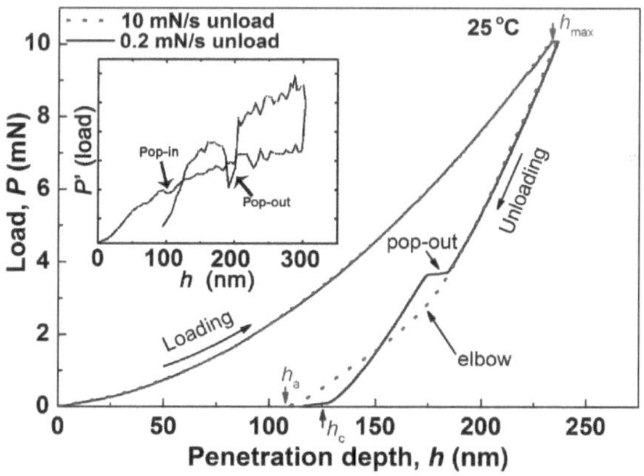

Figure 2 Comparison of *P–h* curves for both fast (10 mN/s) and slow (0.2 mN/s) unloading rates, obtained using nanoindentation with a Berkovich tip on dc-Si at 25 °C. The maximum penetration depth (h_{max}) at 10 mN and the final penetration depths are shown by h_c (for crystalline end phases) and h_a (for an amorphous end structure). Inset shows the derivative of the *P–h* curve obtained using a spherical indenter of ~4.2 μm. *The inset reprinted with permission from Haberl et al. (2006). Copyright (2006), American Institute of Physics.*

unloading, the "sudden" transformation of the dense (β-Sn)-Si phase to less dense r8/bc8 phases results in a rapid volume expansion and a much more obvious discontinuity in the unloading curve termed a pop-out, as shown in Fig. 2 for slow unloading (Domnich et al., 2000). In contrast, the fast unloading rate does not show a pop-out but exhibits a slope change in the unloading curve resembling an elbow (Domnich and Gogotsi, 2002; Domnich et al., 2000; Gogotsi et al., 2000). This has previously been correlated with a more continuous (β-Sn)-Si to a-Si transition on unloading, giving rise to a resultant a-Si end phase (Bradby et al., 2000a; Kailer et al., 1999). It has been suggested that the r8/bc8 crystalline phases need time to nucleate and, since (β-Sn)-Si is highly unstable at low pressures, it can directly transform to an a-Si phase (Ruffell et al., 2006, 2007a). It is interesting to examine the final surface position (penetration depth) on full unloading. As can be seen in Fig. 2, the final position for the a-Si end phase (h_a) is smaller than that for the r8/bc8-Si end phase (h_c), which is consistent with the fact that r8/bc8 phases are around 10% more dense than dc-Si, whereas pure a-Si has a similar density (Haberl, 2011; Roorda et al., 1991). There are several studies where the appearance of a pop-out and the final penetration depth have reliably been used to indicate an r8/bc8 end phase (Fujisawa et al., 2007; Ruffell et al., 2007a), including cyclic loading and unloading under conditions where the probability for the nucleation of these phases is low.

2.2 Raman spectroscopy of end phases

Gogotsi and coworkers (Domnich and Gogotsi, 2002; Gogotsi et al., 2000; Juliano et al., 2004; Kailer, 1999; Kailer et al., 1999) were the first to routinely use Raman spectroscopy to identify metastable end phases following indentation. They suggested that the Raman peak positions observed after slow unloading were consistent with the presence of r8 and bc8 phases of Si. This interpretation was based on previous Raman spectra from these phases obtained in the DAC (Hanfland and Syassen, 1990). However, the closeness of phonon frequencies for these phases together with several overlapping peaks makes it difficult to distinguish between them using Raman spectroscopy, especially when the transformed zone volume is small and the crystallites are of nanometer size. Hence, it is extremely difficult to measure the relative fractions of each phase present after full unloading (Ruffell et al., 2009a). Thus, we hereafter refer to the end phases following indentation as a mixed r8/bc8 phase.

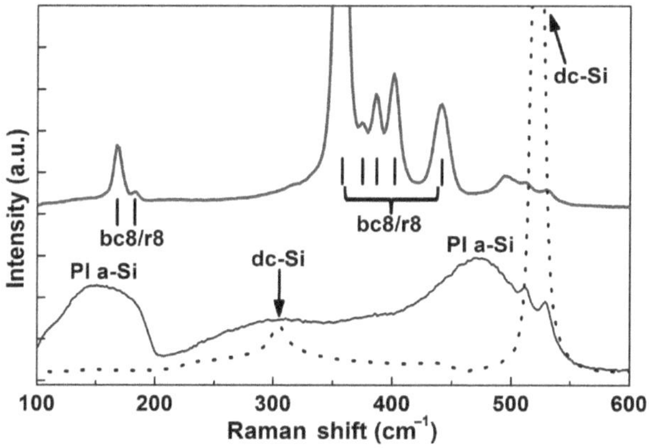

Figure 3 Raman spectra for both slow and fast unloading rates showing different phase transformation pathways, with a Raman spectrum of dc-Si for comparison. Indents were made under the same loading conditions using a 20-μm radii spherical indenter with a maximum load of 700 mN and a loading rate of ~5 mN/s. Unloading was performed with a rate slower than 1 mN/s and faster than 7000 mN/s, respectively. *Adapted from Haberl et al. (2011).*

We show in Fig. 3 typical Raman spectra following both slow and fast unloading using spherical indenter tips (Haberl, 2011). After slow unloading, several Raman peaks between 350 and 480 cm^{-1} are signatures for the mixed r8/bc8 phase. In contrast, after fast unloading, there are no such crystalline phases but evidence for a-Si (pressure-induced a-Si) as indicated by the a-Si Raman signal centered around 480 cm^{-1}. In both cases, a very small contribution of dc-Si from the substrate is observed. The typical Raman data, such as those in Fig. 3, correlate extremely well with the indentation unloading curves that exhibit pop-out and elbow behavior (Fig. 2). Although such *ex situ* Raman can only report on the end phases present after indentation, Gerbig et al. (2011) have recently carried out *in situ* Raman microscopy and reported the formation of r8/bc8 phases during slow unloading from the β-Sn metallic phase. Furthermore, their results suggested that the Raman signatures of these phases did not change further on complete unloading, which is consistent with our suggestion above that a metastable mixed r8/bc8 phase forms during unloading.

In addition to the clear delineation between r8/bc8 and a-Si end phases based on unloading rate, there are some indentation conditions, such as intermediate unloading rates (Ruffell et al., 2007a), the use of sharp Berkovich indenters (Ruffell et al., 2007a, 2009a), and cyclic loading/ unloading (Fujisawa et al., 2007, 2009), where Raman data and unloading curves are not good indicators of the presence of r8/bc8 phases. For example, in cases where the transformed zone only partially transforms to r8/bc8 phases, neither Raman nor unloading curves are sensitive enough to "observe" the small volume of crystalline phases (Ruffell et al., 2007a). In such cases, electron microscopy is needed as shown below.

2.3 Electron microscopy

Typical cross-sectional transmission electron micrographs (XTEMs) following fast and slow unloading using an ~4-μm radius spherical indenter are shown in Fig. 4.

The left side of the figure illustrates the probability of phase transformation as a function of unload rate, adapted from the work of Ruffell et al. (2007a). For the indentation conditions used in this study, unloading rates below about 100 mN/s always resulted in a visible pop-out in indentation load/unload curves, the presence of clear r8/bc8 peaks in Raman spectra and a transformed zone containing r8/bc8 crystallites, as shown in the upper XTEM image in Fig. 4. In contrast, unloading rates above about 1000 mN/s resulted in a totally amorphous transformed zone, as shown in the lower XTEM. In such cases, the load/unload curves exhibited a characteristic elbow and the Raman spectra showed strong evidence for a-Si. In cases where the probability for phase transformation to the r8/bc8 crystalline phases is around 50% (around 300 mN/s in Fig. 4) or when sharp Berkovich

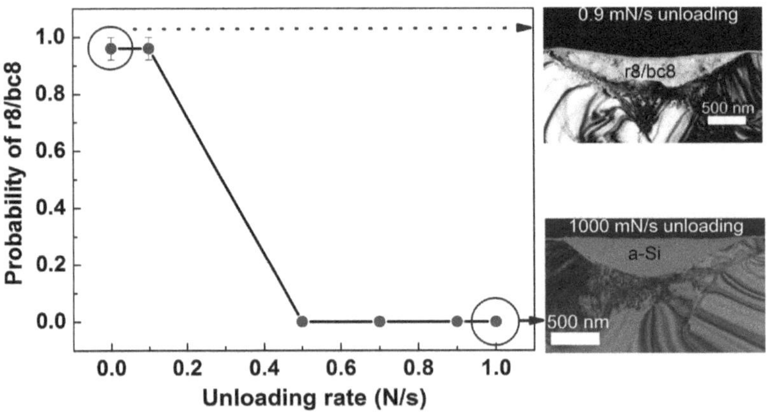

Figure 4 The probability for r8/bc8 formation upon indentation with a 4.3-μm radius spherical indenter in dc-Si with increasing unloading rate. The P_{max} was 80 mN. The bright-field cross-sectional TEM images show the presence of r8/bc8 phases at 0.9 mN/s unloading and a-Si above 1000 mN/s. *Reprinted with permission from Ruffell et al. (2007a). Copyright (2007), American Institute of Physics.*

indenters are used over a wide range of unloading rates, there is often a mixture of a-Si and r8/bc8 end phases. In such cases, TEM is essential to identify the crystalline phases.

Both of the XTEMs in Fig. 4 show the presence of crystalline defects under the transformed zone. This suggests, somewhat surprisingly, that there has been some small defect propagation in dc-Si even at room temperature. Comparing electron microscopy data from the literature (Wong, 2013), it appears that the ratio of the volume of the defective dc-Si region to that of the transformed zone decreases as the sphericality of the tip or hydrostatic stress component under the tip increases. This suggests that more uniform stress fields with less shear component (as resulting from a perfect sphere) favor phase transformation rather than defect propagation. This observation warrants further study to not only better understand the deformation modes in Si as a function of temperature but investigate whether, with large, "perfect" spherical indenters, it is possible to induce a phase transformation without underlying defects in dc-Si.

2.4 Temperature-dependent studies

Early studies of the temperature dependence of phase transformation in Si used microindentation techniques mainly with Vickers indenters under very fast loading and high loads. For example, Gridneva et al. (1972) showed that the dc-Si to β-Sn phase transformation appeared to persist using electrical resistance measurement under the microindenter up to 370 °C. Eremenko and Nikitenko (1972) observed twinning under a diamond pyramidal microindenter along the (111) plane between 350 and 450 °C along with some evidence for the hexagonal diamond structure (hd-Si). Indeed, Pirouz et al. (1990) also reported the formation of hd-Si during microindentation in the temperature range of 450–650 °C and suggested that it resulted from the twinning of dc-Si. However, the presence of hd-Si in the twin bands is still under debate (Cayron et al., 2009; Hertog et al., 2012). There is further discrepancy in the early literature with Suzuki and Ohmura (1996) reporting that phase transformation persisted up to 500 °C (with Vickers indentation), but beyond that temperature plastic deformation via dislocation activity (dislocation glide) occurred. In addition, Gilman (1993) argued that phase transformation and fracture are the only dominating failure modes in dc-Si up to the Debye temperature, θ_D (~372 °C) but plastic deformation by dislocation activity can occur beyond θ_D. Somewhat consistent with the proposal of Gilman (1993), more recent work of Domnich et al. (2008) reported the dc-Si to β-Sn phase transformation during large-load Vickers indentation up to 300 °C but dislocation glide as the dominant mode of deformation beyond 350 °C.

The disagreement in the above literature for elevated temperature indentation could be a result of ill-defined and very different indentation conditions using large indenters at high load and very fast loading rates. There have only been a few recent studies in the literature of controlled instrumented indentation at elevated temperatures that have attempted to map out the phase transformation pathways as a function of temperature. For example, Singh et al. (2008) investigated the effect of temperature on phase transformation in the range 20–135 °C using a Berkovich indenter. Whereas phase transformations occurred over this whole temperature range, they tentatively proposed a tetragonal form of Si on unloading and subsequent annealing for indentation at 135 °C and suggested that this originated from a new unidentifiable phase formed at 135 °C. Furthermore, Ruffell et al. (2009b) performed similar studies in the 25–150 °C temperature range on dc-Si and found only the expected r8/bc8 or a-Si end phases. However, the probability of formation of the bc8/r8 end phase (as opposed to an a-Si end phase) was enhanced with increasing temperature during fast unloading. They also reported some unexpected discrepancies between fast and slow unloading that may have resulted from temperature drift during the indentation. Finally, Bhuyan et al. (2012) suggested that the β-Sn phase may be unstable within a dc-Si matrix at elevated temperatures and that it may directly transform to dc-Si during the early stages of unloading, particularly between 150 and 200 °C. However, this latter study was also subject to temperature drift issues. Nevertheless, it is clear from the above few nanoindentation studies that the deformation mechanism under the indenter at elevated temperature appears to differ significantly from the previous larger-load indentation results.

In Fig. 5, we plot some very recent results of Kiran et al. (2015) who have improved and extended the elevated temperature indentation work of Ruffell et al. (2009b). Figure 5 shows that the probability of an r8/bc8 end phase increases up to 150 °C. The end phase was identified by both pop-out in unloading curves and Raman spectra. The observation of phase transformation was confirmed by XTEM data in selected cases, in which the volume of the transformed zone decreased with increasing temperature, whereas the volume of defective dc-Si underneath the transformed zone increased. These elevated temperature data were obtained for Berkovich indentation under unloading rate conditions in which

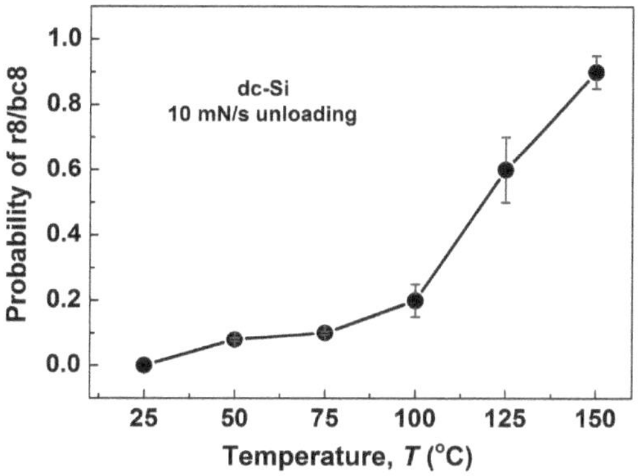

Figure 5 Probability of detecting r8/bc8 phases by Raman measured from 25 indents made at each temperature at an unloading rate of 10 mN/s in dc-Si. *Adapted from Kiran et al. (2015).*

the probability of r8/bc8 formation was extremely low at room temperature where a-Si was the dominant end phase. It has been suggested that if (β-Sn)-Si forms on loading at elevated temperature, then increasing temperature favors the probability of r8/bc8 phase nucleation on unloading, as might be expected from kinetic considerations. Furthermore, the study of Kiran et al. (2015) has shown that no phase transformations were observed at an indentation temperature of 200 °C (under the indentation conditions used); rather, deformation of the dc-Si occurred by the propagation of crystalline defects. It is suggested that at room temperature the dominant deformation mode is phase transformation but a small zone of defected dc-Si is also present. On an increase in temperature, the former volume decreases, but the latter volume of defected dc-Si increases until it totally dominates by 200 °C. This latter study indicates that further controlled elevated temperature work is needed to map out the phase transformation versus defect propagation behavior under widely varying loading and unloading conditions.

2.5 Annealing studies

Gogotsi and coworkers (Domnich and Gogotsi, 2002; Ge et al., 2004; Kailer et al., 1999) were the first to carry out annealing experiments on the r8/bc8 end phases resulting from indentation of Si. They found that these phases transformed to dc-Si in the temperature range of 175–200 °C. Furthermore, it was reported that an intermediate phase, possibly hd-Si, may have formed prior to dc-Si formation at a temperature close to 200 °C (Domnich and

Gogotsi, 2002). Ge et al. (2004) provided more definitive evidence from Raman spectroscopy that an unknown intermediate phase, called Si-XIII, was clearly evident prior to dc-Si formation but hd-Si may also be present. Ge et al. (2004) further suggested that a possible annealing sequence was r8 to bc8 to Si-XIII to hd-Si and/or a-Si to dc-Si at temperatures from 175 to 200 °C. However, the r8 to bc8 transition was based on r8/bc8 pellets made in a DAC (Ge, 2004) as a result of difficulties obtaining clearly distinct r8 and bc8 peaks from Raman spectra in the indentation case. In addition, the inclusion of hd-Si as an intermediate phase, when only sparse evidence for its presence was obtained for indentation results, was also expected from DAC results. Indeed, early annealing studies of the bc8 end phase from DAC experiments reported that bc8-Si annealed to hd-Si (Wentorf and Kasper, 1963), which upon further annealing produced dc-Si at 750 °C (Brazhkin et al., 1992). This latter temperature is clearly well above the temperatures employed in the indentation studies, inhibiting any direct correlation of the DAC and indentation cases.

Ruffell et al. (2007b) attempted to measure the annealing kinetics of r8 and bc8 phases at temperatures up to 300 °C. They found that it was not possible to fully isolate individual r8 and bc8 phases from Raman spectra and could not obtain any strong evidence that r8-Si annealed to bc8-Si prior to dc-Si formation. They therefore measured the combined kinetics for the r8/bc8 to dc-Si transformation and obtained an activation energy of 0.67 eV over the temperature range 125–300 °C. In a further work, Ruffell et al. (2009a) carried out careful annealing experiments on r8/bc8-Si formed in both dc-Si and a-Si with different size spherical indenters (to provide zones of different volumes). Similar to the results of Ge et al. (2004), they found clear evidence for the intermediate Si-XIII phase in Raman spectra and the possibility that hd-Si may also have been present. Again they could not clearly separate the r8 and bc8 phases and proposed several possible transformation pathways during annealing involving both r8 and bc8. However, if r8/bc8 is considered a single mixed (hybrid) phase, then the annealing sequence arrived at by Ruffell et al. (2009a) would be r8/bc8 to Si-XIII (possibly with some hd-Si) and then to dc-Si. Furthermore, this study also found that annealing to dc-Si occurred more rapidly for small r8/bc8 zones in a dc-Si matrix than for similar sized zones in an a-Si matrix. Larger zones formed with large spherical indenters were also found to be more stable and behaved independent of the surrounding matrix. This behavior was attributed to the larger volume/interface ratio in the large indents and residual strain differences in these zones.

From the above results, it is clear that the annealing sequence for the r8 and bc8 phases formed in a DAC [bc8 to hd-Si to dc-Si and r8 to dc-Si (Haberl et al., 2015)] differs considerably to the annealing of the mixed r8/bc8 phase obtained from indentation. In particular, no intermediate phases are found in the DAC case and this may suggest that the indentation end phase (a mixed or hybrid r8/bc8 phase) is different to the r8 and bc8 phases formed during unloading in a DAC.

3. AMORPHOUS AND DEPOSITED Si STUDIES

3.1 Relaxed and unrelaxed ion-implanted amorphous Si

The properties of a-Si vary significantly depending on the thermal history and preparation conditions (Roorda et al., 1991). For example, when an ion-implanted a-Si film is annealed (at 450 °C, for 30 min), it undergoes "structural relaxation" or short-range ordering (Haberl et al., 2004) which is suggested to involve repairing of dangling bonds and other defects that are present in the unrelaxed ion-implanted Si (Roorda et al., 1991). Structurally, relaxed a-Si is considered to be close to a fully coordinated continuous random network (Haberl et al., 2009; Laaziri et al., 1999; Roorda et al., 1991). Interestingly, it has been shown that these two states of a-Si, unrelaxed and relaxed, exhibit a very different behavior under indentation loading (Haberl et al., 2004, 2006). For example, in relaxed a-Si films at room temperature, a pop-out is observed in the unloading curve, and both Raman microspectroscopy and XTEM studies of the residual indent impressions revealed that the relaxed state undergoes a phase transformation under loading above ~11 GPa. Indeed, like dc-Si, a mixed r8/bc8 phase is obtained under appropriate "slow" unloading conditions (Haberl et al., 2004). However, no evidence of phase transformation was found in the unrelaxed a-Si indented under identical conditions but, rather, the material appeared to deform plastically (Haberl et al., 2006).

The above observations are summarized in Fig. 6, where the load/unload curve in (A) shows the typical behavior for a plastically deforming material (unrelaxed a-Si) and the load/unload curve in (B) shows the case of a phase transforming material (relaxed a-Si). The a-Si films were prepared by Si ion implantation (at energies of 400/90 keV and 600/90 keV, respectively) at liquid nitrogen temperature to create 650-nm and 920-nm a-Si layers in dc-Si (Haberl et al., 2004). The indentation was performed using a spherical indenter with an unloading rate that favors the formation of the mixed r8/bc8 phase. The Raman spectra in Fig. 6C confirm that the unrelaxed film does

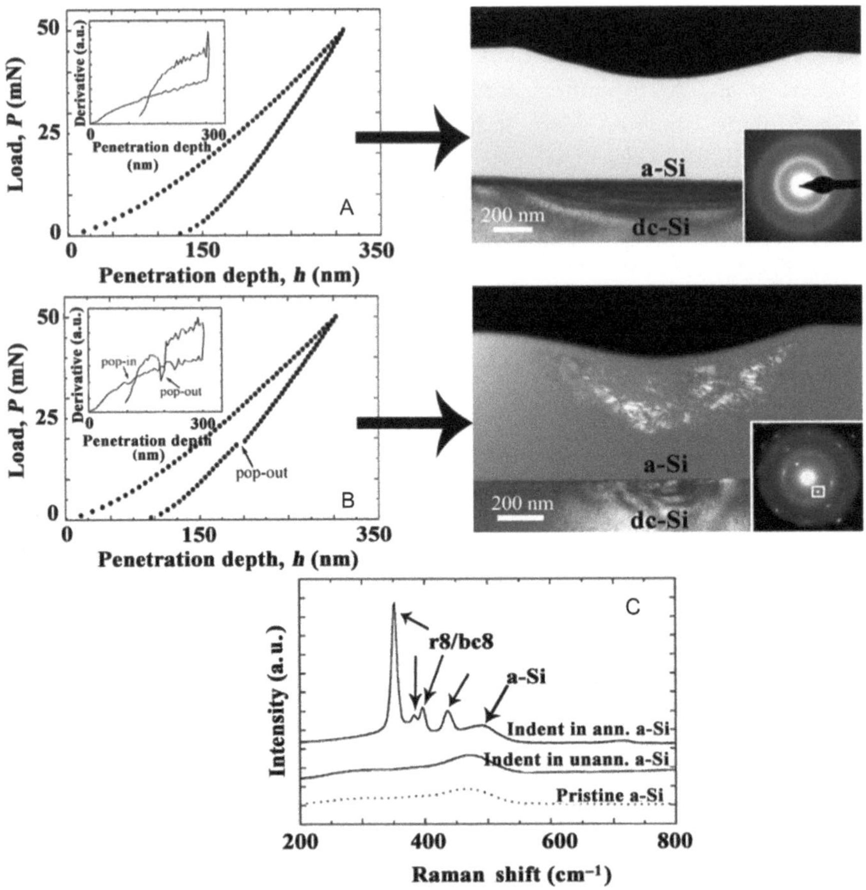

Figure 6 Indentation *P–h* curves (with insets as derivatives of load) and typical XTEM images of residual indents of (A) unrelaxed a-Si and (B) relaxed ~920-nm-thick a-Si. The indentation was performed in ~920-thick a-Si film for the load -displacement data and ~650 nm film for the XTEM images. The XTEM image in (A) is a bright-field image, whereas that in (B) is a dark-field (DF) image and the diffraction spot used to form the DF image is marked with a box in the inset. Indents were made using an ~2-μm radius indenter to a maximum load of 20 mN. (C) Raman spectra of pristine a-Si and residual indents in unrelaxed and relaxed ~650-nm-thick a-Si. Spectra are offset for clarity. In this case, the indentation was performed using an ~5-μm radius indenter applying loads up to 50 mN. *Reprinted with permission from Haberl et al. (2006). Copyright (2006), American Institute of Physics.*

not phase transform but the relaxed film shows clear evidence for a mixed r8/bc8 phase. Finally, the XTEMs in Fig. 6A and B (bright field for the indented, unrelaxed a-Si in (A) and dark field for the relaxed a-Si in (B)) show the untransformed deformation and transformed zone, respectively. Note that the dark-field image has been taken using an r8/bc8 diffraction spot and indicates the typical crystallite size of 20–40 nm. It has been suggested by Haberl et al. (2004) that the mechanical behavior of a-Si is controlled by the dangling bond density and short-range order.

It has also been shown by Haberl et al. (2006) that it is possible to phase transform unrelaxed a-Si films under certain conditions. For example, with spherical indenters a-Si films can transform if the penetration depth exceeds about half of the film thickness on loading. Under such conditions the film is sandwiched between a much harder indenter tip and underlying substrate, thus allowing the applied pressure to build up above the transformation pressure of $>$11 GPa.

3.2 Deposited Si

Again, there is a paucity of data for deposited Si films which is further complicated by the wide variability of deposited films in terms of their morphology and impurity (particularly hydrogen) content. The studies that have indented such films (Danesh et al., 2011; Haberl, 2011; Haberl et al., 2011; Kulikovsky et al., 2008; Pantchev et al., 2010; Ruffell et al., 2009c; Williams et al., 2005), however, have indicated that it is extremely difficult to phase transform them under a wide range of indentation conditions. For example, studies on films with high hydrogen contents (a-Si:H) prepared by chemical vapor deposition show no evidence of phase transformation (Danesh et al., 2011; Haberl, 2011; Pantchev et al., 2010; Williams et al., 2005) even if the films are fully polycrystalline (Haberl, 2011). Thermal annealing (to reduce the H content) also did not enable phase transformation, although the hardness of the films decreased (Danesh et al., 2011). This absence of phase transformation in these cases was attributed to the presence of H (Haberl, 2011; Williams et al., 2005). Indeed, a study of relaxed ion-implanted films enriched with H by implantation revealed a significant decrease in bc8/r8-Si formation for H contents of 1 at.% and above. Furthermore, films deposited via sputtering (that results in lower H contents than chemical vapor deposition techniques) were also not readily observed to phase transform. Haberl et al. (2011) have attributed this to the presence of nanopores that were not removed by thermal annealing. However, Kulikovsky et al. (2008) report occasional phase transformations in indentations performed on sputtered films. No dependence on rate or maximum load was detected, however, and the reason for the presence or absence of phase transformation remained unclear. Therefore, such films deform predominantly by plastic deformation like unrelaxed a-Si films.

3.3 Temperature dependence of indentation and annealing

There have been only two recent studies of the temperature dependence of indentation-induced phase transformations in pure a-Si. The findings are roughly consistent with the behavior of dc-Si outlined in Section 2.4. For example, Ruffell et al. (2009a) found that phase transformations occurred with high probability up to about 150 °C in relaxed a-Si films of 150 nm thickness subjected to indentation with a Berkovich indenter. More recently, Kiran et al. (2014) studied thick ~2 μm films of a-Si prepared by multiple-energy Si ion implantation. Both unrelaxed and relaxed films were indented with a Berkovich indenter at temperatures up to 200 °C. For such thick films, phase transformations were obtained at temperatures up to 100 °C for relaxed films. For unrelaxed films at all temperatures and relaxed films above 100 °C, plastic deformation was the prime deformation mode. Figure 7 summarizes these results.

The probability for phase transformation of relaxed films, as determined by either a pop-out or elbow in nanoindentation unloading curves and the observation of r8/bc8 peaks in Raman spectra, is close to 1 for temperatures up to 100 °C but 0 for temperatures of 150 and 200 °C. The transformation probability is zero at all temperatures for unrelaxed films. The insert bright-field XTEM images and selected area diffraction patterns for relaxed films are

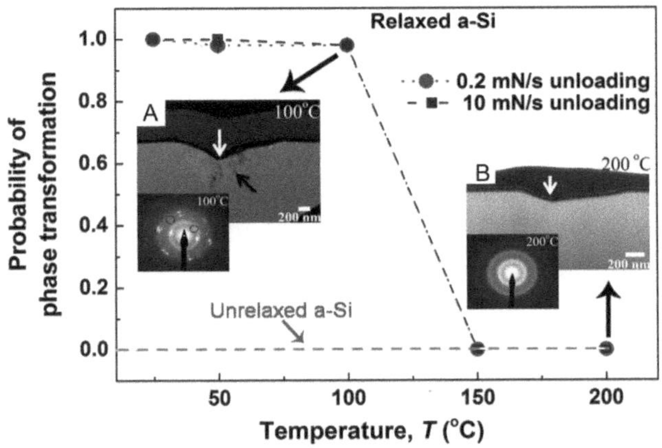

Figure 7 Probability of phase transformation detected by both Raman micro-spectroscopy and indentation curves as a function of temperature for both fast and slow unloading rates for relaxed and unrelaxed a-Si samples. (A) Represents the bright-field XTEM image of relaxed a-Si indented at 100 °C at a load of 10 mN with an unloading rate of 0.2 mN/s and (B) for relaxed a-Si at 200 °C. The Selected Area Diffraction (SAD) pattern shows r8/bc8 and the a-Si structures, respectively. Arrows represent the indentation direction. *Reprinted with permission from Kiran et al. (2014). Copyright (2014), American Institute of Physics.*

shown in Fig. 7A and B for 100 and 200 °C, respectively, and confirm these conclusions. The work of Kiran et al. (2015) also carried out *in situ* electrical measurements to confirm that the indentations at and above 150 °C did not phase transform to the metallic (conducting) (β-Sn)-Si phase on loading, and inadvertently back to insulating a-Si on unloading, a result consistent with *ex situ* Raman and XTEM techniques shown in Fig. 6.

4. ELECTRICAL MEASUREMENTS

4.1 *In situ* electrical measurements

The first *in situ* study that used changes in electrical resistivity to directly monitor the formation of the (β-Sn)-Si phase during indentation was that of Gridneva et al. (1972). Using a large Vickers indenter, the resistance of Si between two metal contacts was observed to decrease sharply at the point of phase transformation and return to its original value on unloading. Subsequent work by Clarke et al. (1988), Pharr et al. (1992) and Mann et al. (2000) used the fact that metallic (β-Sn)-Si caused a Schottky to Ohmic contact on low-doped Si to monitor the onset of the phase transformation. Furthermore, Bradby et al. (2003) fabricated a test structure on epitaxial Si wafers to quantitatively measure the current flow (again using a Schottky to Ohmic conversion at phase transformation to turn on current). The current was measured as a function of applied load and changes in current correlated with discontinuities in the load/unload curve for spherical indentation. In this study, the pop-in event on loading was shown to be related to the load at which the (β-Sn)-Si phase extended beyond the contact radius, and the pop-out event could be correlated with a sudden increase, then decrease, of current during unloading. It was found that current increased roughly as the contact radius of the spherical indenter, but there were a number of unknown parameters such as contact area and contact resistance, spreading resistance and circuit resistances that prohibited extraction of the extent of the (β-Sn)-Si phase from the measured current. Recently, Sprouster et al. (2014) have used a conducting vanadium carbide tip to probe Si during phase transformation and quantitatively measure the various resistances in the circuit, including the change in spreading resistance with tip penetration depth. This latter parameter allowed the electrical contact area to be calculated as a function of load and penetration depth.

Figure 8A shows the through tip-wafer method of electrical measurement used by Sprouster et al. (2014) to obtain the resistance components of the circuit as a function of penetration depth during Si indentation.

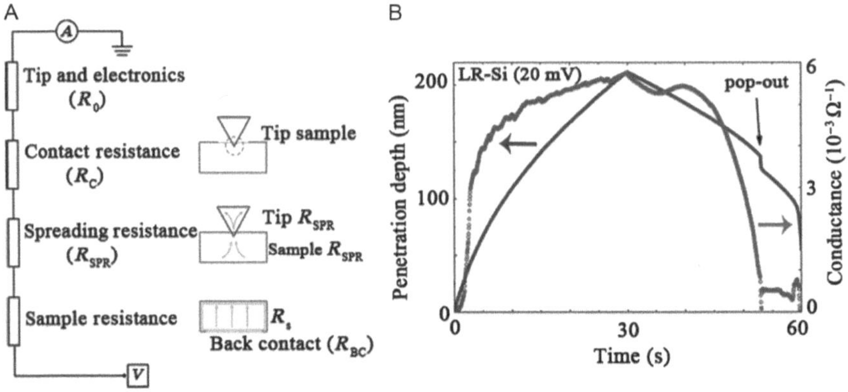

Figure 8 (A) The equivalent circuit diagram of the series arrangement of resistances for a through tip-wafer measurement of Si. (B) The mechanical (penetration depth) and electrical behavior (conductivity) of low resistivity ($9.10 \times 10^{-3}\ \Omega$ cm) Si during *in situ* indentation. *Reprinted with permission from Sprouster et al. (2014). Copyright (2014), Elsevier.*

A typical plot of penetration depth versus time and corresponding conductance versus time is shown in Fig. 8B, for indentation of low resistivity Si using a conducting vanadium carbide (Berkovich) tip and an applied voltage of +20 mV. Note that the rise in current (conductivity) does not occur immediately, but it rises rapidly once the penetration depth corresponds to the initiation of the dc–(β-Sn) transformation. On unloading the conductivity drops rapidly, with often a slight increase at pop-out when the r8/bc8-Si phase is fully nucleated.

In terms of the resistive components of the circuit after the phase transformation (Fig. 8A), the total circuit resistance R_T is given by

$$R_T = R_{SPR} + R_C + R_{BC} + R_S + R_0, \quad (1)$$

where R_{SPR} is the sum of the spreading resistances from the point of contact into the tip and into the Si sample under the contact, R_C is the contact resistance between indenter tip and sample, R_{BC} is the contact resistance of the back contact to Si, R_S is the sample resistance, and R_0 is the sum of the bulk tip resistance and the external circuit resistance. R_0 is typically a few ohms but can be measured by probing an Au thin film standard as outlined by Sprouster et al. (2014). R_S can often be found from the known material resistivity ρ_s but the $R_S + R_{BC}$ terms can also be determined by measuring the cross-sample resistance after forming sintered metal contacts on the top and bottom of the sample (Sprouster et al., 2014). However, usually for thick samples, R_S is small compared to other terms and can be neglected. The spreading resistance and tip-sample contact resistance terms usually dominate the total circuit resistance, as is the case for semiconductors with a low resistance sintered back contact, and hence the other terms in Eq. (1) can often be neglected although in the treatment below we consider the sum $R_{BC} + R_0$ as an unknown constant of small value. R_{SPR} originates from the current flow constriction through the small conducting contact between tip and sample and each component of R_{SPR} is typically inversely proportional to the radius of the electrical contact region, whereas R_C, which is particularly important for small tip-semiconductor contacts, scales inversely as the area of the electrical contact under the tip (Sprouster et al., 2014). For quantitative electrical measurements, it is important that the contact between tip and Si is Ohmic (as is the case for penetration depths that induce the metallic (β-Sn)-Si under the indenter tip).

The spreading resistance terms for semiconductor wafers can be approximated by the Holm equation (Sprouster et al., 2014) as

$$R_{SPR} = \frac{(\rho_s + \rho_t)}{4r} \approx \frac{\rho_s}{4r}, \tag{2}$$

where ρ_t is the tip resistivity and r is the electrical contact radius. In the treatment below, we assume that ρ_s is unknown and the tip resistivity can be neglected since it is more than an order of magnitude smaller than the resistivity of the lowest resistivity Si used. Finally, for Ohmic contacts, the contact resistance can be written as

$$R_C = \frac{R'_C}{\pi r^2}, \tag{3}$$

where R'_C is the specific contact resistance which is also an unknown parameter. Thus, Eq. (1) can now be approximated for indentation of Si as

$$R_T = \frac{\rho_s}{4r} + R'_C/\pi r^2 + (R_{BC} + R_0) \tag{4}$$

Hence, in the fitting procedure described below, there are three "unknowns" or floating parameters: ρ_s, R'_C, and $(R_{BC} + R_0)$. The first term in Eq. (4) varies as $1/r$, the second term as $1/r^2$, and the third term is linear. It is then possible to fit the measured R_T (or conductance) as a function of contact radius for each of the three terms in Eq. (4) and extract the parameters of best fit according to the method described in Sprouster et al. (2014), noting that the conversion of tip penetration depth to contact radius requires an accurate tip-area function calibration to be obtained.

In Fig. 9, we show the measured R_T as a function of contact radius for the Si indentation data in Fig. 8A as well as curves of best fit for the three terms of

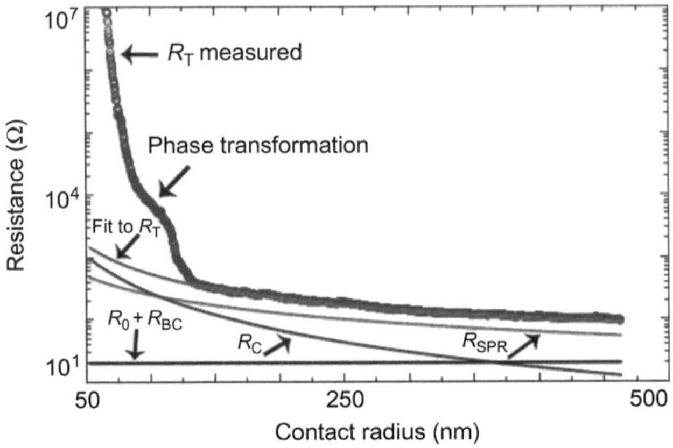

Figure 9 The measured resistance (R_T) as a function of contact radius for the low resistivity Si, together with the fitted circuit resistances. *Reprinted with permission from Sprouster et al. (2014). Copyright (2014), Elsevier.*

Eq. (4) using ρ_s, R'_C, and $(R_{BC} + R_0)$ as floating parameters. It can be seen that the fit to R_T is excellent for all contact radii above about 125 nm. It is interesting that this corresponds to the point of phase transformation and hence Ohmic contact behavior between tip and Si. Below this value, the tip to dc-Si contact is not Ohmic and the quantitative analysis fails. As indicated in Sprouster et al. (2014), the values obtained for ρ_s, R'_C, and $(R_{BC} + R_0)$ are $\sim 9 \times 10^{-3}\ \Omega$ cm, $\sim 10^{-7}\ \Omega\ \text{cm}^2$, and 10 Ω, respectively, which are in extremely good agreement with the values for these parameters measured independently. This demonstrates not only that it is possible to carry out quantitative *in situ* electrical measurements on nanoindented Si but, after the phase transformation occurs, the lateral extent of the metallic (β-Sn)-Si phase under the Berkovich indenter corresponds closely with the contact radius as a function of increasing load. This is not entirely surprising based on previous qualitative *ex situ* XTEM measurements but provides the first quantitative measure of the extent of the metallic phase as a function of increasing load (contact area).

4.2 *Ex situ* electrical characterization of r8/bc8 phase

Prior to more recent indentation studies of mixed r8/bc8-Si, little was known about the electrical properties of r8 or bc8 phases of Si. For example, an early experimental DAC study in which the dominant end phase following pressure release was bc8-Si gave results indicating that bc8-Si was a p-type semimetal with a room temperature carrier concentration of around

5×10^{20} cm^{-3} (Besson et al., 1987). This result was basically in agreement with density functional theory (DFT) calculations (Malone et al., 2008b; Pfrommer et al., 1997; Piltz et al., 1995). However, there were no experimental studies of r8-Si since this was not a stable end phase following DAC depressurization. Nevertheless, there have been DFT predictions that this phase is either a semimetal or a narrow bandgap semiconductor (Malone et al., 2008a; Pfrommer et al., 1997; Piltz et al., 1995). In particular, Malone et al. (2008a,b) predicted a narrow bandgap semiconductor with an energy gap of 0.24 eV.

In light of the fact that indentation can produce a mixed r8/bc8-Si end phase, Ruffell et al. (2011a,b) carried out a number of *ex situ* electrical studies of this end phase to establish its properties. These studies are summarized in part in Fig. 10. In Fig. 10A, a through-tip method was used to measure the *I–V* characteristics and conductivity of the mixed r8/bc8-Si phase for different boron (B) doping levels after indentation. The left panel shows the

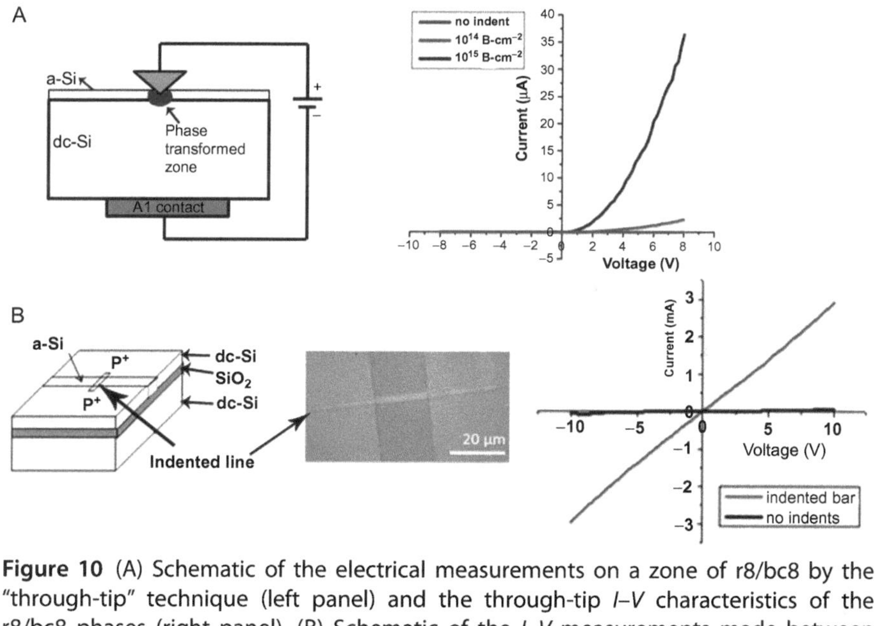

Figure 10 (A) Schematic of the electrical measurements on a zone of r8/bc8 by the "through-tip" technique (left panel) and the through-tip *I–V* characteristics of the r8/bc8 phases (right panel). (B) Schematic of the *I–V* measurements made between two heavily doped contact pads that are electrically connected by a bar of r8/bc8 material across an electrically insulating strip of a-Si (left panel). The bar is formed by a series of overlapping indents (middle panel). The corresponding *I–V* measurements (right panel) clearly show the low resistivity of r8/bc8 phases. *Reprinted panels (A) and (B) with permission from Ruffell et al. (2011b), Copyright (2011), American Physical Society, and Ruffell et al. (2011a), Copyright (2011), American Institute of Physics, respectively.*

device structure which consists of a 50-nm thick insulating a-Si layer made in a p-type dc-Si wafer (B-doped to $\sim 10^{15}$ cm^{-3}) by Si ion implantation and then relaxed by annealing. Parts of the a-Si layer were further implanted with B (fluences of 10^{14} and 10^{15} cm^{-2}) to give B concentrations of $\sim 10^{19}$ or 10^{20} cm^{-3}, respectively. Each region was then indented with a Berkovich tip to 10 mN load to induce phase transformations (as shown in Fig. 10A) and a residual r8/bc8-Si zone extending completely through the a-Si layer and into the underlying dc-Si substrate on full unloading. Each of the three indents was then probed following indentation with a conducting diamond tip at a low pressure of <1 mN to produce the *I–V* characteristics shown on the right panel of Fig. 10A. Under forward bias, there is a clear B effect consistent with activation in a semiconductor and indicating that the residual mixed r8/bc8-Si phase is acting as a semiconductor that can be doped with B during the indentation process at room temperature. The Schottky behavior of the *I–V* curves was shown to arise from the contact between the doped r8/bc8-Si zone and the underlying low-doped dc-Si substrate.

A second structure used to measure conductivity by Ruffell et al. (2011a,b) is shown in the left panel of Fig. 10B. It consists of a dc-Si-on-insulator wafer with two conducting p^+ pads between which was an insulating a-Si layer that electrically isolated the conducting pads. A line of overlapping indents was then made with a 10-μm radius spherical indenter to a load of 200 mN which resulted in a conducting line 4 μm in width and extending down 200 nm to the insulating SiO_2 layer. This bar structure allowed the conductivity of the indented line (see middle panel of Fig. 10B) to be measured as illustrated in the *I–V* characteristics in the right panel for different levels of B doping in the r8/bc8-Si phase. As well as conductivity, carrier mobility was measured by fabricating a microsized van der Paux structure by a series of overlapping indents on the same Si-on-insulator wafer (Ruffell et al., 2011a). Such measurements confirmed p-type carrier conduction and the measured mobility was around 17 cm^2/V s for the 10^{19} B/cm^3 doping concentration. In the same work, n-type doping of r8/bc8-Si with phosphorus (P) was also demonstrated. These results clearly showed that the mixed r8/bc8 semiconducting phase could be doped at room temperature with both B and P.

Further studies (Ruffell et al., 2011a) showed that, as well as "writing" conducting regions in insulating a-Si by overlapping indentation zones of r8/bc8-Si, it was possible to "write" insulating a-Si patterns in conducting dc-Si by overlapping indents where fast unloading had resulted in an a-Si end phase.

5. Ge UNDER INDENTATION

Although indentation with both sharp or spherical diamond tips can induce phase transformations under appropriate conditions in both dc-Ge and a-Ge, similar to the DAC studies mentioned above, there are substantial differences in the literature as to the deformation mechanisms and possible phase transformation pathways that can occur. For dc-Ge, there is clear evidence in some cases of a phase transformation to a metallic Ge phase under pressure (Pharr et al., 1992), but a variety of phases have been suggested on unloading such as possible st12-Ge (Kailer, 1999), potential traces of r8-Ge (Kailer, 1999), and bc8-Ge (Jang et al., 2005a), as well as a-Ge and dc-Ge end phases (Jang et al., 2005a; Oliver et al., 2009a). However, other studies (Bradby et al., 2002; Gogotsi et al., 2000; Oliver et al., 2008) have shown that dc-Ge is difficult to phase transform under indentation loading and that "slow" loading, particularly with spherical indenters, favors deformation by slip, twinning, and dislocation motion rather than inducing a phase change. In terms of an a-Ge starting material, Patriarche et al. (2004) clearly showed that a phase transformation can take place under indentation loading. Furthermore, the indentation and electron microscopy data of Oliver et al. (2009b) showed that it is easier to induce a phase transformation in a-Ge compared with dc-Ge as a result of the ease of deformation by slip and twinning in the latter case. These authors observed a fully crystalline transformed zone within an a-Ge matrix that they interpreted as predominantly dc-Ge. In the sections below, we briefly review the deformation modes (slip vs. phase transformation) in dc-Ge under indentation and then outline some more recent studies of phase transformations under indentation of a-Ge that help to unravel the different end phases reported in the literature.

5.1 Crystalline Ge

Figure 11 is a dark-field image showing the typical deformation processes that occur in dc-Ge with an ~4-μm radius spherical indenter tip (Bradby et al., 2002). There is clear evidence for defect propagation via twinning and slip processes under the indenter but no evidence for phase transformation. In such cases, the loading curve can exhibit a series of small pop-ins that reflect the sudden onset of defect formation and propagation. This is the normal mode of deformation for spherical indentation under all but the very fastest loading conditions as well as under moderate loading rates using sharp (Berkovich) indenters.

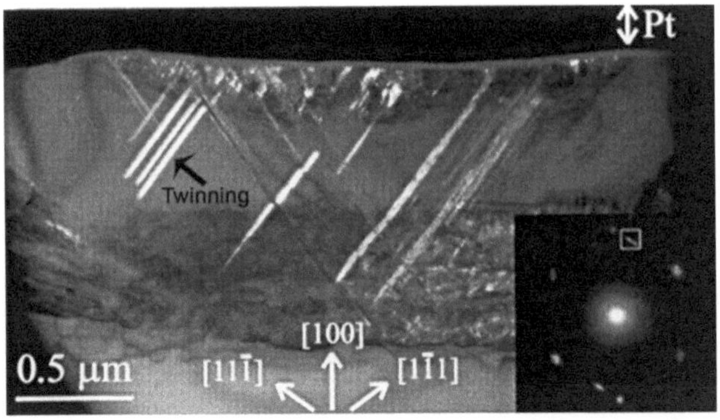

Figure 11 Dark-field XTEM image of a spherical indent in Ge at a maximum load of 50 mN and a loading and unloading rate of ~7 mN/s. Inset shows the SAD pattern indicating twin reflection spots. *A* ~ 4.2-μm radius spherical indenter was used in this study. *Reprinted with permission from Bradby et al. (2002). Copyright (2002), American Institute of Physics.*

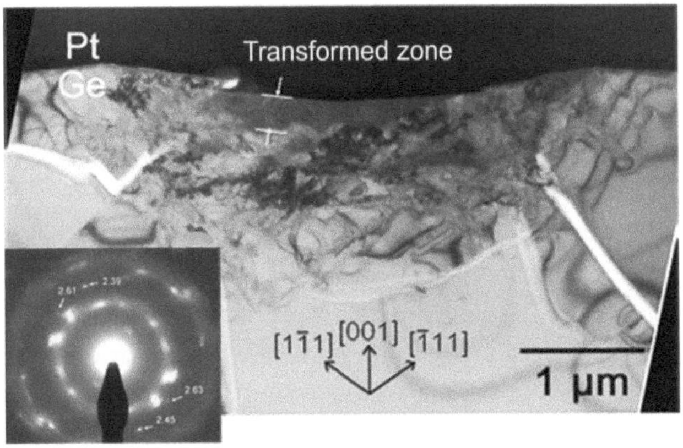

Figure 12 Bright-field <110> zone axis XTEM of ultrarapid loading with a spherical indenter of ~4 μm radius to ~80 mN, at ~165 mN/s. Inset shows SAD pattern taken from the transformed zone. *Reprinted with permission from Oliver et al. (2009a). Copyright (2009), American Institute of Physics.*

However, when the loading rate is extremely fast, it would appear that defect formation and propagation cannot occur fast enough to prevent the pressure under the indenter exceeding the transformation pressure (Oliver et al., 2009a). In such cases, phase transformations can occur and it would appear that the rapidly changing stress gradients in such cases can lead to more than one transformation pathway as well as severe cracking (Jang et al., 2005a; Oliver et al., 2009a). In Fig. 12, we show an example of a typical deformation zone following ultra-fast loading (Oliver et al., 2009a).

The bright-field XTEM in Fig. 12 was obtained from a residual indent (~4-μm radius spherical indenter tip) in dc-Ge following loading to 80 mN at a loading rate of 165 mN/s. Several features are evident in this micrograph: a small transformed zone under the indenter, a much larger region surrounding the transformed zone that contains a dense array of crystalline defects including twins, and severe cracking. The insert shows a selected area diffraction pattern taken from the transformed zone which contains several diffraction spots corresponding to the st12-Ge phase and also some diffuse amorphous rings suggesting the presence of a smaller amount of a-Ge, although ion beam thinning for sample preparation can also give rise to a-Ge. Oliver et al. (2009a) suggest that phase transformations occur during indentation above a critical loading rate since the critical stress required for shear-induced defect propagation rises with loading rate, whereas the critical stress for phase transformation is load-rate independent in the case of Ge. This observation helps explain the difficulties reported by many authors in detecting phase transformations under indentation. Furthermore, the work of Oliver et al. (2008, 2009a) indicates that, if the loading conditions are such as to induce a transformation in dc-Ge under indentation, the tip geometry and details of loading and unloading rate can dictate the phase evolution and the end phase observed. We illustrate these features more clearly in the next section that focuses on indentation into a-Ge, where deformation via slip and twinning processes are not operative and much more controlled indentation can be employed to induce phase transformations.

5.2 Amorphous Ge

The recent work of Deshmukh et al. (2014) has revealed distinctly different transformation pathways in a-Ge depending on the indentation loading/unloading conditions and the morphology of the initial metallic (β-Sn)-Ge formed under the indenter during loading. This work is summarized below. Typical load–unload curves for an a-Ge film (thickness 700 nm) are shown in Fig. 13A. In this case, the a-Ge was made by Ge self-ion implantation into dc-Ge and an ~4 μm spherical indenter was used for a series of 50 indents up to a maximum load of 100 mN at a loading rate of around 0.5 mN/s. In all indents in the series, a clear pop-in was observed at close to 58 mN in each case, but the nature of the pop-in event fell into one of two types.

For "type I" behavior, the pop-in extent (penetration at pop-in) was smaller and the slope of the loading curve after pop-in was similar to the slope before pop-in. For "type II" behavior, the initial pop-in was larger

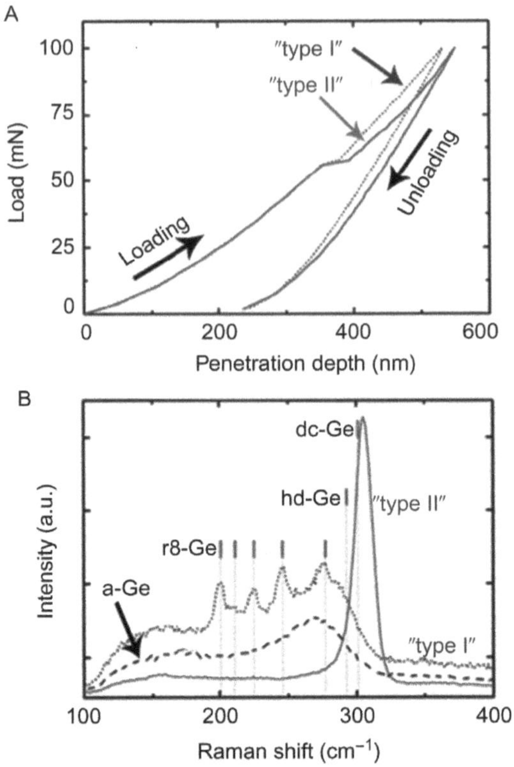

Figure 13 (A) *P–h* curves of indentations made in a 700-nm-thick a-Ge films using an ~4.3-μm radius tip showing two deformation pathways, so-called "type I" and "type II." (B) The corresponding Raman spectra taken from "type I" and "type II" indents. A Raman spectrum from unindented a-Ge is shown for comparison. Spectra are offset for clarity. The experimental peak positions for r8-Ge, hd-Ge, and dc-Ge are also indicated. *Reprinted with permission from Deshmukh et al. (2014). Copyright (2014), American Institute of Physics.*

and the slope after pop-in was clearly smaller than before pop-in. For the indentation conditions used for the data in Fig. 13A, roughly 85% of the indents fall into the "type I" category. Raman spectra from residual indents corresponding to "type I" and "type II" behavior, as well as the starting a-Ge sample, are shown in Fig. 13B. A spectrum taken within 1 day of the indentation for a "type I" indent reveals several clear crystalline Raman peaks which are not characteristic of dc-Ge. Careful fitting of these peaks and comparison of their positions with DFT calculations (Johnson et al., 2013) revealed that these peaks corresponded to an almost pure r8-Ge phase with a small amount of hd-Ge (see calculated r8 and hd-Ge Raman frequencies marked on Fig. 13A). This strongly suggested that a transformation of dc-Ge to (β-Sn)-Ge occurred on loading (presumably at the point of pop-in in Fig. 13A) and that r8-Ge formed from (β-Sn)-Ge on unloading. Indeed, it was subsequently found that the r8-Ge end phase corresponding to "type I" behavior was highly unstable at room temperature and transformed entirely into hd-Ge after about 1 week (Williams et al., 2013). The Raman spectrum in Fig. 13A, corresponding to typical "type II" indents, reveals, quite surprisingly, a signature that is almost entirely dc-Ge (see calculated dominant dc-Ge Raman frequency marked on the figure). To help understand this unexpected behavior XTEM was carried out as below.

Figure 14 shows XTEMs taken from "type I" and "type II" indents of Fig. 13A. A clear deformed/transformed zone in a-Ge is shown that extends right through the thickness of the original 700 nm thick a-Ge film. There are also clear twins and other defects in the underlying dc-Ge substrate but no phase transformation. Analysis of the selected area diffraction patterns indicates that the residual zone in the "type I" material consists of nanocrystalline hd-Ge (Fig. 14A): the fact that no r8-Ge was observed is attributable to the instability of this phase at room temperature and the high probability that it

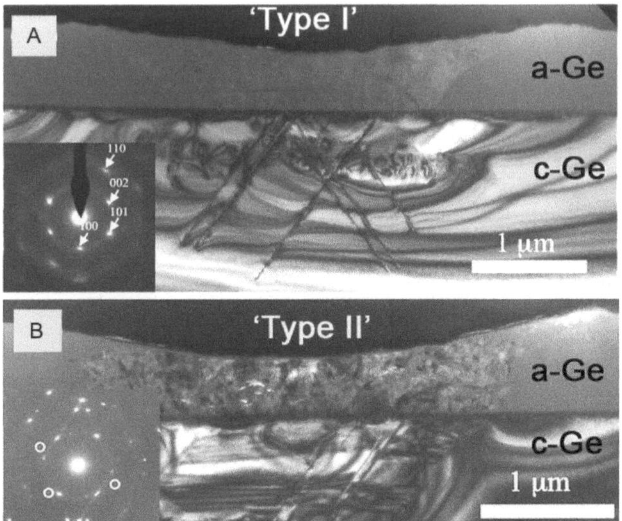

Figure 14 Bright-field XTEM images of (A) "type I" and (B) "type II" indents in an ~700-nm film indented to 100 mN. While the SAD pattern of "type I" indexed to hd-Ge, the "type II" SADP indexed to dc-Ge. *Reprinted with permission from Deshmukh et al. (2014). Copyright (2014), American Institute of Physics.*

will transform to hd-Ge during sample thinning for XTEM. In contrast, the "type II" material in Fig. 14B is almost entirely nanocrystalline dc-Ge with a small amount of st12-Ge, as shown by the circled diffraction spots in the inset diffraction pattern. Hence, these data are evidence that both "type I" and "type II" indents have undergone phase transformation on loading but different transformation pathways were followed, leading to different end phases after full unloading. The possible reasons for this behavior are discussed in the following section.

5.3 Phase evolution

Deshmukh et al. (2014) used the different transformation pathways characterized by "type I" and "type II" indentation behavior in a-Ge to explain many of the inconsistencies in the literature related to pressure-induced phase transformations in Ge. A summary of "type I" and "type II" behavior is shown schematically in Fig. 15. It was proposed that initially some plastic deformation occurs in a-Ge prior to a catastrophic phase transformation when the pressure substantially exceeds 10 GPa. This behavior is suggested to be due to the nucleation of the (β-Sn)-Ge phase and results in a prominent pop-in corresponding to the large volume contraction of about 22% when the more dense metallic (β-Sn)-Ge phase is formed under the indenter tip.

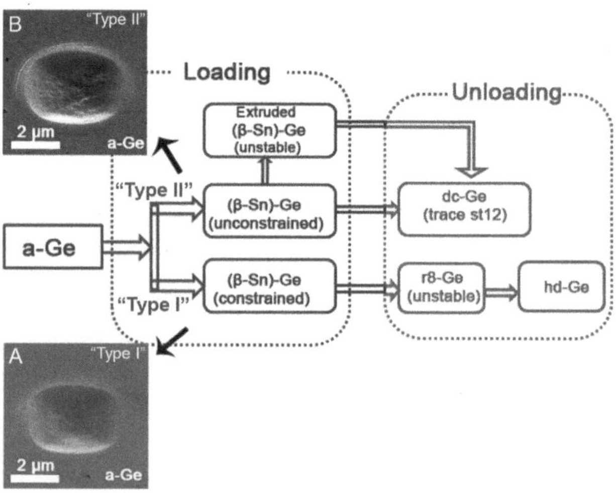

Figure 15 Schematic showing the deformation pathways associated with the indentation of a-Ge. The SEM micrographs in (A) and (B) show typical residual indent impressions (~4-μm radius spherical indenter) of type I and type I behavior, respectively. *Reprinted with permission from Deshmukh et al. (2014). Copyright (2014), American Institute of Physics.*

Two separate types of behavior ensue. In the "type I" case, there is a smaller pop-in and the metallic phase is totally constrained under the indenter. This confinement has been confirmed (Deshmukh et al., 2014) by examining the "type I" indent impressions by scanning electron microscopy which show no pile up outside of the radius of contact of the tip (see inset (A) in Fig. 14). This confinement of the (β-Sn)-Ge phase under the indenter tip is consistent with no change in slope of the loading curve after pop-in (slope determined by the mechanical properties of the surrounding untransformed a-Ge). In the case of "type II" behavior, the large pop-in and slope change of the loading curve is consistent with an unconfined soft (β-Sn)-Ge phase that exudes from under the indenter tip. This behavior is confirmed by examining the residual indent impressions of "type II" cases, shown in inset (B) of Fig. 15, which reveals considerable pile up around the contact area. Thus, it is proposed that the two types of behavior relate to the probability that the catastrophic transformation to (β-Sn)-Ge remains confined under the indenter tip at pop-in. This probability is very much tip shape and load-rate dependent. For example, it has been shown that Berkovich tips at moderate loading rates (Deshmukh et al., 2014) almost always result in "type II" behavior and excessive pile up outside the indenter contact. Indeed, Jang et al. (2005a) related such pile up observations for Berkovich indentation of dc-Ge to evidence for a phase transformation.

For the "type I" case where (β-Sn)-Ge is confined, it is proposed that (β-Sn)-Ge progressively transforms to the r8 phase on unloading, which is unstable and further transforms at room pressure and temperature to the more stable hd-Ge phase. This observation is consistent with the recent DAC results of Haberl et al. (2014) which indicated that (β-Sn)-Ge transformed to the r8-Ge phase under depressurization of a cell prepared under hydrostatic conditions. In terms of early indentation studies in dc-Ge, some works suggest the possibility of r8-Ge (Kailer, 1999) and/or report bc8-Ge (Jang et al, 2005a) as possible end phases in residual indent impressions as obtained from Raman data. It is noted that r8 and bc8 are difficult to distinguish from Raman data (Johnson et al., 2013) and hence it may well be that it was traces of the r8 phase that both these works observed. Indeed, Kailer et al. (1999) and Gogotsi et al. (2000) have observed hd-Ge from Raman spectra under particular indent conditions, noting that bc8/r8-Ge is unstable at room temperature and transforms to hd-Ge. In addition, the main Raman peak of hd-Ge is around six to seven wave numbers lower than the dominant dc-Ge peak at 301 cm^{-1}, and consequently, it is easy to misinterpret Raman spectra when dc-Ge is also present. Hence, the work of Jang et al. (2005a) and Oliver et al (2009a,b) may well have had hd-Ge present in the end phase. Thus, these prior indentation studies could well have observed "type I" behavior as described in Fig. 15.

In the case of "type II" behavior, it is suggested that the unconstrained (β-Sn)-Ge at pop-in suddenly experiences low pressure and undergoes a transformation directly to dc-Ge (with possibly some a-Ge) within the extruded material near the edge of the indenter contact area. This sudden transformation to dc-Ge seeds nucleation of further dc-Ge on pressure release. The rapid pressure release can, however, also favor a-Ge formation on the one hand (if it is too rapid to nucleate dc-Ge) and st12-Ge formation on the other hand (if the pressure release is more gradual but the stress gradients are large). This behavior may well explain the observations of a largely polycrystalline dc-Ge end phase (with some a-Ge present) following rapid loading and unloading experiments in the literature (Gogotsi et al., 2000; Jang et al., 2005a; Oliver et al., 2009a). It is also consistent with both observations of st12-Ge in Fig. 12 under indentation conditions of ultrarapid loading/unloading under high shear stress and the recent DAC studies (Haberl et al., 2014), with depressurization under high shear. These results strongly suggest that high-pressure gradients and shear stress during unloading can favor st12-Ge.

Summing up the high-pressure indentation studies on Ge, it is clear that the situation is particularly complex. Under moderate loading rates with spherical indenters, plastic deformation via slip and twinning is favored for dc-Ge. However, under appropriate choices of indenter tip shape, maximum load, and/or high loading rate, a phase transformation to the (β-Sn)-Ge phase can be induced in dc-Ge. In a-Ge, the slip and twinning mechanism of deformation of the crystal is not relevant, although the material can plastically deform without phase transition (Deshmukh et al., 2014). Nevertheless, even moderate loading rates applied to a-Ge can still induce a sudden phase transformation to (β-Sn)-Ge. Once (β-Sn)-Ge is formed, there are several possible transformation pathways that can ensue. If the (β-Sn)-Ge phase is confined under the indenter and shear stresses directly beneath the tip are small (such as with spherical indenters), the formation of r8-Ge can dominate on unloading. This phase is unstable and appears to transform to hd-Ge at room temperature. When the (β-Sn)-Ge phase is mostly confined, rapid loading/unloading occurs or shear stresses are large, (β-Sn)-Ge can transform to the metastable st12-Ge phase. However, when the (β-Sn)-Ge phase is unconfined, it can extrude from under the indenter and the sudden pressure release can induce transformation directly to either dc-Ge or to a-Ge. Intriguingly, in some cases under rapid loading/unloading with pointed indenters, all of these transformation pathways could coexist in different parts of a single indent impression.

6. SUMMARY AND CONCLUSIONS

This chapter has reviewed the deformation of Si and Ge under indentation loading. Whereas both materials can undergo phase transformations under appropriate indentation loading conditions, there are significant differences. First, both dc-Si and relaxed a-Si readily transform to the metallic (β-Sn)-Si phase at around 11 GPa and this phase transforms to either a metastable mixed r8/bc8-Si phase or a pressure-induced a-Si structure on unloading. This behavior depends on the unloading rate, where sluggish nucleation of the crystalline phases determines the end phase. On annealing, the r8/bc8 structure transforms to dc-Si via at least one intermediate phase (Si-XIII). Both the end phase (r8/bc8) and its annealing behavior are significantly different to DAC results where r8 transforms to mainly bc8 on unloading and this phase anneals to dc-Si via the hd-Si phase, with no evidence for a Si-XIII intermediate phase. The electrical properties of the r8/bc8 end phase are interesting: it is found to be a semiconductor that can be doped both p- and n-type, consistent with theory which has predicted the r8 phase to be a narrow bandgap semiconductor. Thus, indentation has been very useful for the characterization of this interesting phase and its formation via indentation pressure may prove the key for any exploitation of it or other pressure-induced metastable phases.

For dc-Ge, the ability to phase transform the material under indentation loading is compromised by a competing plastic deformation process involving slip and twinning under the indenter tip. However, it can undergo transformation to the (β-Sn)-Ge phase at high loading rates. The a-Ge phase, where slip and twinning processes are not relevant, is easier to phase transform over a wide range of indentation conditions. Following (β-Sn)-Ge formation, the subsequent phase evolution and the end phases are quite complicated compared with Si. For example, the initial transformation to (β-Sn)-Ge does not appear to occur gradually but via a sudden catastrophic process in which a large volume of the soft metallic phase forms on transformation. Several subsequent transformation pathways are possible depending on the tip geometry and load/unload conditions, particularly whether the metallic phase is confined under the indenter tip at the initial transformation. These pathways can lead to a final end phase which is

(i) dc-Ge with significant a-Ge when the metallic phase is not confined under the indenter tip, (ii) st12-Ge when the unloading involves high shear stress, or (iii) r8-Ge (which is unstable and anneals at room temperature to hd-Ge) when the metallic phase is constrained under the indenter tip and the unloading is controlled, particularly with low shear. Although the confinement of the metallic (β-Sn)-Ge phase is not an issue with DAC loading, the DAC behavior is somewhat consistent with indentation results in that there is evidence for either st12-Ge or the r8, bc8, and hd-Ge phases during DAC unloading under high shear and low shear, respectively. There have been no studies to date on the annealing kinetics and properties of the Ge end phases from indentation. It would be of interest to study the r8/bc8 to hd to dc-Ge and the st12 to dc-Ge transitions on annealing in some detail to ascertain the metastability regions for these phases. Moreover, the properties of r8-Ge, st12-Ge, and hd-Ge are of interest in the search, for example, for new semiconductor and superconducting behaviors. Finally, indentation allows for (somewhat) controlled formation of these different phases, thus making their potential exploitation conceivable.

ACKNOWLEDGMENTS

The authors acknowledge the Australian Research Council or financial support of parts of this project. J. E. B. acknowledges an ARC Future fellowship for support and B. H. acknowledges current funding through an Alvin M. Weinberg Fellowship (ORNL) and the Spallation Neutron Source (ORNL), sponsored by the U.S. Department of Energy, Office of Basic Energy Sciences.

REFERENCES

Barkalov, O.I., Tissen, V.G., McMillan, P.F., Wilson, M., Sella, A., Nefedova, M.V., 2010. Pressure-induced transformations and superconductivity of amorphous germanium. Phys. Rev. B 82, 1–4, 020507 (R).

Besson, J.M., Mokhtari, E.H., Gonzalez, J., Weill, G., 1987. Electrical properties of semimetallic silicon III and semiconductive silicon IV at ambient pressure. Phys. Rev. Lett. 59, 473–476.

Bhuyan, S.K., Bradby, J.E., Ruffell, S., Haberl, B., Saint, C., Williams, J.S., 2012. Phase stability of silicon during indentation at elevated temperature: evidence for a direct transformation from metallic Si-II to diamond cubic Si-I. Mater. Res. Commun. 2, 9–12.

Bradby, J.E., Williams, J.S., Wong-Leung, J., Swain, M.V., Munroe, P., 2000a. Transmission electron microscopy observation of deformation microstructure under spherical indentation in silicon. Appl. Phys. Lett. 77, 3749–3751.

Bradby, J.E., Williams, J.S., Wong-Leung, J., Swain, M.V., Munroe, P., 2000b. Mechanical deformation of crystalline silicon during nanoindentation. MRS Proc. 649, Q8.10 (1–6).

Bradby, J.E., Williams, J.S., Wong-Leung, J., Swain, M.V., Munroe, P., 2001. Mechanical deformation in silicon by micro-indentation. J. Mater. Res. 16, 1500–1507.

Bradby, J.E., Williams, J.S., Wong-Leung, J., Swain, M.V., Munroe, P., 2002. Nanoindentation-induced deformation of Ge. Appl. Phys. Lett. 80, 2651–2653.

Bradby, J.E., Williams, J.E., Swain, M.V., 2003. In situ electrical characterization of phase transformations in Si during indentation. Phys. Rev. B 67, 1–9, 085205.

Brazhkin, V.V., Lyapin, A.G., Popova, S.V., Voloshin, R.N., 1992. Solid-phase disordering of bulk Ge and Si samples under pressure. JETP Lett. 56, 152–156.

Brazhkin, V.V., Lyapin, A.G., Popova, S.V., Voloshin, R.N., 1995. Nonequilibrium phase transitions and amorphization in Si, Si/GaAs, Ge, and Ge/GaSb at the decompression of high-pressure phases. Phys. Rev. B 51, 7549–7554.

Cayron, C., Hertog, M.D., Romain, L.L., Mouchet, C., Secouard, C., Rouviere, J.L., Rouviere, E., Simonato, J.P., 2009. Odd electron diffraction patterns in silicon nanowires and silicon thin films explained by microtwins and nanotwins. J. Appl. Crystallogr. 42, 242–245.

Cicco, A.D., Congeduti, A., Coppari, F., Chervin, J.C., Baudelet, F., Polian, A., 2008. Interplay between morphology and metallization in amorphous-amorphous transitions. Phys. Rev. B 78, 1–4, 033309.

Clarke, D.R., Kroll, M.C., Kirchner, P.D., Cook, R.F., Hockey, B.J., 1988. Amorphization and conductivity of silicon and germanium induced by indentation. Phys. Rev. Lett. 60, 2156–2159.

Coppari, F., Chervin, J., Congeduti, A., Lazzeri, M., Polian, A., Principi, E., Cicco, A.D., 2009. Pressure-induced phase transitions in amorphous and metastable crystalline germanium by Raman scattering, x-ray spectroscopy, and *ab initio* calculations. Phys. Rev. B 80, 1–9, 115213.

Daisenberger, D., Wilson, M., McMillan, P.F., Cabrera, R.Q., Wilding, M.C., Machon, D., 2007. High-pressure X-ray scattering and computer simulation studies of density-induced polyamorphism in silicon. Phys. Rev. B 75, 1–11, 224118.

Danesh, P., Pantchev, B., Wiezorek, J., Schmidt, B., Grambole, D., 2011. Effect of hydrogen on hardness of amorphous silicon. Appl. Phys. A 102, 131–135.

Deshmukh, S., Haberl, B., Ruffell, S., Munroe, P., Williams, J.S., Bradby, J.E., 2014. Phase transformation pathways in amorphous germanium under indentation pressure. J. Appl. Phys. 115, 1–10, 153502.

Domnich, V., Gogotsi, Y., 2002. Phase transformations in silicon under contact loading. Rev. Adv. Mater. Sci. 3, 1–36.

Domnich, V., Gogotsi, Y., Dub, S., 2000. Effect of phase transformations on the shape of the unloading curve in the nanoindentation of silicon. Appl. Phys. Lett. 76, 2214–2216.

Domnich, V., Aratyn, Y., Kriven, W.M., Gogotsi, Y., 2008. Temperature dependence of silicon hardness: experimental evidence of phase transformations. Rev. Adv. Mater. Sci. 17, 33–44.

Eremenko, V.G., Nikitenko, V.I., 1972. Electron microscope investigation of the microplastic deformation mechanisms of silicon by indentation. Phys. Status Solidi A 14, 317–330.

Follstaedt, D.M., Knapp, J.A., Myers, S.M., 2004. Mechanical properties of ion-implanted amorphous silicon. J. Mater. Res. 19, 338–346.

Fujisawa, N., Williams, J.S., Swain, M.V., 2007. On the cyclic indentation behavior of crystalline silicon with a sharp tip. J. Mater. Res. 22, 2992–2997.

Fujisawa, N., Ruffell, S., Bradby, J.E., Williams, J.S., Haberl, B., Warren, O.L., 2009. Understanding pressure-induced phase-transformation behavior in silicon through in situ electrical probing under cyclic loading conditions. J. Appl. Phys. 105, 1–3, 106111.

Garg, N., Pandey, K.K., Shanavas, K.V., Betty, C.A., Sharma, S.M., 2011. Memory effect in low-density amorphous silicon under pressure. Phys. Rev. B 83, 1–6, 115202.

Ge, D., 2004. TEM Investigation of Contact Loading Induced Phase Transformation in Silicon. Ph.D Thesis, Drexel University, Philadelphia, PA. http://adsabs.harvard.edu/abs/2004PhDT.......14G.

Ge, D., Dominich, V., Gogotsi, Y., 2004. Thermal stability of metastable silicon phases produced by nanoindentation. J. Appl. Phys. 95, 2725–2731.
Gerbig, Y.B., Stranick, S.J., Cook, R.F., 2011. In situ observation of the indentation-induced phase transformation in silicon thin films. Phys. Rev. B 85, 1–9, 104102.
Gerk, A.P., Tabor, D., 1978. Indentation hardness and semiconductor–metal transition of germanium and silicon. Nature 271, 732–733.
Gilman, J.J., 1993. Why silicon is hard. Science 261, 1436–1439.
Gogotsi, Y.G., Domnich, V., Dub, S.N., Kailer, A., Nickel, K.G., 2000. Cyclic-nanoindentation and Raman microspectroscopy study of phase transformations in semiconductors. J. Mater. Res. 15, 871–879.
Gridneva, V., Milman, Y.V., Trefilov, V.I., 1972. Phase transition in diamond-structure crystals during hardness measurements. Phys. Status Solidi. A 14, 177–182.
Haberl, B., 2011. Structural Characterization of Amorphous Silicon. PhD Thesis, Australian National University, Canberra. http://hdl.handle.net/1885/8937.
Haberl, B., Bradby, J.E., Swain, M.V., Williams, J.S., Munroe, P., 2004. Phase transformations induced in relaxed amorphous silicon by indentation at room temperature. Appl. Phys. Lett. 85, 5559–5561.
Haberl, B., Bradby, J.E., Ruffell, S., Williams, J.S., Munroe, P., 2006. Phase transformations induced by spherical indentation in ion-implanted amorphous silicon. J. Appl. Phys. 100, 1–9, 013520.
Haberl, B., Liu, A.C.Y., Bradby, J.E., Ruffell, S., Williams, J.S., Munroe, P., 2009. Structural characterization of pressure-induced amorphous silicon. Phys. Rev. B 79, 1–8, 155209.
Haberl, B., Bogle, S.N., Li, T., McKerracher, I., Ruffell, S., Munroe, P., Williams, J.S., Abelson, J.R., Bradby, J.E., 2011. Unexpected short- and medium-range atomic structure of sputtered amorphous silicon upon thermal annealing. J. Appl. Phys. 110, 1–3, 096104.
Haberl, B., Aji, L.B.B., Williams, J.S., Bradby, J.E., 2012. The indentation hardness of silicon measured by instrumented indentation: what does it mean? J. Mater. Res. 27, 3066–3072.
Haberl, B., Guthrie, M., Sprouster, D., Williams, J.S., Bradby, J.E., 2013. New insight into pressure-induced phase transitions of amorphous silicon: the role of impurities. J. Appl. Crystallogr. 46, 758–768.
Haberl, B., Guthrie, M., Malone, B.D., Smith, J.S., Sinogeikin, S.V., Cohen, M.L., Williams, J.S., Shen, G., Bradby, J.E., 2014. Controlled formation of metastable germanium polymorphs. Phys. Rev. B 89, 1–6, 144111.
Haberl, B., Guthrie, M., Sinogeikin, S.V., Shen, G., Williams, J.S., Bradby, J.E., 2015. High Pressure Research (submitted).
Hanfland, M., Syassen, K., 1990. Raman modes of metastable phases of Si and Ge. High Press. Res. 3, 242–244.
Hertog, M.I.D., Cayron, C., Gentile, P., Dhalluin, F., Oehler, F., Baron, T., Rouviere, J.L., 2012. Hidden defects in silicon nanowires. Nanotechnology 23, 1–10, 025701.
Imai, M., Mitamura, T., Yaoita, K., Tsuji, K., 1996. Pressure-induced phase transitions of crystalline and amorphous silicon and germanium at low temperatures. High Press. Res. 15, 167–189.
Jamieson, J.C., 1963. Crystal structures at high pressures of metallic modifications of silicon and germanium. Science 139, 762–764.
Jang, J., Lance, M.J., Wen, S., Pharr, G.M., 2005a. Evidence for nanoindentation-induced phase transformations in germanium. Appl. Phys. Lett. 86, 1–3, 131907.
Jang, J., Lance, M.J., Wen, S., Tsui, T.Y., Pharr, G.M., 2005b. Indentation-induced phase transformations in silicon: influences of load, rate and indenter angle on the transformation behavior. Acta Mater. 53, 1759–1770.

Johnson, B.C., Haberl, B., Deshmukh, S., Malone, B.D., Cohen, M.L., McCallum, J.C., Williams, J.S., Bradby, J.E., 2013. Evidence for the R8 phase of germanium. Phys. Rev. Lett. 110, 1–5, 085502.

Juliano, T., Domnich, V., Gogotsi, Y., 2004. Examining pressure-induced phase transformations in silicon by spherical indentation and Raman spectroscopy: a statistical study. J. Mater. Res. 19, 3099–3108.

Kailer, A., 1999. Lokale Schädigung von Oberflächen und Phasenumwandlungen in harten, spröden Materialen verursacht durch mechanischen Kontakt. PhD Thesis, Shaker Verlag, Aachen. ISBN: 3826561147.

Kailer, A., Nickel, K.G., Gogotsi, Y.G., 1999. Raman microspectroscopy of nanocrystalline and amorphous phases in hardness indentations. J. Raman Spectrosc. 30, 939–946.

Kiran, M.S.R.N., Haberl, B., Williams, J.E., Bradby, J.E., 2014. Temperature dependent deformation mechanisms in pure amorphous silicon. J. Appl. Phys. 115, 1–10, 113511.

Kiran, M.S.R.N., Tuan, T.T., Smillae, L., Haberl, B., Subianto, D., Williams, J.S., Bradby, J.E., 2015. Unpublished.

Kulikovsky, V., Vorlicek, V., Bohac, P., Stranyanek, M., Ctvrtlík, R., Kurdyumov, A., 2008. Mechanical properties of amorphous and microcrystalline silicon films. Thin Solid Films 516, 5368–5375.

Laaziri, K., Kycia, S., Roorda, S., Chicoine, M., Robertson, J.L., Wang, J., Moss, S.C., 1999. High-energy x-ray diffraction study of pure amorphous silicon. Phys. Rev. B 60, 13520–13533.

Lyapin, A.G., Brazhkin, V.V., Popova, S.V., Sapelkin, A.V., 1996. Nonequilibrium phase transformations in diamond and zincblende semiconductors under high pressure. Phys. Status Solidi B 198, 481–490.

Malone, B.D., Cohen, M.L., 2012a. Prediction of a metastable phase of silicon in the *Ibam* structure. Phys. Rev. B 85, 1–5, 024116.

Malone, B.D., Cohen, M.L., 2012b. Electronic structure, equation of state, and lattice dynamics of low-pressure Ge polymorphs. Phys. Rev. B 86, 1–7, 054101.

Malone, B.D., Sau, J.D., Cohen, M.L., 2008a. Ab initio study of the optical properties of Si-XII. Phys. Rev. B 78, 1–4, 161202 (R).

Malone, B.D., Sau, J.D., Cohen, M.L., 2008b. Ab initio survey of the electronic structure of tetrahedrally bonded phases of silicon. Phys. Rev. B 78, 1–7, 035210.

Mann, A.B., van Heerden, D., Pethica, J.B., Weihs, T.P., 2000. Size dependent phase transformations during point loading of silicon. J. Mater. Res. 15, 1754–1758.

McMillan, P.F., Wilson, M., Daisenberger, D., Machon, D., 2005. A density-driven phase transition between semiconducting and metallic polyamorphs of silicon. Nat. Mater. 4, 680–684.

Minomura, S., 1981. Pressure-induced transitions in amorphous silicon and germanium. J. Phys. Colloq. 42, C4.181–C4.188.

Mujica, A., Rubio, A., Munoz, A., Needs, R.J., 2003. High-pressure phases of group-IV, III-V, and II-VI compounds. Rev. Mod. Phys. 75, 863–912.

Needs, R.J., Mujica, A., 1995. First-principles pseudopotential study of the structural phases of silicon. Phys. Rev. B 51, 9652–9660.

Nelmes, R.J., McMahon, M.I., Wright, N.G., Allan, D.R., Loveday, J.S., 1993. Stability and crystal structure of BC8 germanium. Phys. Rev. B 48, 9883–9886.

Oliver, D.J., Bradby, J.E., Williams, J.S., Swain, M.V., Munroe, P., 2008. Thickness-dependent phase transformation in nanoindented germanium thin films. Nanotechnology 19, 1–8, 475709.

Oliver, D.J., Bradby, J.E., Williams, J.S., Swain, M.V., Munroe, P., 2009a. Rate-dependent phase transformations in nanoindented germanium. J. Appl. Phys. 105, 1–3, 126101.

Oliver, D.J., Bradby, J.E., Ruffell, S., Williams, J.S., Munroe, P., 2009b. Nanoindentation-induced phase transformation in relaxed and unrelaxed ion-implanted amorphous germanium. J. Appl. Phys. 106, 1–6, 093509.

Pandey, K.K., Garg, N., Shanavas, K.V., Sharma, S.M., Sikka, S.K., 2011. Pressure induced crystallization in amorphous silicon. J. Appl. Phys. 109, 1–7, 113511.

Pantchev, B., Danesh, P., Wiezorek, J., Schmidt, B., 2010. Nanoindentation-induced pile-up in hydrogenated amorphous silicon. J. Phys. Conf. Ser. 253, 1–6, 012054.

Patriarche, G., Le Bourhis, E., Khayyat, M.M.O., Chaudhri, M.M., 2004. Indentation-induced crystallization and phase transformation of amorphous germanium. J. Appl. Phys. 96, 1464–1468.

Pfrommer, B.G., Cote, M., Louie, S.G., Cohen, M.L., 1997. Relaxation of crystals with the quasi-Newton method. J. Comput. Phys. 131, 233–240.

Pharr, G.M., 1992. The anomalous behavior of silicon during nanoindentation. Mater. Res. Soc. Symp. Proc. 239, 301–312.

Pharr, G.M., Oliver, W.C., Cook, R.F., Kirchner, P.D., Kroll, M.C., Dinger, T.R., Clarke, D.R., 1992. Electrical resistance of metallic contacts on silicon and germanium during indentation. J. Mater. Res. 7, 961–972.

Piltz, R.O., Maclean, J.R., Clark, S.J., Ackland, G.J., Hatton, P.D., Crain, J., 1995. Structure and properties of silicon XII: a complex tetrahedrally bonded phase. Phys. Rev. B 52, 4072–4085.

Pirouz, P., Chaim, R., Dahmen, U., Westmacott, K.H., 1990. The martensitic transformation in silicon-I experimental observations. Acta Metall. Mater. 38, 313–322.

Principi, E., Decremps, F., Cicco, A.D., Datchi, F., Panfilis, S.D., Filipponi, A., Polian, A., 2005. Pressure induced phase transitions in amorphous Ge. Phys. Scripta T115, 381–383.

Roorda, S., Sinke, W.C., Poate, J.M., Jacobson, D.C., Dierker, S., Dennis, B.S., Eaglesham, D.J., Spaepen, F., Fuoss, P., 1991. Structural relaxation and defect annihilation in pure amorphous silicon. Phys. Rev. B 44, 3702–3725.

Ruffell, S., Bradby, J.E., Williams, J.S., 2006. High pressure crystalline phase formation during nanoindentation: amorphous versus crystalline silicon. Appl. Phys. Lett. 89, 1–3, 091919.

Ruffell, S., Bradby, J.E., Williams, J.S., Munroe, P., 2007a. Formation and growth of nanoindentation-induced high pressure phases in crystalline and amorphous silicon. J. Appl. Phys. 102, 1–8, 063521.

Ruffell, S., Bradby, J.E., Williams, J.S., 2007b. Annealing kinetics of nanoindentation-induced polycrystalline high pressure phases in crystalline silicon. Appl. Phys. Lett. 90, 1–3, 131901.

Ruffell, S., Haberl, B., Koenig, S., Bradby, J.E., Williams, J.S., 2009a. Annealing of nanoindentation-induced high pressure crystalline phases created in crystalline and amorphous silicon. J. Appl. Phys. 105, 1–8, 093513.

Ruffell, S., Bradby, J.E., Williams, J.S., Paniagua, D.M., Tadayyon, S., Coastworth, L.L., Norton, P.R., 2009b. Nanoindentation-induced phase transformations in silicon at elevated temperatures. Nanotechnology 20, 1–5, 135603.

Ruffell, S., Vedi, J., Bradby, J.E., Williams, J.S., 2009c. Effect of hydrogen on nanoindentation-induced phase transformations in amorphous silicon. J. Appl. Phys. 106, 1–6, 123511.

Ruffell, S., Sears, K., Bradby, J.E., Williams, J.S., 2011a. Room temperature writing of electrically conductive and insulating zones in silicon by nanoindentation. Appl. Phys. Lett. 98, 1–3, 052105.

Ruffell, S., Sears, K., Knights, A.P., Bradby, J.E., Williams, J.S., 2011b. Experimental evidence for semiconducting behaviour of Si-XII. Phys. Rev. B 83, 1–6, 075316.

Shimomura, O., Minomura, S., Sakai, N., Asaumi, K., Tamura, K., Fukushima, J., Endo, H., 1974. Pressure-induced semiconductor–metal transitions in amorphous Si and Ge. Philos. Mag. 29, 547–558.
Singh, R.K., Munroe, P., Hoffman, M., 2008. Effect of temperature on metastable phases induced in silicon during nanoindentation. J. Mater. Res. 23, 245–249.
Sprouster, D.J., Ruffell, S., Bradby, J.E., Stauffer, D.D., Major, R.C., Warren, O.L., Williams, J.S., 2014. Quantitative electromechanical characterization of materials using conductive ceramic tips. Acta Mater. 71, 153–163.
Suzuki, T., Ohmura, T., 1996. Ultra-microindentation of silicon at elevated temperatures. Philos. Mag. A 74, 1073–1084.
Wentorf Jr., R.H., Kasper, J.S., 1963. Two forms of silicon. Science 139, 338–339.
Weppelmann, E.R., Field, J.S., Swain, M.V., 1993. Observation, analysis, and simulation of the hysteresis of silicon using ultra-micro-indentation with spherical indenters. J. Mater. Res. 8, 830–840.
Williams, J.S., Chen, Y., Wong-Leung, J., Kerr, A., Swain, M.V., 1999. Ultra-micro-indentation of silicon and compound semiconductors with spherical indenters. J. Mater. Res. 14, 2338–2343.
Williams, J.S., Haberl, B., Bradby, J.E., 2005. Nanoindentation of Ion implanted and deposited amorphous silicon. Mater. Res. Soc. Symp. Proc. 843, R10.3.1–R10.3.5.
Williams, J.S., Haberl, B., Deshmukh, S., Johnson, B.C., Malone, B.D., Cohen, M.L., Bradby, J.E., 2013. Hexagonal Germanium formed via a pressure-induced phase transformation of amorphous germanium under controlled nanoindentation. Phys. Status Solidi Rapid Res. Lett. 7, 355–359.
Wippermann, S., Voros, M., Rocca, D., Gali, A., Zimanyi, G., Galli, G., 2013. High-pressure core structures of Si nanoparticles for solar energy conversion. Phys. Rev. Lett. 110, 1–5, 046804.
Wong, S., 2013. Optimized Indentation Conditions for Large-Area Phase Transformations in Silicon. M. Phil Thesis, Australian National University, Canberra.
Wosylus, A., Prots, Y., Schnelle, W., Hanfland, M., Schwarz, U., 2008. Crystal structure refinements of Ge (tP12), physical properties and pressure-induced phase transformation Ge(tP12) ↔ Ge(tI4). Z. Naturforsch. B 63, 608–614.

CHAPTER SIX

Analytical Techniques for Electrically Active Defect Detection

Eddy Simoen[1], Johan Lauwaert, Henk Vrielinck
Department Solid State Sciences, Ghent University, Gent, Belgium
[1]Corresponding author: e-mail address: Eddy.Simoen@UGent.be

Contents

1. INTRODUCTION

The room temperature resistivity of semiconductors can be varied over several decades by doping with impurities, which give rise to shallow levels in the band gap and possess a high solid solubility. At the same time, the doping type can be varied from n to p over a short distance, using modern ion-implantation and short-time annealing techniques, suppressing the diffusion of dopants. This has been key to the development of a wide range of basic components, which are integrated in present-day ultra-large scale integrated circuits. These shallow doping levels are thus generally characterized by a high electrically active concentration, a low diffusivity, and an

Semiconductors and Semimetals, Volume 91
ISSN 0080-8784
http://dx.doi.org/10.1016/bs.semsem.2014.12.001

energy level in the band gap close to the corresponding band edge. This means in practice that at room temperature the doping atoms in a semiconductor material are ionized, giving up their free carriers to the conduction (n-type) or valence band (p-type), which compensate the fixed dopant charge.

Deep level defects and impurities, on the other hand, are characterized by a significantly larger energy separation from the band edges than the shallow dopant impurities, i.e., at least more than several kT, with k the Boltzmann constant and T the absolute temperature. The corresponding defect potential, perturbing the periodic lattice potential is much more localized than in the case of shallow-level impurities, which results in a much deeper ground state in the band gap. While their impact on the resistivity is generally rather limited, in many cases, they will govern the nonradiative carrier lifetime by acting as Shockley–Read–Hall (SRH) generation–recombination (GR) centers. As such, deep levels will have a strong impact on the operation of devices, since they determine the generation current (reverse leakage current) or the forward recombination current. Most applications rely on a high carrier lifetime and low leakage currents, so that from a viewpoint of device operation and yield it is of extreme importance to control deep level defects. This, of course, requires the possibility to detect such electrically active states, which are generally present in low to very low concentrations, due to the much lower solid solubility of deep level centers—i.e., a high sensitivity is mandatory.

This chapter aims to review analytical techniques for the detection of electrically active defects in semiconductor materials. In all cases, the operation principles, the strengths, and the weaknesses will be outlined and illustrated for state-of-the-art examples. After defining the basic parameters of a deep level GR center in section 2, the operation principle of techniques based on resistance measurements will be outlined. In this case, no device structure needs to be prepared, except for the fabrication of two (four) ohmic contacts. In section 4 deep level transient spectroscopy (DLTS) is described and adaptations for special experimental conditions will be outlined. It is shown that besides the trap energy level (E_T), the electron (σ_n), or hole (σ_p) capture cross section and the concentration (N_T), also the concentration profile can be determined. The energy resolution can be increased to ~1 meV by implementing Laplace-transform DLTS. In section 5, the use of admittance spectroscopy (AS) will be highlighted, whereby in some cases, information complementary to DLTS can be obtained. Finally, carrier lifetime analysis, based on device current–voltage

(I–V) characteristics is discussed and compared with the spectroscopic techniques described in the previous paragraphs. The chapter will be concluded by a summary, providing also some outlook with respect to deep level studies in nanometric structures.

2. BASIC PROPERTIES OF DEEP LEVEL DEFECTS

A deep level is characterized by its energy level E_T in the band gap, as indicated in Fig. 1. This position determines the electron occupancy of the defect, through Fermi–Dirac statistics, namely:

$$N_T^- = N_T \frac{1}{1 + g_n^{-1} \exp\left(\frac{E_T - E_F}{kT}\right)} \tag{1}$$

and for the empty traps:

$$N_T^0 = N_T \frac{g_n^{-1} \exp\left(\frac{E_T - E_F}{kT}\right)}{1 + g_n^{-1} \exp\left(\frac{E_T - E_F}{kT}\right)} \tag{2}$$

with E_F the Fermi level, g_n the ratio of the degeneracy in the initial bound state and the final free state (usually, $g_n = 2$ for a shallow donor, ¼ for a shallow acceptor) and $N_T^- + N_T^0 = N_T$. Equations (1) and (2) are valid for an acceptor; similar expressions hold for a donor, changing between the neutral (0) and the positive charge state by emission of an electron.

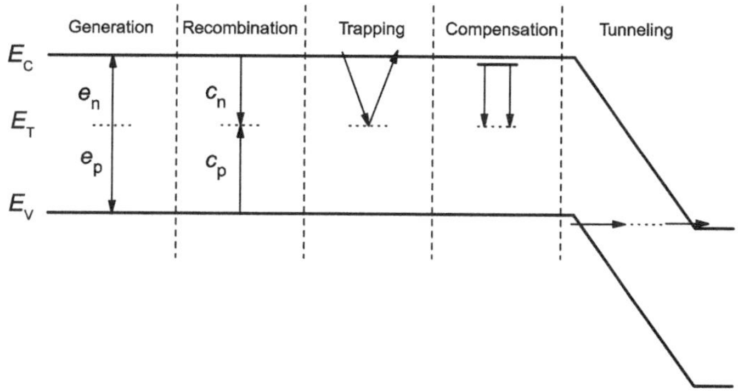

Figure 1 Basic interactions of a deep level in a semiconductor, with from left to right: carrier generation, carrier recombination, carrier trapping, carrier compensation, and carrier tunneling.

Depending on the position of the deep level in the band gap, it may affect the semiconductor properties in different ways, as sketched in Fig. 1. First, a deep level may act as a carrier generation center, whereby the thermal emission rate for electrons is given by:

$$e_{\mathrm{n}} = \frac{N_{\mathrm{c}}\sigma_{\mathrm{n}}v_{\mathrm{thn}}}{g_{\mathrm{n}}}\exp\left(-\frac{(E_{\mathrm{C}} - E_{\mathrm{T}})}{kT}\right) \tag{3}$$

with N_{C} the density of states in the conduction band, σ_{n} the capture cross section for electrons and v_{thn} the thermal velocity for electrons. A symmetrical relationship holds for the hole emission rate e_{p} toward the valence band. Equation (3) assumes detailed balance between carrier emission and carrier capture toward the respective bands. For direct band gap materials, like III–V compounds, the carrier generation may be dominated by optical generation (photoionization) of the deep level, adding to the thermal emission rate. Quite often, Eq. (3) has been rewritten in the form:

$$e_{\mathrm{n}} = K_{\mathrm{T}}T^{2}\exp\left(-\frac{(E_{\mathrm{C}} - E_{\mathrm{T}})}{kT}\right) = \frac{1}{\tau_{e_{\mathrm{n}}}} \tag{4}$$

with (K_{T}, E_{T}) termed the trap signature, which can be derived from an Arrhenius diagram representing e_{n} versus $1/kT$.

The carrier recombination in Fig. 1 is determined by an electron and hole capture rate, c_{n} and c_{p}, defined by:

$$\tau_{c_{\mathrm{n}}}^{-1} = n\sigma_{\mathrm{n}}v_{\mathrm{thn}} = nc_{\mathrm{n}} \qquad \tau_{c_{\mathrm{p}}}^{-1} = p\sigma_{\mathrm{p}}v_{\mathrm{thp}} = pc_{\mathrm{p}} \tag{5}$$

with n and p the free electron and hole density, respectively.

According to Fig. 1, a deep level can act as a generation center, when its position is close to the mid gap or intrinsic Fermi level position E_{i}; this gives rise to electron and hole emission rates of the same order of magnitude. When $e_{\mathrm{n}} \gg e_{\mathrm{p}}$ (or vice versa), the deep level will primarily interact with the conduction band (valence band) and can be considered as an electron (hole) trap (third case in Fig. 1). At the same time, when deep levels trap an electron from the conduction band, they remove a free carrier from the system and behave like compensating centers, reducing the free carrier concentration from the value determined by the (net) shallow donor concentration $N_D^+ = N_D$ to $N_D^+ - N_T^-$, in the case of a deep acceptor. Finally, in the case a large electric field exists, e.g., considering a reversely biased p–n junction or Schottky barrier, an electron can tunnel directly from the top of the valence band to the bottom of the conduction band (band-to-band tunneling). At the same time, a deep level may act as a stepping stone between the valence and conduction band in a trap-assisted tunneling (TAT) event, thus enhancing the carrier generation rate at large electric field (Simoen et al., 2011a).

When an excess free carrier density $\Delta n = \Delta p$ is created in a semiconductor material by an electrical or optical pulse, this will decay by SRH recombination (Shockley and Read, 1952) with a time constant given by:

$$\tau_{\mathrm{SRH}} = \frac{1}{\left[c_{\mathrm{n}}(n + n_{\mathrm{T}}) + c_{\mathrm{p}}(p + p_{\mathrm{T}})\right]} \tag{6}$$

In Eq. (6), n_{T} (p_{T}) corresponds with the free electron (hole) density when the Fermi level coincides with the trap level, i.e.:

$$n_{\mathrm{T}} = n_{\mathrm{i}} \exp\left(\frac{E_{\mathrm{T}} - E_{\mathrm{i}}}{kT}\right) \tag{7}$$

with E_{i} the intrinsic level, close to mid gap. The intrinsic carrier concentration n_{i} is given by:

$$n_{\mathrm{i}}^2 = np = N_{\mathrm{c}} N_{\mathrm{v}} \exp\left(-\frac{E_{\mathrm{G}}}{kT}\right) \tag{8}$$

and depends exponentially on the band gap energy E_{G}. In the case of low-level injection, where the excess carrier density Δn is much smaller than the doping density p, the minority carrier lifetime in the bulk semiconductor (neglecting surface GR) is given by:

$$\tau_{\mathrm{r}} = \frac{1}{\sigma_{\mathrm{n}} v_{\mathrm{thn}} N_{\mathrm{T}}} \tag{9}$$

when a single dominant recombination center with *minority* capture cross section σ_{n} and concentration N_{T} is present in the semiconductor material. At the same time, it has been shown that the relationship between the generation and recombination lifetime under low-injection conditions is given by (Schroder, 1982; Simoen and Claeys, 1999):

$$\tau_{\mathrm{g}} = 2\tau_{\mathrm{r}} \sqrt{\frac{c_{\mathrm{min}}}{c_{\mathrm{maj}}}} \cosh\left[\frac{E_{\mathrm{T}} - E_{\mathrm{i}}}{kT} + 0.5 \ln\left(\frac{c_{\mathrm{maj}}}{c_{\mathrm{min}}}\right)\right] \tag{10}$$

which is often further simplified to (Schroder, 1997):

$$\tau_g = \tau_r \exp\left(\frac{|E_T - E_i|}{kT}\right) \tag{11}$$

Given the importance of the carrier lifetime in device operation, it is clear from Eqs. (9) to (11) that a profound knowledge of the deep level parameters is necessary to model the latter quantity. In the following, different analysis methods assessing deep levels in semiconductors will be described.

3. RESISTIVITY-BASED DEEP LEVEL ANALYSIS

As shown in Fig. 1, a charged deep level can compensate for the shallow doping, thereby reducing the available free carrier density and increasing the resistivity $\rho_n = (qn\mu_n)^{-1}$ ($\rho_p = (qp\mu_p)^{-1}$), with q the electron charge (absolute value) and μ_n (μ_p) the low-field electron (hole) mobility. For a uniformly doped piece of semiconductor material, the free carrier density is given by the charge neutrality equation (Sze, 1981):

$$p + \sum_{i=1}^{l} N_{Di} \frac{1}{1 + g_{Di}\exp\left(-\frac{E_{Di} - E_F}{kT}\right)} = n + \sum_{j=1}^{m} N_{Aj} \frac{1}{1 + \frac{1}{g_{Aj}}\exp\left(\frac{E_{Aj} - E_F}{kT}\right)} \tag{12}$$

with N_{Di} and N_{Aj} the concentration of the i-th donor with level E_{Di} and the j-th acceptor with level E_{Aj}, respectively; g_{Di} and g_{Aj} are the corresponding degeneracy factors. In a p- (n-)type semiconductor, n (or p) is generally negligible, so that the free majority carrier density is a function of the temperature, through the Fermi level and the effective density of states in the conduction or valence band (Sze, 1981). Changing the temperature modifies in principle the Fermi level position, so that each time E_F crosses a (deep) donor or acceptor level it will contribute 50% of its total concentration to the sum terms in Eq. (12), when g_A or $g_D = 1$. In other words, by measuring the sample resistance as a function of T, one expects a step in the sample resistivity due to the so-called freeze-out of a deep level. In order to eliminate the effects of contact resistivity, a four-point probe (4PP) measurement is usually adopted, where the current is measured through two contacts and the potential applied to the two other contacts. However, as

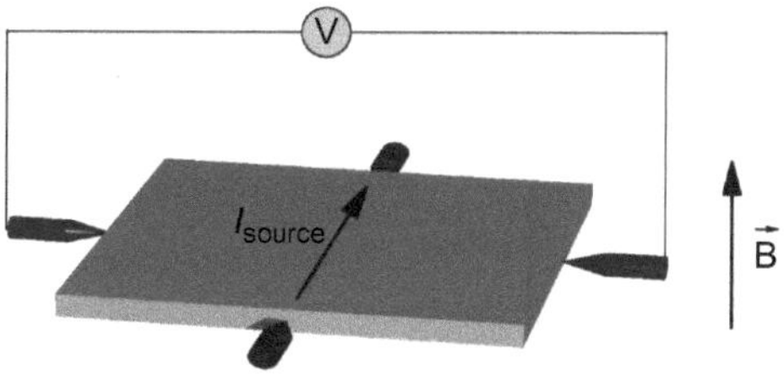

Figure 2 Principle of a Hall measurement on a uniform semiconductor sample. Due to the transverse magnetic field **B**, a potential difference builds up perpendicular to the direction of the current flow and of the magnetic field. This enables to determine the conduction type (p or n), the Hall mobility, and the Hall resistivity.

the carrier mobility μ_n or μ_p is also a strong nonmonotonous function of the temperature, one has to determine this parameter independently.

The most elegant way to do this is by performing Hall effect measurements in de van der Pauw configuration (Van der Pauw, 1958), illustrated in Fig. 2. The application of a magnetic field B perpendicular to the current flow generates a Hall voltage in the transverse direction. The sign of the Hall coefficient R_H gives the conduction type, while the free carrier concentration and the Hall mobility for, e.g., a p-type semiconductor are found from:

$$p_{\mathrm{Hall}} = \frac{1}{q \times R_H} \text{ and } \mu_{\mathrm{Hall}} = \frac{R_H}{\rho_p} \tag{13}$$

Under low magnetic field strength, p_{Hall} and μ_{Hall} differ from the true values p and μ_p by a factor close to unity, called the effective Hall factor (r_H):

$$\mu_{\mathrm{Hall}} = \mu_p \times r_H \text{ and } p_{\mathrm{Hall}} = p / r_H \tag{14}$$

At room temperature, r_H depends on the carrier type, the band structure, the carrier scattering mechanisms, as well as on the magnetic field (Szmulowicz, 1986). At the same time, the measured μ_{Hall} depends slightly on the sample geometry and on the contact layout (Paul and Cornils, 2009; Veirman et al., 2011).

An example of a 4PP resistivity measurement on a p-type germanium sample is given in Fig. 3, indicating the presence of several semi-shallow acceptor states, identified by the different ionization steps (freeze-out) as a function of *T*. Until the advent of DLTS in 1974 (Lang, 1974), Hall effect and resistivity (conductivity) measurements have been systematically used to identify impurity levels in semiconductors (an overview on the early results can be found in Sze and Irvin, 1968). The method has recently been applied to determine the compensating effect of oxygen thermal donors (OTDs) in

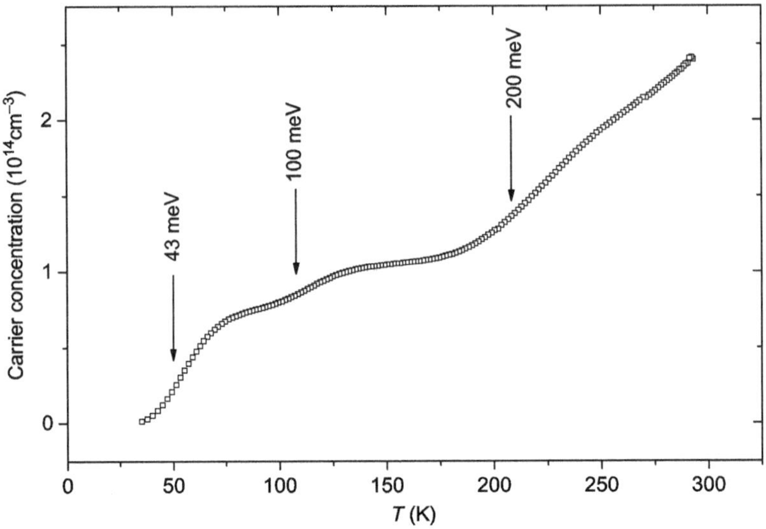

Figure 3 4PP resistivity as a function of temperature obtained on a piece of p-type germanium, showing the presence of several semi-shallow level centers.

p-type Czochralski (CZ) silicon for photovoltaic (PV) applications (Tanay et al., 2014; Veirman et al., 2011). The introduction of so-called micro-four-point probe (M4PP) Hall effect analysis (Petersen et al., 2008) has given a new impulse to the popularity of the technique for resistance profiling, as it solves several of the short-comings of the standard way of working. No dedicated contact preparation is required and a high spatial resolution can be obtained in M4PP. For example, 19 nm deep $n^{++}p^{+}$ junctions have been successfully characterized in InGaAs by combining M4PP and micro Hall effect measurements (Petersen et al., 2010).

Before the advent of M4PP resistivity measurements, dopants, and impurities in semiconductors have been profiled by two-point-probe (2PP) spreading resistance (SR) probing on beveled samples, as sketched in Fig. 4. The method was introduced in 1966 (Mazur and Dickey, 1966) and has since then evolved tremendously (Vandervorst et al., 2002). Spreading resistance probing (SRP) relies on the notion that the ideal resistance of a small-diameter conductive probe in perfect contact with the plane surface of a semi-infinite semiconductor sample is given by (Mazur and Dickey, 1966):

$$R_s = \frac{\rho}{4a} \tag{15}$$

with a the radius of the circular contact area. The SR R_S originates from the radial flow of current from the probe tip into the semiconductor. The actual

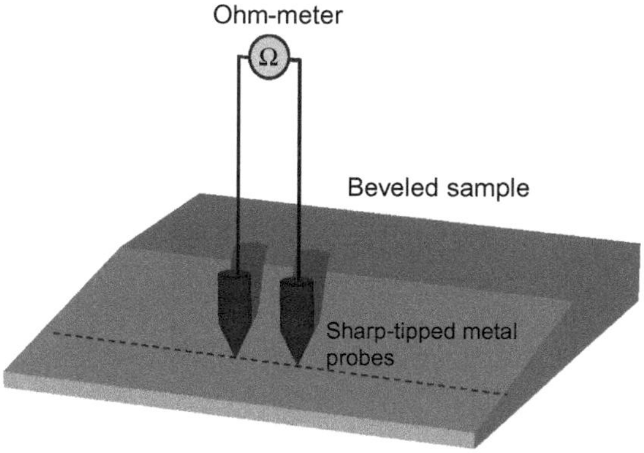

Figure 4 Principle of a two-point spreading resistance measurement. The sharp metal probes are pressed into the semiconductor material, locally transforming the semiconductor lattice into a highly conductive, high-pressure metallic phase, ensuring good ohmic contact. When moving the probes along the bevel, a carrier density profile along the depth of the sample can be obtained.

sampling volume (or the range of potential drop) is about 5*a*. In practical SRP the contact radius can be made as small as a few μm, leading to a correspondingly small probing volume. In practice, a 2PP arrangement is used, whereby the interprobe distance needs to be larger than 5*a*. In addition, the contact is far from circular so that a correction for the contact resistance is required. This is achieved by means of repeated calibration using homogeneously doped samples of the same material, conductivity type, and surface orientation of the sample under test. These calibration samples cover a broad range of resistivities, independently determined by the 4PP technique. In recent years, a so-called variable–temperature (VT) SRP method (150–400 K range) has been developed (Voss et al., 2002) to study not only shallow doping profiles but diffusion profiles of deep level impurities, e.g., S and Se donor levels in silicon.

The advent of scanning probe microscopy (SPM) has opened the route to high spatial resolution two-dimensional nanocontact resistivity profiling (Eyben et al., 2010). On the other hand, based on noncontact scanning tunneling microscopy (STM), the presence of electrical charges at dislocations in GaAs has been investigated (Ebert et al., 2001). At the same time, temperature-dependent scanning capacitance microscopy (SCM) has been developed, allowing for the detection of a hole trap at 0.4 eV above the valence band edge in an SiO_2/p-Si sample (Kim et al., 2001). More recently, the transport through a single donor in silicon has been studied by STM (Miwa et al., 2013), revealing that it is dominated by a minority carrier recombination process with the surrounding p-type matrix. STM has also been applied to investigate the Coulomb energy of a single silicon dangling bond (Nguyen et al., 2010). A simulation framework for the interpretation of three-dimensional carrier and impurity profiling by STM has been proposed by Fukuda et al. (2014). Finally, using Kelvin probe force microscopy on a lateral pn junction in a silicon-on-insulator (SOI) wafer, single dopant levels have been investigated, showing a deeper ground state compared to bulk silicon (Nowak et al., 2013). It is evident that the high spatial resolution of SPM offers the chance for studying individual defects and to investigate new defect physics in nanostructured semiconductor materials.

While a detailed treatment is out of the scope of this chapter, it should be pointed out that techniques based on the electrical detection electron paramagnetic resonance (EDEPR) have recently shown great potential for the detection of single defects/impurities in semiconductor materials (see, e.g., Bagraev et al., 2009; Morello et al., 2010; Pla et al., 2012). The main advantage of such spin-sensitive methods is that they have a real spectroscopic nature, as they allow the chemical identification of the involved atoms and their surroundings. The main drawback is that only paramagnetic defects with an unpaired electron (dangling bond) can be detected.

4. DEEP LEVEL TRANSIENT SPECTROSCOPY

DLTS was originally derived from space-charge-based methods, whereby the charge state of a deep level in the depletion region of a p–n junction or Schottky barrier (SB) diode is modified by a bias pulse (Figs. 5 and 6). The relaxation of the trapped charge after the pulse gives rise to a transient in the junction capacitance, which can then be analyzed to extract information on the deep level parameters. In this paragraph, after a brief introduction to the basic operation principle (Blood and Orton, 1992; Lang, 1974; Schroder, 1998), adapted forms of DLTS will be described, extending the capabilities of the technique beyond the standard approach. In addition, application of DLTS to defects at or beyond interfaces will be outlined. Finally, it will be shown that for the case of nonideal diodes, the device parasitics (series resistance, leakage current, contact resistance) may affect the DLTS signal; ways to address such issues will be discussed.

4.1 Principle of DLTS

The measurement principle of DLTS can be best explained by considering an asymmetrically doped p–n junction, i.e., one where either the p or the n region is at least 10 × higher doped than the complementary region. In that

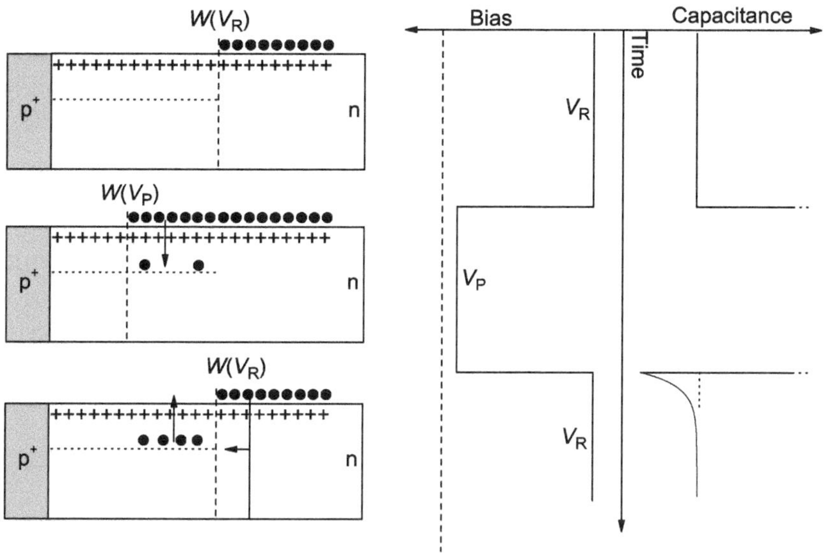

Figure 5 Bias pulsing sequence in DLTS to change the electron occupancy of electron traps in the depletion region of a p^+n junction.

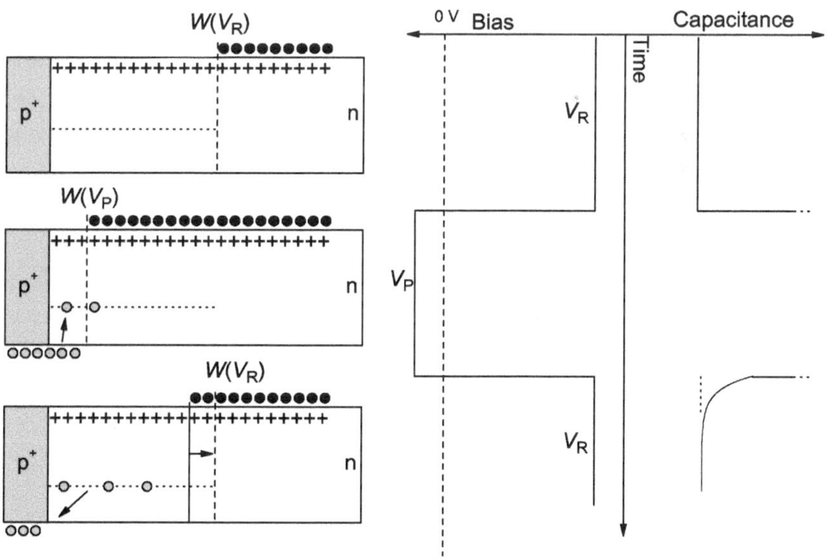

Figure 6 Forward-bias pulsing sequence in DLTS to change the hole occupancy of minority carrier (hole) traps in the depletion region of a p^+n junction.

case, the depletion region with width $W(V_R)$ at a reverse bias V_R will mainly extend in the lower doped n-type region of a p^+n junction, as indicated in Fig. 5 and is in first approximation given by:

$$W(V_R) = \sqrt{\frac{2\varepsilon_S(V_R + V_{bi})}{qN_D}} \tag{16}$$

with ε_S the permittivity of the semiconductor and V_{bi} the built-in potential, which can be derived experimentally from the intercept of a $1/C^2$ versus V_R plot. It is assumed that $N_D \ll N_A$ (the doping density on the p^+ side) and that the junction is abrupt, with a uniform doping profile in the substrate (N_D constant). The corresponding capacitance is given by $C(V_R) = A_J\varepsilon_S/W(V_R)$, with A_J the junction area.

Reducing the reverse bias from V_R to V_P during a pulse time t_p, also reduces the depletion region to $W(V_P)$ (middle diagram of Fig. 5). This allows free electrons (majority carriers) to enter the region between $W(V_R)$ and $W(V_P)$, where they can be captured by the empty deep level centers with activation energy E_T. This trap filling occurs exponentially with time, with a capture rate given by Eq. (5) in first approximation. For a sufficiently long pulse duration, all traps with a concentration N_T, assumed uniform in depth for the moment, will be filled in the probed depletion region. At the end of the pulse, the original V_R is applied to the reversely biased p^+n junction. While the free carriers originating from the shallow donors respond very fast to this bias step, the electrons trapped in the deep levels will be reemitted with a rate constant given by Eq. (3). The negative charges still trapped in the deep levels compensate the positive charge of the ionized shallow donors, resulting in a space charge density $+q(N_D - N_T)$, so that immediately after the pulse, the depletion width $W(V_R)$ will be larger and the capacitance will be smaller than the corresponding steady-state or quiescent values at V_R. As indicated in the bottom diagram of Fig. 5, the thermal emission of the trapped charge from the deep levels will give rise to a capacitance transient, approximately given by:

$$C(t) - C(\infty) = \Delta C\left[1 - \exp\left(-\frac{t}{\tau_{en}}\right)\right] \tag{17}$$

with a negative capacitance transient amplitude given by $\Delta C = C(\infty) - C(0)$. Symmetrical relationships hold for an n^+p junction. Because for most practical cases $e_n \gg e_p$ for levels in the upper half of the band gap and vice versa, only so-called *majority carrier spectra* can be obtained in standard DLTS, using a *majority carrier bias pulse*. This implies that electron traps will be probed in the upper half of the band gap for an n-type semiconductor and hole traps in the bottom half of the band gap for a p-type semiconductor.

In the case of a symmetrical p–n junction, one has to account for the fact that the depletion region extends in both sides and that both holes and electrons will be captured in the p- and n-side, respectively. While this enables the simultaneous study of electron traps in the upper half of the band gap and hole traps in the lower half, interpretation of the resulting spectra is not easy. Therefore, one generally relies on an asymmetric high–low junction or a Schottky barrier at the expense of addressing only the majority carrier traps in their respective half of the band gap. A combined study on both an n^+p and p^+n junction or equivalent Schottky barrier should in principle resolve this issue. Alternatively, for a p^+n diode one can also rely on a forward-bias minority carrier injection pulse, as illustrated in Fig. 6. In that case, minority carriers will be injected from the p^+ anode into the n-type substrate, where they can be captured by hole traps close to the junction. Consequently, the trapped positive charge will add to the fixed donor charge, i.e., $q(N_D + N_T)$ after the pulse, so that the capacitance will be higher, equal to $C(\infty) + \Delta C$. The resulting transient will have the opposite sign (decay), as well as the corresponding DLTS peak. SBs, on the other hand, are majority carrier devices, so that minority carrier traps can in principle only be detected by the creation of minority carriers through an optical pulse—an example is given in Fig. 7 for the case of an irradiated Sn-doped n-type float-zone (FZ) silicon sample, revealing the four deep levels of the Sn–V (V=vacancy) complex (Claeys et al., 2001). However, it has been shown that when the barrier height is sufficient to invert the semiconductor surface, minority carrier peaks can also be detected in DLTS (Simoen et al., 1987, 2006). An example is given in Fig. 8 for a Ni, Pt, or Pd Schottky barrier on n-type Ge, showing both positive peaks (electron traps) and negative peaks due to minority carrier hole traps (Pd and Pt cases).

While the foregoing is common to all junction-based spectroscopies developed in the sixties and early seventies, the typical feature of DLTS is the use of the rate window concept which has originally been implemented by a double boxcar filtering, indicated in Fig. 9 (Lang, 1974). Considering the thermally activated nature of the emission time constant τ_{e_n} in Eq. (17), one expects a strong variation of the nature of the capacitance transient, when varying the temperature. As shown in Fig. 9, at low T, the emission rate of Eq. (4) will be low and the capacitance will hardly change after the pulse: the captured charge is frozen on the deep levels. Conversely, at high

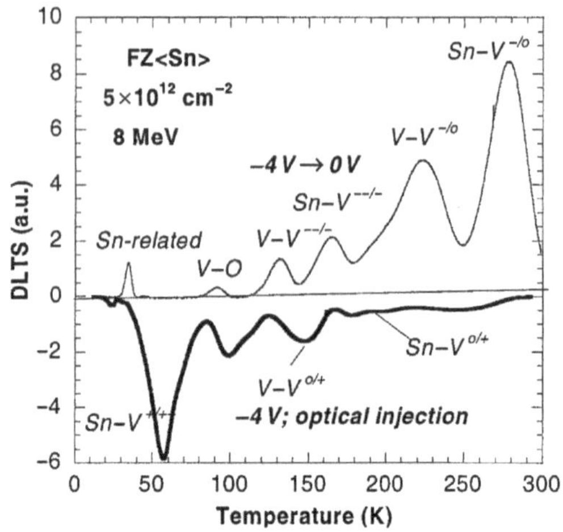

Figure 7 DLT-spectrum of an 8 MeV proton-irradiated n-type FZ–Si sample, doped with Sn and corresponding with a bias pulse from −4→0 V (positive spectrum) or with an optical pulse produced by a laser diode at −4 V reverse bias (negative spectrum). An Au Schottky Barrier was deposited on the front side, while the backside was covered with InGa eutectic and In. *After Claeys et al. (2001), reproduced with permission of the Electrochemical Society.*

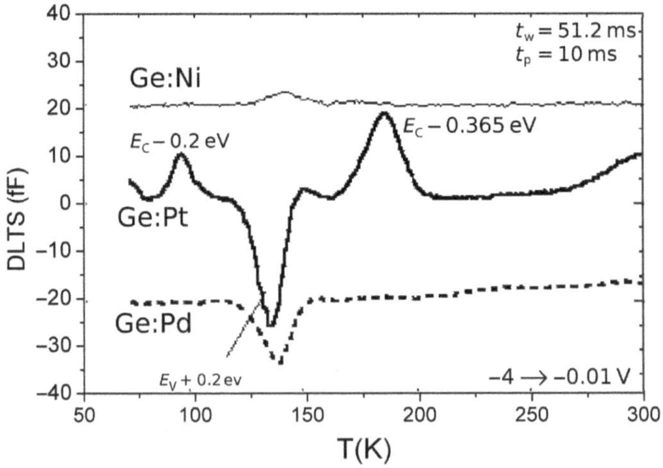

Figure 8 DLT-spectra on Ni, Pt, and Pd Schottky barriers on n-type Ge, whereby the 30 nm metal layer has been deposited by Ar-ion sputtering. A reverse bias pulse of −4→ −0.01 V was applied. The high SB in the range of 0.6 eV enables the detection of a hole trap at E_V+0.3 eV, created by the sputtering damage of the Pt and Pd deposition on Ge. *After Simoen et al. (2006), reproduced with permission from AIP Publishing LLC, Copyright (2006).*

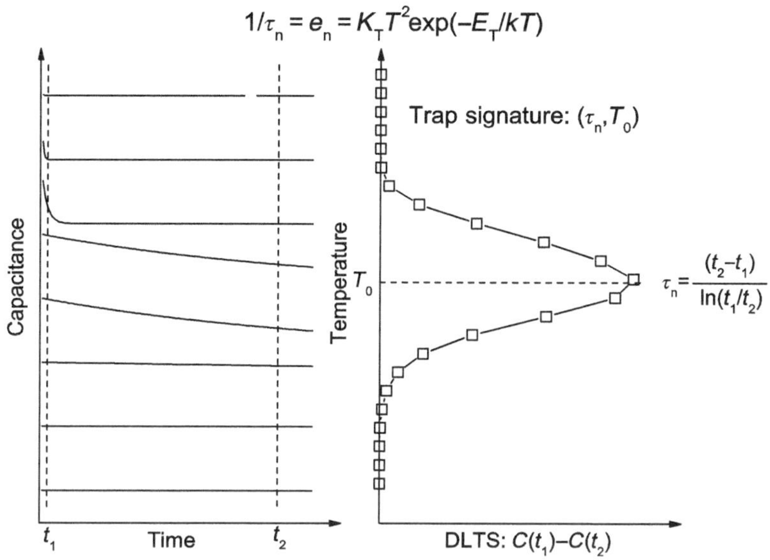

Figure 9 Principle of rate window by the double boxcar technique and resulting spectrum as a function of temperature.

temperatures, charge emission from the traps will be fast, resulting in a fast capacitance decay. As shown schematically in Fig. 9, by taking the capacitance difference at two fixed moments t_1 and t_2 after the end of the pulse, the signal $C(t_1) - C(t_2)$ gives rise to a peak-shaped curve as a function of the temperature and this, for every deep level present with a sufficiently high concentration of filled traps in the depletion region of the junction or SB. At the same time, the DLTS peak maximum corresponds with a fixed emission rate (or time constant) defined by t_1 and t_2 only and given by (Lang, 1974):

$$\tau_n = \frac{t_2 - t_1}{\ln\left(\frac{t_1}{t_2}\right)} \tag{18}$$

Determining the temperature T_m belonging to the maximum of the peak enables to derive one data point of the Arrhenius plot for each deep level (Fig. 10). Repeating the temperature scan for different values of the rate window (different t_1 and t_2 pairs) allows the construction of an Arrhenius diagram, as outlined in Fig. 10. According to Eq. (4), from the slope of a least-squares linear fit to $\ln(\tau_n T^2)$ versus $1/kT$ one can derive the so-called activation energy, while the intercept gives K_T, which is proportional to the majority capture cross section for a majority carrier peak. The

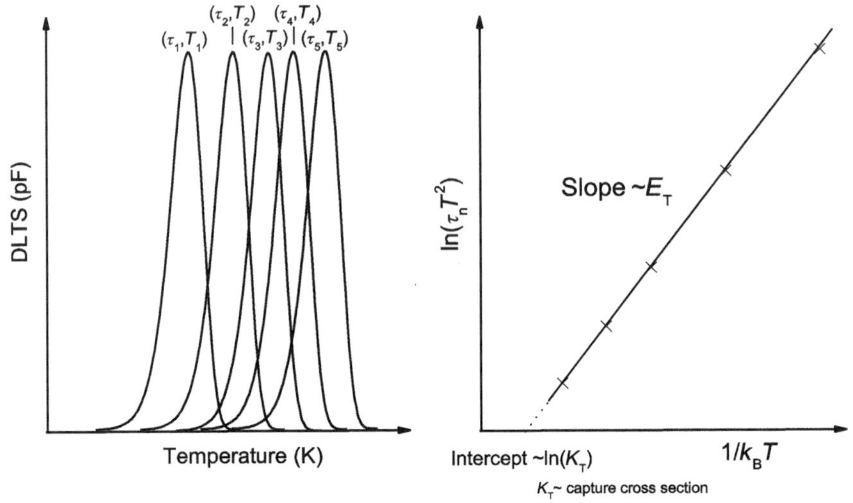

Figure 10 Principle of Arrhenius plot measurement.

derived activation energies will be with respect to the valence band maximum (p-type) or conduction band minimum (n-type)—of course, the opposite holds for the negative minority carrier peaks, which are referenced to their corresponding band edge.

Finally, the trap concentration can be calculated from the peak height, which is proportional to the capacitance transient amplitude ΔC, through (Lang, 1974):

$$\frac{N_\mathrm{T}}{N_\mathrm{dop}} = \frac{1}{2}\frac{\Delta C}{C} \tag{19}$$

which is valid for a trap concentration which is smaller than 10% of the doping concentration ($N_\mathrm{T} \ll N_\mathrm{dop}$) and provided the peak amplitude is saturated by a sufficiently long pulse. Equation (19) implies that DLTS has a relative detection limit or sensitivity, related to the background doping concentration. Typical values quoted in the literature are $10^{-4} \times N_\mathrm{dop}$, enabling the detection of very low impurity or defect concentrations, which are far below the typical detection limits of most chemical analysis techniques. The doping density can in principle be derived from the slope of a $1/C^2$ versus V_R plot.

4.2 Refinements of the basic DLTS principle

Over the years, the DLTS technique has been further refined in order to enhance the measurement accuracy and cope with some physical complications, like inhomogeneous trap concentration profiles, the presence of electric-field-assisted carrier generation (Fig. 1), the limits of the depletion approximation, etc. A first obvious improvement regards the direct determination of the majority capture cross section. While the K_T intercept of an Arrhenius plot provides a value for σ_n or σ_p, this is at best an order of magnitude estimate, neglecting a possible temperature dependence. The intercept value in fact corresponds with a temperature extrapolated to infinity. The best way to measure the capture cross section is by monitoring at a fixed temperature the DLTS amplitude versus the bias pulse duration t_p. In the first instance, the capacitance amplitude will follow an exponential filling, represented by:

$$C(t_p) = C(0) + C(t_p \to \infty)\left[1 - \exp\left(-\frac{t_p}{\tau_{cp}}\right)\right] \tag{20}$$

According to the measurement of Fig. 11, the exponential filling is only valid at short t_p, while there is a long tail at large t_p of so-called slow filling instead of the saturation to a constant amplitude predicted by Eq. (20). This is related to the presence of a free carrier tail at the edge of the depletion region, with a decay length given by the Debye length:

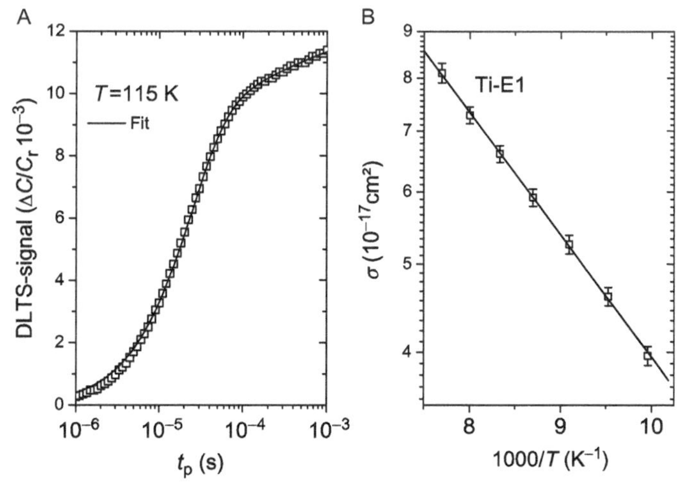

Figure 11 (A) DLTS signal versus t_p for an electron trap belonging to substitutional Ti (Ti_S) in n-type Ge at 115 K and (B) corresponding electron capture cross section versus temperature.

$$\lambda_D = \sqrt{\frac{kT\varepsilon_s}{q^2 N_{dop}}} \; (\lambda \mathrm{D} \rightarrow \lambda_\mathrm{D}) \tag{21}$$

A correction procedure has been proposed in the literature (Pons, 1984), which was recently refined by Lauwaert et al. (2006), enabling an accurate extraction of the capture cross section by fitting a theoretical model including slow capture to the experimental data, like in Fig. 11A.

According to Fig. 11B, the electron capture cross section of the Ti_s electron trap in n-type Ge exhibits an activated temperature dependence, obeying:

$$\sigma_\mathrm{n} = \sigma_\infty \exp\left(-\frac{\Delta E_\sigma}{kT}\right) \tag{22}$$

and is typical for the σ_n of a repulsive center, i.e., for electron capture by a deep acceptor, changing its charge state from single negative (−) or double negative (2−) to a more negative state. Thermal activation indicates dissipation of the excess carrier energy upon capture by multiple phonon emission by the local coupling to the lattice (Henry and Lang, 1977). The thermal activation can be explained by considering the existence of a Coulomb repulsive barrier around the deep level center which is more easily overcome by a free carrier having a higher thermal energy. An activated capture cross section can thus be used as a criterion to distinguish repulsive centers from neutral or attractive ones. In the latter case, the capture cross section generally reduces with T (Darken, 1992).

Another important consequence of the thermal activation of the capture cross section is that the interpretation of the slope of an Arrhenius plot requires a careful discussion. Thermodynamically speaking, the Arrhenius plot reveals the Gibbs free energy change upon emission from a localized state to its respective band (Ricksand and Engström, 1991). The latter can be written as $\Delta G(T) = \Delta H - T\Delta S$, with ΔH the enthalpy change and ΔS the entropy change involved in the transition from bound to free. This implies that the thermal emission rate for electrons can be written as:

$$e_\mathrm{n}(T) = \sigma_\mathrm{n}(T) v_\mathrm{thn}(T) N_\mathrm{c}(T) \exp\left(-\frac{\Delta G(T)}{kT}\right) \tag{23}$$

Combined with Eq. (22), this means that the emission rate can be rewritten as:

$$e_{\mathrm{n}}(T) = \sigma_{\infty} v_{\mathrm{thn}}(T) N_{\mathrm{c}}(T) \exp\left(\frac{\Delta S}{k}\right) \exp\left(-\frac{\Delta H + \Delta E_{\sigma}}{kT}\right) \quad (24)$$

In other words, from the slope of an Arrhenius plot, one can derive a so-called apparent activation enthalpy of the deep level, which generally consists of the sum of the true enthalpy and the activation energy of the cross section (when applicable). A guideline to determine the different parameters of a deep level is then: first, derive $\Delta H + \Delta E_{\sigma}$ from the slope of an Arrhenius plot; second, measure the capture cross section at different T to find ΔE_{σ} and σ_{∞} and, finally, use these values to calculate ΔS and ΔG from the Arrhenius plot. An example of such an analysis for the substitutional Ti and Cr impurity levels in Ge can be found in (Lauwaert et al., 2009).

The trap filling not always obeys an exponential law, followed by some slow increase at larger t_{p}, but may exhibit a DLTS amplitude which increases proportional with $\ln(t_{\mathrm{p}})$ over many decades, as shown in Fig. 12 (Simoen et al., 2011b). This points to a different kind of defects, either in an oxide layer on top of the semiconductor (border traps, BTs; see next section 4.3) or is also typical for extended defects or point defects trapped in the strain field of an extended defect (dislocation, stacking fault, twin boundary, precipitate) (Kimerling and Patel, 1979; Schröter and Cerva, 2002). This filling kinetics can be explained by considering the existence

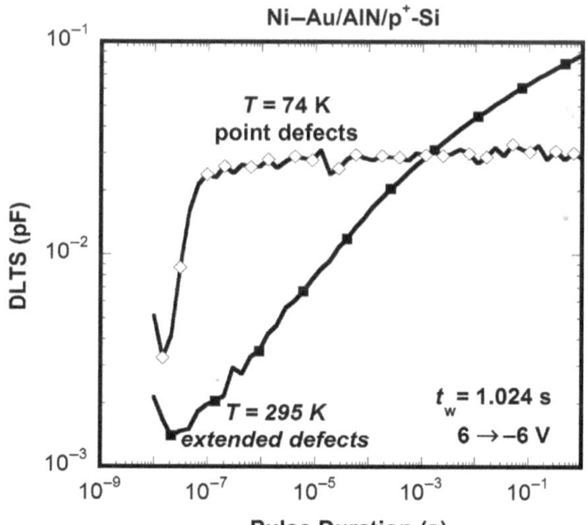

Figure 12 Trap filling kinetics for point defects compared with extended defects. The latter is following an $\ln(t_{\mathrm{p}})$ law.

of a potential barrier around the extended defect, which increases with increasing occupation. In other words, the capture cross section reduces with filling time t_p due to the increase of the charge on the extended defect, creating an increasing potential barrier and thus making it harder to capture another free carrier on the coupled defect sites.

While DLTS is called "spectroscopy", since each deep level species gives rise to a specific peak as a function of temperature, it does not directly provide chemical or structural information on the associated defects. In order to get some idea about the nature of the defects, one has to combine it with real spectroscopic techniques, like electron spin resonance (ESR) or prepare dedicated samples, like the ones in Fig. 11, where Ge was implanted with Ti, followed by a thermal activation anneal to study the Ti_S-related deep acceptors. One can also start from semiconductor material with different doping densities or grown with a known impurity content. Also additional external perturbations (uniaxial stress) may reveal information on the structure and orientation of the defect. Investigating the thermal stability of defect complexes by isothermal and isochronal annealing is common practice, especially in the context of irradiation- or implantation-induced defects and helps to establish defect reaction models. One of the issues related with DLTS is the fairly moderate energy resolution, which scales with the temperature and usually amounts to several kT. As can be seen in Figs. 7 and 8, the full width at half maximum (FWHM) of a DLTS peak increases with its maximum position, so that defects with closely spaced deep levels may give rise to overlapping peaks or at best to a shoulder, if their majority capture cross section is significantly different. The advent of numerical DLTS, whereby the recorded capacitance transients are handled numerically to generate a spectrum, has opened the doors for "high-energy-resolution" DLTS. The best known approach is based on Laplace-transform DLTS (Dobaczewski et al., 2004), providing an energy resolution in the range of 1 meV and whereby the transients are measured at constant temperature and analyzed through a Laplace transformation. This requires, however, a tight temperature control and stability over a long measurement period and stable samples with low-noise ohmic contacts and defect concentrations in the range of 10% of the free carrier concentration.

One aspect which has been overlooked so far is the occurrence of electric-field-assisted thermal emission from a deep level center (Vincent et al., 1979). At lower fields ($F \sim 10^4$ V/cm) and for an attractive center, i.e., a deep donor for electrons in n-type or a deep acceptor for holes in p-type material, the so-called Poole–Frenkel (PF) effect (Frenkel, 1938) can occur, giving rise to an electric-field-dependent emission rate. The presence of the electric field in the depletion region will cause a lowering of the barrier for carrier emission, given by:

$$\Delta E_T(F) = q\sqrt{\frac{ZqF}{\pi\varepsilon_S}} \tag{25}$$

which is valid in the one-dimensional approximation (Hartke, 1968). In Eq. (25), Z is the charge number of the attractive center, i.e., $Z=1, 2, \ldots$ The PF effect causes a DLTS peak shift toward lower T for a higher reverse bias (a higher maximum field in the depletion region of the junction or SB). As a result, the Arrhenius plot and corresponding activation energy becomes V_R dependent or in the case of Fig. 13, doping dependent. In order to determine the PF barrier lowering of Eq. (25), spectra with different V_R have to be recorded. Representing the measured activation energies versus $\sqrt{F}$, with F the average field corresponding with V_R and V_P, should yield a straight line, whereby the intercept with the vertical axis provides the true activation energy (as argued above in fact the apparent activation enthalpy) at zero field. The example in Fig. 13 is interesting as it corresponds with a double donor situation in n-type silicon ($Z=2$) (Rafi et al., 2005). The experimental value for the barrier lowering of $2.2\times10^{-4}\times q\times F^{1/2}$ matches very well the

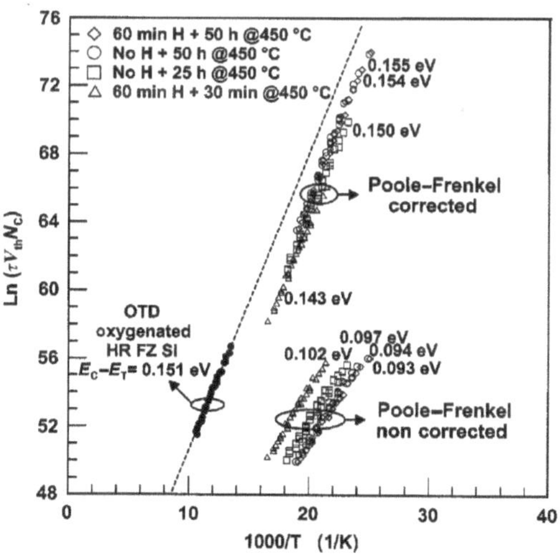

Figure 13 Arrhenius plots of n-type Czochralski Si treated at 450 °C for different times, after different plasma hydrogenation treatments, introducing oxygen-related double thermal donors. The Arrhenius plots correspond with the double donor state OTD$^{+/++}$. *After Rafí et al. (2005), reproduced with permission of the Electrochemical Society.*

theoretical one of Eq. (15) for a charge of +2q in the empty state (Rafi et al., 2005). As the electric field in the depletion region of a junction varies linearly with position with respect to the junction (or surface in case of a SB), a more correct procedure consists of performing two DLTS measurements with slightly different reverse biases V_{R1} and V_{R2} and subtracting the respective spectra (Fleming et al., 2014).

In case the PF effect occurs, it can be used to identify deep attractive centers in a semiconductor, i.e., deep donors in n- and deep acceptors in p-type material. This requires a sufficiently large doping density and/or reverse bias. However, it has been noted that in some cases, the PF shift is modest if not absent, even when the high capture cross section ($>10^{-14}$ cm^2) value indicates an attractive center (Lauwaert et al., 2012a). At higher electric fields, deep level-assisted tunneling effects may occur in the depletion region (Vincent et al., 1979), even for repulse centers, so that a peak shift toward lower temperatures for increasing V_R is not a full-proof method for identifying attractive deep level states.

As mentioned before, the depletion approximation, assuming no free carriers at the edge of the depletion region should be abandoned for a correct evaluation of the capture cross section in DLTS. The same applies for the deep level concentration N_T and more in particular for the determination of concentration profiles in case of a nonuniform trap distribution. This occurs for example for implantation or irradiation-induced deep levels which usually exhibit a peak concentration near the end-of-range of the implanted species or particle. Equation (19) is a first-order approximation, valid for small concentrations and neglecting the free carrier profile at the edge of the depletion region. As a consequence, also the trap occupation shows up a profile, varying smoothly from unoccupied (by an electron) at the depletion region edge to 0.5 where the trap level crosses the Fermi level (for a degeneracy factor 1). This situation is sketched in Fig. 14, for a reverse bias V_R and V_P. As a result, the traps will only be filled in the region between W_R–λ_R and W_P–λ_P during the bias pulse (Zohta and Watanabe, 1982), with λ_R and λ_P the distance from the crossing point of E_F with E_T and the edge of the depletion region and given by:

$$\lambda_R = \sqrt{\frac{2\varepsilon_S(E_F - E_T)}{q^2 N_{dop}}} \ (\lambda R \rightarrow \lambda_R) \tag{26}$$

As a consequence, the trap concentration has to be corrected by a factor, defined by V_R, V_P, and λ, as follows (Zohta and Watanabe, 1982):

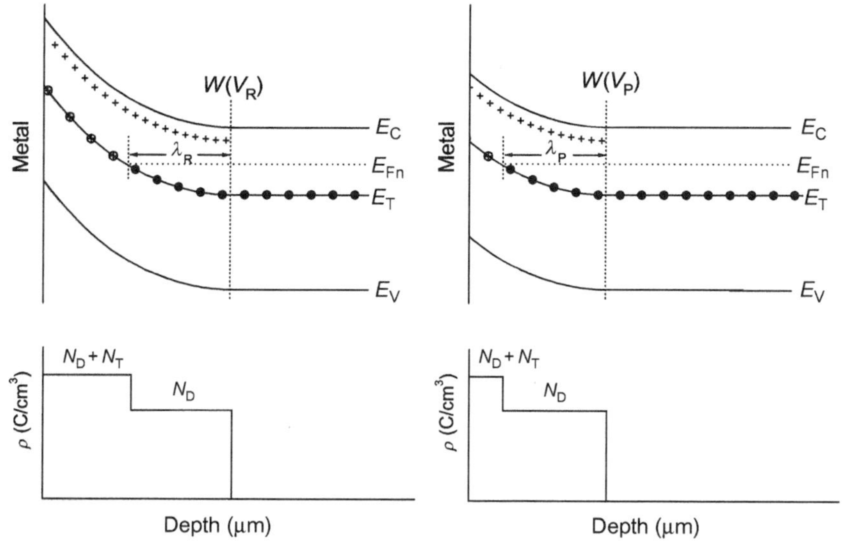

Figure 14 Band diagram in an SB and corresponding charge density profile at a reverse bias V_R (left) and V_P (right). Also indicated is the λ_R and λ_P region confined by the edge of the depletion region and the crossing point between the deep level E_T and the Fermi level E_F.

$$\frac{N_T(W_m - \lambda_m)}{N_{dop}} = \frac{1}{\left(1 - \frac{\lambda_R}{W_R}\right)^2 - \left(\frac{W_P - \lambda_P}{W_R}\right)^2} \frac{2\Delta C(0)}{C} \tag{27a}$$

with:

$$W_m - \lambda_m = \frac{1}{2}[(W_R - \lambda_R) + (W_P - \lambda_P)] \tag{27b}$$

The so-called "λ-effect" becomes particularly important for the determination of trap concentration profiles (Chasin et al., 2014; Zohta and Watanabe, 1982). A further step is the use of constant-capacitance (CC-) DLTS profiling in combination with the double-correlation technique, which should be suitable for the accurate determination of deep level profiles with a high relative concentration (Johnson et al., 1979). In CC-DLTS, the capacitance is kept constant during the transient by a feed-back loop controlling the reverse bias. According to Eq. (16), the space charge and reverse bias are both under the square root sign, so that the V_R transient will always be proportional to the charge transient, irrespective of the trap concentration. In double correlation, two different bias pulses are employed consecutively and the difference spectrum is determined in order to single out a narrow spatial or energy window.

State-of-the-art DLTS systems no longer rely on time-consuming analog filtering, but generate the spectra directly from the stored capacitance or current transients, using for example a fast Fourier transform (FFT) algorithm (Weiss and Kassing, 1988). Different spectra are calculated by a combination of different FFT coefficients, so that Arrhenius plots can be generated after one temperature scan. Another variant of DLTS considers a variation of the bias pulse frequency *f* at constant temperature. Figure 15 compares the standard T-scan DLT-spectrum with an isothermal f-scan spectrum for the same Pt/Al_2O_3/n-GaAs metal–oxide–semiconductor (MOS) capacitor (diameter $\Phi = 500$ µm). The figure illustrates the larger dynamic range of the isothermal DLTS: while in the first case, only part of the broad peak can be measured between 77 and 320 K at a fixed rate window of about 230 ms, the full peak can be accessed in the isothermal spectrum, up to emission rates of several s or more. An Arrhenius plot can then be constructed by combining the emission time constant corresponding with the DLTS peak position for every f-scan at different fixed temperatures. Another advantage of this approach—particular in the case of a continuous density-of-states (DOS) distribution—is the fact that the capture cross section at each activation energy corresponds with the same fixed *T*.

4.3 Application of DLTS to MOS structures: Interface and border traps

Besides SB and p–n junctions, other semiconductor (SC) device structures like MOS capacitors (MOSCAPs) and MOS field-effect transistors (MOSFETs) can be employed for deep level studies. In this case, not only defects in the depletion region but also at the oxide/semiconductor interface and even in the gate oxide can be assessed. The same basic operation principle, relying on a gate voltage pulse from depletion into accumulation, filling the available traps, and monitoring the resulting capacitance transient is used (Fig. 16). The main difference is that besides deep levels in the semiconductor depletion region, also traps at the Semiconductor/Insulator interface are filled, which both can contribute to the capacitance transient. In order to distinguish interface from bulk traps, it has been shown that one can rely on the dependence of the DLTS peak on the bias pulse height (Yamasaki et al., 1979). As shown in Fig. 17, the broad peak at lower temperatures exhibits a pronounced bias pulse dependence, whereby the low-temperature flank of the peak increases when going more into accumulation during the pulse. The deeper peak, on the other hand, stays at the same position while the amplitude reduces with pulse amplitude, in agreement with

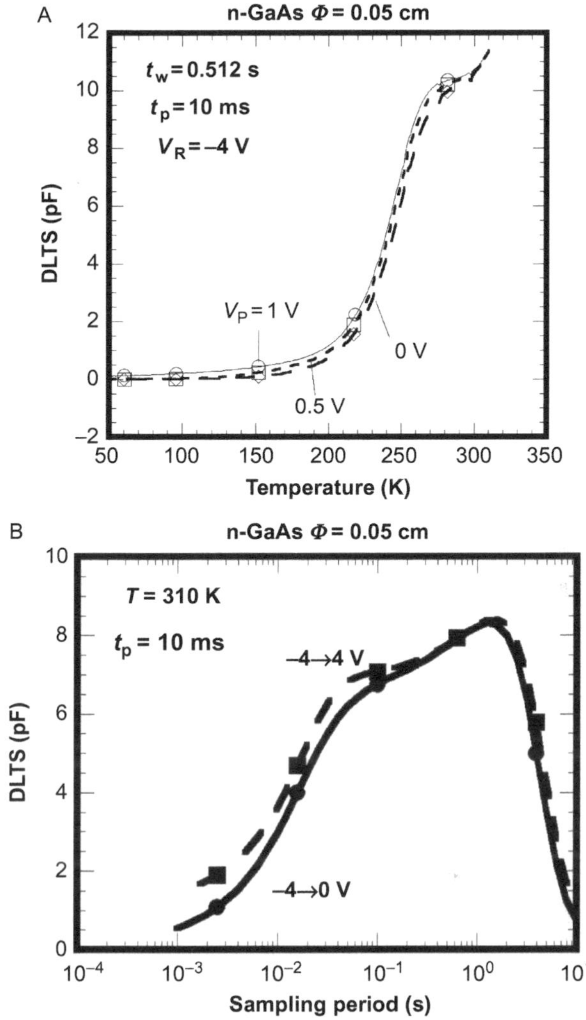

Figure 15 (A) DLTS spectra corresponding with a Pt/Al_2O_3/n-GaAs MOS capacitor with diameter 500 μm. A gate pulse from −4 to 0 V (◊), 0.5 V (□) or 1 V (○) has been applied, respectively, for a pulse time of 10 ms and a sampling period of 0.512 s. (B) DLTS frequency scan of the same n-type GaAs capacitor. The measurement temperature was 310 K.

Eqs. (27a) and (27b). The first behavior is typical for interface states and can be explained as follows: going more into accumulation during the voltage pulse enables to fill shallower interface traps closer to the band edge and promotes their contribution to the spectrum. As shallower traps appear at lower T in DLTS, this will give rise to a stronger increase of the amplitude at lower

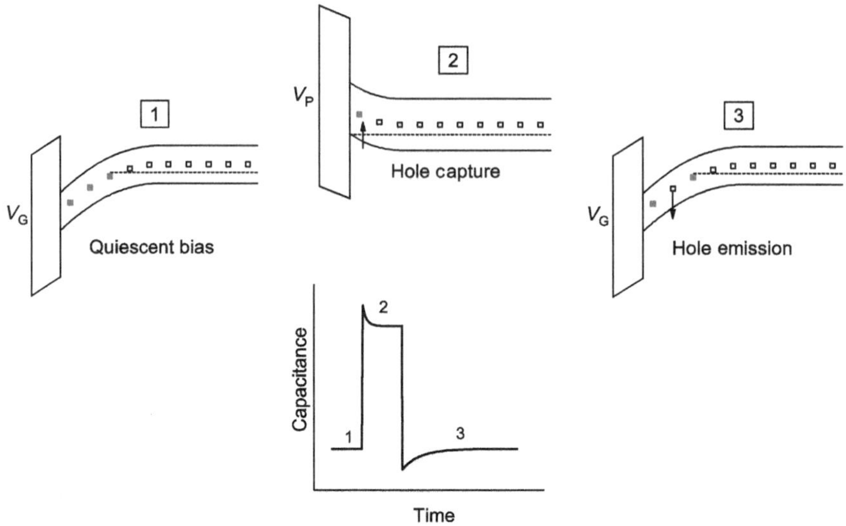

Figure 16 Principle of capacitance DLTS on an MOS capacitor (MOSCAP) on a p-type semiconductor. (1) The MOSCAP is biased in depletion. (2) Traps are filled by pulsing into accumulation. (3) The original reverse bias is restored whereby the filled traps relax back to their steady state by electron emission, resulting in a capacitance transient.

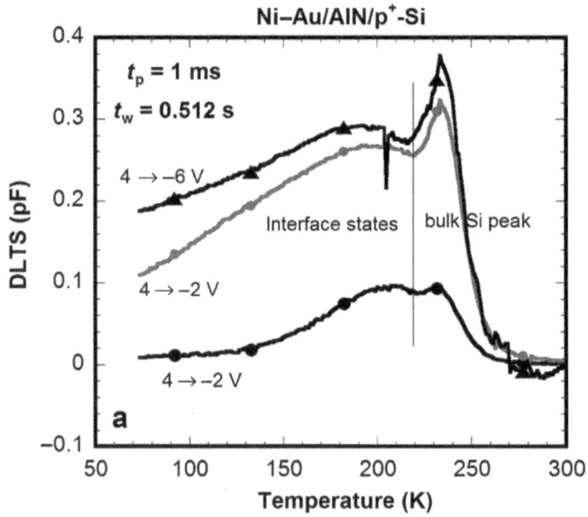

Figure 17 DLT-spectra obtained on an Ni–Au/AlN/p^{++}-Si MIS capacitor ($\Phi = 1$ mm) corresponding with different bias pulses, at the same depletion bias of −6 V. *After Simoen et al. (2011b), reproduced with permission of IOP.*

temperatures, like in Fig. 17. Bulk traps in the depletion region, on the other hand, generally do not exhibit a peak shift upon a change of the pulse height; only the peak amplitude reduces at constant maximum temperature for a neutral or repulsive trap.

One issue regarding interface states is that the energy distribution is usually spread over the entire band gap, while they are spatially localized at the interface. When applying a large-amplitude bias pulse to the gate of a MOSCAP, a large part of the states will contribute to the transient signal, since the Fermi level will be swept over some part of the band gap during the charge relaxation. This obscures the derivation of the DOS and the energy dependence of the capture cross section. Therefore, energy-resolved DLTS has been proposed (Johnson, 1982), whereby CC-DLTS is combined with a double-correlation pulsing scheme, employing two different pulses consecutively, with the same reverse bias and a pulse bias $V_{P1} > V_{P2}$, whereby the $V_{P1} - V_{P2}$ difference defines a small energy interval at the interface, contributing to the DLT-spectrum. In this way, more accurate parameters for interface states can be obtained. Alternatively, one can measure DLT-spectra using a bias pulse with a small amplitude (order 0.1–0.2 V) which again singles out a small energy interval at the interface. An example is given in Fig. 18 for a metal–insulator–semiconductor (MIS) capacitor on n-type Si, showing the shift of the small-pulse DLTS peaks toward lower T (lower activation energy, closer to E_C), for increasing V_R. This is in agreement with the previous statement that when the MOSCAP is biased more into accumulation, shallower interface traps contribute to the spectrum. From the Arrhenius plot corresponding with each small-pulse peak, an activation energy and

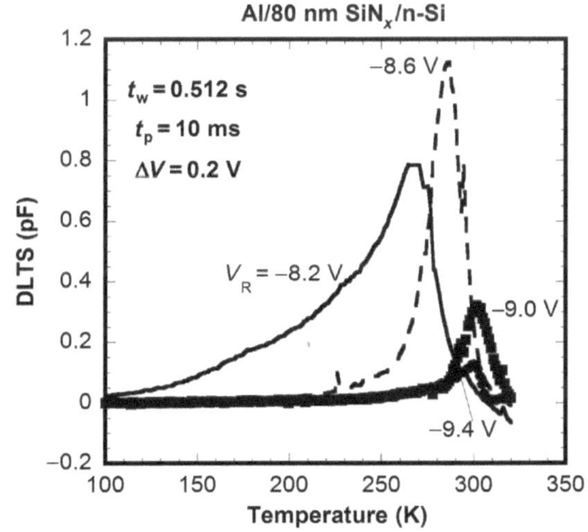

Figure 18 FT-DLT-spectra of an Al/80 nm SiN$_x$/n-Si(100) capacitor, corresponding with small-step (0.2 V) bias pulses. From an Arrhenius plot for each peak, the activation energy and corresponding electron capture cross section can be derived.

corresponding electron capture cross section have been derived showing a strongly increasing σ_n for higher activation energy (Gong et al., 2010).

An additional feature which can occur at higher temperatures (RT and above) when applying a filling pulse to a MOSCAP in deep depletion is the so-called generation or inversion response (Pearce et al., 1987; Volkos et al., 2006). Besides the discharging of the filled traps, minority carrier generation by deep levels in the depletion region and surface states and minority carrier diffusion from the neutral bulk will cause a capacitance increase after the voltage pulse. This brings the capacitance from deep depletion into inversion. The thermally activated behavior of minority carrier generation also yields a DLTS peak with close-to-mid-gap activation energy typically (Pearce et al., 1987; Volkos et al., 2006).

In MOSCAPs, a third kind of deep levels is frequently observed, which is related to traps in the dielectric in communication with the semiconductor interface (Bauza, 1998; Lakhdari et al., 1988; Van Staa et al., 1983; Vuillaume et al., 1986). These have been termed "border traps" (Fleetwood, 1992) and have typical response times in the range of 1 ms to 1 s. They give rise to the well-known $1/f$ noise in MOSFETs (Simoen et al., 2013) and are locally separated from the free carriers, so that capture occurs through an (in)elastic tunneling event. In general, the capture time constant is given by (Simoen et al., 2014):

$$\tau_c = \tau_0 \exp\left(\frac{E_\sigma}{kT}\right)\exp(\alpha_t z) \tag{28}$$

with τ_0 the SRH recombination time constant at the semiconductor/insulator interface ($z=0$), z the tunneling depth in the dielectric from the interface and α_t the tunneling parameter, given by:

$$\alpha_t = \frac{2}{\hbar}\sqrt{2qm_{ox}\Phi_{it}} \tag{29}$$

In Eq. (29), m_{ox} is the tunneling effective mass into the dielectric, Φ_{it} is the potential barrier at the semiconductor/insulator interface and $\hbar$ is the reduced Planck constant. According to Eq. (28) and assuming that the traps are distributed uniformly over the thickness of the dielectric, the tunneling (capture) front z proceeds proportional with $\ln(\tau_c)$. Consequently, it is expected that both the filling kinetics and the DLTS amplitude follow an increase with $\ln(\tau_c)$, as is evidenced in Fig. 19 for a $GdGaO_x$/n-GaAs MOSCAP. This filling behavior resembles closely the one from extended defects, the difference being that border traps will only be seen when

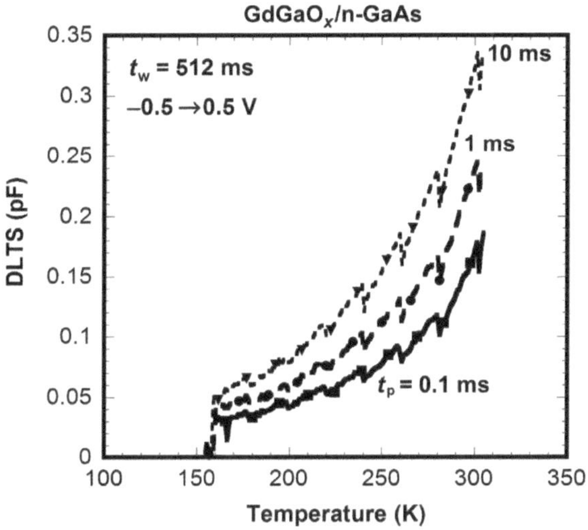

Figure 19 DLT-spectrum for a $GdGaO_x$/n-GaAs MOS capacitor, pulsed from −0.5 to 0.5 V, with a sampling period $t_w = 512$ ms and different filling pulse duration.

applying a bias pulse in accumulation, while extended defects can be probed for a V_P in depletion (Simoen et al., 2011b). A detailed analysis of the border trap parameters can be found in the literature (Bauza, 1998; Lakhdari et al., 1988; Van Staa et al., 1983; Vuillaume et al., 1986) and will not further be discussed here.

There are only few reports on DLTS on transistors, although the possibility to study single interface or oxide traps has been clearly demonstrated, relying on transients in the drain current and not in the (too small) gate capacitance (Schulz and Karmann, 1991). Recently, however, the feasibility of performing capacitance DLTS on silicon nanowires (Sato et al., 2012) and ZnO microwires has been demonstrated (Schmidt et al., 2013), showing the potential application for deep level analysis in future nanoscale device structures.

4.4 DLTS on nonideal junction devices

So far, ideal SB or p–n junctions have been considered in the analysis of the DLTS signal, whereby the junction or barrier is represented by a pure capacitance C_J with zero conductance in parallel and a perfect ohmic contact at the back, with zero potential drop (scheme A in Fig. 20). In reality, a finite reverse current (leakage current) is running through the device, giving rise to a parallel resistance (or conductance R_J). In addition, the resistance of the neutral region can be nonnegligible, resulting in the series resistance r.

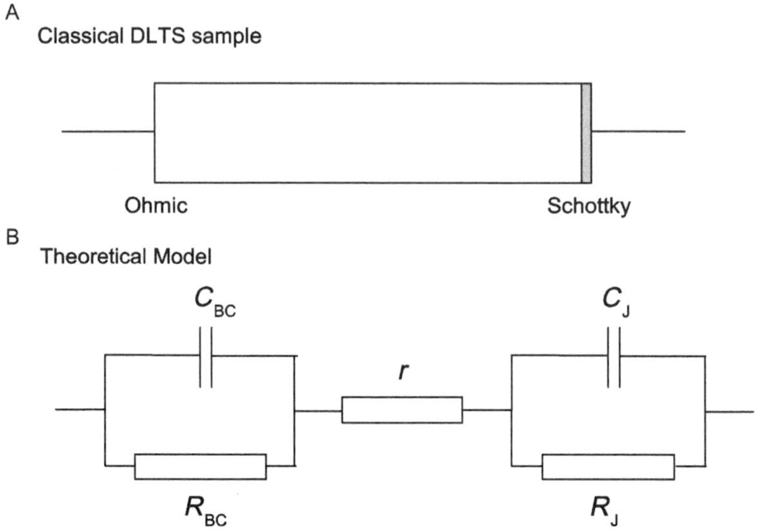

Figure 20 (A) Semiconductor sample with an ideal SB (right) and ohmic contact (left) and (B) circuit representation for a sample with a nonideal (leaky) SB (or p–n junction) and a nonideal back ohmic contact.

Finally, the ohmic contact may not be ideal and in cases a barrier is present, it can be represented by the parallel of a capacitance C_{BC} and a resistance (R_{BC}) in Fig. 20. In principle, these parasitic elements have to be accounted for in the modeling of the capacitance transient following a bias pulse. Fortunately, for most devices used in DLTS, these effects can be largely neglected. In this paragraph, however, the impact of the parasitic circuit elements and response will be discussed and ways to identify and control them are pointed out.

When dealing with a high-resistivity (lowly doped or wide band gap) semiconductor, the series resistance can be considerable (Broniatowski et al., 1983; Simoen et al., 1985). In addition, r is a temperature-dependent parameter, which generally increases when reducing T. In most DLTS measurements, the series connection of C_J and r is converted into a parallel equivalent, consisting of C_p and G_p ($=R_p^{-1}$). The measured C_p is related to the junction capacitance through (Simoen et al., 1985):

$$C_p = \frac{C_J}{(1 + Q^2)} \tag{30}$$

with the Q-factor given by $\omega r C_J$ ($\omega = 2\pi f$ with f the measurement frequency of the capacitance bridge). It is clear that when Q approaches 1, the effect of the series resistance becomes pronounced. As a consequence, the measured capacitance C_p reduces with increasing temperature, which may be considered as a possible finger print of the series resistance effect (Simoen et al., 1985). Moreover, it has been shown that the measured transient amplitude ΔC_p is in that case given by (Simoen et al., 1985):

$$\Delta C_p = \Delta C_J \frac{(1 - Q^2)}{(1 + Q^2)} \tag{31}$$

It is clear from Eq. (31) that for $Q \geq 1$, the transient signal changes its sign, resulting in a DLTS peak becoming negative (Broniatowski et al., 1983; Simoen et al., 1985), rendering the analysis of the DLT-spectra quite unreliable and cumbersome. In order to suppress or avoid the Q-effect in highly resistive semiconductor materials, one can try to minimize the width of the neutral series resistance region either by applying an as high as possible reverse bias or by using an as thin as possible sample (Simoen et al., 1985). Another way out is to perform the capacitance measurements at a lower a.c. signal frequency (Anand et al., 1992), since Q is proportional with f. This, of course, may reduce significantly the range of accessible emission rate windows to lower frequencies. Finally, for very resistive samples, it may turn out more advantageous to perform current transient analysis (Blondeel and Clauws, 1999; Blondeel et al., 1997), which is based on the transient in the reverse current of the junction by the carrier generation from the filled deep levels (Meijer and Grimmeiss, 1992).

It is generally recommended to perform DLTS on device structures with a low parallel leakage current, as this may distort the DLTS measurements (Chen et al., 1984; Dmowski et al., 1993). It has been shown that when the emission rate becomes smaller than the generation rate of the reverse current, the DLTS peak becomes truncated at the high temperature side and the activation energy derived from an Arrhenius plot is dominated by the activation energy of the leakage current (Chen et al., 1984). In the case of high electric fields, the excess carriers can also assist in so-called impact ionization of deep levels, thereby increasing the emission rate. A correction procedure for a proper analysis of the DLT-spectra in the case of leaky Schottky diodes on GaAs has been proposed (Dmowski et al., 1993).

The fabrication of good ohmic contacts on many semiconductor materials may be problematic for several reasons. Generally, when a metal is deposited on a semiconductor, a potential barrier is formed, hindering the flow of carriers through the structure. This potential barrier gives rise to a depletion region at the M/S contact, resulting in the scheme of Fig. 20B. This issue has recently been identified in DLTS of copper–indium–gallium–selenide (CIGS) solar cells (Lauwaert et al., 2012b, 2013). A proper analysis of the effect has been derived for model diodes connected in series with a resistor and a capacitor (Lauwaert et al., 2011).

It has been convincingly shown that in such a case, the DLTS signal can either be positive or negative, depending on the back and front contact parameters. But since the back contact is often a small perturbation on the main junction the situation with a negative conventional signal is more likely. In this case, four distinctive properties of the DLTS signal were found: Negative signal for a conventional bias pulse ($V_r < V_p < 0$), the time constants converge to the same value for small-pulse amplitudes, the signal for the inverted pulse ($V_p < V_r < 0$) has the largest amplitude, and the DLTS amplitude scales with the reverse capacitance to the power 4. Those properties are different for signals originating from defects where a positive peak is expected for a conventional bias pulse and a negative peak for an inverted pulse as shown in Fig. 21 for $Ti_s^{(-/2-)}$ in n-Ge. In the latter case, the negative peak occurs from the capture of carriers at the edge of the depletion region.

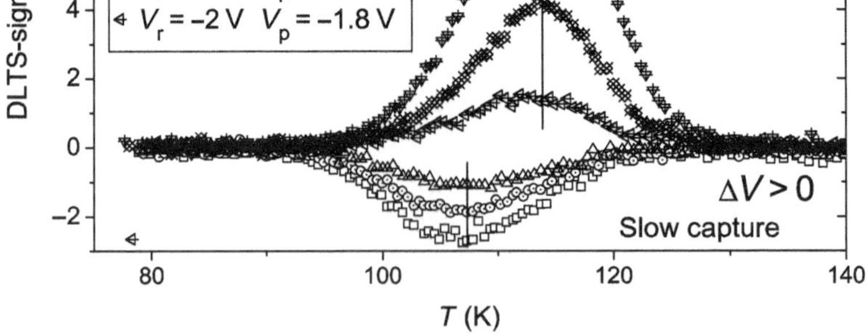

Figure 21 Conventional DLT-spectrum with $\Delta V < 0$ showing emission and with $\Delta V > 0$ showing the slow capture in a Ti-doped n-type Ge SB.

5. AS OF DEEP LEVELS

In AS, the small-signal a.c. response of a semiconductor device under a fixed d.c. bias and temperature T is measured as a function of the frequency f (Losee, 1975; Macdonald, 1987). The a.c. signal has normally a sinusoidal shape with an amplitude of a few tens of mV, which modulates the Fermi level around its steady-state position. In turn, this will provoke a response of the occupation of the deep levels around their crossing point with E_F, causing an oscillation in the charge density, which can be measured as a capacitance signal. The resulting complex admittance (including phase angle) can be expressed as:

$$Y = Z^{-1} = |Y|\exp(j\arg(Y)) = G + jB \tag{32a}$$

while, equally:

$$Z = Y^{-1} = |Z|\exp(j\arg(Z)) = R + jX \tag{32b}$$

with

$$j = \sqrt{-1},$$

Z the impedance, G the conductance, and B the susceptance. The relationship between resistance R, reactance X and G and B is given by:

$$R = \frac{G}{G^2 + B^2} \tag{33a}$$

$$X = \frac{-B}{G^2 + B^2} \tag{33b}$$

$$G = \frac{R}{R^2 + X^2} \tag{33c}$$

$$B = \frac{-X}{R^2 + X^2} \tag{33d}$$

The conductance and susceptance are, by convention, divided by the angular frequency $\omega = 2\pi f$ before plotting, rendering the interpretation of the graphs easier. An example of the capacitance of a MOSCAP at 0 V gate bias is given in Fig. 22.

In order to relate the measured R, X or G, B pairs to different equivalent circuit elements of the device structure (see, e.g., Fig. 19), one can follow two different approaches: the microscopic and the macroscopic analysis. In the first case, a numerical model is established for the device under test,

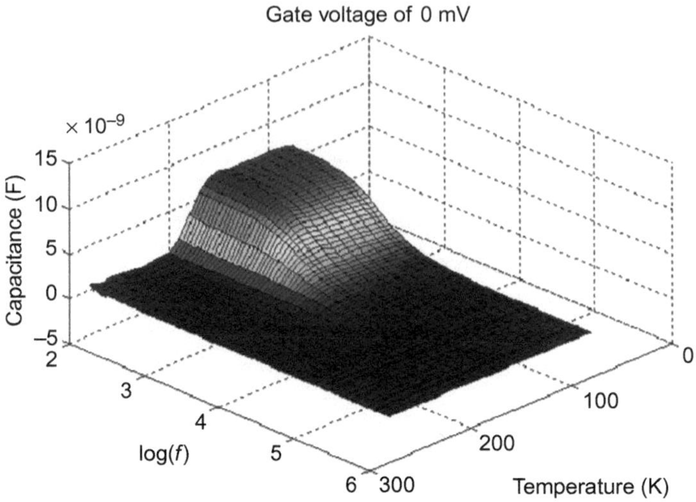

Figure 22 Capacitance versus temperature and frequency for a Ni/9 nm Al_2O_3/~2 nm GeO_x/30 nm $Ge_{0.93}Sn_{0.07}$/1 μm Ge MOSCAP at a gate voltage of 0 V.

solving the drift-diffusion and Poisson equations and also the response to a small-signal a.c. perturbation. In the second approach, an equivalent circuit model is established, such that the admittance matches well the measured data. The circuit must be based on a profound understanding of the system and should not be more complex than required. This enables then the extraction of C_J, R_J, r, C_{BC}, and R_{BC} in Fig. 19 for example.

In order to extract trap responses from the measured G/ω or capacitance, one should keep in mind that at a given temperature and frequency, only traps below an energy limit E_a, given by:

$$E_a = kT[\ln(2\nu_0) - \ln(\omega)] \tag{34}$$

will be able to respond to the a.c. signal, with ν_0 a preexponential rate factor related to the speed of the trap response. The pairs of temperatures and frequencies where the capacitance is dropping off can be investigated by taking the derivative of C with respect to either T or f, due to the $T-\omega$ duality (Li and Levi, 2011). It has been shown that this is advantageous for deep defects (TD-AS), while the frequency-derivative AS method (FD-AS) is better suited for shallow levels. An example of the TD-AS is given in Fig. 23, while the FD-AS is illustrated by Fig. 24. In general, AS can be employed to study shallow levels without the need for a background doping, while DLTS requires a background doping level in order to establish a depletion region.

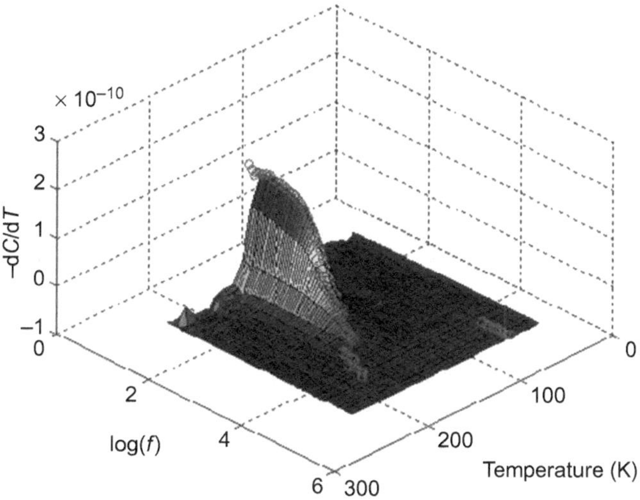

Figure 23 TD-AS plot corresponding with the MOSCAP of Fig. 22 at a gate bias of −1 V, showing the presence of a deep level with activation energy of 0.38 eV above the valence band. This is in reasonable agreement with the value of 0.36 eV found in DLTS.

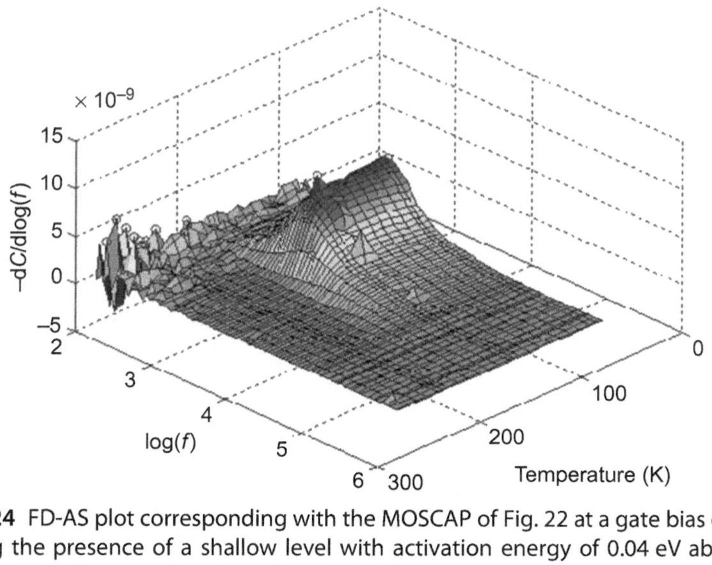

Figure 24 FD-AS plot corresponding with the MOSCAP of Fig. 22 at a gate bias of −1 V, showing the presence of a shallow level with activation energy of 0.04 eV above the valence band. The same level was found in DLTS.

6. DIODE LIFETIME ANALYSIS

As lifetime is a crucial parameter for device operation, a whole plethora of techniques have been developed in the course of the years (Schroder, 1998). However, what is perhaps less known is that the p–n junction current–voltage (I–V) characteristics can be utilized for lifetime extraction as well (Simoen et al., 2007), when proper account is made for certain geometrical and other effects. This offers the possibility to directly correlate the deep level parameters derived from DLTS with the generation and recombination lifetime on the same p–n diodes and enables identifying the dominant SRH center, as will be illustrated in this paragraph. At the same time, such a diode analysis can be applied to diodes of any size, which could be in the μm range or below for source and drain junctions in a MOSFET, where standard DLTS fails because of the too low capacitance. Ultimately, the device characteristics are more sensitive to SRH centers than can be detected by DLTS, so that these form a very simple yet powerful method to study the effects of deep levels in semiconductor materials.

6.1 Standard SRH lifetime analysis

The I–V characteristics of a large-area (A_J) ideal p–n junction without SRH centers and neglecting series resistance can be expressed as (Sze, 1981):

$$I = I_0\left[\exp\left(\frac{qV}{kT}\right) - 1\right] \tag{35}$$

with the saturation current I_0 given by:

$$I_0 = I_d = qn_i^2 A_J\left(\frac{D_n}{L_n N_A} + \frac{D_p}{L_p N_D}\right) \tag{36}$$

The diffusion current I_d is composed of the minority carrier diffusion current in the p- and in the n-side of the junction, with D_n, L_n and D_p, L_p the diffusivity and diffusion length of minority electrons (holes), respectively. The diffusivity is given by Einstein's relationship ($D_n = \mu_n/kT$), while the recombination (or minority carrier) lifetime corresponds with:

$$\tau_{rn} = \frac{L_n^2}{D_n} \tag{37}$$

A symmetrical relationship holds for τ_{rp}. In reverse operation ($V<0$), Eq. (35) reduces to $I \approx I_d$ and from Eq. (36) one can derive that the diffusion current is independent on the bias and exponentially dependent on temperature, whereby the activation energy is determined by n_i^2 and corresponds with the band gap of the semiconductor material. In practice, I_d will dominate the leakage current at sufficiently high temperatures (e.g., >45 °C for a good quality silicon diode; >30 °C for Ge). In the case of an asymmetric p^+n or n^+p junction, the term corresponding with the lower doped side of the junction dominates in Eq. (36), so that one can derive the minority carrier lifetime from the diffusion current component.

Alternatively, one can determine the diffusion current from the saturation current in forward operation as well (Vanhellemont et al., 1995a,b), whereby one should account for the geometry of the diode, correcting for the current flowing through the diode periphery and for the diode non-ideality factor m (>1). In general, the forward saturation current gives less accurate values, as one has to extrapolate an exponential curve (Eq. 35) to 0 V. The advantage is that the lifetime is derived at low values of the electric field in the junction.

In practice, p–n junctions generally contain SRH centers in the depletion region. In that case, the reverse bulk or area current can be expressed as (Simoen et al., 2007):

$$I_R = I_d + I_{gen} = I_d + \frac{qn_i A_J W}{\tau_{gen}} \tag{38}$$

with the depletion width W given by Eq. (16). Equation (38) predicts a $\sqrt{V_R + V_{bi}}$ bias-dependence and—in case the SRH centers have an energy level close to the intrinsic level E_i, corresponding with the most efficient SRH generation centers—an activation energy of $E_G/2$. This value is indeed often found from an Arrhenius plot at lower temperatures, e.g., 0.56 eV for silicon diodes and 0.3–0.4 eV for Ge diodes (Gonzalez et al., 2014; Ioannou-Sougleridis et al., 2013), while at higher T, the diffusion current starts to dominate (activation energy E_G). In case of a small-area diode, with a large impact of the current through the periphery (perimeter P), one can separate the area current density, given by Eq. (38) by a combination of diodes with different area (Simoen et al., 2007; Vanhellemont et al., 1995a) or by using a gated-diode structure (Czerwinski et al., 2000).

One can go one step further in the lifetime analysis by combining a reverse I–V and C–V characteristic (Murakami and Shingyouji, 1994), thereby eliminating the bias dependence, i.e., by plotting I_R versus $W_R = (A_J \varepsilon_S / C_R)$. According to Eq. (38), this should give a straight line, as schematically represented in Fig. 25. From the slope of a linear fit, one

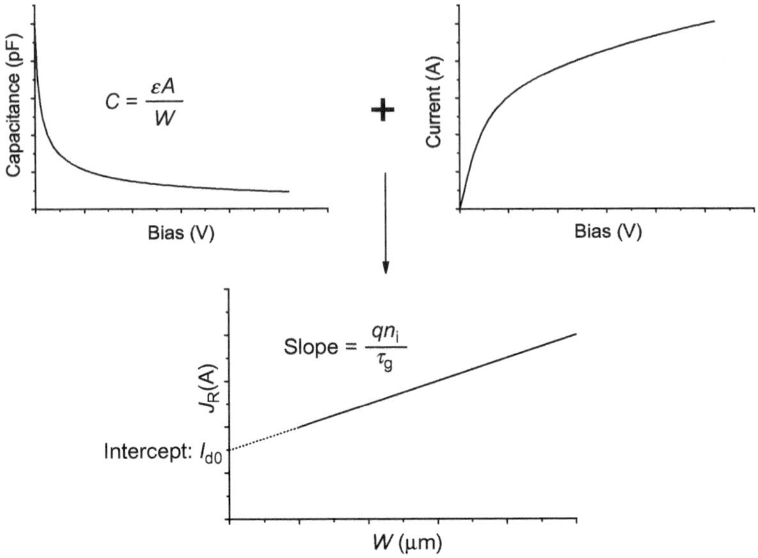

Figure 25 Lifetime analysis for a diode with a reverse current composed of a diffusion component and an SRH term.

can derive then τ_g, while the intercept yields a value for I_d and, hence τ_r. Likewise, both lifetimes can be extracted from the initial part of the forward current, based on the model for the area current density (Poyai et al., 2003):

$$J_f = \frac{qn_i W}{2\tau_r \exp\left(\frac{qV}{2kT}\right) + \tau_g}\left[\exp\left(\frac{qV}{kT}\right) - 1\right] = J_{f0}\left[\exp\left(\frac{qV}{kT}\right) - 1\right] \quad (39)$$

Representing $qn_i W/J_{f0}$ versus $\exp(qV/2kT)$ like in Fig. 26 should yield a straight line over a certain bias range, where the slope gives τ_r and the intercept τ_g. As mentioned before, the generation lifetime is (much) higher than the recombination lifetime, suggesting a dominant SRH center at some distance from mid gap.

A correlation between DLTS and diode lifetimes has been investigated in a number of cases. It has for example been shown that there is a strong dependence of the lifetime of n^+p junctions fabricated in silicon substrates on the initial interstitial oxygen content and on the details of the preheat treatment (nucleation treatment, internal gettering steps) (Vanhellemont et al., 1995b). Moreover, on the same diodes, an electron trap at $E_C - 0.43$ eV has been detected in DLTS, which is responsible for the lifetime degradation in high-oxygen Czochralski silicon diodes. Another

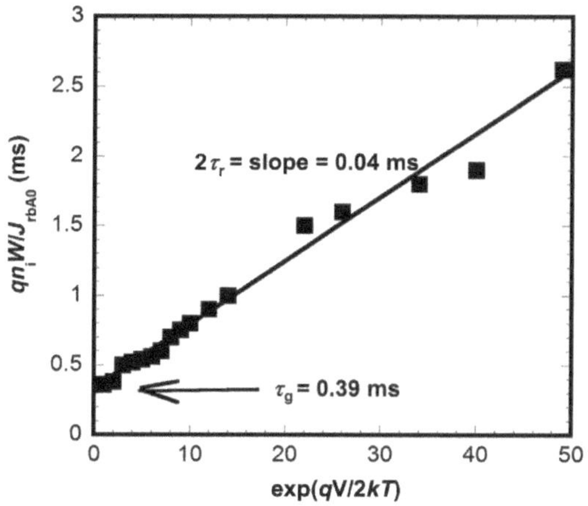

Figure 26 Analysis of τ_g and τ_r from the forward current–voltage characteristic of a silicon p–n junction diode.

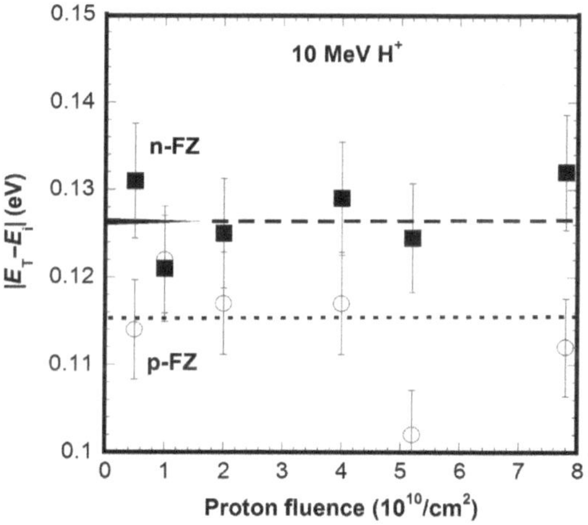

Figure 27 Effective activation energy extracted from the τ_g/τ_r ratio for 10 MeV proton-irradiated n- and p-type FZ diodes as a function of the proton fluence.

example is illustrated by Fig. 27, showing the effective activation energy derived from the diode lifetime ratio (Eq. 11) for proton-irradiated n^+p diodes on FZ–Si substrates. Typical values are in the range of 0.12 eV from mid gap, which coincides with the $V_2^{0/-}$ level at $E_C - 0.42$ eV detected in DLTS (Simoen et al., 1996).

In many realistic cases, the pn junctions are just too small for DLTS analysis (<100 μm^2), so that one has to rely mainly on the analysis of the diode I–V and C–V characteristics to obtain information on the SRH centers in the material. An additional complication arises from the fact that in CMOS compatible diodes, a high doping density is present in the substrate, resulting in a high electric field (>10^4 V/cm). This triggers the so-called field-assisted generation mechanisms in the depletion region, like the Poole–Frenkel effect, trap-assisted-tunneling (TAT), or even band-to-band-tunneling (Simoen et al., 2011a). The impact of the electric field on the carrier generation rate can be implemented by introducing an enhancement factor $(1+\Gamma)$ in the numerator of Eq. (38). Based on this, TAT has been successfully analyzed for the case of Ge pn junctions (Bargallo Gonzalez et al., 2011; Eneman et al., 2008; Simoen et al., 2009). Moreover, state-of-the-art diodes may consist of heterojunctions imparting stress to the underlying substrate (Simoen et al., 2011a). In order to take account of all these different factors, one has to combine diode measurements and analysis with 2D or 3D process (for strain and doping profiles) and device simulations. A result of such a detailed analysis is represented in Fig. 28 (Luque Rodríguez et al., 2011), showing the evolution of the SRH generation lifetime in the silicon substrate for p^+-SiGe/n-Si strained heterojunctions, used as stressors in pMOSFETs. The stress results from the lattice mismatch between silicon and SiGe, increasing with the Ge content. As can be seen, there appears

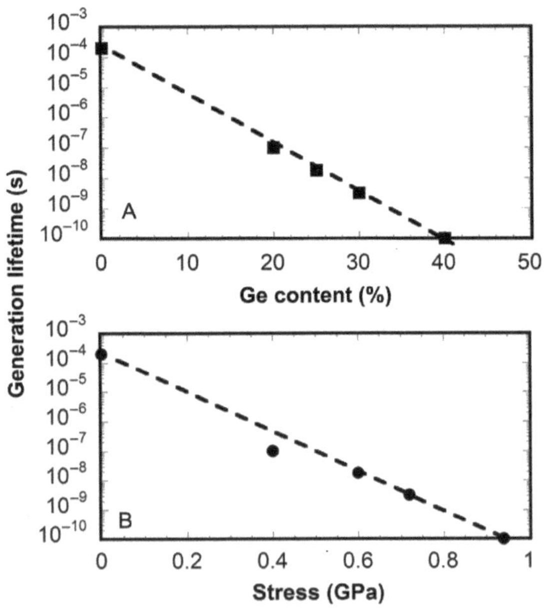

Figure 28 Generation lifetime as a function of (A) Ge content in the SiGe p^+ stressors and (B) as a function of the maximum stress level in the underlying silicon n-well region.

to be an exponential dependence between the generation lifetime in the silicon substrate and the Ge content/stress created by the heterojunction. This leads to the conclusion that the processing of the stressors by selective epitaxial deposition results in an increasing density of deep levels (Luque Rodríguez et al., 2011). In addition, the energy level and capture cross section may be stress dependent as well, which can also contribute to the degradation of the generation lifetime.

7. SUMMARY AND OUTLOOK

Without any doubt, DLTS has pushed the limits of our understanding of electrically active defects in semiconductors to unprecedented levels, providing a high sensitivity and reasonable energy and spatial resolution. At the same time, the technological strive toward the nanoscale indicates that future device operation, yield, and reliability can become sensitive to single-defect effects. Although it is tempting to extrapolate the data obtained on large diode structures toward the nanoscale, this may not be the proper way to go. In recent years, it has been shown that not only the transport through such miniaturized structures is affected by localization and quantization effects, also the properties of well-known defects are modified, so that dedicated analysis techniques are required. The limitations of standard DLTS to diode size have been clear from the start: when the capacitance, mainly dictated by the device area becomes too small (a few pF), it becomes more and more difficult to resolve transient signals form the noise. However, as shown in the past, current transients do not suffer from this limitation—as long as the current through the device is on the order of (or larger than) 1 nA. Detection of single-defect related emission events has been demonstrated in the past. In addition, the analysis of the transport characteristics of nanometric devices and its low-frequency noise may provide an alternative (or complementary) way to study electrically active defects. One may expect also a further development of AFM-based techniques, which yield complementary information on the surface topography, the presence of surface charge and the local transport. In combination with 3D device simulations, this can form a formidable arsenal for future defect studies. Finally, one should consider the fact that the size of nanometric device structures becomes of the same order as typical cluster sizes in density functional theory (DFT) calculations. Although one should be careful with the quantitative information regarding defect energy levels and capture cross sections, it certainly offers a powerful tool for identifying the underlying physical trends.

REFERENCES

Anand, S., Subramanian, S., Arora, B.M., 1992. Use of low-frequency capacitance in deep level transient spectroscopy measurements to reduce series resistance effects. J. Appl. Phys. 72, 3535–3538.

Bagraev, N.T., Gehlhoff, W., Gets, D.S., Klyachkin, L.E., Kudryavtsev, A.A., Malyarenko, A.M., et al., 2009. EDEPR of impurity centers embedded in silicon microcavities. Physica B 404, 5140–5143.

Bargallo Gonzalez, M., Eneman, G., Wang, G., De Jaeger, B., Simoen, E., Claeys, C., 2011. Analysis of the temperature dependence of trap-assisted tunneling in Ge pFET junctions. J. Electrochem. Soc. 158, H955–H960.

Bauza, D., 1998. Near interface oxide trap capture kinetics in metal-oxide-semiconductor transistors: modeling and measurements. J. Appl. Phys. 84, 6178–6186.

Blondeel, A., Clauws, P., 1999. Photoinduced current transient spectroscopy of deep defects in n-type ultrapure germanium. J. Appl. Phys. 86, 940–945.

Blondeel, A., Clauws, P., Vyncke, D., 1997. Optical deep level transient spectroscopy of minority carrier traps in n-type high-purity germanium. J. Appl. Phys. 81, 6767–6772.

Blood, P., Orton, J.W., 1992. The Electrical Characterization of Semiconductors: Majority Carriers and Electron States. Academic, New York, NY.

Broniatowski, A., Blosse, A., Srivastava, P.C., Bourgoin, J.C., 1983. Transient capacitance measurements on resistive samples. J. Appl. Phys. 54, 2907–2910.

Chasin, A., Simoen, E., Bhoolokam, A., Nag, M., Genoe, J., Gielen, G., Heremans, P., 2014. a-IGZO Schottky diode as a material characterization tool: DLTS measurements. Appl. Phys. Lett. 104, 082112/1–082112/5.

Chen, M.C., Lang, D.V., Dautremont-Smith, W.C., Sergent, A.M., Harbison, J.P., 1984. Effects of leakage currents on deep level transient spectroscopy. Appl. Phys. Lett. 44, 790–792.

Claeys, C., Simoen, E., Neimash, V.B., Kraitchinskii, A., Kras'ko, M., Puzenko, O., et al., 2001. Tin doping of silicon for controlling oxygen precipitation and radiation hardness. J. Electrochem. Soc. 148, G738–G745.

Czerwinski, A., Simoen, E., Poyai, A., Claeys, C., 2000. Peripheral current analysis of silicon p-n junction and gated diodes. J. Appl. Phys. 88, 6506–6514.

Darken, L.S., 1992. Theoretical model for hole capture at acceptors in germanium. Phys. Rev. Lett. 69, 2839–2842.

Dmowski, K., Lepley, B., Losson, E., El Bouabdellati, M., 1993. A method to correct for leakage current effects in deep level transient spectroscopy measurements on Schottky diodes. J. Appl. Phys. 74, 3936–3943.

Dobaczewski, L., Peaker, A.R., Bonde Nielsen, K., 2004. Laplace-transform deep-level spectroscopy: the technique and its application to the study of point defects in semiconductors. J. Appl. Phys. 96, 4689–4728.

Ebert, Ph., Domke, C., Urban, K., 2001. Direct observation of electrical charges at dislocations in GaAs by cross-sectional scanning tunneling microscopy. Appl. Phys. Lett. 78, 480–482.

Eneman, G., Wiot, M., Brugère, A., Brunco, D., De Jaegere, B., Satta, A., et al., 2008. Impact of donor concentration, electric field and temperature effects on the leakage current of Ge p^+-n junctions. IEEE Trans. Electron Devices 55, 2287–2295.

Eyben, P., Clemente, F., Vanstreels, K., Pourtois, G., Clarysse, T., Duriau, et al., 2010. Analysis and modeling of the high vacuum scanning spreading resistance microscopy nanocontact on silicon. J. Vac. Sci. Technol. B 28, 401–406.

Fleetwood, D., 1992. "Border traps" in MOS devices. IEEE Trans. Electron Devices 39, 269–271.

Fleming, R.M., Myers, S.M., Wampler, W.R., Lang, D.V., Seager, C.H., Campbell, J.M., 2014. Field dependent emission rates in radiation damaged GaAs. J. Appl. Phys. 116, 013710/1–013710/7.

Frenkel, J., 1938. On pre-breakdown phenomena in insulators and electronic semiconductors. Phys. Rev. 54, 647–648.

Fukuda, K., Nishizawa, M., Tada, T., Bolotov, L., Suzuki, K., Satoh, S., et al., 2014. Three-dimensional simulation of scanning tunneling microscopy for semiconductor carrier and impurity profiling. J. Appl. Phys. 116, 023701/1–023701/11.

Gong, C., Simoen, E., Posthuma, N., Van Kerschaever, E., Poortmans, J., Mertens, R., 2010. A deep-level transient spectroscopy study of silicon interface states using different silicon nitride surface passivation schemes. Appl. Phys. Lett. 96, 103507/1–103507/3.

Gonzalez, M.B., Simoen, E., Eneman, G., De Jaeger, B., Wang, G., Loo, R., et al., 2014. Defect assessment and leakage control in Ge pFET junctions. Microelectron. Eng. 125, 33–37.

Hartke, J.L., 1968. The three-dimensional Poole-Frenkel effect. J. Appl. Phys. 52, 4871–4873.

Henry, C.H., Lang, D.V., 1977. Nonradiative capture and recombination by multiphonon emissions in GaAs and GaP. Phys. Rev. B 15, 989–1016.

Ioannou-Sougleridis, V., Poulakis, N., Dimitrakis, P., Normand, P., Patsis, G.P., Dimoulas, A., et al., 2013. Room temperature analysis of Ge p^+/n diodes reverse characteristics fabricated by platinum assisted dopant activation. Solid-State Electron. 81, 19–26.

Johnson, N.M., 1982. Measurement of semiconductor-insulator interface states by constant-capacitance, deep-level transient spectroscopy. J. Vac. Sci. Technol. 21, 303–314.

Johnson, N.M., Bartelink, D.J., Gold, R.B., Gibbons, J.F., 1979. Constant-capacitance DLTS measurement of defect-density profiles in semiconductors. J. Appl. Phys. 50, 4828–4833.

Kim, C.K., Yoon, I.T., Kuk, Y., Lim, H., 2001. Variable-temperature scanning capacitance microscopy: a way to probe charge traps in oxide or semiconductor. Appl. Phys. Lett. 78, 613–615.

Kimerling, L.C., Patel, J.R., 1979. Defect states associated with dislocations in silicon. Appl. Phys. Lett. 34, 73–75.

Lakhdari, H., Vuillaume, D., Bourgoin, J.C., 1988. Spatial and energetic distribution of $Si\text{-}SiO_2$ near-interface states. Phys. Rev. B 38, 13124–13132.

Lang, D.V., 1974. Deep-level transient spectroscopy: a new method to characterize traps in semiconductors. J. Appl. Phys. 45, 3023–3032.

Lauwaert, J., Van Gheluwe, J., Clauws, P., 2006. An accurate analytical approximation to the capacitance transient amplitude in deep level transient spectroscopy for fitting carrier capture data. Rev. Sci. Instrum. 79, 093902.

Lauwaert, J., Van Gheluwe, J., Vanhellemont, J., Simoen, E., Clauws, P., 2009. Electronic properties of titanium and chromium impurity centers in germanium. J. Appl. Phys. 105, 073707/1–073707/6.

Lauwaert, J., Khelifi, S., Decock, K., Burgelman, M., Vrielinck, H., 2011. Signature of a back contact barrier in DLTS spectra. J. Appl. Phys. 109, 063721/1–063721/6.

Lauwaert, J., Vanhellemont, J., Simoen, E., Vrielinck, H., Clauws, P., 2012a. Electronic properties of iron and cobalt impurity centres in germanium. J. Appl. Phys. 111, 113713/1–113713/7.

Lauwaert, J., Callens, L., Khelifi, S., Decock, K., Burgelman, M., Chirila, A., et al., 2012b. About RC-like contacts in deep level transient spectroscopy and Cu(In, Ga)Se2 solar cells. Prog. Photovolt. Res. Appl. 20 (5), 588–594. http://dx.doi.org/10.1002/pip.2166.

Lauwaert, J., Van Puyvelde, L., Lauwaert, J., Thybaut, J.W., Khelifi, S., Burgelman, M., et al., 2013. Assignment of capacitance spectroscopy signals of CIGS solar cells to effects of non-ohmic contacts. Sol. Energy Mater. Sol. Cells 112, 78–83.

Li, J.V., Levi, D.H., 2011. Determining the defect density of states by temperature derivative admittance spectroscopy. J. Appl. Phys. 109, 083701/1–083701/6.

Losee, D.L., 1975. Admittance spectroscopy of impurity levels in Schottky barriers. J. Appl. Phys. 46, 2204–2214.

Luque Rodríguez, A., Bargallo Gonzalez, M., Eneman, G., Claeys, C., Kobayashi, D., Simoen, E., et al., 2011. Impact of Ge content and recess depth on the leakage current in strained $Si_{1-x}Ge_x/Si$ heterojunctions. IEEE Trans. Electron Devices 58, 2362–2370.

Macdonald, J.R., 1987. Impedance Spectroscopy: Emphasizing Solid Materials and Systems. Wiley-Interscience, New York, NY.

Mazur, R.G., Dickey, D.H., 1966. A spreading resistance technique for resistivity measurements on silicon. J. Electrochem. Soc. 113, 255–259.

Meijer, E., Grimmeiss, H.G., 1992. Transient current measurements for the characterization of deep defects in semiconductors. Semicond. Sci. Technol. 7, 188–197.

Miwa, J.A., Mol, J.A., Salfi, J., Rogge, S., Simmons, M.Y., 2013. Transport through a single donor in p-type silicon. Appl. Phys. Lett. 103, 043106/1–043106/4.

Morello, A., Pla, J.J., Zwanenburg, F.A., Chan, K.W., Tan, K.Y., Huebl, H., et al., 2010. Single-shot readout of an electron spin in silicon. Nature 467, 687–691.

Murakami, Y., Shingyouji, T., 1994. Separation and analysis of diffusion and generation components of pn junction leakage current in various silicon wafers. J. Appl. Phys. 75, 3548–3552.

Nguyen, T.H., Mahieu, G., Berthe, M., Grandidier, B., Delerue, C., Stiévenard, D., et al., 2010. Coulomb energy determination of a single Si dangling bond. Phys. Rev. Lett. 105, 226404/1–226404/4.

Nowak, R., Moraru, D., Mizuno, T., Jablonski, R., Tabe, M., 2013. Effects of deep-level dopants on the electronic potential of thin Si pn junctions observed by Kelvin probe force microscopy. Appl. Phys. Lett. 102, 083109/1–083109/4.

Paul, O., Cornils, M., 2009. Explicit connection between sample geometry and Hall response. Appl. Phys. Lett. 95, 232112/1–232112/3.

Pearce, N.O., Hamilton, B., Peaker, A.R., Craven, R.A., 1987. Application of deep level transient spectroscopy to metal-oxide-semiconductor relaxation transients. J. Appl. Phys. 62, 576–581.

Petersen, D.H., Hansen, O., Lin, R., Nielsen, P.F., 2008. Micro-four-point probe Hall effect measurement method. J. Appl. Phys. 104, 013710/1–013710/10.

Petersen, D.H., Hansen, O., Bøggild, P., Nielsen, P.F., Lin, D., Adelmann, C., et al., 2010. Electrical characterization of InGaAs ultra-shallow junctions. J. Vac. Sci. Technol. B 28, C1C41–C1C47.

Pla, J.J., Tan, K.Y., Dehollain, J.P., Lim, W.H., Morton, J.J.L., Jamieson, D.N., et al., 2012. A single-atom electron spin qubit in silicon. Nature 489, 541–545.

Pons, D., 1984. Accurate determination of the free carrier capture kinetics of deep traps by space charge methods. J. Appl. Phys. 55, 3644–3657.

Poyai, A., Simoen, E., Claeys, C., Gaubas, E., Gräf, D., Huber, A., 2003. Extraction of the carrier generation and recombination lifetime from the forward characteristics of advanced diodes. Mater. Sci. Eng. B 102, 189–192.

Rafí, J.M., Simoen, E., Claeys, C., Huang, Y.L., Ulyashin, A.G., Job, R., et al., 2005. Impact of direct plasma hydrogenation on thermal donor formation in n-type CZ silicon. J. Electrochem. Soc. 152, G16–G24.

Ricksand, A., Engström, O., 1991. Thermally activated capture of charge carriers into irradiation induced Si/SiO_2 interface states. J. Appl. Phys. 70, 6927–6933.

Sato, K., Castaldini, A., Fukata, N., Cavallini, A., 2012. Electronic level scheme in boron- and phosphorus-doped silicon nanowires. Nano Lett. 12, 3012–3017.

Schmidt, F., Müller, S., von Wenckstern, H., Dietrich, C.P., Heinhold, R., Kim, H.-S., et al., 2013. Comparative study of deep defects in ZnO microwires, thin films and bulk single crystals. Appl. Phys. Lett. 103, 062102/1–062102/3.

Schroder, D.K., 1982. The concept of generation and recombination lifetimes in semiconductors. IEEE Trans. Electron Devices 29, 1336–1338.

Schroder, D., 1997. Carrier lifetimes in silicon. IEEE Trans. Electron Devices 44, 160–170.

Schroder, D., 1998. Semiconductor Material and Device Characterization. Wiley, New York, NY.

Schröter, W., Cerva, H., 2002. Interaction of point defects with dislocations in silicon and germanium: electrical and optical effects. Solid State Phenom. 85–86, 67–144.

Schulz, M., Karmann, A., 1991. Individual, attractive defect centers in the Si-SiO_2 interface of μm-sized MOSFETs. Appl. Phys. A 52, 104–111.

Shockley, W., Read Jr., W.T., 1952. Statistics of the recombination of holes and electrons. Phys. Rev. 87, 835–842.

Simoen, E., Claeys, C., 1999. On the impact of the capture rates on the generation/recombination lifetime ratio of a single deep level. IEEE Trans. Electron Devices 46, 1487–1488.

Simoen, E., Clauws, P., Vennik, J., 1985. The determination of deep level concentrations in high resistivity semiconductors by DLTS, with special reference to germanium. J. Phys. D. Appl. Phys. 18, 2041–2058.

Simoen, E., Clauws, P., Huylebroeck, G., Vennik, J., 1987. DLTS of gold impurities in germanium. Semicond. Sci. Technol. 2, 507–512.

Simoen, E., Vanhellemont, J., Claeys, C., 1996. The effective generation-recombination parameters in high-energy proton irradiated Si diodes. Appl. Phys. Lett. 69, 2858–2860.

Simoen, E., Opsomer, K., Claeys, C., Maex, K., Detavernier, C., Van Meirhaeghe, R.L., et al., 2006. Deep level transient spectroscopy study of Pd and Pt sputtering damage in n-type germanium. Appl. Phys. Lett. 89, 202114/1–202114/3.

Simoen, E., Claeys, C., Vanhellemont, J., 2007. Defect analysis in semiconductor materials based on p-n junction diode characteristics. Defect Diffus. Forum 261–262, 1–24.

Simoen, E., De Stefano, F., Eneman, G., De Jaeger, B., Claeys, C., Crupi, F., 2009. On the temperature and field dependence of trap-assisted-tunneling current in Ge p^+n junctions. IEEE Electron Device Lett. 30, 562–564.

Simoen, E., Eneman, G., Bargallo Gonzalez, M., Kobayashi, D., Luque Rodríguez, A., Jiménez Tejada, J.-A., et al., 2011a. High doping density/high electric field, stress and heterojunction effects on the characteristics of CMOS compatible p-n junctions. J. Electrochem. Soc. 158, R27–R36.

Simoen, E., Visalli, D., Van Hove, M., Leys, M., Borghs, G., 2011b. A deep-level analysis of Ni-Au/AlN/(111) p^+-Si metal-insulator-semiconductor capacitors. J. Phys. D. Appl. Phys. 44, 475104/1–475104/8.

Simoen, E., Lin, H.-C., Alian, A., Brammertz, G., Merckling, C., Mitard, J., Claeys, C., 2013. Border traps in Ge/III-V channel devices: analysis and reliability aspects. IEEE Trans. Device Mater. Reliab. 13, 444–455.

Simoen, E., Lee, J.W., Claeys, C., 2014. Assessment of the impact of inelastic tunneling on the frequency-depth conversion from low-frequency noise spectra. IEEE Trans. Electron. Devices 61, 634–637.

Sze, S., 1981. Physics of Semiconductor Devices. John Wiley and Sons, New York, NY.

Sze, S., Irvin, J.C., 1968. Resistivity, mobility and impurity levels in GaAs, Ge, and Si at 300° K. Solid-State Electron. 11, 599–602.

Szmulowicz, F., 1986. Calculation of the mobility and the Hall factor for doped p-type silicon. Phys. Rev. B 34, 4031–4047.

Tanay, F., Dubois, S., Veirman, J., Enjalbert, N., Stendera, J., Périchaud, I., 2014. Oxygen-related thermal donor formation in dopant-rich compensated Czochralski silicon. IEEE Trans. Electron Devices 61, 1241–1245.
Van der Pauw, L.J., 1958. A method of measuring specific resistivity and Hall effect of discs of arbitrary shape. Philips Res. Rep. 13, 1–9.
Van Staa, P., Rombach, H., Kassing, R., 1983. Time-dependent response of interface states in indium phosphide metal-insulator-semiconductor capacitors investigated with constant-capacitance deep-level transient spectroscopy. J. Appl. Phys. 54, 4014–4021.
Vandervorst, W., Clarysse, T., Eyben, P., 2002. Spreading resistance roadmap towards and beyond the 70 nm technology node. J. Vac. Sci. Technol. B 20, 451–458.
Vanhellemont, J., Simoen, E., Claeys, C., 1995a. On the extraction of minority carrier recombination lifetime from forward diode characteristics. Appl. Phys. Lett. 66, 2894–2896.
Vanhellemont, J., Simoen, E., Kaniava, A., Libezny, M., Claeys, C., 1995b. On the impact of oxygen-related extended defects on silicon diode characteristics and material properties. J. Appl. Phys. 77, 5669–5676.
Veirman, J., Dubois, S., Enjalbert, N., Garandet, J.P., Lemiti, M., 2011. Electronic properties of highly-doped and compensated solar-grade silicon wafers and solar cells. J. Appl. Phys. 109, 103711/1–103711/10.
Vincent, G., Chantre, A., Bois, D., 1979. Electric field effect on the thermal emission of traps in semiconductor junctions. J. Appl. Phys. 50, 5484–5487.
Volkos, S.N., Efthymiou, E.S., Bernardini, S., Hawkins, I.D., Peaker, A.R., 2006. The impact of negative-bias-temperature-instability on the carrier generation lifetime of metal-oxynitride-silicon capacitors. J. Appl. Phys. 100, 124103/1–124103/9.
Voss, S., Stolwijk, N., Bracht, H., 2002. Spreading resistance profiling of silicon and germanium at variable temperature. J. Appl. Phys. 92, 4809–4819.
Vuillaume, D., Bourgoin, J.C., Lannoo, M., 1986. Oxide traps in Si-SiO_2 structures characterized by tunnel emission with deep-level transient spectroscopy. Phys. Rev. B 34, 1171–1183.
Weiss, S., Kassing, R., 1988. Deep level transient Fourier spectroscopy (DLTFS)—a technique for the analysis of deep level properties. Solid-State Electron. 31, 1733–1742.
Yamasaki, K., Yoshida, M., Sugano, T., 1979. Deep level transient spectroscopy of bulk traps and interface states in Si MOS diodes. Jpn. J. Appl. Phys. 18, 113–122.
Zohta, Y., Watanabe, M.O., 1982. On the determination of the spatial distribution of deep centers in semiconducting thin films from capacitance transient spectroscopy. J. Appl. Phys. 53, 1809–1811.

CHAPTER SEVEN

Surface and Defect States in Semiconductors Investigated by Surface Photovoltage

Daniela Cavalcoli[1], Beatrice Fraboni, Anna Cavallini
Department of Physics and Astronomy, University of Bologna, Bologna, Italy
[1]Corresponding author: e-mail address: cavalcoli@bo.infn.it

Contents

1. INTRODUCTION

This chapter addresses the surface photovoltage (SPV) spectroscopy method, applied to the study of surface, interface, and defect states in semiconductors.

The SPV method is a well-established contactless technique for the detection of surface and defect states in semiconductors, which has been used since the early 1970s by Gatos and Lagowski (1973) as an extensive source of surface and bulk information on various semiconductors and semiconductor surfaces and interfaces. It has been used to obtain information relevant to semiconductor surfaces (surface potential and flat band potential), to oxides deposited on the surfaces (oxide thickness, oxide integrity, oxide charge, and plasma damage), and to bulk parameters (minority carrier lifetimes and diffusion lengths, doping densities), depending on the light

Semiconductors and Semimetals, Volume 91
ISSN 0080-8784
http://dx.doi.org/10.1016/bs.semsem.2014.11.004

wavelength and on the photon flux used to induce the SPV (Kronik and Shapira, 1999, 2001; Schroder, 2001).

Nowadays, the renewed interest and the advantage of this method rely on the fact that it allows for the detection of electronic transitions (band-to-band, defect-band, and surface state-bands) on a huge range of semiconductors without the need of cumbersome sample preparation, junction formation, etc. The detection of such transitions can be obtained also on buried layers, deposited thin films, heterostructures, multiphase, and nanostructures, where standard transmission optical spectroscopies are not applicable. Such materials are becoming more and more interesting also for many different applicative purposes.

This chapter will be devoted to the description of the physical basis of the SPV spectroscopy, in which changes in the semiconductor surface potential upon illumination are monitored as a function of incident photon energy. The probe wavelength usually spans from below to above bandgap spectral regions, in order to obtain electronic transitions at localized (surface and defects) and extended states (band-to-band) in semiconductors.

In Section 2, we will describe the basic physical principles of the method and the relevant physical quantities will be defined in n-type and p-type semiconductors. In Section 3, we will discuss the experimental set-up. In Section 4, we will report selected applications of the method to the characterization of semiconductor surfaces and defect states in different systems: bulk semiconductors (CdTe), thin layers (heterostructures based on III-nitrides), and Si-based nanostructures (Si nanocrystals (NCs) and nanowires (NWs)). The potentiality of the method will thus be demonstrated spanning from bulk three-dimensional systems to lower dimensionalities. Some conclusions (Section 5) will end the chapter.

2. PHYSICAL BASIS OF THE METHOD

In order to introduce the concept of SPV, the electrical properties of a surface and the definition of surface potential are presented. A crystalline semiconductor is made by a periodic atomic pattern connected by chemical bonds. This pattern can be interrupted at the surface or at any defect, resulting in unsaturated (dangling) bonds, which can either rearrange themselves (surface or defect reconstruction), or be saturated by adatoms (surface) or capture impurities (defects). This results in a change of both the surface crystal structure and of the allowed energies that depend sensitively on the materials and bulk crystal structures involved. The surface or the defects break the translational symmetry, and the resulting states are localized at the surface or at the defects, with the wave function amplitude that decays exponentially over a few lattice constants. It turns out that the energies of these states are usually located inside the band gap, forming a separate band of surface or defect states. A surface can, thus, be seen as a boundary between media with different physical properties. The termination of the periodic structure of a semiconductor at its free surface forms surface-localized electronic states within the band gap and/or a double layer of charge, known as a surface dipole. These localized states can be generated by dangling bonds, surface reconstruction, or relaxation, i.e., a change in the position and/or chemical bonding configuration of surface atoms that minimizes the surface energy, steps and kinks at the surface, impurity atoms adsorbed on the surface, etc.

Surface-localized states induce charge transfer between bulk and surface in order to establish thermal equilibrium between the two (Schroder, 2001). In a p-type semiconductor, we can assume that Q, the usually positive surface charge density, induces an equivalent charge density in the semiconductor, Q_S, such that:

$$Q + Q_S = 0.$$

The semiconductor charge Q_S, in depletion, is provided by the ionized space charge density $qN_A w$, with q electronic charge, N_A acceptor density, and w the space-charge region (SCR) width. Neglecting free electrons and holes in the SCR:

$$w = \frac{Q_S}{qN_A} = \sqrt{\frac{2\varepsilon_S \varepsilon_0 V_S}{qN_A}}$$

with ε_S the semiconductor dielectric constant and ε_0 the vacuum dielectric constant. Solving for the surface potential V_S gives:

$$V_S = \frac{Q_S{}^2}{2\varepsilon_S \varepsilon_0 qN_A} = \frac{qN^2}{2\varepsilon_S \varepsilon_0 N_A},$$

where N represents the surface charge atom density (cm^{-2}). Many semiconductor samples that are characterized by SPV techniques are oxidized and contain surface charges (N) and thus possess a surface potential (V_S). Usually semiconductor surfaces contain positive charges and thus they are in depletion conditions. The equilibrium band diagrams of p-type and n-type

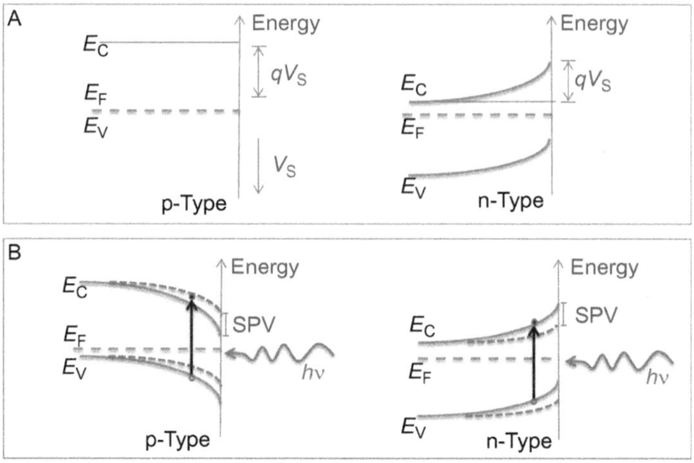

Figure 1 Band diagram of a p-type (left) and n-type (right) semiconductor close to the surface under equilibrium (A) and nonequilibrium/above bandgap illumination (B) conditions. Band edges E_C and E_V, Fermi energy E_F, and surface potential V_S are indicated in the diagram.

semiconductors close to the surfaces are shown in Fig. 1. It must be noted that, as sketched in Fig. 1, energy is defined positive, thus surface potential is by definition positive for downward band bending.

The surface photovoltaic effect represents a specific variant of the photovoltaic effect. The SPV is an illumination-induced change in the equilibrium surface potential V_S, which is followed by a charge transfer and/or redistribution within the device. The driving force for charge transfer is the built-in electric field generated by the surface SCR. The SPV is thus defined as the difference between the surface potential under illumination and the surface potential in dark:

$$\text{SPV} = V_S(\text{ill}) - V_S(\text{dark}).$$

The SPV signal can be detected by illuminating the surface with above or below bandgap light, usually a tunable source is used in order to obtain a continuous spectrum of SPV versus the incident photon energy.

We will examine the above bandgap light first. When photons with energy equal or larger than the band gap hit the semiconductor surface, electron-hole pairs are generated and collected by the surface barrier, a significant amount of charge may transfer in opposite directions under the built-in electric field and/or redistribute within the surface or the bulk so that the net surface charge density (N) changes and the surface potential is consequently reduced. In the case of a p-type semiconductor, the built-in electric field existing across the SCR drives photogenerated electrons toward the surface or interfacial region and holes toward the interior of material or the bulk. The reverse process takes place at an n-type semiconductor. Super-bandgap illumination decreases the surface band bending, as shown schematically in Fig. 1B for p-type and n-type semiconductors. When the photon energy equals the band gap, the resulting SPV increases and becomes negative in p-type and positive in n-type semiconductors. Therefore, the conductivity type of the semiconductor can be obtained by observing the sign of the SPV signal at photon energies close to the band gap. The variation recorded in the SPV spectrum when the incident photon energy approaches band-to-band transitions constitutes the most significant feature in the SPV spectrum, from which the optical band gap E_G can be obtained.

If we now consider sub-bandgap illumination, the probability of band-to-band absorption is negligible since photons do not have sufficient energy for inducing such transitions. However, below bandgap transitions at defect or surface states can be revealed. If we assume that defect or surface states are associated to a trap state E_T, we can represent the possible electronic transitions from the trap state to the band and from the band to the trap states in Fig. 2A and B, respectively. In the first case (Fig. 2A), photons with energy higher than the energy difference between the conduction band E_C and a trapped surface state E_T may promote electron transitions from the trapped surface state at E_T into the conduction band E_C (Fig. 2A), this transition would reduce the band bending in both p-type and n-type semiconductors, resulting in increase or decrease of the SPV signal, respectively. In the second case (Fig. 2B), electronic transitions from the valence band E_V to the trap level E_T would increase band bending in both p-type and n-type semiconductors, resulting in decrease or increase of the SPV signal, respectively. Therefore, if we consider a p-type semiconductor, a positive slope change in the SPV spectrum corresponds to an electron transition from an occupied defect state at E_T to the conduction band. On the other hand, a negative slope changes indicates photoexcitation from the valence band to an unoccupied defect state located at E_T (Fig. 3).

In the spectral region close to the band gap, photons can induce free charge to transfer between shallow states extending from the band gap (also known as "tail states") to one of the bands. The amplitude of the *tail states* can also be evaluated from a SPV spectrum (Fig. 3).

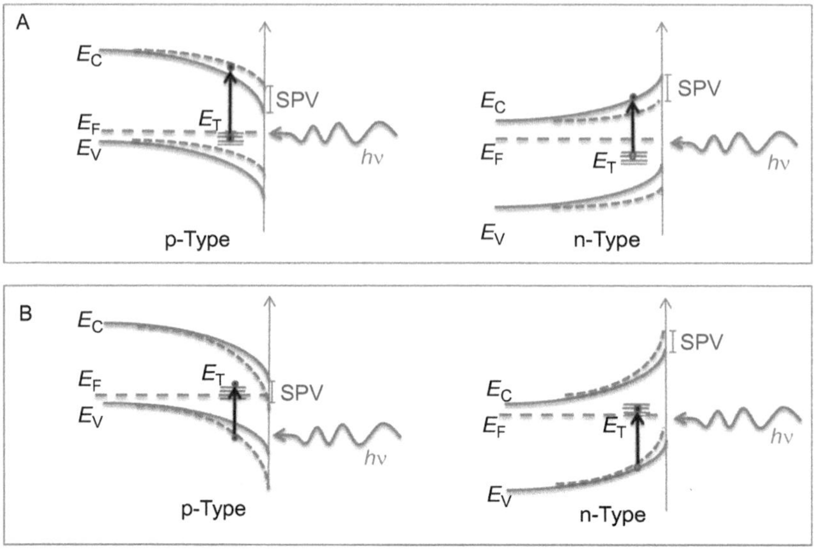

Figure 2 Band diagram of a p-type (left) and n-type (right) semiconductor close to the surface. Below bandgap illumination conditions with photon energies $h\nu = E_C - E_T$ (A) and $h\nu = E_T - E_V$ (B) are shown. Band edges E_C and E_V, Fermi energy E_F, and trap level E_T are indicated in the diagram.

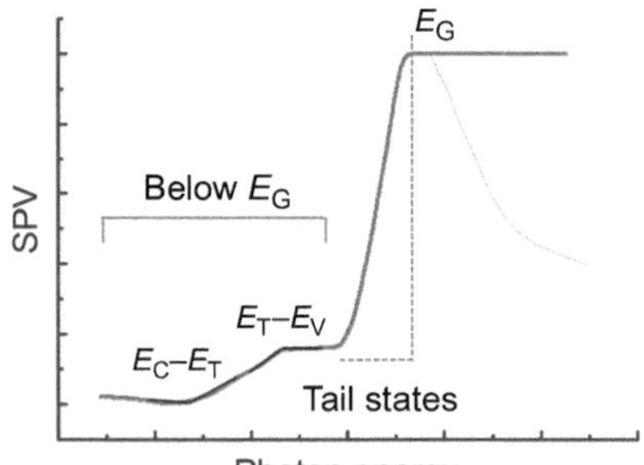

Figure 3 Ideal SPV spectrum for a p-type semiconductor. The features in the spectrum are related to below bandgap electronic transitions at trap level E_T and energy gap E_G. The decreasing trend at energies higher than E_G shows the decrease induced by surface recombination effects.

For photon energies larger than the band gap, a saturation value should be reached by the SPV signal, but often strong electron-hole recombination at the surface significantly reduces the collected SPV signal. Further features that can be detected in the above bandgap region of the spectrum can be related to phase segregations or Van Hove singularities (Kronik and

Shapira, 1999, 2001). Finally, it is worth mentioning that, as the SPV signal is proportional to the optical absorption coefficient α (Kronik and Shapira, 1999, 2001), the sensitivity of the SPV spectroscopy in the detection of bulk defect states is similar to the sensitivity of optical absorption spectroscopy, while SPV spectroscopy is more sensitive than optical absorption for surface states.

Figure 3 summarizes the different features that can appear in an ideal SPV spectrum of a p-type semiconductor.

3. EXPERIMENTAL DETAILS

The SPV experiment requires injection of free carriers into the semiconductor sample, separation of carriers by an external or internal electric field, and detection and amplification of the relevant signal. A convenient source for carrier injection can be provided by a photon source, i.e., a lamp (quartz–tungsten–halogen or Xenon lamp, depending on the spectral range of the experiment) coupled with a grating monochromator. Suitable cut-off filters must be used to avoid high-order diffraction peaks. Free carrier separation can be obtained by the electric field generated by a pn or Schottky junction (in contact mode) or by the surface SCR (in noncontact mode). The detection is usually accomplished by a probe capacitively coupled with the sample, and the signal is usually amplified through phase-sensitive amplification.

In this chapter, we will focus on the noncontact SPV method, the experimental setup of which is sketched in Fig. 4. In order to pickup the signal in noncontact mode, two configurations are usually employed: metal-insulator semiconductor (MIS) and Kelvin probe. In MIS configuration, the signal is capacitively coupled via a semitransparent conductive probe placed in the close proximity of the sample. The MIS device is made up by the sample surface itself, a very thin layer of air or mica spacer and the probe. The SPV probe is made by a thin semitransparent conductive layer, usually ITO sputtered on quartz. In MIS configuration, the light coming from the lamp is chopped and a phase-sensitive amplifier is locked to the chopping light frequency. Thus, the photon flux impinging on the sample causes a variation of the MIS capacitance which is picked up by the probe, the signal goes through an FET preamplifier and is detected by the lock-in amplifier (Fig. 4). The signal measured by the probe V_P is just the SPV signal

$$V_P = \mathrm{SPV}$$

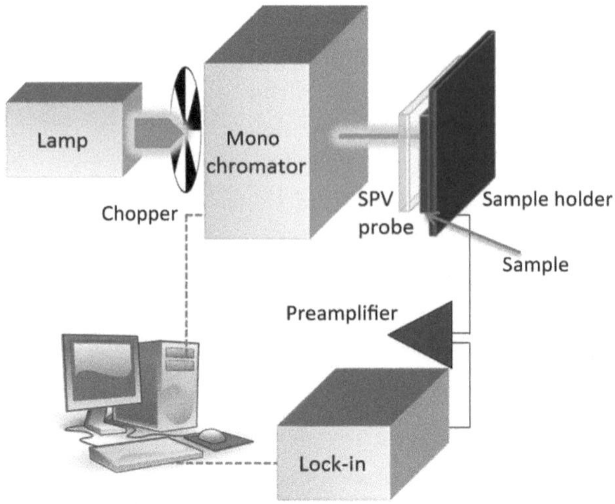

Figure 4 Sketch of the experimental setup for surface photovoltage spectroscopy measurements.

as the SPV is the illumination-induced variation of the surface potential V_S. Therefore, under such conditions, a measurement of surface potential changes is completely equivalent to a direct SPV measurement (in absolute value terms).

In Kelvin probe configuration, the work function difference between the probe and the sample, known as contact potential difference (CPD), is measured via a null method: a periodically vibrated plate is brought in proximity to the sample surface, so that a steady-state AC current develops in the effective AC capacitor formed. This current can be zero if, and only if, the capacitor is discharged. Thus, the CPD may be found easily, even automatically, by determining the DC bias V_B that must be applied to the capacitor for which the external AC current is nullified:

$$I = (\mathrm{CPD} - V_B)\mathrm{d}C/\mathrm{d}t = 0,$$

therefore

$$\mathrm{CPD} = V_B = V_P.$$

It can be demonstrated that:

$$\mathrm{CPD} = -\mathrm{SPV}.$$

When the signal is detected via a Kelvin probe method, the frequency of the lock-in amplifier is locked by the vibrating capacitor frequency.

The different physical principles behind the Kelvin probe and MIS approaches result in several relative strengths and weaknesses, which make the method of choice application dependent. The Kelvin probe is more suitable for the following phenomena with large time constants (surface states with long relaxation times); it does not perturb the true band bending of the free surface, and the extra information provided by the work function is extremely useful in experiments involving surface chemistry and/or film deposition. On the contrary, the MIS technique is more suitable for following fast relaxation phenomena relevant to bulk-related optical transitions. The SPV method has been implemented also with high spatial resolution by using the tip of an atomic force microscope as Kelvin probe force microscopy (Meoded et al., 1999).

Notwithstanding the different methods that can be employed to pick up the signal, it is clear that SPV spectroscopy does not require any cumbersome surface preparation or junction formation, and this is an important advantage when materials with rough surfaces or buried layers must be studied.

In the following sections, applications of SPV to different materials and structures will be reported; all the experiments have been carried out using the MIS method. The photon flux used varies in intensity because of the spectral variation of the optical components of the system (lamp, grating monochromator, etc.); thus, the SPV signal is usually normalized to the photon flux. Such a procedure is applicable only if the linear dependence of the SPV signal versus the photon flux can be demonstrated, and this usually happens in low injection condition (Castaldini et al, 2000).

4. APPLICATIONS OF THE SPV METHOD

In this section, some applications of the method to the study of defect and surface states in different semiconductor systems are presented. In detail, the determination of defect states and optical gap in bulk semiconductor like CdTe and related compounds, interface states in III-N-based heterostructures, optical band gap in Si NCs and NWs, and optical gap in nanoporous Ge will be discussed.

4.1. Surface and defect states in CdTe and $Cd_{1-x}Zn_xTe$

CdTe and $Cd_{1-x}Zn_xTe$ (CZT) single crystals have been intensively studied for applications as radiation detectors (Schleslinger and James, 1995). The here reported samples have been grown by EURORAD by the traveling heater method and high-pressure Bridgman. The samples were analyzed with two different methods: photocurrent (PC) and SPV spectroscopy, in order to compare defect sensitivity and spectral resolution of the two systems. Further details on sample structures and defective states can be found in elsewhere (Cavalcoli et al., 2008; Cavallini et al., 2007; Fraboni et al., 2009). In all the samples examined (CdTe and CZT), negative SPV signals were measured, indicating p-type conductivity.

Surface effects are known to play an important role in the performance of CdTe-based detectors, as the electrical properties of the electrode/CdTe interface are controlled by the interface, surface, and subsurface defects. If such defects are present in substantial concentration, they can cause a significant deterioration of the detector performance or even catastrophic failure of the device. For this reason, the analysis of surface-related states in such materials/devices is of major importance.

The two SPV spectra of CdTe:Cl reported in Fig. 5A, represented by solid and dashed lines, have been measured by illuminating as-received and air-cleaved surfaces, respectively. As the SPV is proportional to the optical absorption coefficient α, the energy gap value E_G could be determined by the variation of α as a function of the photon energy. Nevertheless, as in above bandgap photon energies, the SPV still possesses a significant below bandgap contribution, the extrapolation procedure usually employed for bandgap determination in absorption spectroscopy analyses cannot be applied. This clearly emerges from the diagrams reported in Fig. 5A, where a gradual increase instead of a steep one is observed for energies close to E_G. At photon energies close to the band gap, the SPV should increase rapidly up to a saturation value and E_G should be determined by the energy position of the knee. Sometimes, as in the present case and as also shown in Fig. 3, a peak rather than a knee is observed, due to electron-hole recombination at the surface strongly reducing the SPV signal for $E > E_G$. The main marked peak in each spectrum of Fig. 5A can, thus, be related to the energy gap values E_G of the compound. However, it must be noted that the E_G value measured on as-received surface is significantly lower than those measured on cleaved surfaces. This effect can be explained taking into account the strong electron-hole recombination at the as-received surface.

When the photon energy approaches E_G, α strongly increases and the light penetration depth α^{-1} strongly decreases. Thus, at photon energies close to E_G, the electron-hole pairs are generated in the near-surface layer and mainly recombine there through the high density of surface states. The resulting SPV signal thus decreases. When the surface is cleaved, i.e., with a much lower density of surface states, the surface recombination plays a

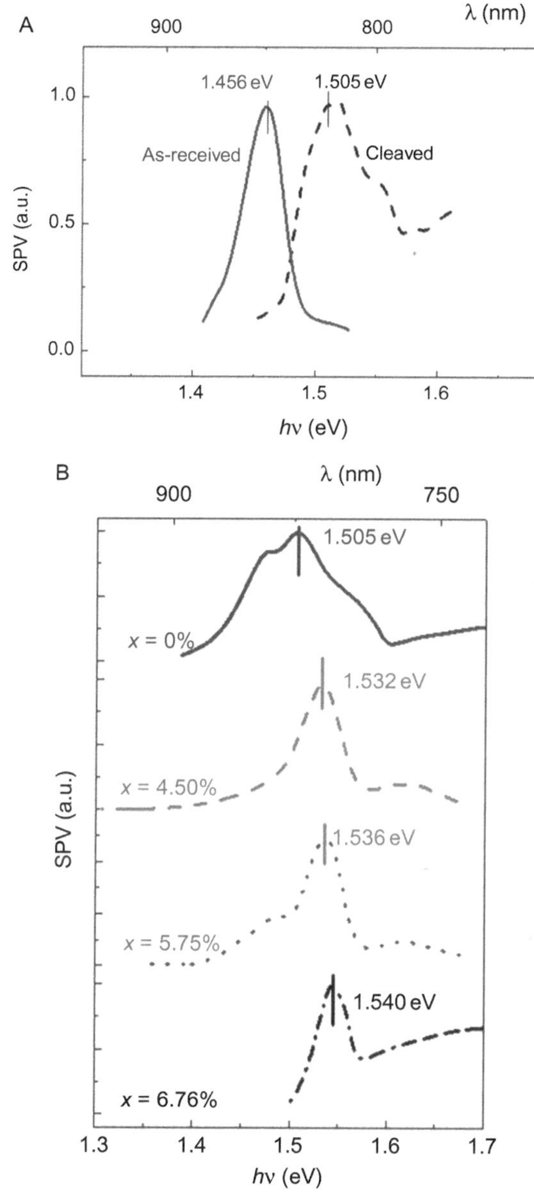

Figure 5 SPV spectra CdTe:Cl (solid line: as-received surface, dashed line: air-cleaved surface) (A). Surface photovoltage spectra of $Cd_{1-x}Zn_xTe$ for different Zn concentration *x*, determined by energy dispersive X-ray spectroscopy (B). *Reprinted from Cavalcoli et al. (2008).*

marginal role. The higher the surface state density, the stronger the recombination-induced decrease of the SPV signal and, thus, the higher the underestimation of E_G. Therefore, only the peaks in the SPV spectra relevant to the cleaved surfaces provide the real value of E_G, which is equal to 1.505 eV for CdTe, in agreement with literature data (Ching-Hua, 2008).

In addition, Cd_{1-x} Zn_xTe (x ranging from 0% to 6.76%) samples have been studied by SPV in order to determine E_G versus the Zn concentration as assessed by energy dispersive X-ray spectroscopy (Cavalcoli et al., 2008). Figure 5B shows SPV spectra of Cd_{1-x} Zn_xTe for different Zn concentrations. It is interesting to observe that the shift in the peak values (corresponding to the estimate of E_G) is in a very good agreement with the predicted Zn% dependence on E_G (Cavalcoli et al, 2008). SPV, thus, represents a valuable method to assess the exact Zn concentration, which significantly affects the transport and lattice parameters of the alloys.

The effect of defect states on the SPV spectrum of CdTe is shown in Fig. 6, where PC and SPV below bandgap spectra are reported. As shown in Fig. 3, the defect states are detected by SPV spectroscopy as slope changes in the below bandgap region of the spectrum. We have observed three major slope changes in the SPV spectrum of Fig. 6: a negative slope change at 0.71 eV indicating the presence of a deep trap located at $E_V+0.71$ eV;

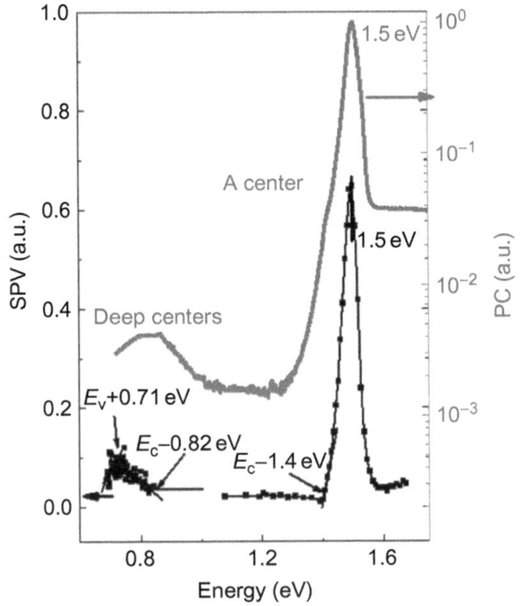

Figure 6 Surface photovoltage (left axis) and photocurrent (right axis) spectra of CdTe:Cl.

two positive slope changes located at 0.82 and 1.4 eV corresponding to two further deep traps located at $E_C - 0.82$ eV and $E_C - 1.4$ eV. It must be noted that SPV spectroscopy not only allows for the detection of defect states but also of their energetic position within the band gap. It is difficult to extract deep energy level values from the corresponding PC spectrum (Fig. 6) as reliabe as from SPV, possibly due to a lower sensitivity of the technique with respect to SPV at room temperature operation. Nonetheless, even in the PC spectrum, it is possible to observe the presence of a peak at 1.4 eV (attributed to the center A, Castaldini et al., 1996) on the left-hand side shoulder of the bandgap peak as assessed by a numerical fit of the spectrum. The broad peak detected around 0.8 eV could be a convolution of two or more deep traps of energy approximately located in the 0.7- to 0.9-eV range. It is evident that more detailed information on the deep trap energy levels can be extracted by SPV spectroscopy. The good agreement between the SPV (sensitive to bulk as well as surface states) and PC (sensitive to bulk states) spectra shows that the detected energy levels ($E_C - 0.82$ eV, $E_C - 1.4$ eV, and $E_V + 0.71$ eV) are mainly related to bulk states.

The comparison between SPV and PC spectra has allowed us to critically discuss the advantages and disadvantages of SPV spectroscopy with respect to other methods.

The main advantage to mention is that SPV spectroscopy is a noncontact method (on the contrary to PC and PICTS, photo-induced current transient spectroscopy, that require electrical contacts), and it is nondestructive (on the contrary, optical absorption spectroscopy requires sample thinning).

The sensitivity of SPV to surface states can be considered as an advantage or a disadvantage, it depends on the applications. By SPV with below bandgap light, both surface and bulk states can be detected. A possible, simple way to distinguish between them is the comparison with other methods: if the same energy levels are revealed by SPV and PC, they can be related to bulk defects (as in the present case), while if they are revealed only by SPV (but not by PC) they can be attributed to surface states.

An important advantage of SPV with respect to other defect state spectroscopies is the capability to reveal the energy level positions with respect to the valence or conduction band edges, while other methods (such as PICTS) do not allow such a distinction.

4.2. Surface and defect states in III-nitride-based alloys and heterostructures

III-Nitrides (III-Ns) semiconductors and their alloys have shown in the last few years high potential for interesting applications in photonics and electronics. III-Ns-based heterostructures have been under wide investigation for different applications such as high-frequency transistors, ultraviolet photodetector, light emitters, etc. (Wu, 2009). In this section, SPV is applied to the investigation of III-Ns-based heterostructures.

The first application concerns nearly lattice-matched $Al_{1-x}In_xN$/AlN/GaN heterostructures. These structures have been widely studied recently due to promising applications as high electron mobility transistors (Gonschorek et al., 2008). SPV spectroscopy allowed for the detection of a shift of the GaN band gap in $Al_{1-x}In_xN$/AlN/GaN heterostructures.

Due to the potential barrier well forming at the interface between $Al_{1-x}In_xN$ and GaN, nearly lattice-matched $Al_{1-x}In_xN$/GaN heterostructures exhibit typical 2D electron gas (2DEG) confined at the heterointerface whose density is of the order of $1.0–2.0 \times 10^{13}$ cm^{-2}. The insertion of AlN layer helps better confinement of electrons and reduction of alloy scattering. It has been demonstrated that the 2DEG density varies as a function of the AlN layer thickness *d*, in particular increases for *d* ranging from 0 to 2.5 nm (Gonschorek et al., 2008; Pandey et al., 2011). Figure 7A shows a sketch of the investigated AlInN/AlN/GaN structures, grown by MOCVD by AIXTRON (Cavalcoli et al, 2011a; Pandey et al., 2011). Structures with different AlN thicknesses have been investigated, and the spectra are reported in Fig. 7B. As already explained in previous sections, the most relevant peak in a SPV spectrum is usually related to band-to-band electronic transitions. Furthermore, by changing the photon energy, the optical absorption coefficient α, and thus the penetration depth α^{-1}, varies. Therefore, considering the spectral variation of α^{-1} in AlInN, AlN, and GaN (Cavalcoli et al., 2011a), it follows that for photon energies 3 eV the carrier generation and collection occurs mainly within the GaN layer, while when the photon energy becomes comparable with the GaN band gap, it occurs mainly at the interface between GaN and AlN, or at the GaN surface in the template. Thus, the peaks in Fig. 7B are all related to band-to-band transitions at the GaN/AlN heterointerface (Fig. 7B) or at the GaN surface (Fig. 7C). While in the GaN template, the maximum value in the spectra expectedly occurs at 3.44 eV (Fig. 7C), in good agreement with the literature value of the optical gap E_G in GaN (Ioffe web site); in the heterostructures, the peaks are detected at higher energy values (Fig. 7B). This shift in the band gap depends on the AlN interlayer thickness *d* and is not revealed by optical transmission measurements, as shown in Fig. 7D, where optical transmittance spectra of GaN template and the 1-nm AlN herostructures are compared.

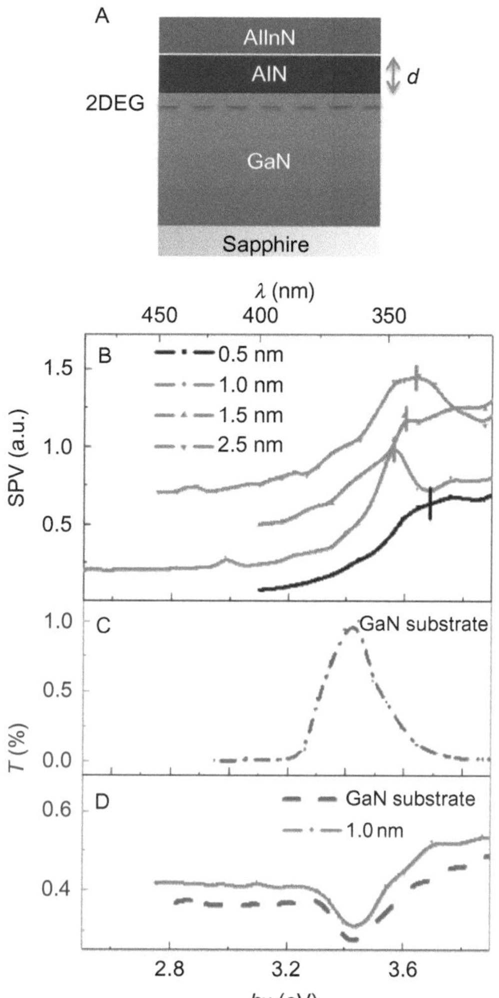

Figure 7 Sketch of AlInN/AlN/GaN heterostructure HS (A), SPV (a.u.) spectra of the HS for different thickness *d* of the AlN layer (B), of the GaN template (C), and Transmission coefficient (T) spectra of GaN template (orange, short dashed dot; light gray in print version) and of AlInN/AlN (1 nm)/GaN heterostructure (red, solid line; dark gray in print version) (D). *Reprinted from Cavalcoli et al. (2011a).*

As the energy gap shift is observed only in the heterostructure, it has been related to the presence of the 2DEG, whose density is related to the AlN layer thickness: in proximity of the heterointerface the semiconductor becomes almost degenerate. The energy gap increases as a function of the free carrier concentration, which is related to the 2DEG density, due to the combined Moss Burnstein and renormalization effects. The comparison between experimental data and theoretical predictions allows for the calculation of a parameter, which quantitatively accounts for the nonparabolic conduction band and to the evaluation the effective mass dependence on free carrier concentration (Cavalcoli et al., 2011a). It is worth noting that in this application SPV allows for the *local* determination of band-to-band transitions, i.e., the energy gap is measured at the interface, where the effect of the 2DEG is significant. The same analyses would be impossible by optical transmission spectroscopy, which is not able to detect the shift (Fig. 7D).

The second example concerning III-N-based structures refers to InGaN/GaN heterostructures. InGaN alloy is particularly interesting as its band gap is tunable, i.e., the energy gap varies in a wide spectral range as a function of the In content. However, some issues on the material properties are still far to be completely understood, in particular the dependence of the band gap as a function of the In concentration and the effectiveness of doping are still debated.

SPV spectroscopy has shown to be able to determine bowing parameter and intragap states in good quality undoped and n-type Si-doped InGaN layers. The bowing parameter b is defined considering that the InGaN band gap E_G(InGaN) varies nonlinearly as a function of In content (x) as follows:

$$E_G(In_xGa_{1-x}N) = (1-x)E_G(GaN) + xE_G(InN) - bx(1-x)$$

with E_G(GaN) and E_G(InN) GaN and InN energy gap, respectively. Up to now, no agreement has been reached on the bowing value and even on the issue if a single bowing parameter can describe the gap over its entire composition range (Moses and Van de Walle, 2010; Wu et al, 2002).

Figure 8 shows SPV spectra of the InGaN/GaN structures sketched in the inset for different In content (In%). The two peaks in the spectra are related to the optical energy gaps E_G of the layers in the structures. While E_G (GaN) does not change for the different samples examined, E_G(InGaN) changes with the In%. It must be noted that SPV allows for the determination of band gap even when the material is buried (GaN). This analysis shows that SPV allows for the extraction of the dependence E_G(InGaN) on In content and thus on the evaluation of the bowing parameter, which results in good agreement with literature data (Pandey et al., 2013).

Below bandgap SPV spectroscopy has been used to determine donor related levels in InGaN. Among the different dopants in InGaN, Si doping particularly interesting is able to mitigate the detrimental effect of the polarization field which strongly reduces the carrier collection efficiency of

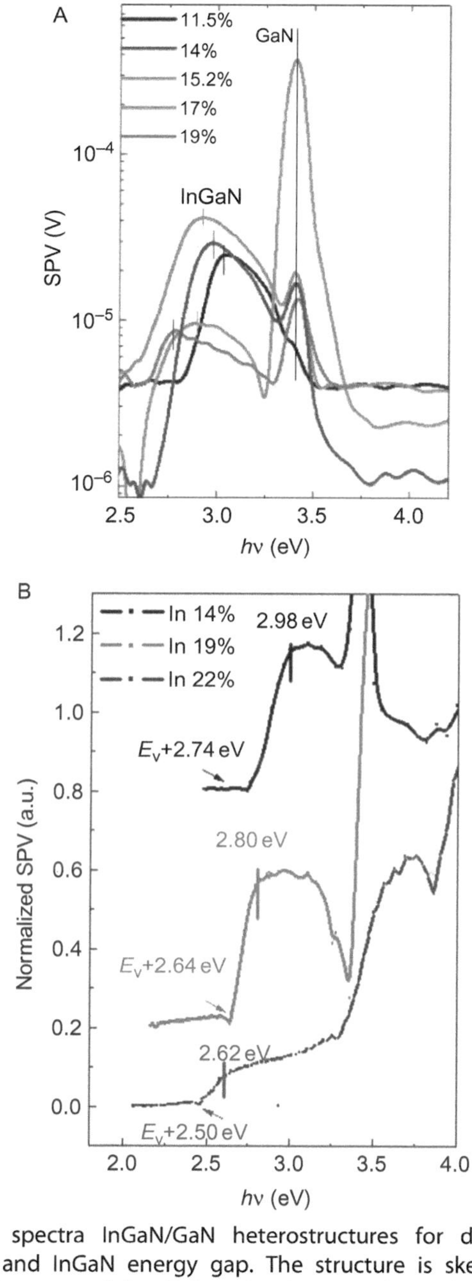

Figure 8 (A) SPV spectra InGaN/GaN heterostructures for different In content showing the GaN and InGaN energy gap. The structure is sketched in the inset. (B) Normalized SPV spectra of the Si-doped InGaN samples with 14%, 19%, and 22% In content. The spectra have been shifted for clarity. Energy gap values and below bandgap transitions are shown in the graph.

(Continued)

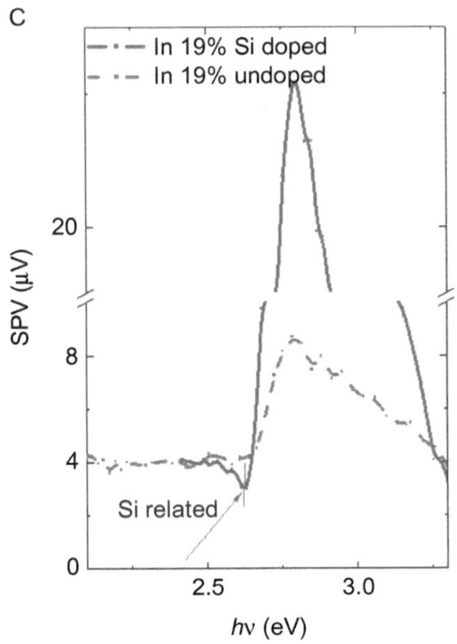

Figure 8—Cont'd (C) Comparison of SPV spectra of doped and undoped structures for 19% indium. *Reprinted from Pandey et al. (2013).*

InGaN-based solar cells and can strongly reduce dislocation lines in InGaN/GaN heterostructures (Hardy et al., 2012; Neufeld et al., 2011).

Figure 8B shows SPV spectra of Si-doped InGaN/GaN heterostructures. For photon energies below the InGaN band-to-band transitions, positive slope changes can be noted in 19% In and 22% In spectra. As this layers show n-type conductivity, positive slope changes correspond to electronic transitions from the valence band, marked as $E_V+2.74$, $E_V+2.64$, and $E_V+2.50$ for In% 14, 19, and 22, respectively, that can be related to donor levels. By comparing SPV spectra of undoped and doped structures with comparable In% (Fig. 8C), we can observe the enhancement of the signal in Si-doped samples, and the presence of these slope changes only in doped samples. Therefore, we can attribute these transitions to Si donor levels, and we can thus measure the activation energy of Si in InGaN for different In%. By this analysis, it can be concluded that when the In% increases, the Si-related energy levels become shallower and, as a consequence the ionized donor concentration increases, thus increasing the efficiency of Si doping in InGaN. This effect can be related to replacement of In by Si (Pandey et al, 2013).

4.3. Surface and defect states in silicon-based nanostructures (NCs and NWs)

SPV spectroscopy has also shown unique potentialities in the case of nanosized semiconductor structures, reviewed by Liqianga et al. (2003). Si-based nanostructures have shown wide potential in different fields and have been extensively studied in the last decades; their characteristics have been recently reviewed by Priolo et al. (2014). Besides Si nanostructures, SPV has been also used for the investigation of surface states in cleaved surfaces of silicon (Craig and Thurgate, 1994), porous silicon (Burstein et al., 1997), and in finely machined silicon surfaces (Yamauchi, 2000).

In this section, the application of SPV spectroscopy to hydrogenated nanocrystalline Si (nc-Si:H) will be presented. nc-Si:H is a multiphase material constituted of an amorphous Si matrix and a dispersion of Si NCs. Crystalline grains, grain boundaries, boundaries between different phases, impurities, and even voids are also present. Such a complicated material has been extensively studied in the last few years due to its promising characteristics for photovoltaic applications (Shah et al., 2003). The experimental results here reported refer to intrinsic nc-Si:H films grown by low-energy plasma-enhanced chemical vapor deposition using different dilution factors d defined as:

$$d = \frac{\phi(\mathrm{SiH_4})}{\phi(\mathrm{SiH_4}) + \phi(\mathrm{H_2})}$$

with ϕ gas flows, thin films grown with different d values showed crystalline fraction ranging from 20% to 80%. SPV spectra of nc-Si:H films have allowed us to characterize defect states in noncontact mode and to determine energy gap and conductivity type (Cavalcoli et al., 2011b). A typical SPV spectrum obtained on high-crystallinity (75%)low-dilution (3%) film is reported in Fig. 9A. The knee in the spectrum at 1.59 eV corresponds to the energy gap (E_G) of the film, which is a typical value of the high-crystallinity films. As explained in Section 2, in the below bandgap region defect states or electronic transitions between spatially localized states can be detected. In this case, as the dimension of the Si NCs range from 5 to 15 nm (Cavalcoli et al., 2011b), free carriers could be localized at Si NCs. Therefore, the spectrum has been fitted with the sum of three Lorentzian functions plus an exponential function whose factor determine the Urbach tail and thus the crystal disorder of the a-Si matrix. Urbach tails in the present samples are about 80 meV, while the electronic transitions occur at 1.31,

1.41, and 1.51 eV. Figure 9B shows a sketch of the proposed band diagram of the material. Therefore, in the present example, the analyses of spectra has allowed for the detection of electronic transitions between localized states, which could be ascribed to quantum confinement (QC) effects at Si NCs. QC occurs if the barriers between amorphous and crystalline phases act as potential energy barriers for moving electrons. However, as stated in Section 2, SPV requires not only carrier generation but also carrier collection, thus to be detected by SPV charge carriers must be separated by the surface junction electric field and must flow within the film. Electronic transitions at fully localized states should not be, in principle, detected by SPV. Therefore, we must conclude that free carriers at Si NCs are only partially

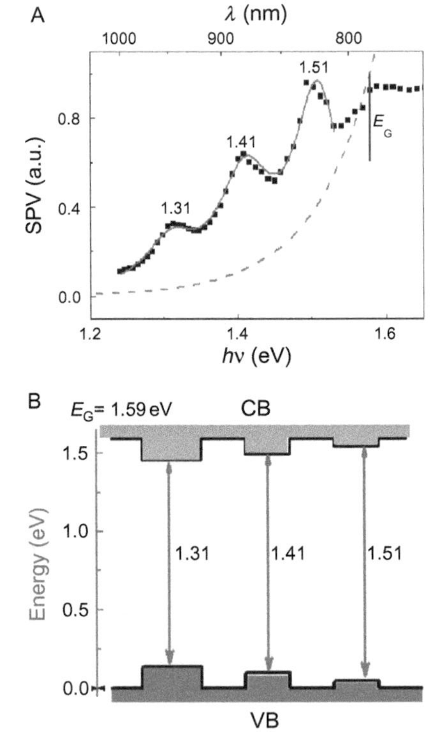

Figure 9 (A) Experimental surface photovoltage spectrum (dark squares) of nc-Si:H film grown at 280 °C, with a rate of 0.37 nm s^{-1} and dilution factor *d* of 3%. The crystallinity fraction is equal to 75%. The features of spectrum have been related to electronic transitions at partially localized states. The fitting function is the sum of three Lorentz functions (centered at 1.31, 1.41, and 1.51 eV) (solid line) plus an Urbach tail contribution due to the amorphous matrix (dashed line). (B) Sketch of the band diagram resulting from the SPV spectrum of Fig. 9A. *Reprinted from Cavalcoli et al. (2011b).*

localized, according to theoretical prediction from Bagolini et al. (2009). Free carriers at Si NCs can be collected as they can be thermally emitted out of the potential barrier, they can tunnel out of the QD, or they can be separated by the surface electric field. In conclusion, SPV spectroscopy has allowed for the identification of electronic transitions at partially localized states at Si NCs.

Another interesting application of SPV spectroscopy is the measurement of band-to-band transitions in Si NWs. Gold-catalyzed silicon NWs are grown on Si substrate at a total pressure ranging from 8 to 10 Torr using a cold wall CVD chamber. B and P doping is obtained using B2H6 and PH3 as n-type and p-type dopant gases, respectively. A scanning electron micrograph of the Si NW mat is visible in the inset of Fig. 10. Further details on NWs' properties are reported by Fabbri et al. (2013). Figure 10 shows the SPV spectra of undoped, B- and P-doped Si NWs compared with SPV spectrum of bulk Si. As already outlined in previous sections, by changing the photon energy the optical absorption coefficient, α, and in turn the

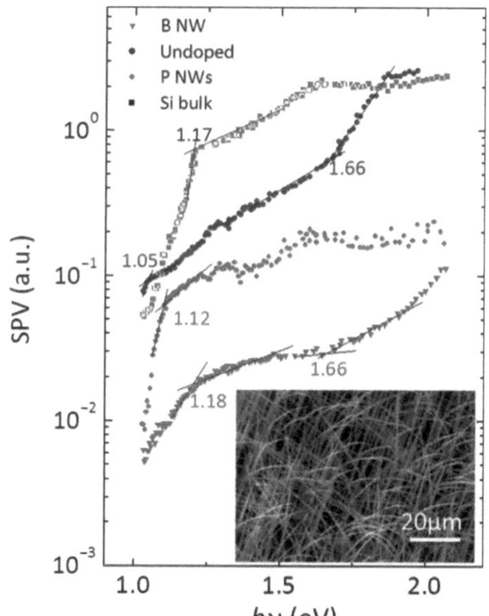

Figure 10 Surface photovoltage spectra of B-doped (red (light gray in the print version) triangle), undoped (green (gray in the print version) dots), P-doped (blue (dark gray in the print version) dots) Si NWs and bulk Si (black square). The slope changes correspond to the estimated optical band gaps, for B-doped and undoped Si NWs is 1.6 ± 0.1 eV, for P-doped NWs and bulk Si is1.1 ± 0.1 eV. The inset shows the scanning electron micrograph of the NW mat.

penetration depth, α^{-1}, varies. For photon energies ranging from 0.8 to 1.4 eV, the penetration depth ranges from 10^6 to 10 μm in bulk Si Dash and Newman (1955). Thus, in this spectral range, the carrier generation and collection occur mainly within the Si substrate, while for photon energies above 1.4 eV they occur mainly within the NW mat. It is clear from the comparison shown in Fig. 10 that, while P-doped NWs and bulk Si show very similar SPV spectra with an onset at the fundamental band gap of Si (around 1 eV), B-doped and undoped NWs show an additional onset at about 1.6 eV. At these photon energies, the penetration depth is within the NW mats; hence, this feature can be safely related to the NWs. Urbach states induced by crystal disorder in the structure give rise to these smooth optical transitions, which do not allow for a more precise determination of the optical band gap which must be associated to an error bar around 0.1 eV. Therefore, the optical gap of B-doped and undoped NWs is equal to 1.6 eV, while the P-doped one is 1.0 eV. This huge difference can be explained by a phase transition induced by doping in Si NWs. Structural and other optical measurements confirm that B and P dopants have different effects on retaining the phase of the grown NWs: the great majority of undoped NWs show totally wurtzite phase, B-doped NWs in the mats predominantly keep the wurtzite phase; meanwhile, P-doped ones have mainly diamond phase (Fabbri et al., 2013). In this example, it has been shown how SPV spectroscopy allows for determining phase changes in Si-based NWs.

4.4. Surface states in porous semiconductors

One of the main advantages of SPV spectroscopy is the possibility to study defective states and band-to-band transitions without the need to realize Schottky or pn junction, as often required by junction spectroscopy methods, like deep level transient spectroscopy (DLTS) or similar techniques. This aspect is interesting and becomes mandatory when very rough surfaces must be studied, as, for example, materials containing pores or voids, where the realization of a Schottky junction could be challenging. In this respect, porous semiconductors represent a useful example to show the potentiality of SPV spectroscopy. In addition, the method allows also determining the optical properties of layers grown on nontransparent substrate, as in the case of several porous thin films. Semiconductor materials containing nanopores have been widely studied in the last few years, thanks to their large surface-to-volume ratio and their ability to adsorb and interact with atoms, ions, and molecules, thus allowing their use in several interesting and emerging applications. Possible applications of porous semiconductors include various novel sensors, solar cells, optoelectronic devices; they can be used for catalysis, biological molecular isolation, and purification or electrodes for microfuel cells (Foll et al., 2006). Regarding optical properties, SPV studies of porous Si and TiO_2 and recently of porous Ge have been reported by Burstein et al. (1997), Dittrich et al. (2006), and Cavalcoli et al. (2014), respectively. Porous semiconductors can also be efficiently decorated with metallic nanoparticles (NPs), as metal NPs have attracted much interest in recent years due to their uses in a wide range of applications such as surface-enhanced Raman scattering (SERS) (Wang et al., 2010) and biosensors (Saha et al., 2012). In the present example, the optical characterization of ion-implanted nanoporous Ge samples will be discussed (Cavalcoli et al., 2014). These substrates were implanted with Ge+ ions at 300 keV being the Ge-projected range about 130 nm (Ziegler et al., 1984). Continuous amorphous layers were generated by implanting the substrates, held at the liquid nitrogen temperature, at a fluence of 4×10^{13} cm^{-2}. In this condition, the dynamic annealing of the defects created by the ion beam is suppressed and the amorphization of the material occurs. Porous layers were obtained by implanting at room temperature the Ge substrates at different fluences in the range from 5×10^{15} to 4×10^{16} cm^{-2}. After ion implantation, some samples were annealed at temperatures ranging from 360 to 500 °C for 1 h. The evolution of the prominent features in the SPV spectra with annealing temperature has been studied; the relevant SPV spectra are reported in Fig. 11. The SPV spectrum of a nanoporous sample obtained with an implant fluence of 1×10^{16} cm^{-2} is compared with the spectra of samples obtained with the same implant fluence but annealed at increasing temperatures. The SPV spectrum of the as-implanted sample shows the main feature at 0.78 eV, typical of the band-to-band transitions in the amorphous phase, while the annealed sample spectra show the gradual appearance of the 0.67- and 0.80-eV features which are typical of band-to-band transitions in crystalline-Ge (Cavalcoli et al., 2014 and ref therein). This result is in good agreement with the micrograph analyses (Impellizzeri et al., 2012). The broadening of the spectra is likely due to large tail states related to crystal disorder.

Besides the effect of the annealing temperature, the effect of decorating nanoporous Ge with Au NPs has been studied by SPV. Some np-Ge samples were decorated with Au NPs by electrochemical deposition at a fixed potential, and their SPV spectra has been studied. Figure 12A–C shows cross-section TEM (X-TEM) images of the 1×10^{16} cm^{-2} implanted

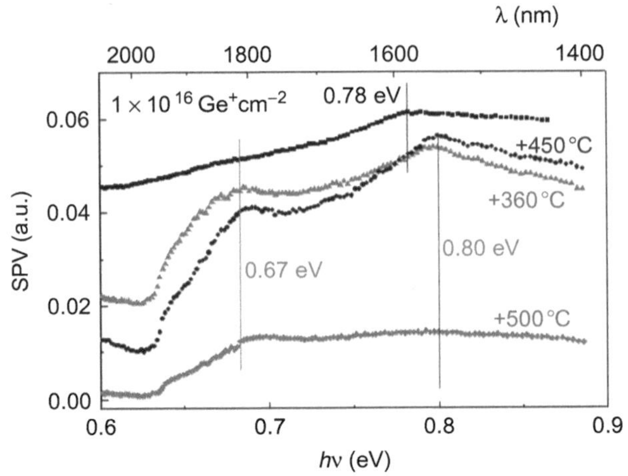

Figure 11 Normalized SPV spectra of nanoporous Ge layers at different annealing temperatures. The spectra have been shifted for clarity. The feature at 0.78 eV corresponds to the main electronic transition in the amorphous phase, the ones at 0.67 and 0.80 eV to the main transition in the crystalline phase. Ge fluence ($Ge+cm^{-2}$) and annealing temperatures (°C) are shown in the plot.

np-Ge samples decorated with Au NPs for 10-, 20-, and 40-s deposition time, respectively, while Fig. 12D shows the relevant SPV spectra (raw data, not normalized to the photon flux spectrum). It is worth noting in Fig. 12D that for Au deposition time of 20 s a significant increase of the SPV signal occurs while deposition times of 10 and 40 s causes a decrease of the SPV signal with respect to the as-implanted. The explanation of this behavior can be found in the X-TEM pictures, as it is evident from the comparison between Fig. 12A–C that only in the case of 20-s deposition time small (diameters ranging from 1 to 5 nm) Au NPs coat the inner walls of the porous structure, while 10- and 40-s deposition time produces large Au NPs (diameters larger than 20 nm), mostly located at the top of the columns. Therefore, these large Au NPs located at the top of the columns seem to be responsible for photovoltage suppression, while small NPs mostly located in the interior of the nanopores are mostly effective in photovoltage enhancement. This can be understood as follows: large NP clusters located at the top surface increase surface recombination, thus decreasing the surface band bending and thus its variation with impinging photons, i.e., the SPV signal; on the contrary, small NPs located in the nanopore interior are effective light scatter centers. They possibly enhance the bouncing of the incoming photon within the nanopore, implementing an effective light trapping process. In

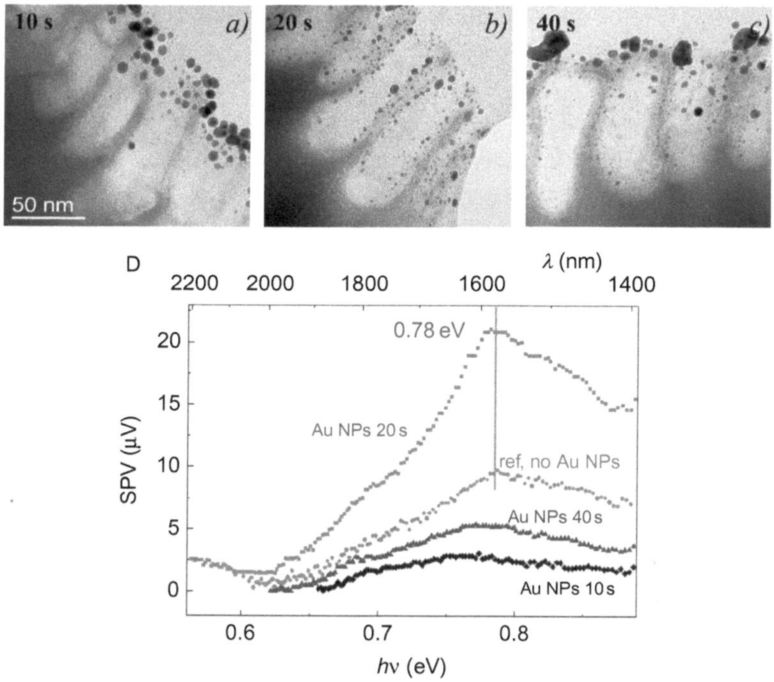

Figure 12 X-TEM micrograph showing the Au nanoparticles embedded within the nanoporous Ge structure deposited with (A) 10 s, (B) 20 s, and (C) 40 s. SPV signal plotted versus photon energy for np-Ge samples decorated with gold nanoparticles for different deposition times, compared with nondecorated sample. Fluence: 1×10^{16} cm^{-2}, as deposited (D). *Reprinted from Cavalcoli et al. (2014).*

addition, the photovoltage enhancement occurs at 1650 nm, wavelengths which are very close to the Ge band-to-band transitions, and far from the spectral region of the surface plasmon resonance of Au NPs (Eustis and El-Sayed, 2006); as a consequence, the observed enhancement of the SPV signal is most likely originated from the light scattering from the metallic NPs, and surface plasmonic resonance effects can be ruled out. This last example shows the potentiality of SPV spectroscopy to reveal phase changes and light trapping effects.

5. CONCLUSIONS

The aim of this chapter was to discuss the characteristics of the SPV spectroscopy method, starting from the definitions, the basic physical principles, going through the experimental details and discussing some relevant applications. It has been shown that the method allows for obtaining several different information on material properties. In more detail, in bulk CdTe and CZT, SPV has been used to determine the energy gap, and its variation with Zn concentration. In addition, an important surface recombination effect has been revealed and below bandgap defect states identified.

SPV spectroscopy has also allowed for optical characterization of buried layers in III-N-based heterostructures, where the effect that the two-dimensional electron gas forming in InAlGaN/GaN has on the energy gap value has been demonstrated. Impurity related levels, gap values, and bowing parameters have been obtained in InGaN as a function of In content.

The potentialities of the method have further been shown on Si-based nanostructures; in this case, the method helped obtaining evidence of QC in Si NCs and doping-induced phase changes in Si NWs. Finally, the application of the method to nanoporous Ge has contributed to the identification of phase transitions induced by thermal treatment and enhanced light trapping effects due to Au NPs embedded in the np-Ge matrix.

In conclusion, the here reported examples evidence the potential, the flexibility, and the ease of application of the SPV method, which can be used to investigate a wide range of different systems allowing to achieve relevant physical information on material properties that can hardly be otherwise obtained.

REFERENCES

Bagolini, L., Mattoni, A., Colombo, L., 2009. Electronic Localization and optical absorption in embedded silicon nanograins. Appl. Phys. Lett. 94, 053115.

Burstein, L., Shapira, Y., Partee, J., Shinar, J., Lubianiker, Y., Balberg, I., 1997. Surface photovoltage spectroscopy of porous silicon. Phys. Rev. B 55, 1930–1933.

Castaldini, A., Cavallini, A.B., Fraboni, B., Fernandez, B., Piqueras, J., 1996. Comparison of electrical and luminescence data for the A center in CdTe. Appl. Phys. Lett. 69, 3510–3513.

Castaldini, A., Cavalcoli, D., Cavallini, A., 2000. Determination of bulk and surface transport properties by photocurrent spectral measurements. Appl. Phys. A 71, 305–310.

Cavalcoli, D., Fraboni, B., Cavallini, A., 2008. Surface photovoltage spectroscopy analyses of $Cd_{1-x}Zn_xTe$. J. Appl. Phys. 103 (4), 043713–043719.

Cavalcoli, D., Pandey, D., Fraboni, B., Cavallini, A., 2011a. Band gap shift in $Al_{1-x}In_xN$/AlN/GaN heterostructures studied by surface photovoltage spectroscopy. Appl. Phys. Lett. 98, 142111–142113.

Cavalcoli, D., Rossi, M., Cavallini, A., 2011b. Defect states in nc-Si: H films investigated by surface photovoltage spectroscopy. J. Appl. Phys. 109, 053719–053724.

Cavalcoli, D., Fraboni, B., Impellizzeri, G., Romano, L., Scavetta, E., Grimaldi, M.G., 2014. Optoelectronic properties of nanoporous Ge layers investigated by surface photovoltage spectroscopy. Micropor. Mesopor. Mat. 196, 175–178.

Cavallini, A., Castaldini, A., Cavalcoli, D., Fraboni, B., 2007. Photocurrent and surface photovoltage spectroscopy investigations of CdTe-based compounds. IEEE Trans. Nucl. Sci. 54, 1719.

Ching-Hua, S., 2008. Energy band gap, intrinsic carrier concentration, and Fermi level of CdTe bulk crystal between 304 and 1067 K. J. Appl. Phys. 103, 084903–084909.

Craig, R.P., Thurgate, S.M., 1994. Two photon processes in surface photovoltage spectroscopy. Surf. Sci. 310, 103–112.

Dash, W.C., Newman, R., 1955. Intrinsic optical absorption in single-crystal germanium and silicon at 77°K and 300°K. Phys. Rev. 99, 1151–1155.

Dittrich, T., Mora-Seró, I., García-Belmonte, G., Bisquert, J., 2006. Temperature dependent normal and anomalous electron diffusion in porous TiO_2 studied by transient surface photovoltage. Phys. Rev. B 73, 045407–045415.

Eustis, S., El-Sayed, M.A., 2006. Why gold nanoparticles are more precious than pretty gold: noble metal surface plasmon resonance and its enhancement of the radiative and nonradiative properties of nanocrystals of different shapes. Chem. Soc. Rev. 35, 209–217.

Fabbri, F., Rotunno, E., Lazzarini, L., Cavalcoli, D., Castaldini, A., Fukata, N., Sato, K., Salviati, G., Cavallini, A., 2013. Preparing the way for doping wurtzite silicon nanowires while retaining the phase. Nano Lett. 13, 5900–5906.

Foll, H., Carstensen, J., Frey, S., 2006. Macroporous semiconductors. J. Nanomater. 91635, 1–10.

Fraboni, B., Cavalcoli, D., Cavallini, A., Fochuk, P., 2009. Electrical activity of deep traps in high resistivity CdTe: spectroscopic characterization. J. Appl. Phys. 105, 073705–073710.

Gatos, H.C., Lagowski, J., 1973. Surface photovoltage spectroscopy—a new approach to the study of high-gap semiconductor surfaces. J. Vac. Sci. Technol. 10, 130–134.

Gonschorek, M., Carlin, J.-F., Feltin, E., Py, M.A., Grandjean, N., Darakchieva, V., Monemar, B., et al., 2008. Two-dimensional electron gas density in $Al_{1-x}In_xN$/AlN/GaN heterostructures ($0.03 \leq x \leq 0.23$). J. Appl. Phys. 103, 093714–093716.

Hardy, M.T., Young, E.C., Hsu, P.S., Haeger, D.A., Koslow, I.L., Nakamura, S., et al., 2012. Suppression of m-plane and c-plane slip through Si and Mg doping in partially relaxed ($20\bar{2}1$) InGaN/GaN heterostructures. Appl. Phys. Lett. 101, 132102–132104.

Impellizzeri, G., Romano, L., Fraboni, B., Scavetta, E., Ruffino, F., Bongiorno, C., Privitera, V., Grimaldi, M.G., 2012. Nanoporous germanium electrodes as a template for nanosized (<5 nm) aggregates. Nanotechnology 23, 395604–395614.

Ioffe web site: http://www.ioffe.ru/SVA/NSM/Semicond/GaN/bandstr.html.

Kronik, L., Shapira, Y., 1999. Surface photovoltage phenomena: theory, experiment, and applications. Surf. Sci. Rep. 37, 1–206.

Kronik, L., Shapira, Y., 2001. Surface photovoltage spectroscopy of semiconductor structures: at the crossroads of physics, chemistry and electrical engineering. Surf. Interface Anal. 31, 954–965.

Liqianga, J., Xiaojunb, S., Jingc, S., Weiminb, C., Zilic, X., Yaoguoc, D., Hongganga, F., 2003. Review of surface photovoltage spectra of nanosized semiconductors and its applications in heterogeneous photocatalysis. Sol. Energy Mater. Sol. Cells 79, 133–151.

Meoded, T., Shikler, R., Fried, N., Rosenwaks, Y., 1999. Direct measurement of minority carrier diffusion length using atomic force microscopy. Appl. Phys. Lett. 75, 2435–2437.

Moses, P.G., Van de Walle, C.G., 2010. Band bowing and band alignment in InGaN alloys. Appl. Phys. Lett. 96, 021908–021910.

Neufeld, C.J., Cruz, S.C., Farrell, R.M., Iza, M., Lang, J.R., Keller, S., et al., 2011. Effect of doping and polarization on carrier collection in InGaN quantum well solar cell. Appl. Phys. Lett. 98, 243507–243509.

Pandey, S., Fraboni, B., Cavalcoli, D., Minj, A., Cavallini, A., 2011. Two-dimensional electron gas properties by current-voltage analyses of $Al_{0.86}In_{0.14}$ N/AlN/GaN heterostructures. Appl. Phys. Lett. 99, 012111–012113.

Pandey, S., Cavalcoli, D., Cavallini, A., 2013. Band bowing and Si donor levels in InGaN layers investigated by surface photo voltage spectroscopy. Appl. Phys. Lett. 102, 142101–142103.

Priolo, F., Gregorkiewicz, T., Galli, M., Krauss, T.F., 2014. Silicon nanostructures for photonics and photovoltaics. Nat. Nanotechnol. 9, 19–32.

Saha, K., Agasti, S.S., Kim, C., Li, X., Rotello, V.M., 2012. Gold nanoparticles in chemical and biological sensing. Chem. Rev. 112, 2739–2779.

Schleslinger, T.E., James, R.B., 1995. Semiconductors for Room Temperature Nuclear Detector Applications. Academic, San Diego, CA.

Schroder, D.K., 2001. Surface voltage and surface photovoltage: history, theory and applications. Meas. Sci. Technol. 12, R16–R31.

Shah, A.V., Meier, J., Vallat-Sauvain, E., Wyrsch, N., Kroll, U., Droz, C., Graf, U., 2003. Material and solar cell research in microcrystalline silicon. Sol. Energy Mater. Sol. Cells 78, 469–491.

Wang, X.T., Shi, W.S., She, G.W., Mu, L.X., Lee, S.T., 2010. High-performance surface-enhanced Raman scattering sensors based on Ag nanoparticles-coated Si nanowire arrays for quantitative detection of pesticides. Appl. Phys. Lett. 96, 053104–053106.

Wu, J., 2009. When group-III nitrides go infrared: new properties and perspectives. J. Appl. Phys. 106, 011101–011121.

Wu, J., Walukiewicz, W., Yu, K.M., Ager III, J.W., Haller, E.E., Lu, H., Schaff, W.J., 2002. Small band gap bowing in $In_{1-x}Ga_xN$ alloys. Appl. Phys. Lett. 80 (4741), 4743.

Yamauchi, K., 2000. Development of SPV (Surface Photo-voltage) spectroscopy to observe finely-prepared surfaces by ultraprecision machining processes. J. Jap. Soc. Precis. Eng. 66, 630–634.

Ziegler, J.F., Biersack, J.P., Littmark, U., 1984. The Stopping and Range of Ions in Solids. In: Stopping and Ranges of Ions in Matter, vol. 1. Pergamon Press, New York.

CHAPTER EIGHT

Point Defects in ZnO

Matthew D. McCluskey[1]
Department of Physics & Astronomy, Washington State University, Pullman, Washington, USA
[1]Corresponding author: e-mail address: mattmcc@wsu.edu

Contents

Semiconductors and Semimetals, Volume 91
ISSN 0080-8784
http://dx.doi.org/10.1016/bs.semsem.2014.11.002

1. INTRODUCTION

1.1. Motivation

Zinc oxide (ZnO) is a venerable wide-band-gap semiconductor that has experienced a resurgence of interest, due to its potential for light-emitting, spintronic, and transparent electronic applications. Its relatively large free-exciton binding energy (60 meV) means that excitons are stable at room temperature, a potentially desirable feature for efficient light emission (Look, 2001). In contrast to wide-band-gap semiconductors such as gallium nitride (GaN), bulk single crystals can be grown relatively easily (Triboulet, 2014). In addition, Zn is an "earth abundant" element, making it economically competitive.

Because of the intense focus on this material, the ZnO literature is vast and often contradictory. Many exciting claims of p-type conductivity and room-temperature ferromagnetism have not withstood the test of time. The purpose of this chapter is to summarize reasonably well-established results on point defects in ZnO. Due to the volume of papers in the field, only an illustrative subset of papers and review articles is presented. The chapter is not intended to be exhaustive or comprehensive. Instead, it presents an overview of the field, with enough references to allow the reader to delve further.

Interested readers will find no shortage of review articles on ZnO. For the "pre-2010" era, there are general reviews by Özgür et al. (2005), Klingshirn (2007), and Janotti and Van de Walle (2009) and a review of defects by McCluskey and Jokela (2009). Other papers include discussions of exciton recombination (Meyer et al., 2004), the theory of impurities (Park et al., 2002) and native defects (Janotti and Van de Walle, 2007a), processing (Pearton et al., 2005), and defects at surfaces and interfaces (Brillson, 2013; Brillson and Lu, 2011).

The field has gone through many recent changes. Through the decade of 2000–2009, the conventional wisdom was that substitutional nitrogen was a relatively shallow acceptor, with a binding energy of a few hundred millielectron volts, around the value for Mg in GaN. This assumption was challenged by a boldly titled theory paper by Lyons et al. (2009a), which claimed that nitrogen was a deep acceptor that could not produce p-type ZnO. Around the same time, experimentalists were finding that much—maybe all—of the ferromagnetism reported in transition-metal-doped ZnO originated from phase-segregated clusters rather than the ZnO material.

These and other developments suggest that a new review of defects in ZnO is in order. Hopefully, this chapter will serve that purpose.

1.2. Properties of ZnO

As shown in Fig. 1, ZnO has the hexagonal wurtzite crystal structure, with lattice constants $a = 3.25$ Å, $c = 5.21$ Å, and $u = 0.38$ (parameters in this section are from Özgür et al., 2005, unless noted otherwise). Similar to diamond and zincblende crystals, each atom is bonded to four neighboring atoms. ZnO has a bulk modulus of 143 GPa (Hanna et al., 2011) and transforms to the rocksalt phase at a pressure of ~9 GPa. Room-temperature Raman spectra of wurtzite ZnO typically show a longitudinal optical (LO) phonon at 588 cm^{-1} (E_1 symmetry). There is also an A_1 LO phonon mode at 576 cm^{-1}. These LO phonon modes, which have energies of ~72 meV, produce phonon replicas in photoluminescence (PL) spectra.

ZnO is a direct-gap semiconductor. Because of the wurtzite symmetry plus spin–orbit coupling, the valence band (VB) splits into three at the Brillouin zone center. These give rise to *A*, *B*, and *C* free excitons, in order of increasing energy. The free-exciton (*A*) binding energy is 60 meV, making it stable at room temperature. This feature may make ZnO desirable for efficient lighting applications. At liquid-helium temperatures, the band gap is 3.437 eV. This gap decreases to 3.37 eV at room temperature. In this chapter, spectroscopy results are quoted for low temperatures (2–12 K) unless noted otherwise.

Bulk ZnO crystals can be grown by several methods (Triboulet, 2014). Although the melting point is high (1975 °C), Cermet, Inc., grew ZnO by a high-pressure (50 atm) melt growth process. A radio frequency field couples to the melt or an iridium crucible. Chemical vapor transport (CVT), at

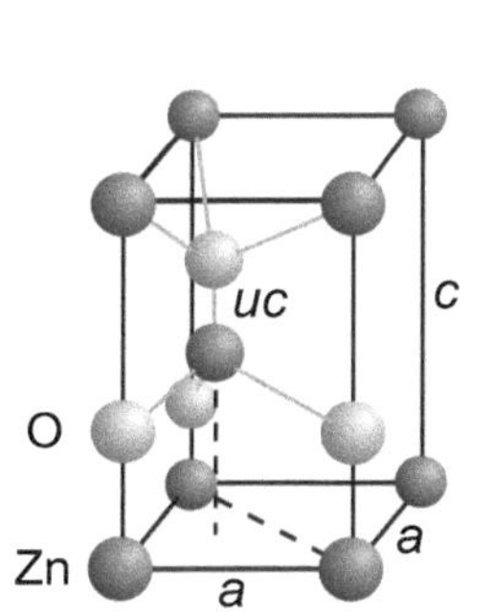

Figure 1 Wurtzite crystal structure for ZnO.

temperatures around 800–1300 °C, involves growth in a sealed silica ampoule (Ntep et al., 1999; Jokela and McCluskey, 2007). ZnO powder, along with a catalyst such as graphite, is placed at one end of the ampoule and a seed crystal is placed at the other end. A thermal gradient causes the transport of Zn and O, which grows on the "cold" seed. Hydrothermal growth (Ohshima et al., 2004) can be performed at temperatures below 400 °C and high pressures (~1000 atm). ZnO dissolved in mineralizers, such as LiOH or KOH, precipitates onto a seed crystal, resulting in crystals as large as several centimeters.

Thin-film ZnO is produced by standard techniques such as sputtering, molecular beam epitaxy, pulsed laser deposition, and metal organic chemical vapor deposition (Norton et al., 2004). High-quality films are grown on sapphire but cheaper substrates such as glass may also be used.

2. DONOR IMPURITIES

Group-III impurities substitute for the Zn site and act as shallow donors, with an effective mass donor binding energy of ~50 meV (Table 1). Together with hydrogen, these elements account for the n-type behavior found in most ZnO samples. Aluminum and gallium, in particular, are used for transparent conducting oxide thin films (King and Veal, 2011; Minami, 2000). With these dopants, electron densities of 10^{20} cm^{-3} can be achieved, with room-temperature mobilities ~60 cm^2/V s. Lower electron densities (10^{16}–10^{17} cm^{-3}) give rise to mobilities ~150 cm^2/V s.

Electrons bound to shallow donors give rise to the commonly observed "effective mass donor" electron paramagnetic resonance (EPR) at g ~1.96 (Table 2). In PL experiments performed at liquid-helium temperatures

Table 1 Donor binding energies and PL energies in ZnO

Donor	Binding energy (meV)	PL energy (eV)
Al_{Zn}	51.6	3.3608
Ga_{Zn}	54.6	3.3598
In_{Zn}	63.2	3.3567
H_O	47	3.3628
H_{BC}	53	3.3601

Data are at liquid-helium temperatures.
From Meyer et al. (2004) and Lavrov et al. (2009b).

Table 2 Magnetic resonance parameters for donors in ZnO, where || and ⊥ are parallel and perpendicular to the *c*-axis, respectively

Donor	*g* Values	Hyperfine parameters	Reference
Shallow (EM) donor	$g_{\|\|}=1.9570$ $g_{\perp}=1.9551$		Carlos et al. (2001)
Al_{Zn} (neutral)	$g_{\|\|}=1.9578$ $g_{\perp}=1.9564$	$A=2.0$ MHz	Herklotz et al. (2011)
Ga_{Zn} (neutral)	$g_{\|\|}=1.957$ $g_{\perp}=1.956$	$A=6.7$ G	Garces et al. (2002a)
In_{Zn} (neutral)	$g_{\|\|}=1.957$ $g_{\perp}=1.956$	$A=36.6$ G	Gonzales et al. (1982)
H_O (neutral)	$g_{\|\|}=1.9569$ $g_{\perp}=1.9552$	$A=1.4$ MHz	Hofmann et al. (2002)
Pb^{3+}	$g_{\|\|}=2.0133$ $g_{\perp}=2.0135$		Laiho et al. (2008)
V_O (1+)	$g_{\|\|}=1.9945$ $g_{\perp}=1.9960$	Axial ^{67}Zn: $A_{\|\|}=57.3$ MHz $A_{\perp}=42.3$ MHz Nonaxial ^{67}Zn: $A_{xx}=76.6$ MHz $A_{yy}=75.9$ MHz $A_{zz}=94.8$ MHz	Meyer et al. (2004)
Zn_i (1+)	$g_{\|\|}=1.9605$ $g_{\perp}=1.9595$		Vlasenko and Watkins (2005)

All spins are $S=½$.

(Meyer et al., 2004), donor-bound excitons give rise to sharp lines. As shown in Fig. 2, peaks are observed for H (3.3628 eV), Al (3.3608 eV), Ga (3.3598 eV), and In (3.3567 eV). In commercial bulk samples with $n \sim 10^{17}$ cm^{-3}, secondary ion mass spectrometry (SIMS) showed that the free-electron density matched the group-III impurity concentration (McCluskey and Jokela, 2007).

A ZnO PL spectrum, over a broad range of photon energy, is shown in Fig. 3. Peaks due to exciton recombination are seen at 3.36 and 3.33 eV, along with their phonon replicas. The peak at 3.33 eV was attributed to an exciton bound to a structural defect (Meyer et al., 2004). Peaks attributed to donor–acceptor pair (DAP) transitions are also observed. A broad "green

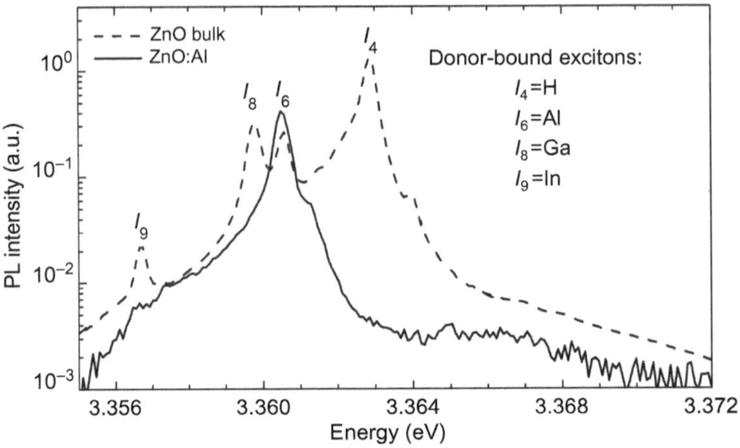

Figure 2 PL spectra (4.2 K) of donor-bound excitons in ZnO. *After Meyer et al. (2004).*

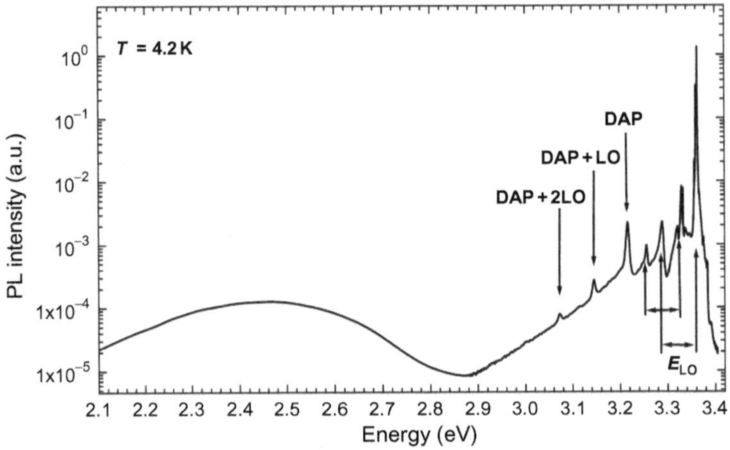

Figure 3 PL spectrum of ZnO. Phonon replicas of bound exciton peaks are indicated by E_{LO}. A donor–acceptor pair (DAP) transition with two phonon replicas are shown. A broad green luminescence peaks near 2.45 eV. *After Meyer et al. (2004).*

luminescence" is centered at 2.45 eV. These and other PL features are discussed further in Section 12.2.

Group-IV elements are sometimes present as impurities. Substitutional lead (Pb_{Zn}) impurities are common in ZnO powders and act as shallow double donors. In the doubly ionized state, the Pb^{4+} ion is EPR inactive. Illumination can convert Pb^{4+} to Pb^{3+}, giving rise to a magnetic resonance line with g ~2.013 (Laiho et al., 2008). Silicon impurities have been observed with SIMS (McCluskey and Jokela, 2007), but it is not known whether they exist as point defects or a segregated phase. According to theory, Si and Ge should occupy the substitutional Zn site and act as shallow double donors (Lyons et al., 2009b).

3. HYDROGEN DONORS

Hydrogen is considered to be an important contributor to n-type conductivity in oxides (McCluskey et al., 2012). After the pioneering studies of Mollwo (1954) and Thomas and Lander (1956), Van de Walle (2000) performed calculations showing that hydrogen could act as a shallow donor in ZnO. This prediction stimulated studies of muonium, an exotic analog of hydrogen, in ZnO, which showed behavior consistent with a shallow donor (Cox et al., 2001). Electron-nuclear resonance techniques showed hydrogen has a delocalized (i.e., shallow) electron wavefunction (Hofmann et al., 2002).

There are two forms in which hydrogen acts as a shallow donor: interstitial and substitutional hydrogen. Interstitial hydrogen forms an O–H bond that gives rise to a bond-stretching peak in the infrared (IR) absorption spectrum (Table 3). There are two different interstitial hydrogen donors, with peaks at 3326 cm^{-1} (McCluskey et al., 2002) and 3611 cm^{-1} (Lavrov et al., 2002). It has been proposed that the 3326 cm^{-1} line (Fig. 4) may be an O–H pair adjacent to a Ca impurity (Bang et al., 2009; Li et al., 2008; McCluskey and Jokela, 2007). The atom sits in an antibonding ($AB_{\perp}$) location. The line at 3611 cm^{-1}, denoted H_{BC}, arises from a hydrogen donor in a bond-centered ($BC_{\parallel}$) position, between a Zn and O. Subscripts refer to O–H orientations that are nearly perpendicular ($\perp$) or parallel ($\parallel$) to the *c*-axis.

Interstitial hydrogen donors are unstable at room temperature, decaying away after a few weeks (Jokela and McCluskey, 2005). Some of the hydrogen atoms form H_2 molecules, which are electrically neutral and "hidden" from IR spectroscopy (Lavrov et al., 2009a; Shi et al., 2004). Since interstitial hydrogen donors are unstable, they are not responsible for the prevailing n-type conductivity in ZnO samples.

Substitutional hydrogen (H_O), in contrast, is a shallow donor that is stable to ~500 °C (Shi et al., 2004). The I_4 line in Fig. 2 was assigned to H_O by Lavrov et al. (2009b). First-principles calculations showed that the H atom sits in an O vacancy and forms weak bonds with the neighboring Zn atoms

Table 3 Vibrational frequencies of impurities in ZnO

Frequency (cm^{-1})	Defect	References
274 (RT, Raman)	Zn cluster	Gluba et al. (2013)
742, 792	H_O	Koch et al. (2012)
2306 (RT, Raman)	N_2	Soudi et al. (2009)
2782.9	Unknown	Lavrov and Weber (2009) and Li et al. (2009)
3150.6	NH	Jokela and McCluskey (2010)
3191.8	CuH	Gärtner and Mollwo (1978)
3304	NaH	Parmar et al. (2013)
3312.2, 3349.6	$V_{Zn}H_2$	Lavrov et al. (2002)
3329.0, 3348.4	$V_{Zn}H_2$*	Bastin et al. (2011)
3326.3	CaH	Jokela and McCluskey (2005), Li et al. (2008), and McCluskey and Jokela (2007)
3346.9, 3373.9	CuH_2	Lavrov et al. (2008)
3577.3	LiH	Halliburton et al. (2004)
3611.3	H_{BC}	Lavrov et al. (2002)
4145 (Raman)	H_2	Lavrov et al. (2009a)

Data are from low-temperature IR spectroscopy unless noted otherwise (RT = room temperature).

(Janotti and Van de Walle, 2007b). Because the bonds are weak, the calculated vibrational frequency is low, lying in a spectral region with strong IR two-phonon and free-carrier absorption. For this reason, detecting H_O with IR spectroscopy presented a technical challenge. Koch et al. (2012) circumvented the problem by measuring the IR photoconductivity of a ZnO:H sample. Dips in the photoconductive response at 742 and 790 cm^{-2} correspond to the A_1 and E vibrational modes of H_O (Fig. 5). The corresponding deuterium (D) vibrations could not be observed because they lie in the *Reststrahlen* band.

Regarding hydrogen diffusion, experiment and theory are at odds. Experimentally, a hydrogen migration barrier of 1.0 eV has been found by Mollwo (1954), Nickel (2006), and Thomas and Lander (1956). This value concurs with the temperature-dependent decay rate of interstitial hydrogen donors, which has an activation energy of 0.96 eV (Jokela and McCluskey, 2005). However, local density approximation (LDA)

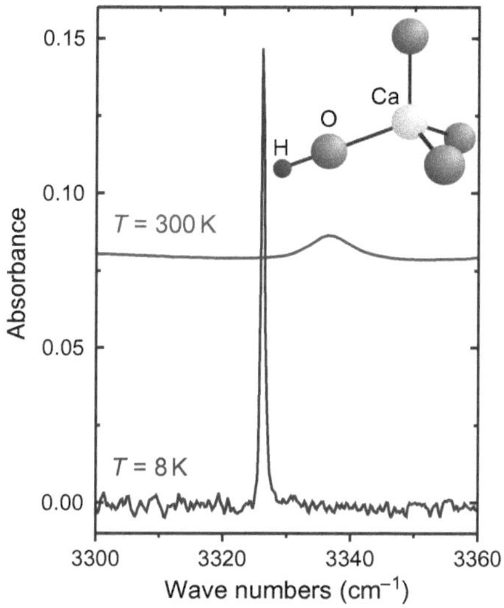

Figure 4 IR absorption spectra of a hydrogen donor in ZnO, attributed to a CaH complex. The peak corresponds to an O–H bond-stretching vibration. For a discussion of the temperature dependence, see McCluskey et al. (2006).

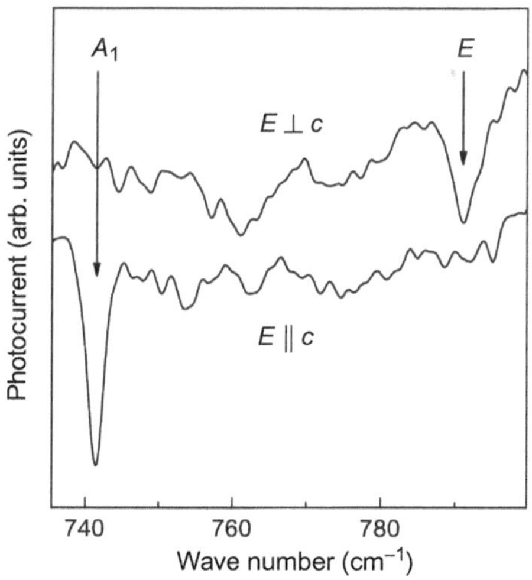

Figure 5 Photoconductivity spectrum of substitutional hydrogen (H_O) in ZnO. The dips correspond to H vibrational modes, perpendicular (*E*) and parallel (*A*) to the *c*-axis. *After Koch et al. (2012).*

calculations (Bang and Chang, 2008; Wardle et al., 2006) predict a much lower migration barrier of 0.4–0.5 eV. It is conceivable that the theoretical approach does not choose realistic paths for the diffusion process.

4. ISOELECTRONIC CENTERS

A PL line labeled I^* was detected in bulk samples at 3.3621 eV (Chen et al., 2012). This line was attributed to an exciton bound to an isoelectronic center. The hole is tightly bound to the center, while the electron is "effective mass" like and orbits at a large, hydrogenic radius. A line at 3.2765 eV was attributed to an exciton bound to Hg on a substitutional Zn site (Agne et al., 2003). Isoelectronic centers may be important for a number of reasons. As discussed in Section 3, an isoelectronic center such as Ca may bind a H atom, becoming a donor. This "hydrogen activation" of neutral impurities was first observed in Ge, where CH and SiH complexes are acceptors while the OH complex acts as a donor (Haller, 1991).

5. COPPER

5.1. Copper acceptors

Copper is a common impurity in ZnO, where it exists as an interstitial donor (Qiu et al., 2011) or as a substitutional acceptor (Cu_{Zn}). The substitutional Cu^{2+} oxidation state is often referred to as neutral (0) because it has the same charge as the ion that it replaces (Zn^{2+}). Cu^{2+} has one unoccupied d-orbital; i.e., it has a hole in its core. The negative (−) Cu acceptor is denoted Cu^{1+} and has filled d-orbitals. The hole energy is shown in Fig. 6, where the VB is on the top and the conduction band (CB) is on the bottom.

In a PL experiment, a hole can be excited from the Cu^{2+} core to a neighboring O atom. The resulting state is denoted (Cu^{1+}, h). When the hole transitions from (Cu^{1+}, h) to Cu^{2+}, light is emitted. The resultant green PL band is centered at ~2.4 eV (510 nm). Phonon replicas, which arise from the emission of a photon plus n phonons, give rise to its characteristic structure (Dingle, 1969). While other defects may also contribute to the green band (Reshchikov et al., 2007), the phonon-replica structure is unique to Cu_{Zn} and therefore provides a useful spectroscopic "fingerprint."

Samples may contain Cu in the form of interstitials or complexes, but upon annealing, they assume substitutional sites. Garces et al. (2002b) showed that bulk ZnO crystals annealed in air at 900 °C exhibit the Cu

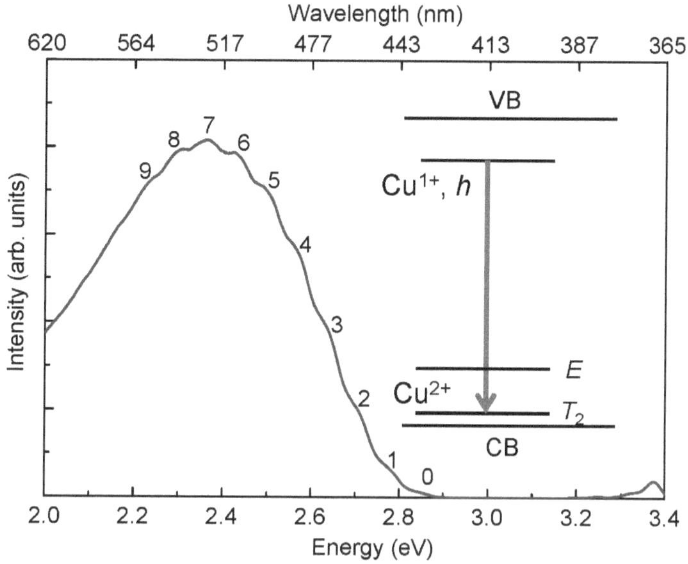

Figure 6 Structured green luminescence due to Cu in ZnO. The PL spectrum was obtained at 8 K. The numerals designate the number of phonons emitted. See McCluskey et al. (2014).

structured green luminescence. McCluskey et al. (2014) observed a similar phenomenon after a 1000 °C anneal in O_2. In the latter study, the phonon replicas were analyzed and found to occur at a regular spacing of 70 meV. This value is similar to the resonant Raman mode energy of 68.83 meV that was attributed to a localized vibration of Cu (Dahan et al., 1998a,b). The regular energy spacing is consistent with Raman experiments on ZnO that showed purely harmonic behavior for LO phonons (Huso et al., 2013).

As mentioned previously, Cu^{2+} contains one hole in its 3d shell. Light can excite the hole from its ground state (T_2) to excited states (E). This process results in sharp peaks at 5782 and 5820 cm^{-1} in low-temperature IR absorption spectra (Dietz et al., 1963). The IR peaks provide an unambiguous signature for Cu^{2+} acceptors in ZnO. N-type samples lightly doped with Cu show structured green PL but, since they only contain Cu^{1+} at low temperatures (absent excitation by a light source), do not show the IR peaks.

First-principles calculations by Yan et al. (2006) indicated the (0/−) acceptor level for Cu_{Zn} lies only 0.7 eV above the VB maximum. (The Ag_{Zn} level was even shallower.) To test this prediction, studies were undertaken on neutron transmutation doped ZnO (Selim et al., 2011). In this approach, samples are exposed to thermal neutrons in a nuclear reactor in order to yield well-controlled concentrations of Cu impurities. Variable-temperature Hall-effect measurements indicated an acceptor level 0.16 eV below the CB, consistent with admittance spectroscopy measurements by Kanai (1991).

In summary, substitutional Cu is a *very deep* acceptor. An acceptor level close to the CB can act as an electron trap but can never produce p-type material. Gallino and Di Valentin (2011) used hybrid density functional theory (DFT) to calculate the acceptor level. The calculations showed an acceptor level 2.7 eV above the VB, more in line with the experimental data. In general, DFT/LDA studies that do not correct for the band-gap underestimation tend to produce incorrect donor/acceptor binding energies.

5.2. Copper–hydrogen complexes

Hydrogen passivates acceptors in semiconductors by providing the needed electron and then binding to the acceptor or to a nearby host atom (McCluskey and Haller, 2012). Using vibrational spectroscopy (IR or Raman), the vibrational frequencies of the hydrogen complex can be determined, providing insight into the structure (McCluskey and Haller, 1999). One of the earliest examples of this was the CuH complex in ZnO. Gärtner and Mollwo (1978) observed an O–H bond-stretching frequency at 3192 cm^{-1}. According to first-principles calculations by Wardle et al. (2005), hydrogen is located in a $BC_{\perp}$ location, between the Cu and an off-axis O atom. Cu atoms with two nearby hydrogens have also been discovered and have vibrational frequencies of 3347 and 3374 cm^{-1} (Lavrov et al., 2008).

6. LITHIUM

6.1. Lithium acceptors

Substitutional lithium (Li_{Zn}) is a single acceptor. In its neutral charge state, its hole is localized on one of the neighboring O atoms. The axial configuration, in which the hole sits on the O atom along the *c*-axis, is energetically preferred, having an energy 15 meV lower than the nonaxial location (Schirmer, 1968). Compensated Li acceptors give rise to a broad yellow PL band, centered at ~2.25 eV (550 nm). The yellow band is caused by the transition of an electron from a shallow donor (or CB) to the acceptor level. From the width of the yellow band, Schirmer and Zwingel (1970) estimated the acceptor level to lie at least 0.8 eV above the VB maximum.

A photoluminescence excitation (PLE) experiment was performed by monitoring the 550-nm emission intensity versus excitation wavelength. The PLE spectrum shows an onset at ~3.1 eV, near the band edge (Fig. 7). The large Stokes shift between excitation and emission, 0.85 eV, is consistent with a deep-level impurity. To estimate the acceptor binding energy, resistivity measurements were performed by McCluskey et al. (2014). The resistivity was quite high (~10^9 Ω cm) and constant up to 400 K. For temperatures above 400 K, the resistivity dropped significantly. The constant and freeze-out regimes were attributed to surface conduction (Allen et al., 2010) and bulk hole conduction, respectively. Modeling the data yielded an acceptor level of 0.86 eV, in agreement with the level estimated by Schirmer and Zwingel (1970).

Diffusion experiments performed by Knutsen et al. (2013) showed that Li diffuses to a depth of 1–10 μm after 10 min at temperatures of 450–600 °C. Concentration versus depth profiles, measured by SIMS, show a sharp drop-off where the Li concentration falls below the donor concentration. The diffusion of mobile interstitial donors (Li_i^+) is fast compared to Li_{Zn} acceptors. When the Li concentration is low, Li_{Zn} acceptors are formed, which compensate the donors. The immobile Li_{Zn} species causes the abrupt decrease in the concentration profile. At much higher temperatures, one can remove Li impurities from a sample by out-diffusion. After annealing hydrothermally grown ZnO in air at 1500 °C, the Li concentration is reduced from ~10^{17} cm^{-3} to 10^{15} cm^{-3} (Vines et al., 2010).

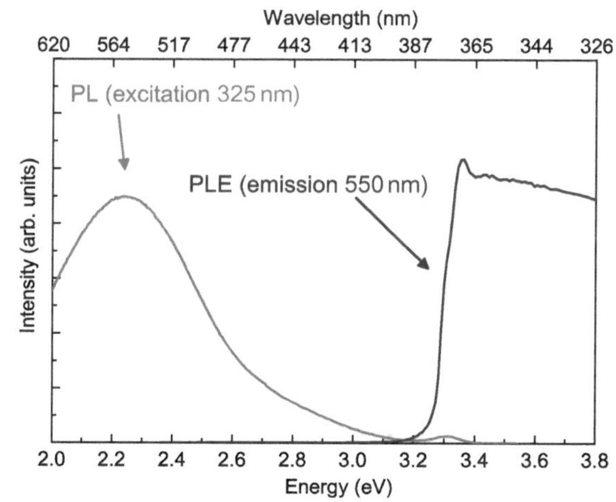

Figure 7 PL and PLE spectra of ZnO:Li at 8 K. The PLE spectrum shows the intensity of the yellow emission (550 nm) versus excitation photon energy. See McCluskey et al. (2014).

6.2. Lithium–hydrogen complexes

Hydrothermal ZnO samples generally contain Li acceptors, some of which are passivated by hydrogen. The LiH complex gives rise to an O–H stretch mode at 3577 cm^{-1} (Halliburton et al., 2004). It is a very stable complex—in bulk samples, annealing temperatures over 1100 °C are required to break up the LiH complexes (Lavrov et al., 2005). Hydrogen occupies a $BC_{\parallel}$ site, between the substitutional Li and axial O atom. The O–H dipole points along the *c*-axis. Li has two stable isotopes, ^{6}Li and ^{7}Li. While the Li isotope splitting cannot be resolved for the O–H mode, the O–D line at 2645 cm^{-1} does show splitting (Shi et al., 2006).

7. SODIUM

7.1. Sodium acceptors

Similar to Li, Na exhibits yellow luminescence near 550 nm. In contrast to Li, the nonaxial hole configuration is preferred by ~25 meV (Zwingel and Gärtner, 1974). Studies on Na-implanted ZnO showed that it interacts with Li already in the sample (Neuvonen et al., 2011). Interstitial Na can kick out a substitutional Li. This trapping mechanism results in substitutional, immobile Na. Positron annihilation measurements showed that Na impurities tend to reduce the concentration of open-volume defects.

Parmar et al. (2013) diffused Na into ZnO, achieving a near-surface Na concentration of $\sim 10^{18}$ cm^{-3}. Resistivity measurements showed that the sample went from n-type to insulating, consistent with the introduction of a deep acceptor. The Na-doped sample showed a room-temperature PL peak near 414 nm (3.0 eV). This peak, similar to that observed by Meyer et al. (2007) in Li- or Na-diffused ZnO, was attributed to the transition of an electron from the CB (or shallow donor) to a shallow (0.3–0.4 eV) acceptor.

7.2. Sodium–hydrogen complexes

Parmar et al. (2013) hydrogenated ZnO:Na by annealing in H_2 gas at 500 °C for 60 h. Using low-temperature (9 K) IR spectroscopy, they observed an O–H stretch mode at 3304 cm^{-1} (Fig. 8). The observed frequency agrees with first-principles calculations for passivated Na, where H attaches to O in an $AB_{\perp}$ configuration. The NaH complex dissociated at 900 °C, lower than the temperature for the LiH complex (1100 °C). This behavior is

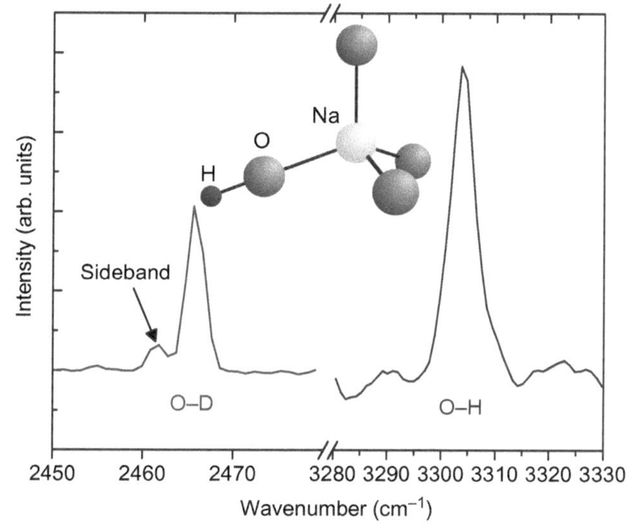

Figure 8 IR spectra (9 K) of NaH and NaD complexes in ZnO. Peaks correspond to O–H or O–D bond-stretching vibrations. The sideband is tentatively assigned to a Fermi resonance. See Parmar et al. (2013).

qualitatively consistent with the calculations of Lee and Chang (2004), which found LiH to be more stable than NaH.

Samples annealed in D_2 gas showed an O–D peak at 2466 cm^{-1}. The deuterated sample also showed a sideband at 2461 cm^{-1}. This sideband was tentatively attributed to a Fermi resonance between the stretch mode and the third harmonic of a wag mode. This is an accidental degeneracy: three times the O–D wag-mode frequency almost exactly equals the O–D stretch-mode frequency. The degeneracy causes the wag mode to acquire some stretch-like character and become IR active.

8. NITROGEN

8.1. Nitrogen acceptors

For a time, nitrogen appeared to be the best hope for ZnO-based devices. As summarized by McCluskey and Jokela (2009), there have been numerous reports that nitrogen doping can produce p-type ZnO. Bierwagen et al. (2008) and Look and Claflin (2004) discussed the difficulty in determining carrier type from Hall-effect measurements on low-quality samples. Erroneous p-type readings may be caused by defects and phase segregation at interfaces, inhomogeneous doping, or other factors.

Lyons et al. (2009a) determined the properties of substitutional nitrogen (N_O), using DFT and hybrid functionals. The results of these calculations show that N_O is a deep acceptor with an acceptor level 1.3 eV above the VB maximum. Lany and Zunger (2010) used generalized Koopmans DFT and also found the level to be very deep (1.6 eV). The deep acceptor level cannot produce significant hole densities at room temperature.

Tarun et al. (2011) tested the theories by performing experiments on bulk ZnO:N crystals. A broad PL emission band near 1.7 eV, with a PLE onset of ~2.2 eV, was observed, in agreement with the deep-acceptor model of the nitrogen defect (Fig. 9). In the PL process, a photon of energy 2.2 eV or higher excites an electron from N^- to the CB. The neutral N^0 acceptor relaxes to a new configuration. When it recaptures an electron, 1.7 eV light is emitted. A photo-EPR investigation by Stehr et al. (2012) showed the absorption onset for converting N^- to N^0 was near 2 eV, consistent with the PL results. These experimental studies point toward an acceptor level 1.4–1.5 eV above the VB maximum (McCluskey et al., 2014).

Like any good mystery, the signs were there before these recent studies. In an EPR study of nitrogen acceptors in ZnO, Carlos et al. (2001) noted their results were consistent with a deep-level defect (Table 4). Visible

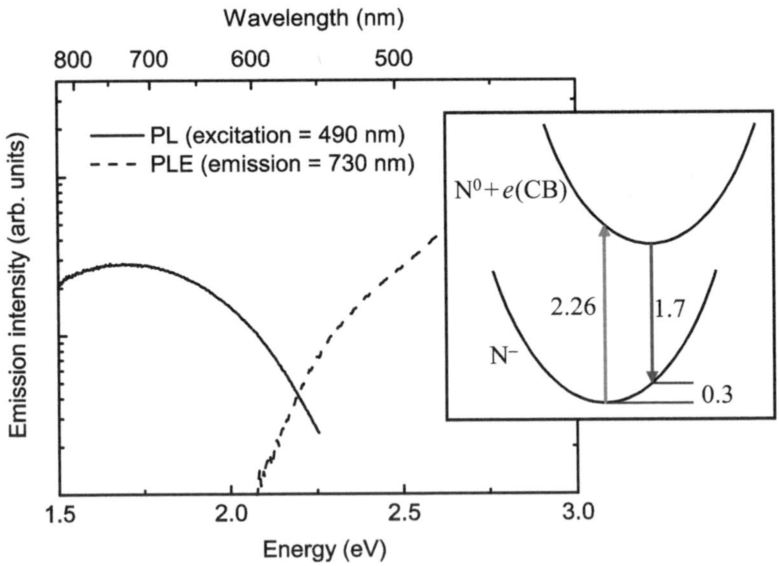

Figure 9 PL and PLE spectra of ZnO:N, at 8 K. The "red" emission peaks at 1.7 eV and the PLE spectrum has an onset near 2.1 eV. A configuration coordinate diagram is shown, where the vertical transition energy of 2.26 eV was obtained by McCluskey et al. (2014).

Table 4 Magnetic resonance parameters for acceptors in ZnO

Acceptor	*g* Values	Other parameters	References
N_O (neutral)	$g_{\|\|}=1.9948$ $g_{\perp}=1.9632$	$A_{\|\|}=81.3$ MHz $A_{\perp}=9.5$ MHz	Garces et al. (2002a)
N_2 (1 −)	$g_{\|\|}=2.0036$ $g_{\perp}=1.9935$	$A_{\|\|}=9.8$ MHz $A_{\perp}=20.1$ MHz	Garces et al. (2003)
Li_{Zn} (neutral)	Axial: $g_{\|\|}=2.0028$ $g_{\perp}=2.0253$ Nonaxial: $g_{xx}=2.0223$ $g_{yy}=2.0254$ $g_{zz}=2.004$	 $A_{\|\|}=0.22$ G $A_{\perp}=1.81$ G $A_{xx}=1.8$ G $A_{yy}=1.8$ G $A_{zz}=0.29$ G	Meyer et al. (2004)
Na_{Zn} (neutral)	Axial: $g_{\|\|}=2.0024$ $g_{\perp}=2.0298$ Nonaxial: $g_{xx}=2.0241$ $g_{yy}=2.0302$ $g_{zz}=2.0032$	 $A_{\|\|}=2.89$ G $A_{\perp}=1.58$ G $A_{xx}=1.61$ G $A_{yy}=1.42$ G $A_{zz}=4.15$ G	Meyer et al. (2004)
V_{Zn} (1 −)	Axial: $g_{\|\|}=2.0024$ $g_{\perp}=2.0193$ Nonaxial: $g_{xx}=2.0173$ $g_{yy}=2.0183$ $g_{zz}=2.0028$		Galland and Herve (1970)
V_{Zn} (neutral), $S=1$	$g_{xx}=2.0132$ $g_{yy}=2.0187$ $g_{zz}=2.0088$	$D=1470$ MHz $E=58$ MHz	Galland and Herve (1970)
$V_{Zn}Al_{Zn}$ (neutral)	$g_{xx}=2.0243$ $g_{yy}=2.0143$ $g_{zz}=2.0045$	$A=26.1$ MHz	Stehr et al. (2014)
$V_{Zn}H$ (neutral)	$g_{xx}=2.0648$ $g_{yy}=2.0023$ $g_{zz}=2.0077$	$A_{xx}=8.4$ MHz $A_{yy}=4.8$ MHz $A_{zz}=-13.2$ MHz	Son et al. (2013)
V_{Zn} complex	$g_{\|\|}=2.0056$ $g_{\perp}=2.0015$		McCluskey et al. (2014)
ODMR center, $S=1$	$g_{\|\|}=1.9710$ $g_{\perp}=2.0224$	$D=763$ MHz	Carlos et al. (2001)

The axial configuration is preferred for Li, while the nonaxial configuration is preferred for Na and $V_{Zn}{}^{-}$. Spins are $S=½$ except where noted. The $S=1$ defect observed by ODMR is listed but might not be an acceptor.

transmission spectra of ZnO:N showed absorption with an onset of ~2.2 eV (Garces et al., 2002a), in agreement with the PLE onset measured by Tarun et al. (2011). Earlier DFT calculations underestimated the ZnO band gap and placed the acceptor level too close to the VB (Park et al., 2002; Yamamoto and Katayama-Yoshida, 1999). However, they correctly showed the hole was localized, as expected for a deep acceptor.

8.2. Nitrogen–hydrogen complexes

IR spectroscopy performed by Jokela and McCluskey (2010) provided unambiguous evidence that hydrogen passivates nitrogen in ZnO, giving rise to a N–H bond-stretching mode at 3151 cm^{-1} (Fig. 10). All stable isotope combinations (^{14}NH, ^{15}NH, ^{14}ND, and ^{15}ND) were measured and showed frequency shifts expected for a N–H bond. The vibrational frequencies are in agreement with first-principles calculations (Li et al., 2005). The calculations found that hydrogen attaches to nitrogen in an $AB_{\perp}$ configuration. In bulk ZnO, the NH pairs dissociate at temperatures around 700 °C, leaving behind substitutional N acceptors.

8.3. Nitrogen pairs

EPR measurements on ZnO:N show, in addition to the isolated N_O acceptor, an axially symmetric N_2^- center (Garces et al., 2003). The O site is energetically unfavorable for an N_2 molecule (Gluba and Nickel, 2009). Instead, first-principles calculations by Lambrecht and Boonchun (2013) showed

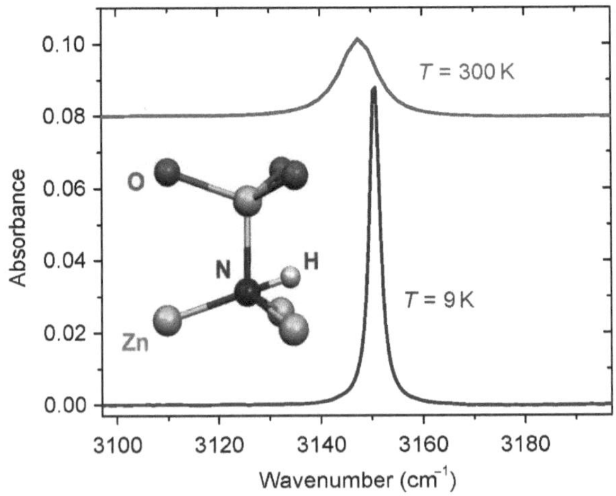

Figure 10 IR spectra of the N–H complex in ZnO. The peak corresponds to a N–H bond-stretching vibration. See Jokela and McCluskey (2010).

that N_2 may occupy a Zn vacancy, with calculated EPR parameters in good agreement with experiment. Photo-EPR measurements by Philipps et al. (2014) indicate a N_2^- photoionization threshold ~0.5 eV lower than that of N_O^-. This observation suggests that the N_2 acceptor level is even deeper than N_O (i.e., closer to the CB).

In addition to the EPR-active N_2^- acceptor, N_2 may exist as a molecule in interstitial locations or voids. Raman spectroscopy on ZnO nanowires showed a room-temperature N_2 vibrational mode at 2306 cm^{-1} (Soudi et al., 2009), a frequency slightly lower than that of atmospheric N_2 (2330 cm^{-1}). X-ray absorption experiments on MBE-grown ZnO:N confirm N occupies a substitutional site, N_O, as grown (Fons et al., 2006). Upon annealing at 800 °C, however, N_2 molecules were observed. The authors proposed that annealing caused the formation of N_2 bubbles. This explanation seems consistent with the observation that N-doped ZnO, grown by a variety of methods, always contain vacancy clusters that are detectable by positron annihilation measurements (Tuomisto et al., 2013).

9. TRANSITION METALS

Besides Cu (Section 5), ZnO may contain various transition-metal impurities. Hydrothermally grown ZnO contains Mn, Fe, Co, and Ni at concentrations up to ~10^{15} cm^{-3} (Jiang et al., 2007). They substitute the Zn^{2+} host atom and act as deep donors, assuming the 2+ (neutral) or 3+ (positive) oxidation states. They give rise to distinct EPR signatures, summarized in Table 5. Since the transition metals have unfilled d-orbitals, there exist internal electronic transitions that give rise to peaks in PL or absorption spectra.

Co^{2+} and Ni^{3+} are similar in that they both have the $3d^7$ electronic core. Intra-d transitions lead to IR absorption/luminescence peaks at 3611 and 3616 cm^{-1} (Co^{2+}); 6091 and 6096 cm^{-1} (Ni^{3+}) (Schulz and Thiede, 1987). In units of energy, these peaks are near 0.45 and 0.76 eV, respectively. Co^{2+} also shows visible peaks near 1.87 eV. Ni^{2+} has IR peaks near 4200 cm^{-1} (0.52 eV). V^{3+}, which is in the $3d^2$ configuration, has sharp IR luminescence peaks near 6880 cm^{-1} (0.85 eV) (Heitz et al., 1991). In the visible spectrum, Fe^{3+} has sharp luminescence peaks near 690 nm (1.79 eV) (Heitz et al., 1992).

Mn^{2+} has been studied in many II–VI semiconductors and has six distinct lines in the EPR spectrum (Table 5). The (0/+) donor level, or $Mn^{2+/3+}$ level, was found to lie close to the VB maximum (Johnson et al., 2010). Electronic transitions from this level to the CB lead to subgap optical absorption and photoconductivity. However, Mn alloying appears to suppress dark

Table 5 Magnetic resonance parameters for transition metals in ZnO

Ion	*g* Values	Other parameters
Mn^{2+}	$g=1.9984$	$A=-74.1\times10^{-4}$ cm^{-1} $D=-235\times10^{-4}$ cm^{-1} $\|a-F\|=5.44\times10^{-4}$ cm^{-1}
Fe^{3+}	$g=2.0060$	$\|A\|=9.02\times10^{-4}$ cm^{-1} $D=-593.7\times10^{-4}$ cm^{-1} $F=4\times10^{-4}$ cm^{-1} $a=39\times10^{-4}$ cm^{-1}
Co^{2+}	$g_{\parallel}=2.2500$ $g_{\perp}=4.5536$	$A=15.30$ G $B=2.82$ G
Ni^{3+}	$g_{\parallel}=2.1426$ $g_{\perp}=4.3179$	
Cu^{2+}	$g_{\parallel}=0.74$ $g_{\perp}=1.531$	$\|A\|=195\times10^{-4}$ cm^{-1} $\|B\|=231\times10^{-4}$ cm^{-1}

After Dietz et al. (1963) and Jiang et al. (2007).

conductivity and green luminescence in ZnO films, presumably by lowering the concentration of other defects (Hlaing Oo et al., 2009).

Intentional transition-metal doping was motivated by the possibility of ferromagnetism at room temperature. As with p-type ZnO, there have been numerous reports of ferromagnetic ZnO, often with Curie temperatures well above room temperature (Pearton et al., 2004). As discussed by McCluskey and Jokela (2009), after the initial excitement, careful studies showed that secondary phases were often present in transition-metal-doped ZnO. For example, Kaspar et al. (2008) used X-ray absorption fine structure to identify ferromagnetic CoZn phases in ZnO:Co thin films. Techniques such as transmission electron microscopy and synchrotron X-ray spectroscopies have identified secondary phases in transition-metal-doped ZnO and other semiconductors (Bonanni and Dietl, 2010). In summary, it has become clear that the reported ferromagnetism was caused by phase-segregated material rather than the ZnO host.

10. NATIVE DEFECTS

10.1. Zinc interstitials

Historically, the zinc interstitial (Zn_i) has been blamed for the prevailing *n*-type conductivity of as-grown ZnO samples. First-principles calculations indicate that it is a shallow, double donor (Janotti and Van de Walle, 2007a). Magnetic resonance experiments performed on electron-irradiated samples (Vlasenko and Watkins, 2005) showed that Zn_i is indeed a shallow donor, with $g \sim 1.96$. However, the signal disappears at temperatures above 170 K, indicating that Zn_i cannot be a source of free electrons at room temperature. These experimental results are consistent with the calculations, which showed a low migration barrier such that Zn_i is mobile at room temperature. Interstitials diffuse until they get trapped at a defect site or the surface.

Although isolated Zn_i is not stable at room temperature, it does appear to form complexes and clusters. Work by Look et al. (2005) suggests that Zn_i may form a complex with a N acceptor. Since Zn_i is a double donor and N is a single acceptor, the Zn_iN complex should act as a single donor. Using isotopically enriched Zn and room-temperature Raman spectroscopy, Gluba et al. (2013) showed that clusters of Zn atoms form and have a vibrational mode at 275 cm^{-1}.

It was commonly assumed that annealing in Zn vapor was a good way to introduce interstitials and raise the n-type conductivity of ZnO. However, positron annihilation studies by Selim et al. (2007) showed that annealing in Zn vapor did not decrease the Zn vacancy concentration. Presumably an in-diffusion of interstitials would have filled the vacancies. Instead, annealing in Zn vapor grows new ZnO on the surface, resulting in oxygen vacancies in the bulk. The increase in oxygen vacancies, perhaps in combination with hydrogen, results in increased conductivity.

10.2. Oxygen vacancies

Along with Zn_i, the oxygen vacancy (V_O) has long been suspected as being a contributor to n-type conductivity. A large number of theoretical studies have reached a consensus that oxygen is a deep double donor (McCluskey and Jokela, 2009). It is a "negative U" center, which means that the 1+ charge state is never stable. Photo-EPR experiments by Evans et al. (2008) indicate the photoionization threshold for neutral V_O is 2.1 eV, consistent with a deep donor. Calculations using hybrid density functionals place the (0/2+) level 1.7–2.3 eV above the VB maximum (Clark et al., 2010). While the donor level may compensate acceptors, it is too deep to contribute free electrons to the CB. (As discussed in Section 3, V_O with one hydrogen atom inside — i.e., substitutional hydrogen — is a shallow donor.)

The neutral and 2+ charge states are EPR inactive. The metastable 1+ charge state, which has one unpaired electron spin, has an EPR resonance at $g \sim 1.995$ (Vlasenko and Watkins, 2005). This signal has been observed in electron-irradiated ZnO at low temperature, following exposure to light. While the green luminescence around 2.4 eV has been attributed to oxygen vacancies, there is no direct evidence for this. Indeed, V_O may not be luminescent. Besides EPR, there are few characterization methods that can detect oxygen vacancies. The small open volume of neutral V_O renders it essentially invisible to positron annihilation measurements (Tuomisto and Makkonen, 2013).

10.3. Zinc vacancies

The zinc vacancy (V_{Zn}) is a double acceptor that can compensate donors in ZnO (Tuomisto et al., 2003). A positron that is trapped in V_{Zn} has a lifetime of 230 ps, as compared to the bulk ZnO lifetime of 169 ps. Crystals that are ion implanted typically show larger open-volume defects. For example, O-implanted (2 MeV) ZnO shows isolated V_{Zn} up to a critical fluence and vacancy clusters for larger doses (Zubiaga et al., 2008). A similar situation is observed for ZnO exposed to high-intensity laser pulses (Khan et al., 2013).

EPR measurements show the singly ionized acceptor V_{Zn}^- has spin $S=½$. It is a deep acceptor with the hole localized on a specific O atom. The complex is referred to as "axial" if the hole is on the O atom along the c-axis and "nonaxial" if it sits on one of the other three O atoms (Fig. 11). For V_{Zn}^-, the nonaxial configuration is preferred, having an energy 20 meV lower than the axial configuration (Galland and Herve, 1974). Theory (Janotti and Van de Walle, 2007a) and experiment (Wang et al., 2009) estimated the ($-/2-$) acceptor level to lie $\sim$1 eV above the VB maximum.

The neutral double acceptor, V_{Zn}^0, has $S=1$. This triplet spin state comes from two holes on neighboring nonaxial O sites. Because of the symmetry of the defect, the $m=-1$ to 0 transition has a different energy than the $m=0$ to 1 transition. For an applied magnetic field along the c-axis, this difference leads to two EPR lines, split by 0.050 T (Evans et al., 2008). Theory estimated the ($0/-$) level at 0.2 eV above the VB maximum (Janotti and Van de Walle, 2007a). Calculations using various hybrid functionals estimated the ($0/-$) and ($-/2-$) acceptor levels to lie 0.7–1.5 and 1.2–2.3 eV above the VB maximum, respectively (Clark et al., 2010).

To study the contribution of V_{Zn} to luminescence, Khan et al. (2013) performed room-temperature PLE experiments on laser-irradiated samples that contained isolated V_{Zn} centers. They monitored the green (2.4 eV) PL versus excitation photon energy. The irradiated samples showed a strong PLE peak at 3.18 eV; i.e., excitation photons of 3.18 eV produced intense

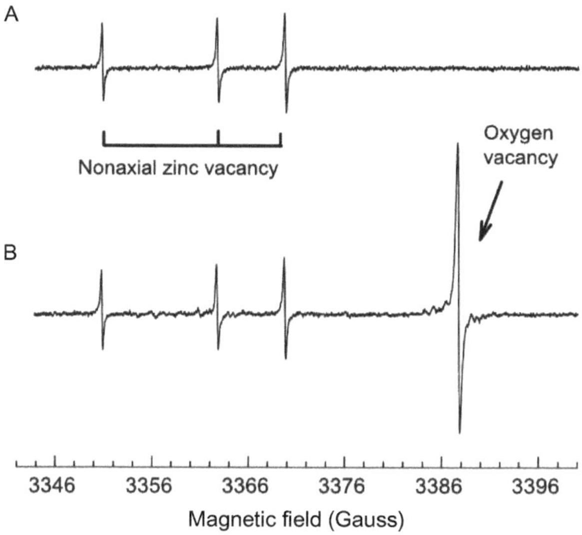

Figure 11 EPR spectra (30 K, 9.48 GHz, $B \| c$) of electron-irradiated ZnO: (A) In the dark. (B) Following exposure to 325 nm light. The EPR lines correspond to singly ionized Zn and O vacancies (1− and 1+, respectively). *After Kappers et al. (2008).*

green PL (Fig. 12). This PLE peak provides a useful "fingerprint" for V_{Zn}. The unique shape appears to be an absorption onset at ~3.1 eV, followed by an abrupt reduction in intensity for photon energies >3.2 eV. This reduction could be due to the generation of electron–hole pairs, which recombine and produce near-band-edge luminescence at the expense of deep-level PL.

The PLE peak was attributed to the excitation of an electron from the (0/−) acceptor level to the CB minimum. The recapture of the electron then produced green PL. However, the origin of green PL is not well understood. Wang et al. (2009) found that the V_{Zn}^- optically detected magnetic resonance (ODMR) signature is peaked at 1.6 eV. Hence, they concluded that the famous green PL is *not* due to Zn vacancies. Studies on electron-irradiated samples have attributed an emission band, peaked at ~1.75 eV, to Zn vacancies (Knutsen et al., 2012).

10.4. Zinc vacancy complexes

A double acceptor, V_{Zn} can form a complex with donors. Hydrogen vibrations observed by Lavrov et al. (2002) were attributed to a V_{Zn} passivated by two H atoms. It is fully passivated and therefore electrically neutral. This $V_{Zn}H_2$ complex consists of a Zn vacancy with one H attached to the axial O ($BC_{\|}$) and the other attached to a nonaxial O ($BC_{\perp}$), resulting in two IR

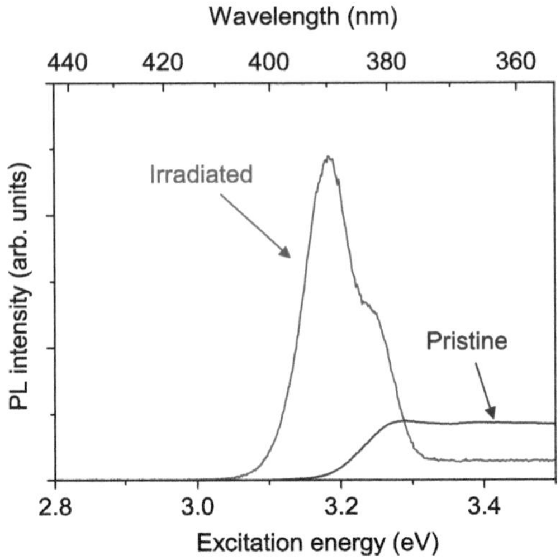

Figure 12 Room-temperature PLE spectra of pristine ZnO and ZnO irradiated with high-intensity laser pulses. The characteristic peak shape of the irradiated sample spectrum is correlated with Zn vacancies. See Khan et al. (2013).

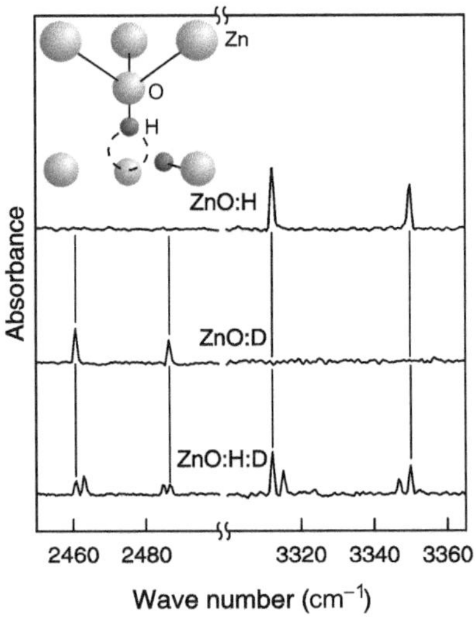

Figure 13 IR spectra (10 K) of fully passivated Zn vacancies in ZnO. Peaks correspond to O–H bond-stretching vibrations of $V_{Zn}H_2$, $V_{Zn}D_2$, and $V_{Zn}HD$ complexes. *After Lavrov et al. (2002).*

absorption peaks (Fig. 13). Deuterated samples show the expected isotopic frequency shift. When the sample is exposed to a mixture of hydrogen and deuterium, new lines appear, corresponding to $V_{Zn}HD$ complexes.

A metastable complex, $V_{Zn}H_2$*, with both H atoms in nonaxial positions, has also been observed (Bastin et al., 2011).

A partially passivated complex, $V_{Zn}H$, has been observed by magnetic resonance on electron-irradiated samples (Son et al., 2013). The H atom resides in a $BC_{\perp}$ configuration, with the O–H bond oriented 78° to the *c*-axis. This complex is a single acceptor, with the hole localized on a non-axial O atom. EPR measurements on electron-irradiated samples containing Al donors showed that a large fraction of the Zn vacancies formed a $V_{Zn}Al_{Zn}$ complex (Stehr et al., 2014). In this complex, the Al_{Zn} donor resides on a substitutional site that is a next-nearest neighbor of V_{Zn}. The $V_{Zn}Al_{Zn}$ complex is a single acceptor, with the hole localized on an axial O atom.

11. NANOCRYSTALS

A wide variety of nanocrystalline ZnO has been synthesized, including nanowires and nanosprings (Huang et al., 2001; Wang, 2004). For a general review of ZnO nanocrystals, the reader is referred to Schmidt-Mende and MacManus-Driscoll (2007). A review of transition-metal-doped II–VI quantum dots is given by Beaulac et al. (2010).

Point defects in nanocrystals are, in principle, no different than the corresponding defects in the bulk. For ZnO diameters above 5 nm, quantum confinement does not play a significant role. However, the large surface-to-volume ratio and particular synthesis conditions may lead to different defect populations and distributions. Annealing ZnO nanocrystals in H_2 produces very large free-electron densities of $\sim 10^{19}$ cm^{-3}, at temperatures of only 300 °C (Hlaing Oo et al., 2007). For Cu doping, the incorporation of Cu^{2+} acceptors increases with nanocrystal size (Hlaing Oo et al., 2010).

Nominally undoped ZnO nanocrystals exhibit IR and PL peaks tentatively assigned to an acceptor ~0.4 eV above the VB maximum (Teklemichael and McCluskey, 2012; Teklemichael et al., 2011). Examples of PL spectra are shown in Fig. 14, for commercial (90 nm diameter) and home-grown (20 nm diameter) powders. The peaks at 3.22 and 3.31 eV have been observed previously in bulk ZnO and are discussed in Section 12.2. The 3.0-eV peak was attributed to the transition of a CB electron to a 0.4-eV acceptor level. Associated with the 3.0-eV peak, high-field EPR spectra revealed an axial center with $g_{\parallel} = 2.0056$ and $g_{\perp} = 2.0015$ (McCluskey et al., 2014). These g values suggest an acceptor that may involve a Zn vacancy (Table 4).

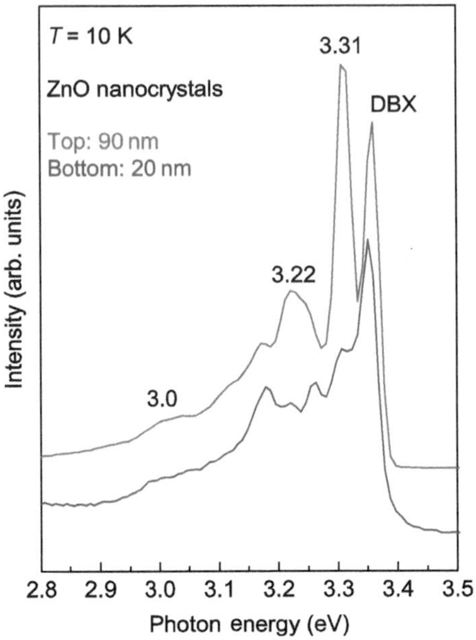

Figure 14 PL spectra of ZnO nanocrystals with average diameters of 90 and 20 nm. Donor-bound excitons (DBX) are shown. Peaks at 3.0, 3.22, and 3.31 eV are discussed in the text.

Teklemichael and McCluskey (2011) explained various experimental results with a "unified model" that considered the interaction of the 0.4-eV acceptor level with surface states. Parashar et al. (2012) proposed a "core–shell model," with donors in the core and acceptors in the shell of the nanocrystal. Both models agree that (1) sufficiently small nanocrystals can have more acceptors than donors and (2) red luminescence comes from the near-surface region (Kaftelen et al., 2012).

12. UNIDENTIFIED DEFECTS

12.1. Deep-level characterization

Using Schottky contacts and deep-level transient spectroscopy (DLTS), researchers have measured several electron traps in ZnO. The most common traps are denoted E1 ($E_c-0.12$ eV), E2 ($E_c-0.11$ eV), E3 ($E_c-0.30$ eV), E4 ($E_c-0.54$ eV), and E5 ($E_c-0.62$ eV), where the energy levels are given in parentheses and E_c refers to the CB minimum. E1 and E3 are prevalent in CVT and melt-grown ZnO, respectively, with concentrations $\sim 10^{16}$ cm^{-3} (Auret et al., 2006). E3 is a dominant defect in hydrothermally grown ZnO (Simpson and Cordaro, 1988). Because the displacement energy for O is greater than Zn, electron energies less than 200 keV damage the Zn sublattice only. Frank et al. (2007) used electron irradiation and DLTS to probe the identities of deep-level defects. Their results suggest that E3 involves a Zn vacancy while E4 involves an O vacancy.

12.2. Luminescent mysteries

PL is a widely used technique to characterize defects in ZnO (Thonke and Feneberg, 2012). In general, it is difficult to assign emission features to specific defects. Different ZnO samples show a rainbow of deep-level luminescence peaks, depending on growth method, processing, and doping (Reshchikov et al., 2007). Few of these peaks have been identified. Studies on ZnO by Vanheusden et al. (1996) and Heo et al. (2005) attributed the green luminescence near 2.45 eV to O and Zn vacancies, respectively. As discussed in Section 10, there is no solid evidence to support these assignments. Structured green PL is clearly due to Cu acceptors (Section 5). The origin(s) of *unstructured* green PL, however, is debatable.

ODMR experiments by Carlos et al. (2001) and Leiter et al. (2001) show that the unstructured green luminescence involves an $S=1$ center with axial symmetry. For a magnetic field applied parallel to the *c*-axis, the magnetic resonance peaks are split by 0.060 T. This center cannot be V_O^-, which is spin ½ (Table 2). Janotti and Van de Walle (2007a) suggested that it may involve V_{Zn}^- next to a neutral donor, where the interaction between the two unpaired spins results in the $S=1$ triplet state.

Several near-band-edge peaks have been assigned to transitions involving shallow acceptors. Cathodoluminescence (CL) measurements showed a 3.31-eV peak in the vicinity of stacking faults (Schirra et al., 2008). This peak was attributed to the recombination of a free electron with a 0.13-eV acceptor. Since the acceptor defects are localized at stacking faults, rather than distributed homogeneously throughout the bulk, they cannot yield p-type ZnO. Similarly, a line at 3.333 eV has been assigned to excitons bound to structural defects (Meyer et al., 2004).

A PL peak near 3.22 eV has been attributed to transitions involving a shallow acceptor (Thonke et al., 2001). A similar line at 3.24 eV was assigned to shallow N acceptors (Lautenschlaeger et al., 2012). However, since substitutional N is known to be a deep acceptor (Section 8.1), this assignment is not straightforward. Nitrogen doping is correlated with Zn vacancy clusters (Tuomisto et al., 2013), which may play a role in these PL features.

As discussed in Section 11, a 3.0-eV line was observed in nanocrystalline ZnO. A similar peak has been observed in the PL spectra of ZnO:Li (Meyer

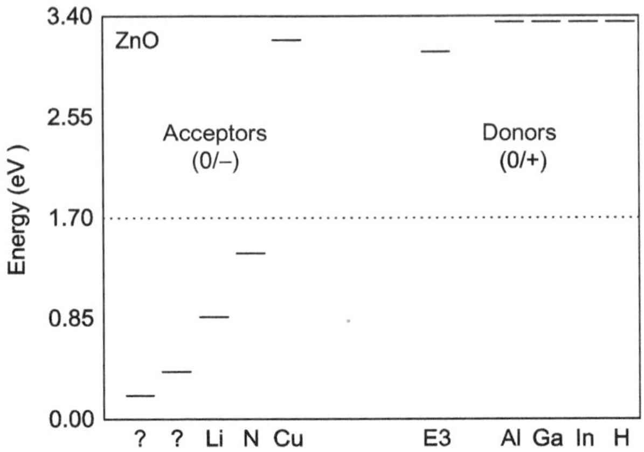

Figure 15 Summary of acceptor and donor levels in ZnO. Question marks (?) indicate levels obtained from PL spectra.

et al., 2007) and ZnO:Na (Meyer et al., 2007; Parmar et al., 2013), and CL spectra of ZnO:Li (Zhang et al., 2012). The fact that it is observed in Na- and Li-doped samples suggests that the peak is not due to either of these dopants. Instead, it may be due to another defect such as a Zn vacancy complex.

13. CONCLUSIONS

While much is known about point defects ZnO, many fundamental problems remain. The origin of green luminescence, observed for many decades, has not been determined. Reproducible p-type doping has not been achieved. This latter problem stems from the low position of the VB as compared to conventional semiconductors (Kiliç and Zunger, 2002; Van de Walle and Neugebauer, 2003). As a summary, Fig. 15 shows reasonably well-established energy levels for point defects in ZnO, along with a couple of acceptor levels that are questionable. Future research will not only clarify the picture but hopefully also produce new breakthroughs.

ACKNOWLEDGMENTS

I would like to acknowledge numerous colleagues and collaborators in this area, including Leah Bergman, Leonard Brillson, Eugene Haller, M. Zafar Iqbal, Anderson Janotti, Bruno Meyer, Michael Reshchikov, Michael Stavola, Filip Tuomisto, Chris Van de Walle, Joerg Weber, Su-Huai Wei, Yanfa Yan, Shengbai Zhang, and the students and postdocs who contributed to this work. Research at WSU was supported by the Department of Energy (DE-FG02-07ER46386) and National Science Foundation (DMR-1004804).

REFERENCES

Agne, Th., Dietrich, M., Hamann, J., Lany, S., Wolf, H., Wichert, Th., 2003. Optical properties of the isoelectronic trap Hg in ZnO. Appl. Phys. Lett. 82, 3448–3450.

Allen, M.W., Swartz, C.H., Myers, T.H., Veal, T.D., McConville, C.F., Durbin, S.M., 2010. Bulk transport measurements in ZnO: the effect of surface electron layers. Phys. Rev. B 81 (075211), 1–6.

Auret, F.D., Nel, J.M., Hayes, M., Wu, L., Wesch, W., Wendler, E., 2006. Electrical characterization of growth-induced defects in bulk-grown ZnO. Superlattice. Microst. 39, 17–23.

Bang, J., Chang, K.J., 2008. Diffusion and thermal stability of hydrogen in ZnO. Appl. Phys. Lett. 92 (132109), 1–3.

Bang, J., Choi, E.-A., Chang, K.J., 2009. The effect of impurities on hydrogen bonding site and local vibrational frequency in ZnO. J. Appl. Phys. 106 (053522), 1–5.

Bastin, D., Lavrov, E.V., Weber, J., 2011. Metastable state of the $V_{Zn}H_2$ defect in ZnO. Phys. Rev. B 83 (195210), 1–5.

Beaulac, R., Ochsenbein, S.T., Gamelin, D.R., 2010. Colloidal transition-metal-doped quantum dots. In: Klimov, V.I. (Ed.), Nanocrystal Quantum Dots, second ed. CRC Press, Boca Raton, FL, pp. 397–453.

Bierwagen, O., Ive, T., Van de Walle, C.G., Speck, J.S., 2008. Causes of incorrect carrier-type identification in van der Pauw–Hall measurements. Appl. Phys. Lett. 93 (242108), 1–3.

Bonanni, A., Dietl, T., 2010. A story of high-temperature ferromagnetism in semiconductors. Chem. Soc. Rev. 39, 528–539.

Brillson, L.J., 2013. Surfaces and interfaces of ZnO. In: Svensson, B.G., Pearton, S.J., Jagadish, C. (Eds.), In: Semiconductors and Semimetals, 88. Elsevier Academic Press, San Diego, CA, pp. 105–157.

Brillson, L.J., Lu, Y.C., 2011. ZnO Schottky barriers and Ohmic contacts. J. Appl. Phys. 109 (121301), 1–33.

Carlos, W.E., Glaser, E.R., Look, D.C., 2001. Magnetic resonance studies of ZnO. Physica B 308–10, 976–9.

Chen, S.L., Chen, W.M., Buyanova, I.A., 2012. Zeeman splitting and dynamics of an isoelectronic bound exciton near the band edge of ZnO. Phys. Rev. B 86 (235205), 1–7.

Clark, S.J., Robertson, J., Lany, S., Zunger, A., 2010. Intrinsic defects in ZnO calculated by screened exchange and hybrid density functionals. Phys. Rev. B 81 (115311), 1–5.

Cox, S.F.J., Davis, E.A., Cottrell, S.P., King, P.J.C., Lord, J.S., Gil, J.M., Alberto, H.V., Vilão, R.C., Duarte, J. Piroto, Ayres de Campos, N., Weidinger, A., Lichti, R.L., Irvine, S.J.C., 2001. Experimental confirmation of the predicted shallow donor hydrogen state in zinc oxide. Phys. Rev. Lett. 86, 2601–2604.

Dahan, P., Fleurov, V., Thurian, P., Heitz, R., Hoffmann, A., Broser, I., 1998a. Isotope shift in semiconductors with transition-metal impurities: experiment and theory applied to ZnO:Cu. Phys. Rev. B 57, 9690–9694.

Dahan, P., Fleurov, V., Thurian, P., Heitz, R., Hoffmann, A., Broser, I., 1998b. Properties of the intermediately bound α-, β-, and γ-excitons in ZnO:Cu. J. Phys.: Condens. Matter 10, 2007–2019.

Dietz, R.E., Kamimura, H., Sturge, M.D., Yariv, A., 1963. Electronic structure of copper impurities in ZnO. Phys. Rev. 132, 1559–1561.

Dingle, R., 1969. Luminescent transitions associated with divalent copper impurities and the green emission from semiconducting zinc oxide. Phys. Rev. Lett. 23, 579–581.

Evans, S.M., Giles, N.C., Halliburton, L.E., Kappers, L.A., 2008. Further characterization of oxygen vacancies and zinc vacancies in electron-irradiated ZnO. J. Appl. Phys. 103 (043710), 1–7.

Fons, P., Tampo, H., Kolobov, A.V., Ohkubo, M., Niki, S., Tominaga, J., Carboni, R., Boscherini, F., Friedrich, S., 2006. Direct observation of nitrogen location in molecular beam epitaxy grown nitrogen-doped ZnO. Phys. Rev. Lett. 96 (045504), 1–4.

Frank, T., Pensl, G., Tena-Zaera, R., Zuniga-Perez, J., Martinez-Tomas, C., Munoz-Sanjose, V., Ohshima, T., Itoh, H., Hofmann, D., Pfisterer, D., Sann, J., Meyer, B., 2007. Energetically deep defect centers in vapor-phase grown zinc oxide. Appl. Phys. A 88, 141–145.

Galland, D., Herve, A., 1970. ESR spectra of the zinc vacancy in ZnO. Phys. Lett. A 33, 1–2.

Galland, D., Herve, A., 1974. Temperature dependence of the ESR spectrum of the zinc vacancy in ZnO. Solid State Commun. 14, 953–956.

Gallino, F., Di Valentin, C., 2011. Copper impurities in bulk ZnO: a hybrid density functional study. J. Chem. Phys. 134 (144506), 1–10.

Garces, N.Y., Giles, N.C., Halliburton, L.E., Cantwell, G., Eason, D.B., Reynolds, D.C., Look, D.C., 2002a. Production of nitrogen acceptor in ZnO by thermal annealing. Appl. Phys. Lett. 80, 1334–1336.

Garces, N.Y., Wang, L., Bai, L., Giles, N.C., Halliburton, L.E., Cantwell, G., 2002b. Role of copper in the green luminescence from ZnO crystals. Appl. Phys. Lett. 81, 622–624.

Garces, N.Y., Wang, L., Giles, N.C., Halliburton, L.E., Cantwell, G., Eason, D.B., 2003. Molecular nitrogen (N_2^-) acceptors and isolated nitrogen (N^-) acceptors in ZnO crystals. J. Appl. Phys. 94, 519–524.

Gärtner, F.G., Mollwo, E., 1978. IR absorption of OH and OD centres and OH/OH, OD/OD, and OH/OD complexes in Cu-doped ZnO single crystals. Phys. Stat. Sol. B 89, 381–388.

Gluba, M.A., Nickel, N.H., 2009. Defects in compound semiconductors caused by molecular nitrogen. Phys. Rev. Lett. 103 (145501), 1–4.

Gluba, M.A., Nickel, N.H., Karpensky, N., 2013. Interstitial zinc clusters in zinc oxide. Phys. Rev. B 88 (245201), 1–8.

Gonzales, C., Block, D., Cox, R.T., Hervé, A., 1982. Magnetic resonance studies of shallow donors in zinc oxide. J. Cryst. Growth 59, 357–362.

Haller, E.E., 1991. Hydrogen-related phenomena in crystalline germanium. In: Pankove, J.I., Johnson, N.M. (Eds.), In: Semiconductors and Semimetals, vol. 34. Academic Press, San Diego, CA, pp. 113–137.

Halliburton, L.E., Wang, L.J., Bai, L.H., Garces, N.Y., Giles, N.C., Callahan, M.J., Wang, B.G., 2004. Infrared absorption from OH^- ions adjacent to lithium acceptors in hydrothermally grown ZnO. J. Appl. Phys. 96, 7168–7172.

Hanna, G.J., Teklemichael, S.T., McCluskey, M.D., Bergman, L., Huso, J., 2011. Equations of state for ZnO and MgZnO by high pressure X-ray diffraction. J. Appl. Phys. 110 (073511), 1–5.

Heitz, R., Hoffmann, A., Hausmann, B., Broser, I., 1991. Zeeman spectroscopy of V^{3+}-luminescence in ZnO. J. Lumin. 48–49, 689–692.

Heitz, R., Hoffmann, A., Broser, I., 1992. Fe^{3+} center in ZnO. Phys. Rev. B 45, 8977–8988.

Herklotz, F., Lavrov, E.V., Weber, J., Mamin, G.V., Kutin, Y.S., Volodin, M.A., Orlinskii, S.B., 2011. Identification of shallow Al donors in ZnO. Phys. Stat. Sol. B 248, 1532–1537.

Heo, Y.W., Norton, D.P., Pearton, S.J., 2005. Origin of green luminescence in ZnO thin film grown by molecular-beam epitaxy. J. Appl. Phys. 98 (073502), 1–6.

Hlaing Oo, W.M., McCluskey, M.D., Huso, J., Bergman, L., 2007. Infrared and Raman spectroscopy of ZnO nanoparticles annealed in hydrogen. J. Appl. Phys. 102 (043529), 1–5.

Hlaing Oo, W.M., Saraf, L.V., Engelhard, M.H., Shutthanandan, V., Bergman, L., Huso, J., McCluskey, M.D., 2009. Suppression of conductivity in Mn-doped ZnO thin films. J. Appl. Phys. 105 (013715), 1–4.

Hlaing Oo, W.M., McCluskey, M.D., Huso, J., Morrison, J.L., Bergman, L., Engelhard, M.H., Saraf, L.V., 2010. Incorporation of Cu acceptors in ZnO nanocrystals. J. Appl. Phys. 108 (064301), 1–3.

Hofmann, D.M., Hofstaetter, A., Leiter, F., Zhou, H., Henecker, F., Meyer, B.K., Orlinskii, S.B., Schmidt, J., Baranov, P.G., 2002. Hydrogen: a relevant shallow donor in zinc oxide. Phys. Rev. Lett. 88 (045504), 1–4.

Huang, M.H., Wu, Y.Y., Feick, H., Tran, N., Weber, E., Yang, P.D., 2001. Catalytic growth of zinc oxide nanowires by vapor transport. Adv. Mater. 13, 113–116.

Huso, J., Morrison, J.L., Bergman, L., McCluskey, M.D., 2013. An harmonic resonant Raman modes in $Mg_{0.2}Zn_{0.8}O$. Phys. Rev. B 87 (125205), 1–5.

Janotti, A., Van de Walle, C.G., 2007a. Native point defects in ZnO. Phys. Rev. B 76 (165202), 1–22.

Janotti, A., Van de Walle, C.G., 2007b. Hydrogen multicentre bonds. Nat. Mater. 6, 44–47.

Janotti, A., Van de Walle, C.G., 2009. Fundamentals of zinc oxide as a semiconductor. Rep. Prog. Phys. 72 (126501), 1–29.

Jiang, Y., Giles, N.C., Halliburton, L.E., 2007. Persistent photoinduced changes in charge states of transition-metal donors in hydrothermally grown ZnO crystals. J. Appl. Phys. 101 (093706), 1–8.

Johnson, C.A., Kittilstved, K.R., Kaspar, T.C., Droubay, T.C., Chambers, S.A., Salley, G.M., Gamelin, D.R., 2010. Mid-gap electronic states in $Zn_{1-x}Mn_xO$. Phys. Rev. B 82 (115202), 1–11.

Jokela, S.J., McCluskey, M.D., 2005. Structure and stability of O-H donors in ZnO from high-pressure and infrared spectroscopy. Phys. Rev. B 72 (113201), 1–4.

Jokela, S.J., McCluskey, M.D., 2007. Hydrogen complexes in ZnO grown by chemical vapor transport. Phys. B 401–402, 395–398.

Jokela, S.J., McCluskey, M.D., 2010. Structure and stability of N–H complexes in single-crystal ZnO. J. Appl. Phys. 107 (113536), 1–5.

Kaftelen, H., Ocakoglu, K., Thomann, R., Tu, S., Weber, S., Erdem, E., 2012. EPR and photoluminescence spectroscopy studies on the defect structure of ZnO nanocrystals. Phys. Rev. B 86 (014113), 1–9.

Kanai, Y., 1991. Admittance spectroscopy of Cu-doped ZnO crystals. Jpn. J. Appl. Phys. 30, 703–707.

Kappers, L.A., Gilliam, O.R., Evans, S.M., Halliburton, L.E., Giles, N.C., 2008. EPR and optical study of oxygen and zinc vacancies in electron-irradiated ZnO. Nucl. Instrum. Methods Phys. Res. B 266, 2963–2967.

Kaspar, T.C., Droubay, T., Heald, S.M., Engelhard, M.H., Nachimuthu, P., Chambers, S.A., 2008. Hidden ferromagnetic secondary phases in cobalt-doped ZnO epitaxial thin films. Phys. Rev. B 77 (201303), 1–4.

Khan, E.H., Weber, M.H., McCluskey, M.D., 2013. Formation of isolated Zn vacancies in ZnO single crystals by absorption of ultraviolet radiation: a combined study using positron annihilation, photoluminescence, and mass spectroscopy. Phys. Rev. Lett. 111 (017401), 1–5.

Kiliç, Ç., Zunger, A., 2002. N-type doping of oxides by hydrogen. Appl. Phys. Lett. 81, 73–75.

King, P.D.C., Veal, T.D., 2011. Conductivity in transparent oxide semiconductors. J. Phys. Condens. Matter 23 (334214), 1–17.

Klingshirn, 2007. ZnO: From basics towards applications. Physica Status Solidi B 244, 3027–3073.

Knutsen, K.E., Galeckas, A., Zubiaga, A., Tuomisto, F., Farlow, G.C., Svensson, B.G., Kuznetsov, A.Yu., 2012. Zinc vacancy and oxygen interstitial in ZnO revealed by sequential annealing and electron irradiation. Phys. Rev. B 86 (121203), 1–5.

Knutsen, K.E., Johansen, K.M., Neuvonen, P.T., Svensson, B.G., Kuznetsov, A.Y., 2013. Diffusion and configuration of Li in ZnO. J. Appl. Phys. 113 (023702), 1–5.
Koch, S.G., Lavrov, E.V., Weber, J., 2012. Photoconductive detection of tetrahedrally coordinated hydrogen in ZnO. Phys. Rev. Lett. 108 (165501), 1–4.
Laiho, R., Vlasenko, L.S., Vlasenko, M.P., 2008. Optical detection of magnetic resonance and electron paramagnetic study of the oxygen vacancy and lead donors in ZnO. J. Appl. Phys. 103 (123709), 1–10.
Lambrecht, W.R.L., Boonchun, A., 2013. Identification of a N-related shallow acceptor and electron paramagnetic resonance center in ZnO:N_2^+ on the Zn site. Phys. Rev. B 87 (195207), 1–5.
Lany, S., Zunger, A., 2010. Generalized Koopmans density functional calculations reveal the deep acceptor state of N_O in ZnO. Phys. Rev. B 81 (205209), 1–5.
Lautenschlaeger, S., Eisermann, S., Haas, G., Zolnowski, E.A., Hofmann, M.N., Laufer, A., Pinnisch, M., Meyer, B.K., Wagner, M.R., Reparaz, J.S., Callsen, G., Hoffmann, A., Chernikov, A., Chatterjee, S., Bornwasser, V., Koch, M., 2012. Optical signatures of nitrogen acceptors in ZnO. Phys. Rev. B 85 (235204), 1–7.
Lavrov, E.V., Weber, J., 2009. Comment on "Infrared absorption spectroscopy on OH–Ni complex in hydrothermally grown ZnO" (J. Appl. Phys. 105, 093516, 2009). J. Appl. Phys. 106, 086104.
Lavrov, E.V., Weber, J., Börrnert, F., Van de Walle, C.G., Helbig, R., 2002. Hydrogen-related defects in ZnO studied by infrared absorption spectroscopy. Phys. Rev. B 66 (165205), 1–7.
Lavrov, E.V., Börrnert, F., Weber, J., 2005. Dominant hydrogen-oxygen complex in hydrothermally grown ZnO. Phys. Rev. B 71 (035205), 1–6.
Lavrov, E.V., Weber, J., Börrnert, F., 2008. Copper dihydrogen complex in ZnO. Phys. Rev. B 77 (155209), 1–10.
Lavrov, E.V., Herklotz, F., Weber, J., 2009a. Identification of hydrogen molecules in ZnO. Phys. Rev. Lett. 102 (185502), 1–4.
Lavrov, E.V., Herklotz, F., Weber, J., 2009b. Identification of two hydrogen donors in ZnO. Phys. Rev. B 79 (165210), 1–13.
Lee, E.-C., Chang, K.J., 2004. Possible p-type doping with group-I elements in ZnO. Phys. Rev. B 70 (115210), 1–4.
Leiter, F.H., Alves, H.R., Hofstaetter, A., Hofmann, D.M., Meyer, B.K., 2001. The oxygen vacancy as the origin of a green emission in undoped ZnO. Phys. Stat. Sol. B 226, R4–R5.
Li, X., Keyes, B., Asher, S., Zhang, S.B., Wei, S.-H., Coutts, T.J., Limpijumnong, S., Van de Walle, C.G., 2005. Hydrogen passivation effect in nitrogen-doped ZnO thin films. Appl. Phys. Lett. 86 (122107), 1–3.
Li, X.B., Limpijumnong, S., Tian, W.Q., Sun, H.B., Zhang, S.B., 2008. Hydrogen in ZnO revisited: bond center versus antibonding site. Phys. Rev. B 78 (113203), 1–4.
Li, Y.J., Zhang, B., Lu, W., 2009. Infrared absorption spectroscopy on OH–Ni complex in hydrothermally grown ZnO. J. Appl. Phys. 105 (093516), 1–6.
Look, D.C., 2001. Recent advances in ZnO materials and devices. Mater. Sci. Eng. B 80, 383–387.
Look, D.C., Claflin, B., 2004. P-type doping and devices based on ZnO. Phys. Stat. Sol. B 241, 624–630.
Look, D.C., Farlow, G.C., Reunchan, P., Limpijumnong, S., Zhang, S.B., Nordlund, K., 2005. Evidence for native-defect donors in *n*-type ZnO. Phys. Rev. Lett. 95 (225502), 1–4.
Lyons, J.L., Janotti, A., Van de Walle, C.G., 2009a. Why nitrogen cannot lead to *p*-type conductivity in ZnO. Appl. Phys. Lett. 95 (252105), 1–3.
Lyons, J.L., Janotti, A., Van de Walle, C.G., 2009b. Role of Si and Ge as impurities in ZnO. Phys. Rev. B 80 (205113), 1–5.

McCluskey, M.D., Haller, E.E., 1999. Vibrational spectroscopy of hydrogen in III–V and II–VI semiconductors. In: Nickel, N.H. (Ed.), In: Semiconductors in Semimetals, vol. 61. Academic Press, San Diego, CA, pp. 373–440.
McCluskey, M.D., Haller, E.E., 2012. Dopants and Defects in Semiconductors. CRC Press, Boca Raton, FL.
McCluskey, M.D., Jokela, S.J., 2007. Sources of *n*-type conductivity in ZnO. Phys. B 401–402, 355–357.
McCluskey, M.D., Jokela, S.J., 2009. Defects in ZnO. J. Appl. Phys. 106 (071101), 1–13.
McCluskey, M.D., Jokela, S.J., Zhuravlev, K.K., Simpson, P.J., Lynn, K.G., 2002. Infrared spectroscopy of hydrogen in ZnO. Appl. Phys. Lett. 81, 3807–3809.
McCluskey, M.D., Jokela, S.J., Hlaing Oo, W.M., 2006. Hydrogen in bulk and nanoscale ZnO. Phys. B 376–7, 690–693.
McCluskey, M.D., Tarun, M.C., Teklemichael, S.T., 2012. Hydrogen in oxide semiconductors. J. Mater. Res. 27, 2190–2198.
McCluskey, M.D., Corolewski, C.D., Lv, J., Tarun, M.C., Teklemichael, S.T., Walter, E.D., Norton, M.G., Harrison, K.W., Ha, S., 2015. Acceptors in ZnO. J. Appl Phys. 117, (112802), 1–6.
Meyer, B.K., Alves, H., Hofmann, D.M., Kriegseis, W., Forster, D., Bertram, F., Christen, J., Hoffmann, A., Strassburg, M., Dworzak, M., Haboeck, U., Rodina, A.V., 2004. Bound exciton and donor–acceptor pair recombinations in ZnO. Phys. Stat. Sol. B 241, 231–260.
Meyer, B.K., Stehr, J., Hofstaetter, A., Volbers, N., Zeuner, A., Sann, J., 2007. On the role of group I elements in ZnO. Appl. Phys. A 88, 119–123.
Minami, T., 2000. New *n*-type transparent conducting oxides. MRS Bull. 25, 38–44.
Mollwo, E., 1954. Die Wirkung von Wasserstoff auf die Leitfähigkeit und Lumineszenz von Zinkoxydkristallen. Z. Phys. 138, 478–488.
Neuvonen, P.T., Vines, L., Venkatachalapathy, V., Zubiaga, A., Tuomisto, F., Hallén, A., Svensson, B.G., Kuznetsov, A.Yu., 2011. Defect evolution and impurity migration in Na-implanted ZnO. Phys. Rev. B 84 (205202), 1–7.
Nickel, N.H., 2006. Hydrogen migration in single crystal and polycrystalline zinc oxide. Phys. Rev. B 73 (195204), 1–9.
Norton, D.P., Heo, Y.W., Ivill, M.P., Ip, K., Pearton, S.J., Chisholm, M.F., Steiner, T., 2004. ZnO: growth, doping & processing. Mater. Today 7, 34–40.
Ntep, J.M., Hassani, S.S., Lusson, A., Tromson-Carli, A., Ballutaud, D., Didier, G., Triboulet, R., 1999. ZnO growth by chemical vapour transport. J. Cryst. Growth 207, 30–34.
Ohshima, E., Ogino, H., Niikura, I., Maeda, K., Sato, M., Ito, M., Fukuda, T., 2004. Growth of the 2-in-size bulk ZnO single crystals by the hydrothermal method. J. Cryst. Growth 260, 166–170.
Özgür, Ü., Alivov, Y.I., Liu, C., Teke, A., Reshchikov, M.A., Doğan, S., Avrutin, V., Cho, S.-J., Morkoç, H., 2005. A comprehensive review of ZnO materials and devices. J. Appl. Phys. 98 (041301), 1–103.
Parashar, S.K.S., Murty, B.S., Repp, S., Weber, S., Erdem, E., 2012. Investigation of intrinsic defects in core-shell structured ZnO nanocrystals. J. Appl. Phys. 111 (113712), 1–7.
Park, C.H., Zhang, S.B., Wei, S.H., 2002. Origin of *p*-type doping difficulty in ZnO: the impurity perspective. Phys. Rev. B 66 (073202), 1–3.
Parmar, N.S., McCluskey, M.D., Lynn, K.G., 2013. Vibrational spectroscopy of Na–H complexes in ZnO. J. Electron. Mater. 42, 3426–3428.
Pearton, S.J., Heo, W.H., Ivill, M., Norton, D.P., Steiner, T., 2004. Dilute magnetic semiconducting oxides. Semicond. Sci. Technol. 19, R59–R74.
Pearton, S.J., Norton, D.P., Ip, K., Heo, Y.W., Steiner, T., 2005. Recent progress in processing and properties of ZnO. Prog. Mater. Sci. 50, 293–340.

Philipps, J.M., Stehr, J.E., Buyanova, I., Tarun, M.C., McCluskey, M.D., Meyer, B.K., Hofmann, D.M., 2014. Recharging behavior of nitrogen-centers in ZnO. J. Appl. Phys. 116 (063701), 1–4.
Qiu, H., Gallino, F., Di Valentin, C., Wang, Y., 2011. Shallow donor states induced by in-diffused Cu in ZnO: A combined HREELS and hybrid DFT study. Phys. Rev. Lett. 106 (066401), 1–4.
Reshchikov, M.A., Morkoç, H., Nemeth, B., Nause, J., Xie, J., Hertog, B., Osinsky, A., 2007. Luminescence properties of defects in ZnO. Phys. B 401, 358–361.
Schirmer, O.F., 1968. The structure of the paramagnetic lithium center in zinc oxide and beryllium oxide. J. Phys. Chem. Solids 29, 1407–1429.
Schirmer, O.F., Zwingel, D., 1970. The yellow luminescence of zinc oxide. Solid State Commun. 8, 1559–1563.
Schirra, M., Schneider, R., Reiser, A., Prinz, G.M., Feneberg, M., Biskupek, J., Kaiser, U., Krill, C.E., Thonke, K., Sauer, R., 2008. Stacking fault related 3.31-eV luminescence at 130-meV acceptors in zinc oxide. Phys. Rev. B 77 (125215), 1–10.
Schmidt-Mende, L., MacManus-Driscoll, J.L., 2007. ZnO—nanostructures, defects, and devices. Mater. Today 10, 40–48.
Schulz, H.J., Thiede, M., 1987. Optical spectroscopy of 3d7 and 3d8 impurity configurations in a wide-gap semiconductor (ZnO:Co, Ni, Cu). Phys. Rev. B 35, 18–34.
Selim, F.A., Weber, M.H., Solodovnikov, D., Lynn, K.G., 2007. Nature of native defects in ZnO. Phys. Rev. Lett. 99 (085502), 1–4.
Selim, F.A., Tarun, M.C., Wall, D.E., Boatner, L.A., McCluskey, M.D., 2011. Cu-doping of ZnO by nuclear transmutation. Appl. Phys. Lett. 99 (202109), 1–3.
Shi, G.A., Saboktakin, M., Stavola, M., Pearton, S.J., 2004. "Hidden hydrogen" in as-grown ZnO. Appl. Phys. Lett. 85, 5601–5603.
Shi, G.A., Stavola, M., Fowler, W.B., 2006. Identification of an OH-Li center in ZnO: infrared absorption spectroscopy and density functional theory. Phys. Rev. B 73 (081201), 1–3.
Simpson, J.C., Cordaro, J.F., 1988. Characterization of deep levels in zinc oxide. J. Appl. Phys. 63, 1781–1783.
Son, N.T., Isoya, J., Ivanov, I.G., Ohshima, T., Janzén, E., 2013. Magnetic resonance identification of hydrogen at a zinc vacancy in ZnO. J. Phys. Condens. Matter 25 (335804), 1–5.
Soudi, A., Khan, E.H., Dickinson, J.T., Gu, Y., 2009. Observation of unintentionally incorporated nitrogen-related complexes in ZnO and GaN nanowires. Nano Lett. 9, 1844–1849.
Stehr, J.E., Hofmann, D.M., Meyer, B.K., 2012. Electron paramagnetic resonance and photo-electron paramagnetic resonance investigation on the recharging of the substitutional nitrogen acceptor in ZnO. J. Appl. Phys. 112 (103511), 1–4.
Stehr, J.E., Johansen, K.M., Bjørheim, T.S., Vines, L., Svensson, B.G., Chen, W.M., Buyanova, I.A., 2014. Zinc-vacancy–donor complex: a crucial compensating acceptor in ZnO. Phys. Rev. Appl. 2 (021001), 1–6.
Tarun, M.C., Iqbal, M.Z., McCluskey, M.D., 2011. Nitrogen is a deep acceptor in ZnO. AIP Adv. 1 (022105), 1–7.
Teklemichael, S.T., McCluskey, M.D., 2011. Acceptor and surface states of ZnO nanocrystals: a unified model. Nanotechnology 22 (475703), 1–4.
Teklemichael, S.T., McCluskey, M.D., 2012. Compensation of acceptors in ZnO nanocrystals by adsorption of formic acid. J. Phys. Chem. C 116, 17248–17251.
Teklemichael, S.T., Hlaing Oo, W.M., McCluskey, M.D., Walter, E.D., Hoyt, D.W., 2011. Acceptors in ZnO nanocrystals. Appl. Phys. Lett. 98 (232112), 1–3.
Thomas, D.G., Lander, J.J., 1956. Hydrogen as a donor in zinc oxide. J. Chem. Phys. 25, 1136–1142.
Thonke, K., Feneberg, M., 2012. Photoluminescence of ZnO: basics and applications. In: Bergman, L., McHale, J.L. (Eds.), Handbook of Luminescent Semiconductor Materials. CRC Press, Boca Raton, FL, pp. 87–124.

Thonke, K., Gruber, T., Teofilov, N., Schönfelder, R., Waag, A., Sauer, R., 2001. Donor-acceptor pair transitions in ZnO substrate material. Phys. B 308–10, 945–948.
Triboulet, R., 2014. Growth of ZnO bulk crystals: a review. Prog. Cryst. Growth Charact. Mater. 60, 1–14.
Tuomisto, F., Makkonen, I., 2013. Defect identification in semiconductors with positron annihilation: experiment and theory. Rev. Mod. Phys. 85, 1583–1613.
Tuomisto, F., Ranki, V., Saarinen, K., Look, D.C., 2003. Evidence of the Zn vacancy acting as the dominant acceptor in *n*-type ZnO. Phys. Rev. Lett. 91 (205502), 1–4.
Tuomisto, F., Rauch, C., Wagner, M.R., Hoffmann, A., Eisermann, S., Meyer, B.K., Kilanski, L., Tarun, M.C., McCluskey, M.D., 2013. Nitrogen and vacancy clusters in ZnO. J. Mater. Res. 28, 1977–1983.
Van de Walle, C.G., 2000. Hydrogen as a cause of doping in zinc oxide. Phys. Rev. Lett. 85, 1012–1015.
Van de Walle, C.G., Neugebauer, J., 2003. Universal alignment of hydrogen levels in semiconductors, insulators, and solutions. Nature 423, 626–628.
Vanheusden, K., Warren, W.L., Seager, C.H., Tallant, D.R., Voigt, J.A., Gnade, B.E., 1996. Mechanisms behind green photoluminescence in ZnO phosphor powders. J. Appl. Phys. 79, 7983–7990.
Vines, L., Monakhov, E.V., Schifano, R., Mtangi, W., Auret, F.D., Svensson, B.G., 2010. Lithium and electrical properties of ZnO. J. Appl. Phys. 107 (103707), 1–7.
Vlasenko, L.S., Watkins, G.D., 2005. Optical detection of electron paramagnetic resonance for intrinsic defects produced in ZnO by 2.5-MeV electron irradiation *in situ* at 4.2 K. Phys. Rev. B 72 (035203), 1–12.
Wang, Z.L., 2004. Zinc oxide nanostructures: growth, properties and applications. J. Phys. Condens. Matter 16, R829–R858.
Wang, X.J., Vlasenko, L.S., Pearton, S.J., Chen, W.M., Buyanova, I.A., 2009. Oxygen and zinc vacancies in as-grown ZnO single crystals. J. Phys. D 42 (175411), 1–8.
Wardle, M.G., Goss, J.P., Briddon, P.R., 2005. Theory of Fe, Co, Ni, Cu, and their complexes with hydrogen in ZnO. Phys. Rev. B 72 (155108), 1–13.
Wardle, M.G., Goss, J.P., Briddon, P.R., 2006. First-principles study of the diffusion of hydrogen in ZnO. Phys. Rev. Lett. 96 (205504), 1–4.
Yamamoto, Y., Katayama-Yoshida, H., 1999. Solution using a codoping method to unipolarity for the fabrication of p-type ZnO. Jpn. J. Appl. Phys. 38, L166–L169.
Yan, Y., Al-Jassim, M.M., Wei, S.-H., 2006. Doping of ZnO by group-IB elements. Appl. Phys. Lett. 89 (181912), 1–3.
Zhang, Z., Knutsen, K.E., Merz, T., Kuznetsov, A.Yu., Svensson, B.G., Brillson, L.J., 2012. Thermal process dependence of Li configuration and electrical properties of Li-doped ZnO. Appl. Phys. Lett. 100, 042107.
Zubiaga, A., Tuomisto, F., Coleman, V.A., Tan, H.H., Jagadish, C., Koike, K., Sasa, S., Inoue, M., Yano, M., 2008. Mechanisms of electrical isolation in O^+-irradiated ZnO. Phys. Rev. B 78 (035125), 1–5.
Zwingel, D., Gärtner, F., 1974. Paramagnetic and optical properties of Na-doped ZnO single crystals. Solid State Commun. 14, 45–49.

CHAPTER NINE

Point Defects in GaN

Michael A. Reshchikov[1]
Department of Physics, Virginia Commonwealth University, Richmond, Virginia, USA
[1]Corresponding author: e-mail address: mreshchi@vcu.edu

Contents

1. INTRODUCTION

GaN and its alloys with InN and AlN are the key components for bright blue and white LEDs and blue lasers. GaN-based light-emitting devices are gaining favor over the mercury-containing fluorescent lamps in the lighting technology. Finally, GaN is a very promising material for high-power electronics.

The thermodynamically stable crystal structure of GaN is wurtzite. Detailed description of GaN properties can be found in a handbook by Morkoç (2008). GaN is mostly used in the form of thin films grown on foreign substrates such as sapphire and SiC. The lattice mismatch between GaN and the substrate leads to a high density of threading dislocations (10^9–10^{10} cm^{-2}). In freestanding GaN templates with thickness of about

Semiconductors and Semimetals, Volume 91
ISSN 0080-8784
http://dx.doi.org/10.1016/bs.semsem.2014.11.003

0.5–1 mm, the dislocation density can be reduced down to $\sim 10^6$ cm^{-2}. The high density of structural defects was the major problem for the development of GaN-based materials and devices in the past two decades, whereas now it has been mostly resolved, and high-efficiency light-emitting devices have become available for consumers.

However, the concentration of point defects in GaN is still relatively high (on the order of 10^{16} cm^{-3} in high-quality material). Point defects include impurities introduced during growth such as O, C, Si, and H, and native defects such as vacancies, antisites, and interstitials, as well as complexes formed from different elementary defects. Point defects affect the performance of light-emitting devices by decreasing their lifetime and limiting their efficiency. They are also the main obstacle hindering the realization of high-power electronic devices. A very high breakdown voltage is predicted for GaN (Boles et al., 2013; Saitoh et al., 2010); however, point defects can reduce this parameter significantly. In spite of many years of research, most of the point defects in GaN remain unidentified and are not well understood.

Several experimental techniques are currently used to detect and quantify point defects in GaN (Table 1). The Hall effect provides limited information because of large ionization energies for almost all defects in GaN and the presence of degenerate layers (e.g., near the GaN/substrate interface). Secondary-ion mass-spectrometry (SIMS) analysis allows detecting and quantifying impurities, but its detection limit is often too high. Positron annihilation spectroscopy (PAS) is efficient in detecting negatively charged defects, especially cation vacancies. Many modifications of capacitance techniques, which include deep-level transient spectroscopy (DLTS), optical DLTS (ODLTS), photoinduced current transient spectroscopy (PICTS), deep-level optical spectroscopy (DLOS), and steady-state photocapacitance (SSPC), can detect point defects with low concentrations. Magnetic resonance techniques, which include electron paramagnetic resonance (EPR) and optically detected magnetic resonance (ODMR), can identify defects in certain charge states. One of the most powerful tools for the investigation of point defects in GaN is photoluminescence (PL). It can detect point defects with both shallow and deep energy levels and determine their concentrations, charge states, the strength of electron–phonon coupling, and many other parameters. In certain cases, usually in combination with other characterization tools and theoretical insight, PL provides the identity of defects.

Table 1 Capability of experimental methods for the investigation of point defects in GaN

Method	Modifications	Energy level	Symmetry, structure	Capture coefficients	Concentration	Comment
Hall effect		Yes	No	No	Yes	Only shallow
SIMS		No	No	No	Yes	Only impurities
Luminescence	PL, CL, ODMR	Yes	Possible	Yes	Yes	Only radiative defects
Capacitance	DLTS, ODLTS, PICTS, SSPC, DLOS	Yes	Possible	Yes	Yes	Requires a diode structure
Magnetic resonance	EPR, ODMR, photo-EPR	Possible	Possible	No	Yes	In certain charge states
PAS		No	Possible	No	Yes	Mostly vacancies

In this chapter, experimental results on point defects in GaN will be compared with first-principles calculations. Only for a few defects, the identification appears to be reliable, whereas for a majority of the defects, further investigation is needed. The readers may note that the author is an expert in luminescence methods, and the critical analysis of other techniques may not be as encompassing. The most interesting and important cases (from the viewpoint of the author) are analyzed in detail, whereas a few classes of defects, such as rare-earth, transition metals, shallow donors, and defects introduced by intentional doping (except for carbon), are left out.

2. THEORETICAL PREDICTIONS

It is difficult to compare experimental and theoretical results for defects in GaN, because the former are often ambiguous, while the latter involve a wide range of assumptions and corrections. Then, by seeking mutual validation, the conclusions of experimentalists may be affected by inaccurate theoretical predictions, and in turn, the theoretical adjustments may be wrong because of unreliable experimental data. As an example, we will briefly review how the theoretical predictions on the location of energy levels for the isolated Ga vacancy (V_{Ga}), the $V_{Ga}O_N$ complex, and the C_N acceptor evolved in the last 25 years. This will help us to understand why experimentalists were often influenced by these predictions while interpreting their experimental results.

Jenkins and Dow (1989) have calculated that the isolated Ga vacancy (V_{Ga}) is a *shallow* acceptor in GaN. In accord with this prediction, Tansley and Egan (1992), while discussing the origin of the shallow (0.2 eV from the valence band) acceptor in GaN, concluded in their experimental critical review: "The only point defect unambiguously located appears to be the gallium vacancy observed in studies of GaN luminescence at 3.26 eV below the conduction band and close to the calculated value of about 3.3 eV." Later, by using density functional theory (DFT), Neugebauer and Van de Walle (1996), Laaksonen et al. (2009), and few other groups have calculated that V_{Ga} is a *deep* acceptor with the 3-/2- transition energy level at ~1.1 eV above the valence band (Fig. 1A). The prediction resulted in the attribution of the ubiquitous yellow luminescence (YL) band peaking at 2.2 eV to V_{Ga} or to the $V_{Ga}O_N$ complex, which was accepted by many experimentalists (Reshchikov and Morkoç, 2005). The transition level for the defect responsible for the YL band (also known as

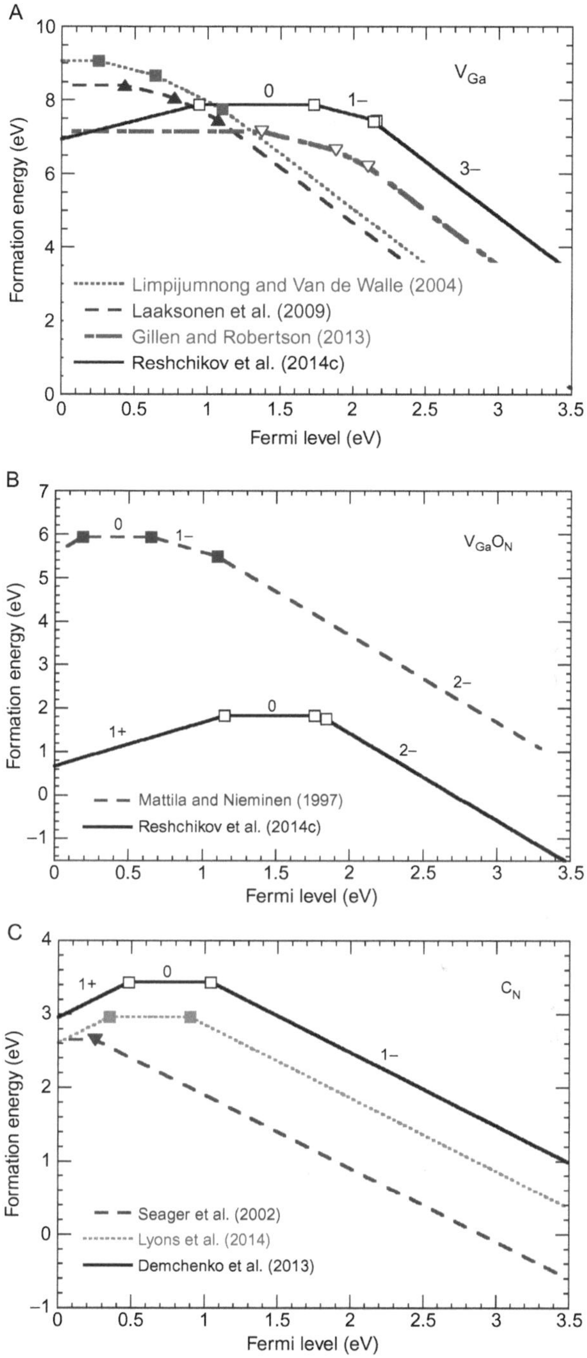

Figure 1 First-principles calculations for V_{Ga} (A), the $V_{Ga}O_N$ complex (B), and the C_N acceptor (C). Symbols indicate transition levels. *Reproduced with permission from Limpijumnong and Van de Walle (2004), Copyright 2004 by The American Physical Society; Laaksonen et al. (2009) and Gillen and Robertson (2013), © IOP Publishing. Reproduced by permission of IOP Publishing. All rights reserved; Reshchikov et al. (2014c), Copyright 2014 by The American Physical Society; Mattila and Nieminen (1997), Copyright 1997 by The American Physical Society; Seager et al. (2002), Copyright 2002, AIP Publishing LLC; and Lyons et al. (2014), Copyright 2014 by The American Physical Society.*

the hole trap H1 in DLTS) was found to be at 0.9 eV above the valence band, which almost matched the calculated values for the V_{Ga} and $V_{Ga}O_N$ defects (Fig. 1A and B). However, recent calculations using hybrid functionals indicate that the energy levels of the isolated V_{Ga} and the $V_{Ga}O_N$ complex are *even deeper* than predicted in early DFT calculations (Fig. 1A and B), and the related PL is now expected in the infrared and not in the yellow part of the spectrum (Gillen and Robertson, 2013; Reshchikov et al., 2014c).

Similarly, the C_N defect was predicted to be a *shallow* acceptor in early DFT calculations (Bogusławski and Bernholc, 1997; Eberlein et al., 2007; Neugebauer and Van de Walle, 1996; Wright, 2002). In "agreement" with this prediction, a shallow acceptor with the energy level of about 0.2 eV has been attributed to the C_N defect in a number of experimental works using luminescence (Armitage et al., 2005; Fischer et al., 1995) and photocapacitance techniques (Kamyczek et al., 2012; Zhang et al., 2012). Even an achievement of p-type conductivity with the concentration of free holes in excess of 10^{17} cm^{-3} has been reported for GaN samples doped with carbon (Abernathy et al., 1995; Hikosaka et al., 2006). However, recent calculations, using hybrid functionals (Demchenko et al., 2013; Lyons et al., 2010), indicate that C_N is a *deep* acceptor, with the energy level at about 1 eV above the valence band (Fig. 1C). This new prediction is in agreement with other experimental observations, according to which the YL band is caused by a deep carbon-related defect (Kucheyev et al., 2002; Ogino and Aoki, 1980; Zhang and Kuech, 1998).

In recent years, significant progress has been achieved in first-principles calculations for defects in semiconductors (Freysoldt et al., 2014). The accuracy of the calculations is markedly improved by more accurate treatment of electronic interactions in solids. The currently used hybrid functionals, for example, the popular HSE (Heyd et al., 2003), offer a practical compromise between the semi/local approximations to DFT and the computationally demanding many-body methods (such as GW). These methods are expected to produce more accurate results that can be compared to experiment. However, one of the main problems impeding the progress in defect physics is that solid experimental data are not always available for comparison with theoretical results. Only a few defects in GaN are reliably identified and well understood. In the following sections, a number of point defects in GaN are analyzed in detail.

3. GROWTH METHODS AND SIMS ANALYSIS

Currently, the main growth techniques for GaN are metalorganic chemical vapor deposition (MOCVD), which is the primary production method for optoelectronic devices such as LEDs and laser diodes; hydride vapor phase epitaxy (HVPE); molecular beam epitaxy (MBE); ammonothermal (AT) bulk growth; and high nitrogen pressure (HNP) bulk growth. Typical contaminants for all growth techniques are O, C, Si, and H, yet each growth method can produce specific point defects due to specific sources or ambient conditions. The impurities can infiltrate the GaN from the substrate, reactor walls, susceptor, or precursors. The formation of point defects and the incorporation of impurities depend on growth conditions such as temperature, pressure, and Ga/N ratio.

Common impurities in MOCVD GaN include C, Si, O, and H. Silicon and oxygen are shallow donors, and their concentrations in unintentionally doped GaN can range from 10^{16} to 10^{19} cm^{-3}. The C concentration strongly depends on growth conditions in MOCVD growth. Ambacher et al. (1997) have shown that the concentration of C, as measured by SIMS, decreases from 1×10^{21} to 2×10^{18} cm^{-3} as the substrate temperature increases from 700 to 1100 °C. The concentration of H decreases by two orders of magnitude in this temperature range. A similar temperature dependence was observed by Koleske et al. (2002) and Chen et al. (2013), who reported that the C concentration decreased from $\sim10^{18}$ to $\sim10^{16}$ cm^{-3} with increasing growth temperature from 950 to 1080 °C. The growth pressure is another factor affecting the incorporation of carbon. Koleske et al. (2002) and Brunner et al. (2011) reported on a decrease in the C concentration in undoped GaN by about two orders of magnitude (from $\sim10^{19}$ to $\sim10^{17}$ cm^{-3}) as the growth pressure was increased from 50 to 150 Torr. In the same conditions, the concentration of Si increased by a factor of 2. The C concentration in MOCVD GaN also strongly depends on the III/V ratio (Chen et al., 2013; Koleske et al., 2002). By optimizing conditions, the concentration of C in MOCVD GaN can be reduced to about 10^{16} cm^{-3} (Chen et al., 2013; Murthy et al., 2007). The typical concentrations of O and Si are between 10^{16} and 10^{17} cm^{-3} in MOCVD GaN. Often, the results of SIMS measurements are inconclusive because the detection limits for C, O, and Si are near 10^{16} cm^{-3}, and the detection limit for H is about 10^{17} cm^{-3}.

In GaN grown by HVPE, the concentrations of C and Si are often close to or below the SIMS detection limit. Fujito et al. (2009) have demonstrated that bulk HVPE GaN crystals have very low concentrations of C, O, H, and Cl impurities, below their detection limits of 2×10^{15}, 7×10^{15}, 2×10^{16}, and 5×10^{14} cm^{-3}, respectively. The concentration of C in a 30-μm-thick GaN layer on a sapphire substrate can be lower than 1×10^{15} cm^{-3} (Reshchikov et al., 2014c). The lowest concentrations of Si and O in HVPE-grown GaN are about 10^{15} and 10^{16} cm^{-3}, respectively (Freitas et al., 2012; Murthy et al., 2007). It is known that oxygen can diffuse from the sapphire substrate to GaN via structural defects and create a thin, degenerate layer near the GaN/sapphire interface (Look et al., 2001). It is also established that Si tends to accumulate at surfaces and subinterfaces (caused by growth interruption) in GaN grown by HVPE (Usikov et al., 2013).

MBE-grown GaN is generally inferior to MOCVD and HVPE GaN. For example, unintentionally doped or Si-doped MBE GaN layers on sapphire substrates had the O concentration in the range of 2×10^{18} to 1×10^{20} cm^{-3} and the C concentration in the range of 3×10^{17} to 8×10^{19} cm^{-3} (Bell et al., 2000). Significantly, lower impurity concentrations can be achieved by optimizing growth conditions. Murthy et al. (2007) have demonstrated very low defect density in MBE layers grown homoepitaxially on high-quality MOCVD or HVPE GaN substrates. The concentration of Si and C impurities was lower than 10^{15} and 10^{16} cm^{-3}, respectively, and the concentration of oxygen was slightly higher ($\sim 3 \times 10^{16}$ cm^{-3}).

Bulk GaN can be grown from solution, and there are several modifications of this method. Bulk AT GaN commonly has a high concentration of oxygen, on the order of 10^{19} cm^{-3} (Bliss et al., 2010; Hashimoto et al., 2013). Bulk HNP GaN contains oxygen with the concentration exceeding $\sim 10^{19}$ cm^{-3} (Leszczynski et al., 1996; Tuomisto et al., 2005). In GaN grown from solution at ~800 °C and near atmospheric pressure, the concentrations of the O, Si, C, and H impurities are near or below their detection limits of 5×10^{16}, 5×10^{15}, 8×10^{15}, and 2×10^{17} cm^{-3}, respectively (Garces et al., 2010).

4. DEFECTS REVEALED BY PL

PL is a powerful tool for the detection and identification of point defects in GaN (Reshchikov and Morkoç, 2005). We will leave out the analysis of the excitonic, or in the more general case, near-band-edge (NBE) emission (typically between 3.2 and 3.5 eV), and will focus on the

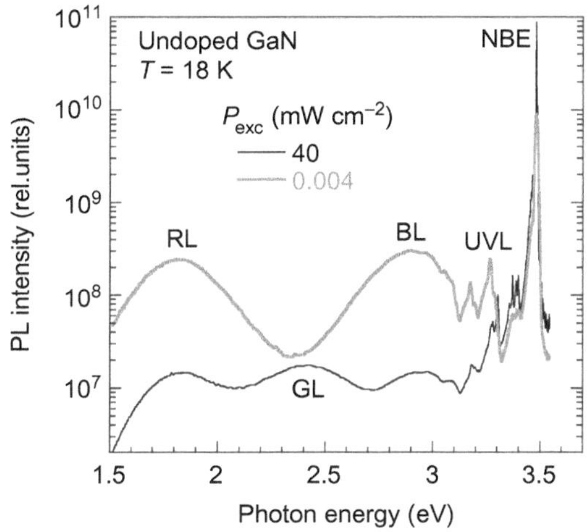

Figure 2 Low-temperature PL spectra at two excitation intensities from undoped GaN grown by HVPE. The PL intensity is divided by the excitation intensity.

defect-related PL in the photon energy range between 1.5 and 3.3 eV. Figure 2 shows low-temperature PL spectra from undoped GaN grown by HVPE on a sapphire substrate.

The defect-related PL bands can be better observed at low excitation intensities, when defects are not saturated with photogenerated carriers and the dependence of the PL intensity on the excitation intensity is linear. For the excitation power density of 4 μW cm^{-2} in Fig. 2, the ultraviolet luminescence (UVL) band is observed in the photon range closest to the NBE emission. The UVL band has the strongest peak at 3.267 eV followed by weaker LO phonon replicas at distance multiples of 91 meV. At lower energies, the blue luminescence (BL) band can be found, which has a sharp onset at 3.1 eV and a maximum at 2.9 eV. The red luminescence (RL) band with a maximum at 1.8 eV is another defect-related PL band. With increasing excitation intensity from 0.004 to 40 mW cm^{-2}, the efficiency of the above-mentioned defect-related PL bands (i.e., the ratio of the PL intensity to the excitation intensity) gradually decreases due to the saturation of the associated defects with photogenerated holes. The quantitative analysis of the quantum efficiency (QE) of PL (Reshchikov, 2012a) shows that the QE of the RL, BL, and UVL bands decreases from 3.4%, 3.8%, and 0.67% to 0.2%, 0.18%, and 0.08%, respectively, as P_{exc} increases from 0.004 to 40 mW cm^{-2} for the sample analyzed in Fig. 2. We can see that a new defect-related band, namely the green luminescence (GL) band with a maximum at 2.4 eV emerges at high excitation intensity, while its observation was impossible at low excitation intensity when it was obstructed by other PL bands. Thus, by varying excitation intensity, new PL bands can be revealed. The variation of temperature, as well as the use of time-resolved PL analysis, further extends the capabilities of the PL technique and can help to discover otherwise hidden PL bands and to study their behavior. In contrast to conventional wisdom that defect-related PL bands are broad because of the spread of the energy levels in the bandgap, the broadness typically arises from strong electron–phonon coupling, and important parameters of the related defect can be determined from the analysis of the PL band shape and position. Below, we will briefly review the basic methods of PL analysis which allow us to recognize the defect-related PL bands and determine their properties.

4.1. Phenomenological model

PL is usually excited with a laser having photon energy higher than the semiconductor bandgap. It is often assumed for simplicity that the laser light penetrates to a depth d defined as $d = \alpha^{-1}$, where α is the absorption coefficient at the wavelength of the laser emission ($\alpha \approx 10^5$ cm^{-1} for GaN and a HeCd laser emitting at 325 nm, Muth et al., 1997), and the electron–hole pairs are created uniformly in this volume. The exponential decrease of the laser intensity inside a semiconductor can also be considered, and the exponential decrease of the electron–hole density with increasing distance from the semiconductor surface can be regarded as an advanced model.

In n-type GaN, photogenerated free holes (with concentration p) annihilate with free electrons (with total concentration n) via different recombination channels, including recombination at point defects with emission of photons (radiative recombination producing PL), and recombination at defects with emission of multiple phonons (nonradiative recombination). The transition rates are described in the Shockley–Read–Hall theory as the product of the concentrations of available carriers and available empty sites, multiplied by a phenomenological parameter called the capture coefficient (Hall, 1951; Shockley and Read, 1952). In particular, holes are captured by defect A, and electrons recombine with the holes bound to this defect at rates $C_{pA}N_A^-\mathrm{p}$ and $C_{nA}N_A^0\mathrm{n}$, respectively, where C_{pA} and C_{nA} are the capture coefficients for holes and electrons by defect A, and the superscripts "−" and "0" indicate the charge states of the defect. The concept of an effective capture cross-section σ is often used with the physical meaning of a target area where a carrier will be trapped. For example,

$C_{pA}=\sigma_{pA}\langle v_p\rangle$, where $\langle v_p\rangle=\sqrt{8kT/\pi m_p}$ is the mean thermal velocity of free holes, k is Boltzmann's constant, T is ambient temperature, and m_p is the effective mass of holes in the valence band. The captured hole may thermally escape to the valence band before the radiative recombination takes place, leading to the quenching of PL. The rate of the thermal emission of holes from defect A to the valence band can be found from detailed balance as

$$Q_A=C_{pA}N_v g^{-1}\exp\left(-\frac{E_A}{kT}\right), \tag{1}$$

where $N_v=2(2\pi m_p kT)^{3/2}/h^3$ is the effective density of states in the valence band, g is the degeneracy of the acceptor level, and E_A is the energy difference between the transition level of the defect and the top of the valence band. Similarly, the rate of the thermal emission of electrons from a donor D to the conduction band is

$$Q_D=C_{nD}N_c g^{-1}\exp\left(-\frac{E_D}{kT}\right), \tag{2}$$

where N_c is the effective density of states in the conduction band, g is the degeneracy of the donor level, and E_D is the ionization energy of the donor D.

The PL intensity caused by the recombination of free electrons with holes bound to defect A is $I^{PL}=C_{nA}N_A^0 n$. We can introduce the absolute internal QE (IQE) of this PL as $\eta=I^{PL}/G$, where G is the electron–hole pair generation rate (in $cm^{-3}\,s^{-1}$). The system of rate equations can be solved numerically. For a few simple cases, analytical solutions can be obtained. For example, for conductive n-type GaN and low excitation intensity, the IQE of PL via defect A has the following temperature dependence (Reshchikov and Korotkov, 2001; Reshchikov et al., 2011):

$$\eta(T)=\frac{\eta_0}{1+(1-\eta_0)\tau Q_A}, \tag{3}$$

where η_0 is the IQE of PL at low temperature and τ is the PL lifetime. Analysis of Eq. (3) shows that at $T<T_0$, where T_0 is found from $(1-\eta_0)\tau Q_A=1$, the IQE of PL is constant and equal to η_0, while at $T>T_0$ the IQE is proportional to $\exp(E_A/kT)$. By fitting an experimental dependence of the PL intensity on temperature with Eq. (3) and taking τ from time-resolved PL measurements, parameters E_A and C_{pA} for defect A can be readily determined if $\eta_0 \ll 1$. Moreover, when E_A and C_{pA} for a particular defect are known, the parameter η_0 can be found for samples in which the IQE is very high (Reshchikov et al., 2012).

The concentration of defects participating in PL can be determined from the observation of PL intensity saturation with increasing excitation intensity. In the simple model (generation rate is constant within an active layer of thickness $d = \alpha^{-1}$), the following expression can be derived for the IQE of PL from defect A (Reshchikov, 2012a; Reshchikov and Korotkov, 2001):

$$\eta(G) \approx \left(\frac{G\tau}{N_A} + \frac{1}{\eta_0}\right)^{-1}, \tag{4}$$

where η_0 is the IQE in the limit of low excitation intensity. According to Eq. (4), the PL intensity increases linearly with G at low excitation intensity and is equal to $\eta_0 G$, and it saturates at the level of N_A/τ at high G. The transition from the linear dependence to the saturation occurs at $G = G_0 = N_A/\eta_0\tau$. In the advanced model (Reshchikov, 2006, 2012a), which accounts for the exponential decrease of the laser light intensity in a thick GaN layer,

$$\eta(P_0) \approx \frac{N_A}{\alpha\tau P_0} \ln\left(1 + \frac{P_0}{P_1}\right), \tag{5}$$

where, instead of the constant generation rate G, we introduced the excitation intensity P_0 expressed as the number of photons entering through the unit area of the sample surface per unit time, and $P_1 = N_A/\alpha\tau\eta_0$. This method is more accurate than the simple method, and it requires only an accurate estimate of the laser intensity and parameters τ and η_0, which can be found from independent experiments (Reshchikov, 2012a).

The PL lifetime in n-type GaN reveals important characteristics of the radiative defects, such as the electron-capture cross-section. Indeed, in time-resolved PL experiments, the PL intensity decays exponentially after a laser pulse, with the characteristic time constant τ (Reshchikov, 2014a):

$$I^{PL}(t) \approx I^{PL}(0)\exp\left(\frac{-t}{\tau}\right), \tag{6}$$

where $\tau = (C_{nA}n)^{-1}$ is the PL lifetime.

For high-resistivity semiconductors, the solutions of rate equations are not straightforward, but at least numerical solutions can be obtained, which successfully simulate the experimental dependences. Sometimes, an abrupt and tunable quenching of PL is observed, and the detailed analysis of which may reveal a number of parameters for radiative and nonradiative defects in a semiconductor sample (Reshchikov, 2012b; Reshchikov et al., 2011).

4.2. Configuration-coordinate model

The commonly used band diagrams do not account for electron–phonon coupling, which is important in the case of deep-level defects. In particular, defect-related PL bands are usually broad, and their maxima correspond to the emission of one photon and several phonons, so that determining the defect energy level from the PL spectrum may be not an easy task. The basis of the theory of multiphonon transitions was developed by Huang and Rhys (1950), and the simplest approach is called the one-dimensional configuration-coordinate (CC) model of a localized defect (Abakumov et al., 1991; Dexter et al., 1955; Rebane, 1970; Stoneham, 1975). The CC diagram represents the potential energy (the sum of electronic and lattice energies) for the ground and excited states of a defect system as a function of a generalized atomic coordinate (Fig. 3). For simplicity, the one-dimensional CC model considers an effective phonon mode, and the energy of a single phonon may correspond to the energy of the dominant local or quasi-local phonon. The potential curves (adiabatic potentials) are assumed to be simple parabolas. For a localized defect with a deep level, the potential minima of the ground and excited states are displaced due to different lattice geometries. The optical transition between the potential minima produces only a photon, and the corresponding feature in the PL spectrum is called the no-phonon or zero-phonon line (ZPL). Several sharp lines can be observed in the PL spectrum at energy multiples of the dominant phonon energy, which are called phonon replicas of the ZPL and correspond to the consecutive emission of one photon and several phonons. However, the line shape of a PL band from deep defects is often structureless, usually Gaussian or asymmetric, with more abrupt high-energy side.

Figure 3 shows examples of possible CC diagrams for defects in GaN with a bandgap of 3.5 eV. Here, the lowest parabola is the ground state corresponding to a defect filled with electrons when the Fermi level lies well above its energy level. The top parabola, which corresponds to the creation of a free electron and a free hole by light, is identical to the ground state displaced vertically by the energy of the bandgap. The middle parabola is the excited state corresponding to the defect after it captures a hole in n-type GaN. A vertical upward arrow AB in Fig. 3A indicates the resonant excitation of the defect, i.e., the excitation of an electron from the defect

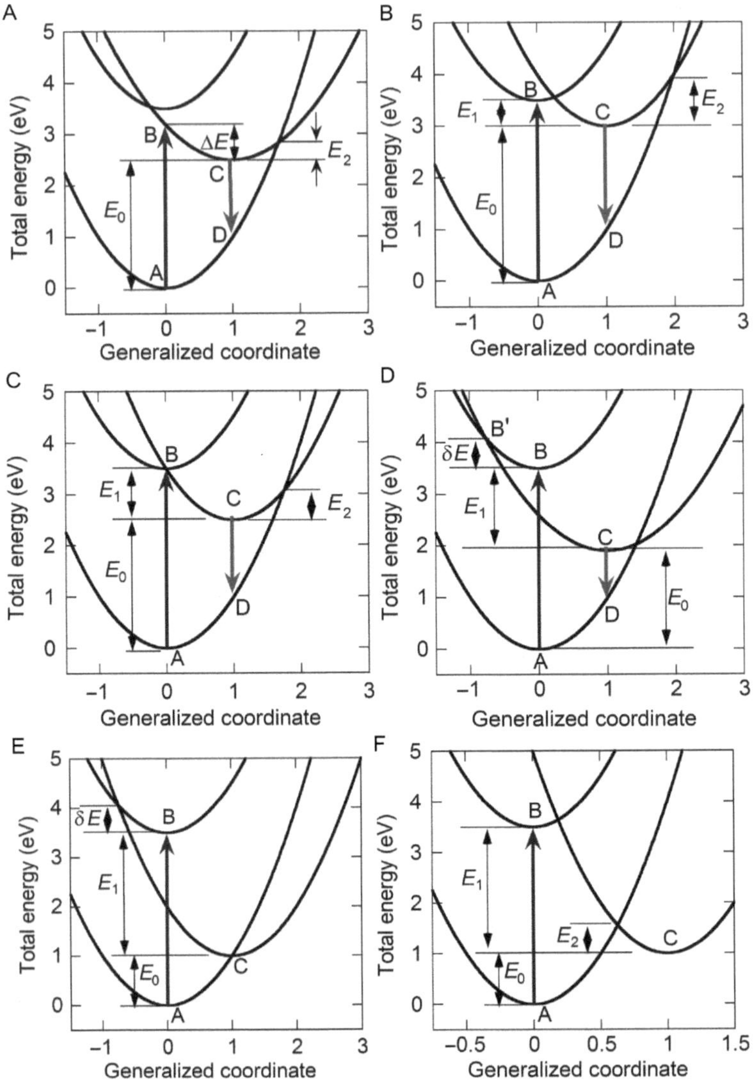

Figure 3 Examples of CC diagrams. (A–D)—radiative defects, and (E) and (F)—nonradiative defects.

level to the conduction band. After the excitation, the defect and surrounding lattice relax to the new equilibrium state (point C), and the excess energy, ΔE, is released in the form of phonons. The dimensionless value $\Delta E/\hbar\Omega$, with $\hbar\Omega$ being the effective phonon energy, is called the Huang–Rhys factor, and it describes the strength of the electron–phonon coupling. The transition CD is the optical transition with the length of the downward arrow equal to the position of the PL band maximum. The ZPL energy is equal to E_0, and the ZPL can be observed as a sharp line on the high-energy side of the PL band. After the optical transition, the defect relaxes into the minimum of the ground state by emitting several phonons (transition DA). Alternatively, the transition from point B to point A may occur nonradiatively, via the crossover of the excited and ground state curves. According to Bartram and Stoneham (1975), at low temperatures, the recombination is purely radiative when $\Delta E < E_2$, and it is only partially radiative (low luminescence efficiency) when $\Delta E > E_2$. When E_2 is close to zero (Fig. 3E) or when the excited state minimum lies outside the ground state curve (Fig. 3F), the electron–hole recombination via the defect is nonradiative.

Figure 3B–D illustrates different cases of PL via a defect after above-bandgap excitation. First, an electron–hole pair is created (transition AB), and then the hole is captured by the defect (transition BC). The energy of the system decreases by E_1, which is equal to the distance from the defect transition level to the top of the valence band. The excess energy is released in the form of several phonons. For some defects, a barrier for the capture of a charge carrier (a hole in our example) exists, such as δE in Fig. 3D and E. For such defects, the capture coefficient (and the capture cross-section) increases with increasing temperature (Alkauskas et al., 2014).

As the temperature increases, the probability that the bound hole will be emitted back to the valence band (transition CB) increases. The average frequency of such an event is $\nu \approx \nu_0 \exp[-(E_1 + \delta E)/kT]$, where ν_0 is the frequency of atomic oscillations (about $10^{13}\ \mathrm{s}^{-1}$). The PL quenching is observed when $\nu > \tau^{-1}$, because the thermally emitted holes can be captured by other defects and recombine with electrons via other channels (the Schön–Klasens mechanism). Alternatively, the PL quenching may be caused by another mechanism (the Seitz–Mott mechanism). In this case, the defect system in point C acquires enough thermal energy to pass over a barrier with height E_2 in Fig. 3A–C and relaxes to point A with the emission of many phonons. Such a defect behaves as a radiative defect at low temperature and a nonradiative defect at high temperatures (Reshchikov, 2014b).

Several parameters of a radiative defect can be obtained from the analysis of the shape of a defect-related PL band and from its temperature behavior. In particular, within a one-dimensional CC model, assuming that the effective phonon energies of the ground and excited states ($\hbar\Omega_g$ and $\hbar\Omega_e$, respectively) and the corresponding Huang–Rhys factors (S_g and S_e, respectively) are not necessarily identical, the following expression can be obtained for the shape of a PL band at low temperature (Reshchikov et al., 2014a):

$$I^{PL} \propto \exp\left[-2S_e\left(\sqrt{\frac{E_0 + 0.5\hbar\Omega_e - \hbar\omega}{E_0 + 0.5\hbar\Omega_e - \hbar\omega_{max}}} - 1\right)^2\right], \tag{7}$$

where $\hbar\omega$ and $\hbar\omega_{max}$ are the photon energy and the PL band maximum position, respectively. For very strong electron–phonon coupling ($S \gg 1$), the shape of a PL band approaches the Gaussian shape, and its full width at half maximum (FWHM), W, in the low-temperature limit is

$$W(0) = \sqrt{8\ln 2}\frac{S_g\hbar\Omega_g}{\sqrt{S_e}}. \tag{8}$$

As the temperature increases, the FWHM increases as

$$W(T) = W(0)\sqrt{\coth\left(\frac{\hbar\Omega_e}{2kT}\right)}. \tag{9}$$

From the analysis of the temperature dependence of the PL band FWHM, the parameter $\hbar\Omega_e$ can be found. A more detailed analysis of the PL band shape and position also allows one to find S_g, S_e, $\hbar\Omega_g$, and E_0 (Reshchikov and Morkoç, 2005; Reshchikov et al., 2014a). Table 2 shows the main parameters for defects in GaN determined by PL methods. The parameters and the PL band nomenclature are taken from Reshchikov (2014c) and Reshchikov and Morkoç (2005) and updated with the latest data.

4.3. Main PL bands in GaN

The most common defect-related PL bands in undoped GaN are shown in Fig. 4. Below we will briefly describe their properties.

4.3.1 RL (1.81 eV) band in undoped GaN

One of the main PL bands in undoped GaN grown by the HVPE method is the RL band with a maximum at 1.81 eV at low temperatures (Fig. 5). The RL band has an unusual shape: a less steep high-energy side with an abrupt drop at ~2.36 eV. The sharp peak at 2.36 eV is identified as the ZPL of the RL band (Reshchikov et al., 2014b). The LO phonon replicas of the ZPL can be seen at distances of 91 and 182 meV. Two other sets of peaks with separations of 38 and 70 meV are attributed to local or quasi-local phonon

Table 2 Parameters of defect-related PL bands in GaN

PL band	$\hbar\omega_{max}$ (eV)	E_0 (eV)	E_A (eV)	C_{nA} ($cm^3\ s^{-1}$)	C_{pA} ($cm^3\ s^{-1}$)	$\hbar\Omega_e$ (meV)	$\hbar\Omega_g$ (meV)	S_e	S_g	Identity
UVL	3.27	3.27	0.2	8×10^{-13}	7×10^{-8}	91				
BL2	3.02	3.34	0.15			36, 91				
BL	2.88	3.10	0.4	2×10^{-13}	7×10^{-7}	36, 91	60	3	4.2	Zn_{Ga}
AL	2.56	~3.0	0.5	1×10^{-13}		38				
GL	2.4	2.9	0.54	3×10^{-12}	2×10^{-7}	41	50	8.5	10.4	C_N
GL2	2.35	2.85	0.43			23	21	27	24	V_N
YL	2.2	2.64	0.86	1×10^{-13}	6×10^{-7}	52	57	7.4	8.1	C_NO_N
YL	2.1	2.57	0.9			52	59	7.4	8.4	C_N
RL	1.81	2.36	1.13	2×10^{-14}		38, 91				
RL2	~1.8	~2.5				~30	~30	~20	~20	

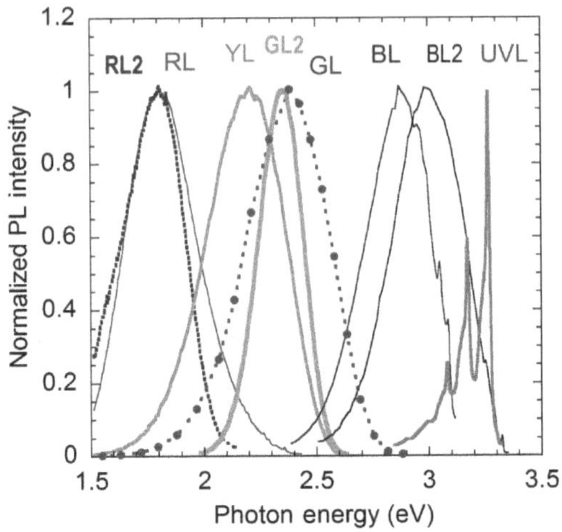

Figure 4 Main defect-related PL bands in undoped GaN at $T = 18$ K.

modes. The ZPL at 2.36 eV has been assigned to transitions from a shallow donor to a deep acceptor level located at 1.13 eV above the valence band. Such an assignment is supported by the transformation of the fine structure with increasing temperature (Fig. 5B): the donor–acceptor pair (DAP)-type lines decrease, while the eA-type lines emerge at $T > 40$ K.

The identity of the deep acceptor responsible for the RL band is unknown. Wang et al. (2007) observed the RL band with the same position and shape (yet without the fine structure) in GaN grown by HVPE, and they associated this band with the presence of C and O (with concentrations of about 10^{19} cm^{-3}). However, their attribution of the RL band to transitions from a deep donor, V_NC_N, to a deep acceptor, $V_{Ga}O_N$, is inconsistent with the first-principles calculations for the $V_{Ga}O_N$ acceptor (Section 2) and contradicts the above conclusion that the donor involved in the DAP transition is shallow.

4.3.2 RL2 (1.8 eV) and GL2 (2.35 eV) bands in Ga-rich, high-resistivity GaN

An RL band with a maximum at about 1.8 eV was observed in high-resistivity GaN grown by MBE in Ga-rich conditions (Reshchikov and Morkoç, 2005). It is labeled RL2, to distinguish it from the RL band described in Section 4.3.1, because the properties of the RL and RL2 bands are very different. The RL2 band commonly appears together with a green PL band, which has a maximum at 2.35 eV and is labeled GL2 (Fig. 6).

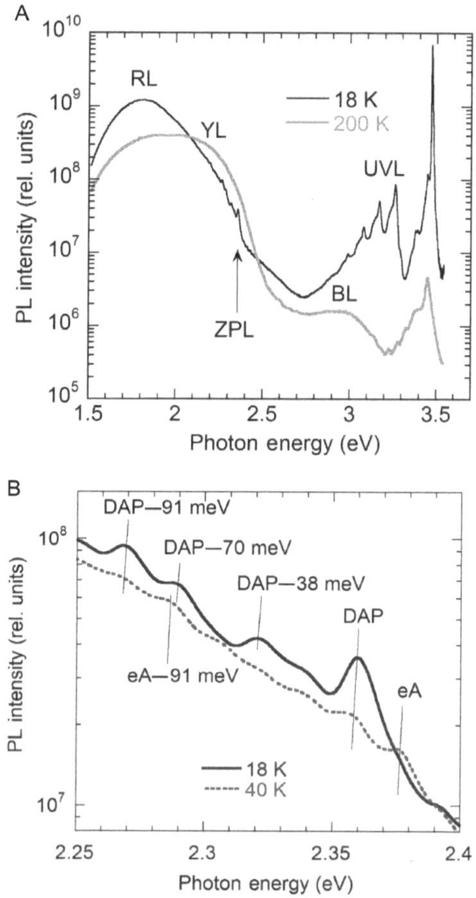

Figure 5 PL spectra from undoped GaN grown by HVPE at $P_{exc} = 1$ mW cm^{-2}. (A) PL spectra at 18 and 200 K. (B) Fine structure of the RL band at 18 and 40 K. *Reproduced with permission from Reshchikov et al. (2014b), Copyright 2014, AIP Publishing LLC.*

McNamara et al. (2013) have found that the RL2 and GL2 bands can also be observed in the near-surface layer of high-purity freestanding GaN mechanically polished and containing a great number of defects in the top 1-μm-thick layer (sample 1412.3 in Fig. 6).

The RL2 band has an FWHM of 300 meV at low temperatures. It is quenched at temperatures above 100 K with an activation energy of about 100 meV, and its intensity decreases by two orders of magnitude by 200 K. The position of the maximum of the RL2 band varies between 1.75 and 1.9 eV in different samples. However, the temperature behavior of PL is reproducible for all the samples containing this PL band. The

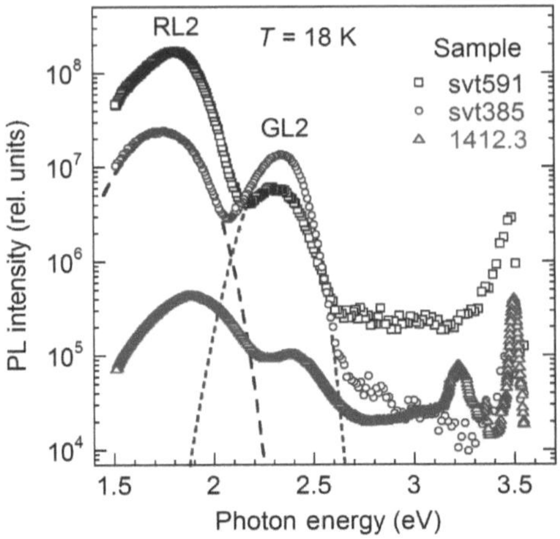

Figure 6 The RL2 and GL2 bands in GaN at $T = 18$ K and $P_{exc} = 1$ mW cm^{-2}. The RL2 band has a maximum at 1.80, 1.75, and 1.90 eV, and the GL2 band has a maximum at 2.34, 2.31, and 2.40 eV in MBE GaN (samples svt591 and svt385), and HVPE GaN (sample 1412.3), respectively. The dashed lines show fits for sample svt385 using Eq. (7) with the following parameters: $S_e = 23$, $\hbar\Omega_e = 30$ meV, $\hbar\omega_{max} = 1.75$ eV, $E_0 = 2.48$ eV (RL2 band), and $S_e = 26.5$, $\hbar\Omega_e = 23$ meV, $\hbar\omega_{max} = 2.34$ eV, $E_0 = 2.84$ eV (GL2 band).

RL2 band can be excited resonantly with photons having energy higher than 2.7 eV. The shape and position of the RL2 band are consistent with the shape and position of its PLE spectrum, provided that the ZPL is located at about 2.5 eV. The activation energy of ~0.1 eV is most probably associated with the thermal emission of electrons from an excited state to the conduction band. The decay of the RL2 band after the laser pulse is exponential at low temperatures, which can be explained by an internal transition from an excited state (a level at 0.1 eV below the conduction band) to the ground state (a level at ~0.9 eV above the valence band).

In high-resistivity GaN grown under Ga-rich conditions, a green PL band with a maximum at 2.35 eV is observed. It has unique properties and is called the GL2 band to distinguish it from the GL band with a maximum at 2.4 eV in high-purity GaN grown by HVPE. The GL2 band is relatively narrow (FWHM is 230 meV) for a deep defect, and its shape can be fit using Eq. (7) with $E_0 = 2.85$ eV and $S_e = 26.5$ (Reshchikov et al., 2014a). Such high Huang–Rhys factors ($S_g = 24$, $S_e = 26.5$) and small characteristic phonon energies ($\hbar\Omega_e = 23$ meV and $\hbar\Omega_g = 21$ meV) are typical for deep donors (Alkauskas et al., 2012).

Reshchikov et al. (2014a) have attributed the GL2 band to the isolated nitrogen vacancy, V_N. While for a majority of PL bands in n-type GaN, the

PL decay after a pulsed excitation is nonexponential at low temperatures because DAP-type transitions dominate, and the separations in these pairs are random, the decay of the GL2 band is exponential for a wide range of temperatures (15–100 K), with a characteristic PL lifetime of about 0.3 ms. To explain this unusual behavior, Reshchikov et al. (2014a) suggested that the observed GL2 band is caused by an internal transition, whereby the weakly localized electron collapses to the localized orbital, and the center converts from V_N^{2+} to V_N^{+}. The PL quenching with an activation energy of ~100 meV in the temperature range of 100–200 K is attributed to the thermal emission of electrons from the 0/+ level of V_N to the conduction band, whereas the PL quenching with an activation energy of ~400 meV at higher temperatures is attributed to the thermal emission of holes from the +/2+ level to the valence band.

4.3.3 YL (2.2 eV) band in MOCVD and MBE GaN

The YL band with a maximum at about 2.2 eV is the most common defect-related PL band in n-type GaN. Recent studies show that there are at least two YL bands originating from different defects. In this section, the YL band in MOCVD or MBE GaN will be considered, while a similar in shape and position YL band observed in HVPE GaN samples will be discussed in Section 4.3.4. Figure 7 shows the YL band in undoped and Si-doped GaN grown by MOCVD and MBE. It has the same shape and position in different samples. The YL band shape is fit using Eq. (7) with $S_e = 7.4$, $\hbar\omega_{max} = 2.21\,\text{eV}$, and $E_0 = 2.64$ eV. The inset of Fig. 7 shows the high-energy side of the YL band, where the onset at 2.61 eV can be identified as the position of the ZPL.

The characteristic excitation band maximum for the YL band was estimated to be at 3.19 eV (Ogino and Aoki, 1980) or 3.32 eV (Reshchikov et al., 2002b) from the analysis of the shape of the PL excitation spectrum. This energy corresponds to transition AB in the CC diagram shown in Fig. 3A. The transition level of the YL-related defect at low temperature can be estimated as 0.85–0.89 eV by taking the difference between the bandgap and the ZPL. It can also be found from the thermal quenching of the YL band.

The thermal quenching of the YL band begins at temperatures above 400 K (Fig. 8) and can be described with Eq. (3). For the majority of GaN samples grown by MOCVD and MBE (undoped and Si-doped), the activation energy of the defect responsible for the YL band is 0.85–0.9 eV (Reshchikov and Morkoç, 2005). However, in some samples,

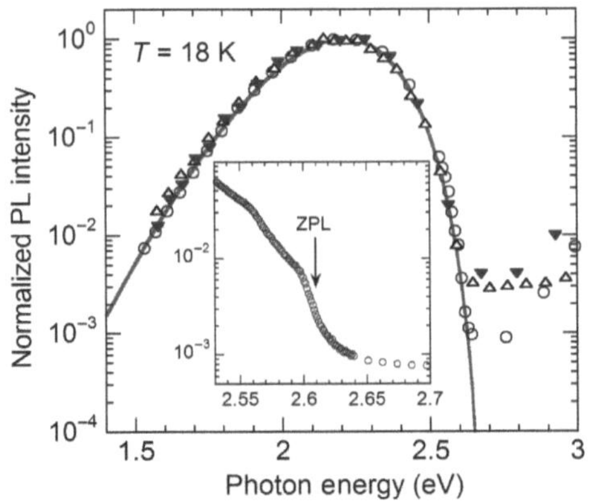

Figure 7 The YL band in undoped and Si-doped GaN grown by MBE and MOCVD. The symbols are experimental points (every 10th point is shown): empty circles—MOCVD GaN:Si ($n = 7 \times 10^{18}$ cm^{-3} at room temperature); empty triangles—MOCVD GaN ($n = 2 \times 10^{16}$ cm^{-3}); filled triangles—MBE GaN ($n = 3 \times 10^{16}$ cm^{-3}). The line is calculated using Eq. (7) with the following parameters: $S_e = 7.4$, $\hbar\Omega_e = 56$meV, $\hbar\omega_{max} = 2.21$ eV, and $E_0 = 2.64$ eV. The inset shows a high resolution of the high-energy edge with an apparent ZPL at 2.61 eV for the MOCVD GaN:Si sample ($n = 7 \times 10^{18}$ cm^{-3}).

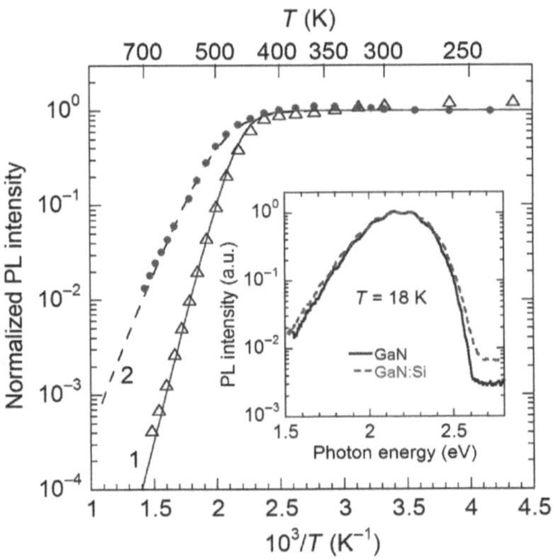

Figure 8 The temperature dependence of the YL intensity for MOCVD GaN, $n = 2 \times 10^{16}$ cm^{-3} (empty triangles) and MOCVD GaN:Si, $n = 1 \times 10^{19}$ cm^{-3} (filled circles). The solid and dashed lines are calculated using Eq. (7) with the following parameters: $E_A = 0.85$ eV, $\sigma_{pA} = 7 \times 10^{-14}$ cm^2, $\eta_0 = 0.05$, and $\tau = 0.35$ ms (curve 1 for undoped GaN); $E_A = 0.45$ eV, $\sigma_{pA} = 4 \times 10^{-17}$ cm^2, $\eta_0 = 0.05$, and $\tau = 14$ μs (curve 2 for Si-doped GaN); and $m_p = 0.8m_0$ and $g = 2$ for both. The inset shows normalized low-temperature PL spectra for these samples.

the activation energy found from the thermal quenching is much smaller (Fig. 8). Armitage et al. (2003) suggested that two types of defects (C-related and V_{Ga}-related) are responsible for the YL band with very similar shapes and positions.

The YL band is often attributed to V_{Ga} or the $V_{Ga}O_N$ complex. However, recent theoretical calculations have shown that the V_{Ga}-related defects are unlikely to be responsible for the YL band (Section 2). Namely, hybrid functional calculations predict that the best candidate for the YL band is either C_N (Lyons et al, 2010) or the C_NO_N complex (Demchenko et al., 2013). Of these two assignments, only in the first case is the saturation of the YL band with photogenerated holes (caused by transitions of electrons to the $-/0$ level of C_N) expected to cause an emergence of another PL band at higher energies (caused by transitions of electrons to the $0/+$ level of C_N). Since such a transformation has not been observed experimentally in GaN grown by MOCVD and MBE, we attribute the YL band to the C_NO_N complex in these samples (Reshchikov et al., 2014c).

In the literature, conclusions on the identity of the YL band are often drawn from seeking a correlation between the YL band intensity and some "independent" factors such as the growth conditions, the presence of structural defects, the concentrations of impurities, and vacancies detected by SIMS, PAS, EPR, and other techniques. Such an approach to identify a PL band is rarely successful, and sometimes it creates long-standing misconceptions. This is because many defects are nonradiative, and the intensity of a particular PL band depends not only on the concentration of the related defect but also on the concentrations of other radiative and nonradiative defects, which change with varying the growth conditions. Moreover, defect-related PL bands are often caused by complexes, not by isolated defects, and the concentration of defects causing a dominant PL band in a sample may be much smaller than the concentrations of several impurities present in this sample. For example, a very strong YL band (with QE of about 3%) is observed in a sample with the concentration of YL-related defects of 3×10^{15} cm^{-3} and the concentrations of Si, C, and O impurities between 3×10^{16} and 5×10^{16} cm^{-3} (Demchenko et al., 2013). Furthermore, in a set of MOCVD GaN samples with the concentration of C between 4×10^{16} and 2×10^{17} cm^{-3}, no correlation between the intensity of the YL band and the concentration of C could be established (Reshchikov et al., 2006a). Most likely, this is because, as the concentration of C increases, other channels of nonradiative recombination change in an uncontrolled way.

4.3.4 YL (2.1–2.2 eV) and GL (2.4 eV) bands in HVPE GaN

The GL band with a maximum at 2.4–2.5 eV is observed in high-quality, thick GaN grown by the HVPE technique (Reshchikov and Morkoç, 2005). A similar PL band is the dominant band in GaN grown from solution (Garces et al., 2010). The GL band emerges only at high excitation intensity, when the YL and RL bands saturate (Fig. 9A). The intensity of the GL band increases as a square of the excitation intensity, indicating that the PL band is caused by a multicharged defect, which may capture two holes before radiative recombination takes place (Reshchikov et al., 2002a). After a laser pulse, the GL intensity decreases exponentially at temperatures between 30 and 100 K, with the characteristic lifetime of 1–2 μs. In a time-resolved PL spectrum, the GL band vanishes at time delays exceeding $\sim 10^{-5}$ s, and subsequently the YL and RL bands can be observed (Fig. 9B).

Based on recent theoretical predictions (Section 2), the GL band can be assigned to transitions of electrons from the conduction band to the 0/+ level of the C_N defect located at ~0.5 eV above the valence band. In this case, the YL band can be caused by transitions via the −/0 level (E_v +1.0 eV) of the same defect. Note that for a long time the YL and GL bands were attributed to the 2−/− and −/0 transition levels of the $V_{Ga}O_N$ complex (Reshchikov and Morkoç, 2005). Such an attribution was mostly based on early theoretical predictions (Section 2) and results of PAS studies (Section 6). While some experimental data are in favor of such an attribution, other experimental data are inconsistent with it. For example, Buyanova et al. (1998) observed that electron irradiation of GaN results in the disappearance of the YL band and the appearance of new PL bands in the infrared region (broad overlapping bands within the 0.7–1.1 eV and a PL band with the ZPL at 0.88 eV followed by phonon replicas). Furthermore, Armitage et al. (2003) have found that the concentration of the V_{Ga}-containing defects in C-doped GaN ([C] = 2×10^{18} cm^{-3}) was below the detection limit ($<10^{16}$ cm^{-3}), whereas in a reference (undoped) GaN sample with [C] = 6×10^{16} cm^{-3}, the concentration of V_{Ga} was in the low 10^{17} cm^{-3} range, but the YL band was weaker.

4.3.5 BL (2.9 eV) band in undoped and Zn-doped GaN

The BL band with a maximum at 2.88 eV is attributed to transitions of electrons from the conduction band (or from shallow donors at low temperature) to the Zn_{Ga} acceptor (Demchenko and Reshchikov, 2013; Reshchikov and Morkoç, 2005). This PL band is often observed in undoped

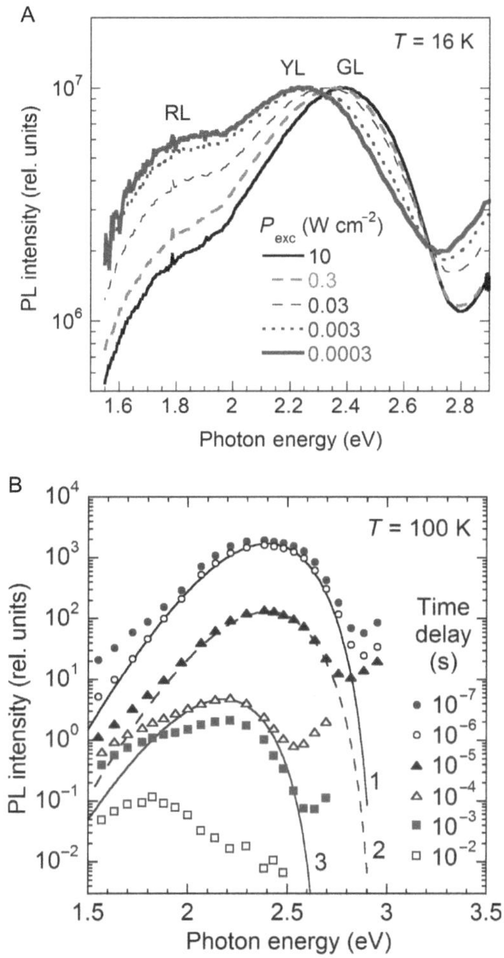

Figure 9 YL and GL bands in HVPE-grown freestanding GaN. (A) Transformation of the YL band into the GL band with increasing excitation intensity at $T = 16$ K. (B) PL spectra at selected time delays after a laser pulse at $T = 100$ K. Lines are calculated using Eq. (7) with the following parameters: $S_e = 8.5$, $\hbar\Omega_e = 41$ meV, $\hbar\omega_{max} = 2.4$ eV, and $E_0 = 2.91$ eV (curves 1 and 2 for the GL band), and $S_e = 7.4$, $\hbar\Omega_e = 52$ meV, $\hbar\omega_{max} = 2.18$ eV, and $E_0 = 2.63$ eV (curve 3 for the YL band).

GaN grown by the HVPE and MOCVD methods where Zn is introduced as a Ga source contaminant or due to a memory effect. The BL band is asymmetric, with a characteristic fine structure on the high-energy side. A sharp peak at 3.10 eV, identified as the ZPL, is followed by two sets of phonon replicas: an LO phonon mode (91 meV) and a local or quasi-local phonon mode (36 meV) (Reshchikov and Morkoç, 2005; Reshchikov

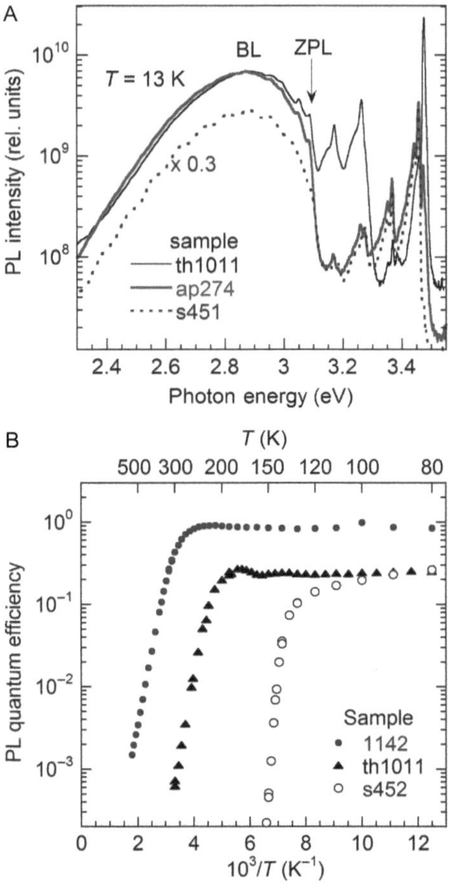

Figure 10 (A) Low-temperature PL spectra from undoped GaN (sample th1011), lightly Zn-doped ([Zn] = 1.7×10^{17} cm^{-3}) GaN (sample ap274), and heavily Zn-doped ([Zn] = 5×10^{18} cm^{-3}) GaN (sample s451). The BL band has a ZPL at 3.10 eV. (B) The temperature dependence of the BL band quantum efficiency for undoped GaN (sample th1011), Zn-doped ([Zn] = 2×10^{19} cm^{-3}) GaN (sample s452), and GaN codoped with Si and Zn ([Si] = 10^{19} cm^{-3}, [Zn] ≈ 6×10^{17} cm^{-3}) GaN (sample 1142).

et al., 2000). In GaN samples lightly doped with Zn, the BL band has the same asymmetrical shape and traces of the phonon structure (Fig. 10A). In GaN heavily doped with Zn, the fine structure disappears, and the PL band broadens. This may be in part due to potential fluctuations in high-resistivity GaN:Zn.

In conductive n-type GaN (both undoped and lightly doped with Zn), the BL band is quenched at temperatures above 200 K, with the activation energy of about 300–350 meV (Fig. 10B). The quenching is explained by the thermal emission of holes from the Zn_{Ga} acceptor to the valence band and their recapture by other recombination channels. When the QE of the BL band is very high (over 90%), the quenching region shifts to higher temperatures, because the Zn_{Ga} acceptors efficiently recapture the emitted holes (Reshchikov et al., 2012). According to Eq. (3), the shift of the quenching region to higher temperatures can also be caused by shorter PL lifetime in GaN with a higher concentration of free electrons.

For high-resistivity GaN:Zn (sample s452 in Fig. 10B), the quenching of PL is very different: the BL band intensity abruptly drops at a characteristic temperature T^*, which shifts to higher temperatures as the excitation intensity increases; i.e., the quenching is tunable by the excitation intensity. This abrupt and tunable quenching is explained by a sudden redirection of the recombination flow from the radiative channel (via the Zn_{Ga} acceptor) to a nonradiative channel (via some unknown deep donor) at temperatures when the thermal emission of holes from the Zn_{Ga} acceptor to the valence band becomes significant (Reshchikov, 2014b; Reshchikov et al., 2011). At $T < T^*$, the nonradiative donor is saturated with photogenerated electrons due to its large electron-capture cross-section. This causes a population inversion in the system (higher-energy levels in the gap are almost completely filled with electrons, while lower-energy levels have a deficit of electrons), a buildup of electrons in the conduction band, and a conversion of conductivity from p-type to n-type under UV illumination (Reshchikov, 2014d). At $T \approx T^*$, the concentration of the thermally emitted holes becomes approximately equal to the concentration of free holes generated by light. The additional holes are captured by the deep donor and unblock the nonradiative channel. A small change in temperature near T^* results in the abrupt reduction of the concentration of free electrons and electrons bound to the deep donor. At $T > T^*$, the system returns to the thermal equilibrium population, and conductive n-type converts into high-resistivity n-type or even p-type. T^* shifts to higher temperatures as the excitation intensity increases, because more thermal holes are needed to keep the "balance" with photogenerated holes. The abrupt and tunable quenching can be observed not only for Zn-doped GaN but also for other high-resistivity semiconductors (Reshchikov, 2014b). Two-step tunable quenching is sometimes observed and is explained in a similar way (Reshchikov, 2012b).

Very low concentrations of Zn can cause the BL band (Mohajerani et al, 2013). The concentration of the Zn_{Ga} acceptors can be controlled by the

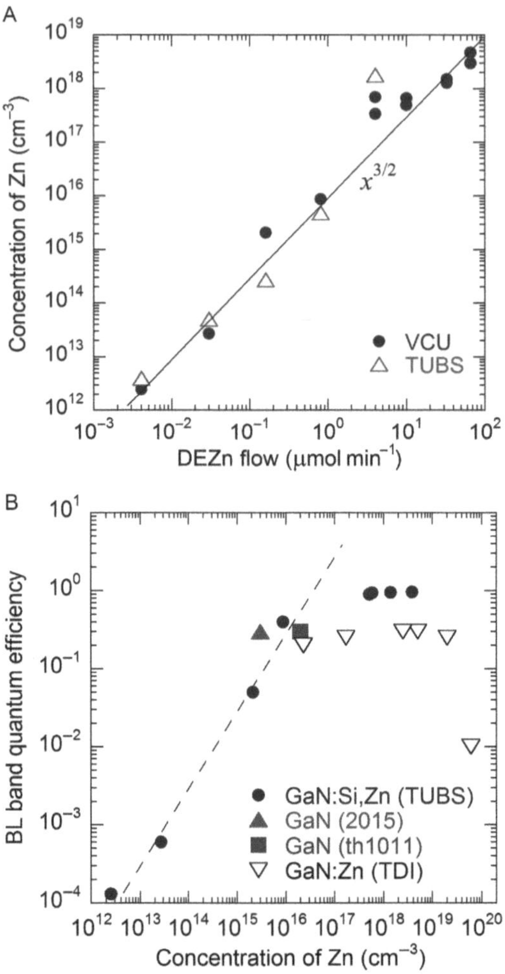

Figure 11 (A) The concentration of the Zn_{Ga} acceptors determined from PL as a function of the diethylzinc flow in the MOCVD growth. The PL measurements and analysis were independently carried out at VCU and at TUBS (Germany). (B) The dependence of the BL band quantum efficiency on the concentration of the Zn_{Ga} acceptors.

diethylzinc (DEZn) flow in the MOCVD growth (Fig. 11A). The BL intensity increases linearly with the concentration of Zn_{Ga} up to 10^{16} cm^{-3}, when it reaches a very high QE (more than 20%). The BL intensity is nearly independent with a further increase in Zn doping and even decreases for very high doping levels (Fig. 11B).

The Zn-related BL band should not be confused with the BL2 band (Section 4.3.6) or with the blue band (labeled BL_{Mg}) in Mg-doped

GaN. The blue band in GaN heavily doped with Mg is attributed to transitions from a deep donor (~0.4 eV below the conduction band) to the shallow Mg_{Ga} acceptor (Kamiura et al., 2005; Kaufmann et al., 1999; Reshchikov et al., 1999). Such an attribution explains the very large shift of the blue band with increasing excitation intensity (from ~2.7 to ~3.0 eV), along with other properties of this band. The deep donor is sometimes attributed to the V_NMg_{Ga} complex. However, it is known that the BL_{Mg} band can be greatly enhanced by thermal annealing, whereas the concentration of the V_NMg_{Ga} complexes significantly decreases after annealing at $T > 500$ °C (Hautakangas et al., 2003). Since the BL_{Mg} band is commonly observed in GaN:Mg grown by MOCVD and rarely observed in the MBE-grown GaN:Mg, one may expect that the deep donor involved in the BL_{Mg} band contains hydrogen as a part of the defect. Such an assumption is supported by the annealing studies in H plasma, according to which the BL_{Mg} band is greatly enhanced by remote plasma treatment, with plasma containing atomic hydrogen (Kamiura et al., 2005).

4.3.6 BL2 (3.0 eV) band in high-resistivity GaN

A broad blue band with a maximum at 3.0 eV (labeled BL2) can be found in PL spectra from high-resistivity GaN. It is often observed for C-doped GaN grown by MOCVD, along with the YL band (Polyakov et al., 1996; Seager et al., 2002, 2004). It can also be seen in undoped or Fe-doped high-resistivity GaN grown by MOCVD or HVPE (Reshchikov and Morkoç, 2006; Reshchikov et al., 2006b). In some GaN samples (undoped, C-doped, and Fe-doped), the BL2 band has a fine structure with the ZPL at 3.33–3.34 eV followed by a few phonon replicas with the dominant phonon modes having energies of 36 and 91 meV (Fig. 12A). From the position of the ZPL, it was suggested that the BL2 band is caused by transitions of electrons from the conduction band or from an excited state near it to a level located at 0.15 eV above the valence band (Reshchikov and Morkoç, 2006). In agreement with this assumption, the BL2 band is quenched above 75 K with an activation energy of about 150 meV (Reshchikov et al., 2006b; Seager et al., 2004).

An important feature of the BL2 band is its bleaching during continuous UV illumination (Reshchikov et al., 2006b). Simultaneously with this bleaching, the YL band intensity rises (Fig. 12B). It was suggested that the BL2 band is associated with a defect complex containing hydrogen, and the bleaching is caused by a recombination-assisted dissociation of this

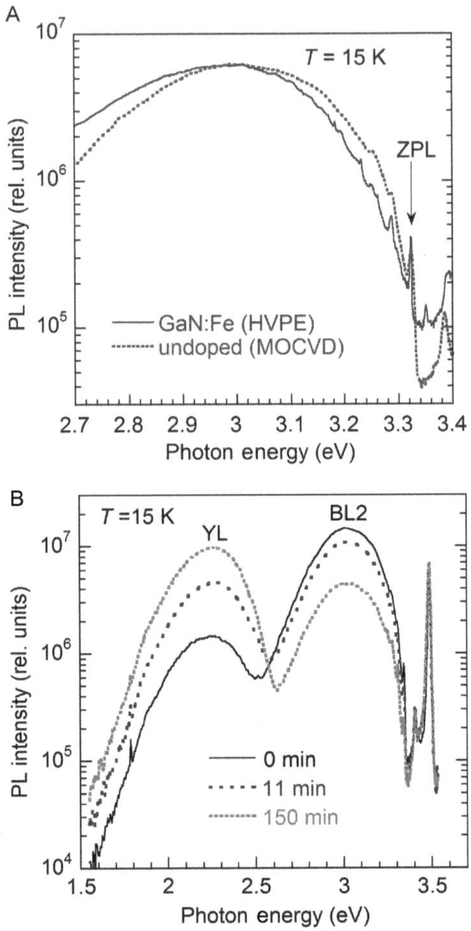

Figure 12 The BL2 band in high-resistivity GaN. (A) The fine structure of the BL2 band with the ZPL at 3.33 eV. (B) Evolution of the PL spectrum in MOCVD-grown GaN measured at $P_{exc} \approx 1$ mW cm^{-2} after exposure of the sample to the laser light with $P_{exc} = 0.3$ W cm^{-2} for selected times.

complex under UV exposure (Reshchikov and Morkoç, 2005). It is tempting to assign the BL2 band to the C_N–H or C_NO_N–H complexes, which dissociate under UV irradiation to form the C_N or C_NO_N defect. The released hydrogen remains in the crystal lattice and can be trapped back by the carbon-containing defects after keeping the sample in dark at high temperatures for an extended period of time. Attribution of the BL2 band to a complex containing carbon and hydrogen is consistent with the fact that the BL2 band is observed in GaN grown by either MOCVD or HVPE, the techniques in which relatively high concentrations of C and H are typical. The fact that the BL2 band in most cases appears in high-resistivity GaN may indicate that the defect responsible for this PL band has a lower formation energy when the Fermi level is far from the conduction band.

4.3.7 UVL (3.27 eV) band

The UVL band has a main peak at about 3.27 eV followed by a few LO phonon replicas with decreasing intensities. At low temperatures ($T < 20$ K), the UVL band is caused by transitions of electrons from shallow donors (Si_{Ga} and O_N) to an unidentified shallow acceptor having an ionization energy of about 0.2 eV. For this reason, the UVL band is often called the shallow DAP band. Due to the random distribution of distances between the DAP, two characteristic effects are commonly observed at low temperatures: the UVL peaks shift to higher photon energies (by up to ~10 meV) with increasing excitation intensity, and they shift to lower photon energies (by 10–15 meV) with increasing time delay in the case of a pulse excitation (Reshchikov and Morkoç, 2005). Another characteristic feature of DAP-type transitions is a nonexponential PL decay after a pulse excitation. For such PL decay, no characteristic PL lifetime can be found, and the time dependence of the PL decay is close to the t^{-1} dependence for a wide range of time delays (Reshchikov et al., 2003).

As the temperature increases from 20 to 50 K, more and more electrons are thermally emitted from the shallow donors to the conduction band. As a result, the DAP-type UVL band transforms into the eA-type UVL band, which has a very similar shape, but is shifted to higher energies by 10–20 meV. The shift is equal to the effective ionization energy of the shallow donors, which decreases with increasing concentration of shallow donors. Since the eA-type UVL band is caused by transitions of electrons from the conduction band to the shallow acceptor, the decay of PL is exponential and can be fit with Eq. (6) in conductive n-type GaN. Then, the concentration of free electrons can be found if the electron-capture coefficient C_{nA} is known (Reshchikov, 2014a). At temperatures above 100 K, the UVL band is quenched with an activation energy of about 180 meV. The quenching is caused by the thermal emission of holes from the shallow acceptor to the valence band. The PL lifetime decreases very similarly to the PL intensity, in agreement with theory (Reshchikov, 2014a).

The identity of the shallow acceptor responsible for the UVL band is uncertain. In some samples, it may be Mg_{Ga} if the contamination with Mg during growth cannot be excluded (Monemar et al., 2002). In contrast to early theoretical predictions and a number of experimental reports (Sections 2 and 5), the shallow acceptor is not related to carbon. Indeed, in MOCVD GaN containing a high concentration of C, the intensity of the UVL band is extremely low, and the concentration of the related defect is on the order of 10^{12} cm^{-3} (Reshchikov et al., 2006a). Armitage et al. (2005) also noticed that the UVL band is surprisingly weak in C-doped GaN samples.

5. POINT DEFECTS REVEALED BY DLTS AND OTHER CAPACITANCE TECHNIQUES

The DLTS method was introduced by Lang (1974). It allows for the detection of electron and hole traps in a semiconductor, the determination of the concentration of traps and their properties such as the energy levels, the capture cross-section, and the presence of a potential barrier for the capture. In contrast to PL, DLTS is capable of detecting both radiative and nonradiative traps.

The traditional DLTS is a high-frequency capacitance technique, which allows detecting defects in the depletion region of a Schottky diode or a p–n junction. For example, electron traps in the depletion region are emptied under a reverse bias. A forward bias is applied to reduce the band bending and fill the traps with electrons (a charging or filling pulse). The capacitance of the diode changes, and this change is proportional to the trap concentration. After the pulse, the electron occupation of traps returns to equilibrium, and the capacitance returns to the initial baseline. The capacitance transients depend on the energy level of a trap and its electron-capture cross-section. According to Lang (1974), the emission rates of electrons from a trap to the conduction band, e_n, and of holes from a trap to the valence band, e_p, are

$$e_n = \sigma_n \langle v_n \rangle N_c g^{-1} \exp\left(\frac{-\Delta E}{kT}\right), \tag{10}$$

and

$$e_p = \sigma_p \langle v_p \rangle N_v g^{-1} \exp\left(\frac{-\Delta E}{kT}\right), \tag{11}$$

respectively. Here, $\Delta E = E_c - E_T$ for the emission of electrons and $\Delta E = E_T - E_v$ for the emission of holes and E_T is the transition level energy for the trap. Equations (11) and (12) are identical to Eqs. (2) and (1), respectively, since $Q_D \equiv e_n$ and $Q_A \equiv e_p$ are just different labels of the same quantities. The characteristic time of the electron (hole) emission is defined as $\tau_n \equiv e_n^{-1}$ ($\tau_p \equiv e_p^{-1}$).

For a wide-bandgap semiconductor such as GaN, only majority carriers are usually detected by the traditional DLTS; i.e., only electron traps are revealed in n-type GaN. The capacitance change has a maximum at a temperature (T_{max}) when τ_n becomes equal to

$$\tau_{max} = (t_1 - t_2)\left[\ln\left(\frac{t_1}{t_2}\right)\right]^{-1}, \tag{12}$$

where the times t_1 and t_2 determine the rate window for a DLTS thermal scan. By changing the t_1 and t_2 constants, the dependence of τ_{max} on T_{max} can be plotted to find parameters of the trap σ_n and E_T. Traditionally, to account for the temperature dependence of $\langle v_n \rangle$ and N_c, the dependence of $\ln(e_n/T^2)$ on T^{-1} is plotted. Then a slope of the dependence gives ΔE, and the position of the dependence gives σ_n. By varying the reverse bias before applying the filling pulse, the trap distribution in depth can be determined, so that one can distinguish between the near-surface and bulk traps. Often the near-surface traps are induced during the diode fabrication, e.g., by electron beam.

A careful analysis of the capacitance transients at fixed temperatures, a method called the isothermal DLTS, may allow more accurate determination of the trap parameters. The resolution and sensitivity of the DLTS method can be improved by using its modifications such as Laplace transform DLTS (Dyba et al., 2011) or deep-level transient Fourier spectroscopy (Asghar et al., 2006). Minority carrier traps in wide-bandgap semiconductors can be detected by ODLTS, also called the minority carrier transient spectroscopy (MCTS). In this method, an optical pulse (below or above bandgap) with an intensity and duration needed to fill the minority carrier traps is applied. Similar techniques, in which the photocurrent is measured instead of photocapacitance, include the optical transient current spectroscopy (Polyakov et al., 1998a), PICTS (Polyakov et al., 1998b), or optical-current DLTS (Calleja et al., 1997). Note that not only above-bandgap light can be used to fill hole traps but also below-bandgap light (Polyakov et al., 2011) or electron beam (Polyakov et al., 1998b). Another capacitance method for the detection of hole traps in n-type GaN is DLOS. In this method, the incident light wavelength is changed and the change of the diode capacitance is plotted as a function of the photon energy. The trap energy level can be estimated in this method from a threshold of the spectral dependence. When the threshold is not abrupt and signals from several defects overlap, the results of the fit may be less reliable.

5.1. Electron traps

Several electron traps in GaN have been detected by many research groups using the DLTS technique. A detailed analysis of the electron traps in GaN can be found elsewhere (Morkoç, 2008). The main electron traps in GaN are E_1 at $E_c-(0.14-0.27)$ eV, E_2 at $E_c-(0.49-0.60)$ eV, E_3 at $E_c-(0.62-0.67)$ eV, E_4 at $E_c-(0.81-0.85)$ eV, and E_5 at $E_c-(1.07-1.44)$ eV. Note that there is very large scatter in the reported parameters and the suggested names for the traps.

Sometimes, a DLTS peak is assumed to be caused by a single trap, whereas a more detailed analysis of its shape at different pulse widths reveals two or more components contributed by different defects. In such a case, significant errors in the defect parameters are possible. For example, Polenta et al. (2000) suggested that the signal observed at about 120 K in the DLTS spectra of GaN irradiated with electrons (the E trap with apparent parameters of $\Delta E = 0.18\,\text{eV}$ and $\sigma_n = 2\times10^{-15}\,\text{cm}^2$) is in fact caused by a superposition of two traps, each with $\Delta E \approx 0.06\,\text{eV}$ and $\sigma_n \approx 10^{-20}\,\text{cm}^2$. The relatively shallow donors have been preliminary assigned to V_N or the V_N-containing complexes.

Significant errors in ΔE and σ_n may originate from the narrow range of time windows, when a defect is recognized in a DLTS spectrum as a peak at T_{max}, but the change of T_{max} with τ_{max} contains large uncertainty due to peak broadness or noise. An error in the determination of the slope in the Arrhenius plot containing only 4–5 points (an error in ΔE) may lead to an error in σ_n by orders of magnitude. This sort of error results in a large dispersion in the reported data for well-identified traps. Morkoç (2008) illustrates this with an example where different research groups reported very different parameters (in the ranges of $\Delta E \approx 0.5-0.8\,\text{eV}$ and $\sigma_n \approx 10^{-16}-10^{-12}\,\text{cm}^2$) for apparently the same electron trap responsible for a DLTS peak at about 350 K. This is because in Eq. (10), T_{max} and e_n are determined, but ΔE and σ_n are both unknown. Then, a small error in ΔE causes a large error in σ_n.

The electron traps are not expected to be optically and electrically active in n-type GaN unless the role of a depletion region is significant. Indeed, the above traps are donors with low concentration and relatively large ionization energy; they are filled with electrons in dark (i.e., neutral) and practically do not affect the electron conductivity. They cannot be observed in PL experiments because defects with levels in the lower half of the bandgap, especially acceptors, capture photogenerated holes much faster than donors near the conduction band. Most likely, all electron traps revealed by DLTS are inefficient nonradiative defects or traps which, being saturated with charge carriers, remain inactive.

5.2. Hole traps

5.2.1 Optical DLTS

Hole traps are important as defects reducing the efficiency of light-emitting devices and affecting the electrical parameters of high-power devices. They are also likely to be detected in PL measurements, because most of the PL bands in n-type GaN are caused by transitions of electrons from the conduction band to different acceptors in the lower half of the bandgap (Reshchikov and Morkoç, 2005). Several hole traps have been detected in n-type GaN, mostly by ODLTS and DLOS techniques. Since there is no consensus about the trap names in the literature, we will choose the notation given by Polyakov et al. (2011) (Table 3).

The dominant hole trap in MOCVD-grown n-type GaN is a trap appearing at about 330 K in the ODLTS spectrum, denoted as H1 by Polyakov et al. (2011). It has an energy level at E_V +0.85 eV, and its appearance and concentration correlate with the YL band (Calleja et al., 1997; Honda et al.,

Table 3 Hole traps in n-type GaN

Trap	Typical T_{max}	Nominal E_A (eV)	Range of E_A (eV)	Other names, similar defects
Shallow acceptor	105 K	0.25	0.2–0.25	H(0.25) (Auret et al., 2004; Polyakov et al., 1998a)
H2	205 K[a]	0.55	0.5–0.6	
H3	235 K[a]	0.65	0.6–0.7	
H4	270 K[a]	0.80	0.7–0.9	
H1	330 K[a]	0.85	0.81–0.95	H1 (Honda et al., 2012; Tokuda et al., 2011), H1a (Kamyczek et al., 2012), H(085) (Auret et al., 2004), and H2 (Polyakov et al., 1998a)
H5	400 K[b]	1.2	1.1–1.2	
H6	400 K[b]	1.05	0.95–1.05	

[a]For time windows close to 1 s.
[b]For time windows close to 10 s.

2012; Kamyczek et al., 2012; Polyakov et al., 1998a,b). This trap was detected in undoped and Si-doped GaN by many research groups, and its energy level in different reports varies between 0.8 and 0.95 eV above the valence band (Auret et al., 2004; Polyakov et al., 2011). The scatter in the value of ΔE is often caused by a broadness of the DLTS peak and by the use of a narrow range of time windows. Figure 13 shows the data for the H1 hole trap from three reports. A fit for the compound data, which encompasses more than four orders of magnitude for the characteristic times of the hole emission, yields $\Delta E = 0.85$ eV and $\sigma_p = 7 \times 10^{-14}$ cm^2 for the H1 trap (the solid line). The determined value for the hole-capture cross-section greatly depends on the value of ΔE, because the latter appears in the power of the exponent in Eq. (11). The calculated σ_p also depends on the values of the hole effective mass m_p and the degeneracy of the trap level g used in the expression for the hole emission rate given by Eq. (11). For example, the use

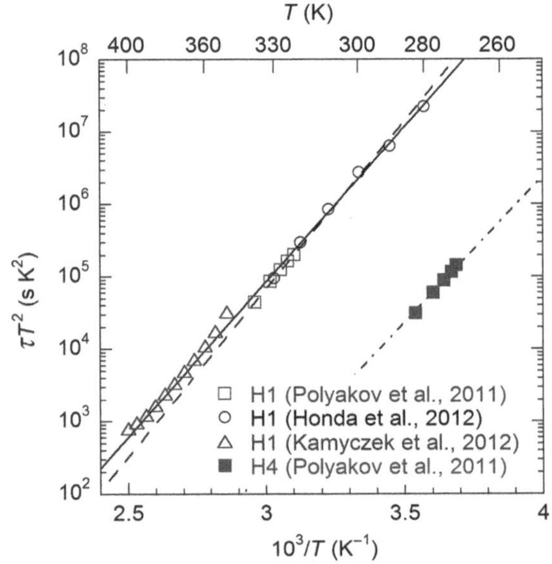

Figure 13 Arrhenius plot for the H1 trap (open symbols) and the H4 trap (closed squares) from the ODLTS (or MCTS) (Honda et al., 2012; Polyakov et al., 2011) and Laplace DLTS (Kamyczek et al., 2012) measurements. The solid and dashed lines are fits using Eq. (11) with the following parameters: $\Delta E = 0.85$eV and $\sigma_p = 7 \times 10^{-14}$cm^2 (solid line), $\Delta E = 0.92$eV and $\sigma_p = 1 \times 10^{-12}$cm^2 (dashed line). The dash-dotted line for the H4 trap is a fit using Eq. (11) with the following parameters: $\Delta E = 0.8$eV and $\sigma_p = 4.6 \times 10^{-12}$cm^2. $m_p = 0.8m_0$, $g = 2$ in all fits. *Reproduced with permission from Polyakov et al. (2011), Copyright 2011, AIP Publishing LLC; Honda et al. (2012), Copyright 2012, The Japan Society of Applied Physics; and Kamyczek et al. (2012), Copyright 2012, AIP Publishing LLC.*

of $m_p = 1.1m_0$ and $g=1$ (Polyakov et al., 1998b) instead of $m_p = 0.8m_0$ and $g=2$ (the fit in Fig. 13), would lead to a higher value of σ_p by a factor of 2.6. Interestingly, the mean values of ΔE and σ_p from several reports (Auret et al., 2004; Honda et al., 2012; Kamyczek et al., 2012; Muret et al., 2002; Polyakov et al., 1998b, 2011; Tokuda et al., 2011; $\Delta E = 0.86$ eV and $\sigma_p = 8 \times 10^{-14}$ cm^2) are very close to the ones found from the fit shown in Fig. 13, whereas ΔE and σ_p in these works vary in the ranges of 0.81–0.92 eV and $1.4 \times 10^{-14} - 1.7 \times 10^{-13}$ cm^2, respectively, and different parameters m_p and g are used.

A hole trap with a similar to the H1 trap energy level but with a different capture cross-section was observed as a peak at 270 K in thick GaN layers grown by HVPE. The new trap is labeled H4 in Lee et al. (2012, 2014) and Polyakov et al. (2011). It is clear from Fig. 13 that the H1 and H4 traps are different defects, while their energy levels are close. From the fit shown in Fig. 13, by taking $\Delta E = 0.8$ eV, we have found that the H4 trap has a very large capture cross-section ($\sigma_p = 4.6 \times 10^{-12}$ cm^2), which explains the large temperature separation between the Arrhenius plots for the two traps with similar ΔE.

It is important to note that in the MOCVD-grown GaN the H1 trap is the dominant hole trap and the H4 trap is absent, whereas in thick GaN layers grown by HVPE the H1 trap is absent or is present with a very low concentration (lower than 10^{13} cm^{-3}), much lower than that for the H4 trap ($\sim 10^{14}$ cm^{-3}; Polyakov et al., 2002, 2011, 2014). Many researchers noted that the H1 trap correlates with the appearance of the YL band in MOCVD-grown samples. From the quenching of the YL band in undoped GaN grown by MOCVD, we have determined exactly the same parameters for the defect responsible for the YL band (Fig. 8). In thick GaN grown by HVPE, the YL band is rarely detected. The YL band can be detected in these samples in a time-resolved PL spectrum (Section 4.3.4) and could be related to the same defect as the H4 trap.

The H5 trap at $T_{max} \approx 400$ K is the dominant trap in freestanding GaN grown by HVPE (Lee et al., 2011a,b, 2014; Polyakov et al., 2011). The concentration of the H5 trap is in the mid 10^{15} cm^{-3} in bulk, undoped GaN grown by HVPE at Kyma Inc., whereas the concentration of other hole traps in these samples (H2, H3, and H4) is an order of magnitude lower (Polyakov et al., 2011). The H5 trap is also observed in MOCVD-grown GaN, but only after irradiation with neutrons and following annealing at 1000 °C (Lee et al., 2011a). The concentration of these defects increased linearly with the neutron fluence and reached 5×10^{15} cm^{-3} for the fluence of 2×10^{16} cm^{-2}.

Polyakov et al. (2011) and Lee et al. (2011a) attributed the H5 trap to the $V_{Ga}O_N$ complex, because Ga vacancies are expected to be created by neutron irradiation, while the isolated V_{Ga} cannot survive annealing at such high temperatures and is expected to form a complex with O_N (Section 6). The photoionization spectrum for the H5 trap has been obtained by measuring the ODLTS signal with below-bandgap illumination (Lee et al., 2012, 2014). It showed a threshold at 2.1–2.2 eV and a plateau for photon energies 2.5–3.1 eV. From the position of the threshold, the transition level for this defect can be estimated to be at about 1.3 eV. However, it is possible that the ODLTS signal with below-bandgap illumination is related to another trap, labeled H6 (Table 3), because it has a slightly different T_{max} than the H5 signal observed with above-bandgap illumination (Lee et al., 2014).

5.2.2 Photoionization spectra

Other popular capacitance-based techniques include SSPC and transient photocapacitance spectroscopy (or DLOS) (Chantre et al., 1981). In the SSPC method, a photoinduced change in the capacitance of a p–n or Schottky diode, ΔC, is plotted as a function of photon energy. For n-type GaN, the photons with below-bandgap energy excite electrons from defect levels to the conduction band and produce capacitance steps in an SSPC spectrum with onsets corresponding to the energy difference between the conduction band minimum and defect energy levels. From the $\Delta C/C_0$ ratio, where C_0 is the dark capacitance, the concentration of a defect can be determined similar to how it is determined in the DLTS method. In the DLOS technique, the spectral dependence of the optical cross-section, $\sigma^{o}(\hbar\omega)$, is obtained from the time derivative of the photocapacitance transient. The SSPC and DLOS spectra are expected to have similar shapes and the same onset or threshold because both these methods, as well as photoconductivity, optical transmission, and PL excitation spectra, measure a photoionization spectrum.

Figure 14 shows examples of the SSPC and DLOS spectra from GaN. To find the defect energy level, a photoionization spectrum is fit with a theoretical model. The commonly used Lucovsky (1965) model predicts that

$$\sigma^{o}(\hbar\omega) \propto \frac{(\hbar\omega - E_0)^{3/2}}{(\hbar\omega)^3}. \tag{13}$$

For defects in n-type GaN, $E_0 = E_g - E_T$, where E_T is the defect transition level energy measured from the valence band maximum. Note that the

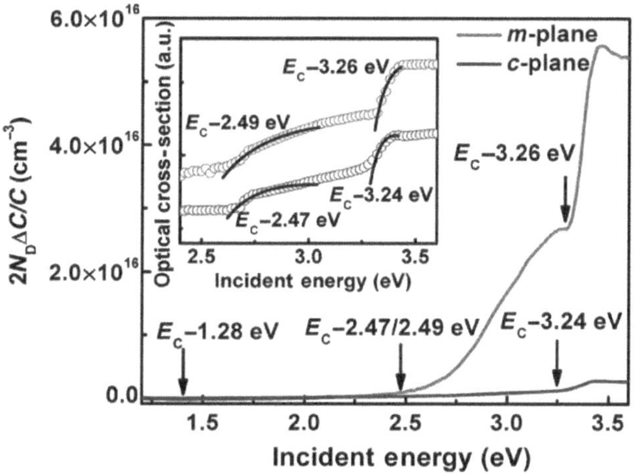

Figure 14 SSPC spectra of *m*-plane and *c*-plane MBE GaN. The concentration of carbon at 0.2 μm depth is 2×10^{17} and $\sim 1 \times 10^{16}$ cm^{-3} for the *m*-plane and *c*-plane samples, respectively. The inset shows the optical cross-section obtained from DLOS, with fits to the model of Chantre et al. (1981). *Reproduced with permission from Zhang et al. (2012), Copyright 2012, AIP Publishing LLC.*

Lukovsky model does not account for electron–phonon coupling and ignores the thermal broadening. A model of Chantre et al. (1981) is sometimes used to account for these effects. In terms of the CC model (Fig. 3A), the change in capacitance is caused by a resonant excitation of a defect (transition AB). The photoionization spectrum is expected to have an onset at $\hbar\omega = E_0$ (or higher energies) and reach the maximum at $\hbar\omega = E_0 + \Delta E$, where ΔE is called the Frank–Condon shift (Fig. 3A). Note that when a photocapacitance spectrum is measured at room temperature, the thresholds are blurred and the transition of electrons from a shallow acceptor to the conduction band can be confused with the band-to-band transition.

One of the main traps identified with the SSPC, DLOS, and DLTS techniques is the H1 trap, which we attributed above to the C_NO_N complex. The defect with the level at E_v +0.9 eV causes an onset at 2.5–2.6 eV in the SSPC and DLOS spectra. The E_c −2.6 eV signal greatly increases with increasing concentration of carbon in GaN samples (Fig. 14) (Armstrong et al., 2004, 2005a,b; Zhang et al., 2012). It would be logical to assign the E_c −2.6 eV signal to a C-related defect. However, Armstrong et al. (2004), Arehart et al. (2008), and Zhang et al. (2012) assigned the E_c −2.6 eV signal to the V_{Ga} defect.

A correlation between the E_c −3.28 eV signal and the concentration of carbon in GaN has been reported in several works (Armstrong et al., 2004,

2005a,b; Zhang et al., 2012). Based on this observation, the authors of these works assigned the shallow acceptor with a level at about 0.16 eV to the C_N defect. However, such an attribution contradicts recent theoretical calculations, which predict that the C_N acceptor is a deep defect (Section 2). Moreover, it is known from PL studies that the intensity of the 3.27 eV band, which is attributed to transitions from the conduction band to an unidentified shallow acceptor, dramatically decreases with carbon doping (Section 4.3.7). It appears that the onset at 3.28 eV in the SSPC and DLOS spectra is just a tail from the band-to-band excitation. Indeed, the SSPC and DLOS signals at $\hbar\omega > E_g$ increase in a similar way with C doping (Fig. 14). The signal appearing to have an onset at 3.26–3.30 eV might be the Urbach tail (Urbach, 1953) in the room-temperature photoionization spectrum. The increase of photocapacitance at $\hbar\omega > E_g$ with increasing C concentration can be attributed, at least partially, to higher resistivity and an increased depletion region width in C-doped GaN.

5.2.3 Identification of hole traps

The H1 trap appears to be the same defect as the one responsible for the YL band in MOCVD-grown GaN. In the past, the YL band and the H1 trap were often attributed to a V_{Ga}-related defect (Section 2). In contrast with this assignment, Auret et al. (2004) have found that irradiation of GaN with 1.8 MeV protons does not increase the concentration of the H1 trap at 0.85 eV. Demchenko et al. (2013) identify this defect as the C_NO_N complex, although Lyons et al. (2010) suggest that it is an isolated C_N defect. The attribution of the H1 trap to a C-containing defect agrees with results of Honda et al. (2012), where the concentration of the 0.86 eV hole trap increased from 1.8×10^{14} to 2.2×10^{15} cm^{-3} with increasing carbon concentration from $(2–5) \times 10^{16}$ to 1×10^{17} cm^{-3} in MOCVD-grown GaN.

The H4 trap could be the isolated C_N defect and be responsible for the YL band in HVPE GaN. Such an attribution can explain why this trap appears only in thick GaN grown by HVPE and why the capture cross-section for this trap is much larger than the cross-section for the H1 trap: the hole-capture cross-section of deep acceptors (C_N) is expected to be larger than the one of deep donors (C_NO_N). The H2 (0.55 eV) or H3 (0.6–0.65 eV) traps are most likely nonradiative defects.

According to the DLTS studies, a hole trap H5 with a level near 1.2 eV can be the $V_{Ga}O_N$ complex because the trap appeared after irradiation with neutrons (Lee et al., 2011a). Also, it is the dominant defect in freestanding

GaN where its concentration is (2–4) × 10^{15} cm^{-3} (Polyakov et al., 2011). According to recent calculations (Section 2), transitions of electrons from the conduction band to the 2 −/− level of the $V_{Ga}O_N$ complex should produce a PL band in the infrared region. An alternative explanation is that the $V_{Ga}O_N$ complex is a nonradiative defect.

6. VACANCY-RELATED DEFECTS REVEALED BY PAS

The PAS technique was successful in detecting V_{Ga} and the $V_{Ga}O_N$ complexes in GaN. Positrons get trapped at neutral and negatively charged vacancies because of the missing positive charge. As a result, the positron lifetime τ increases, and the positron–electron–momentum distribution (described with parameters S and W) narrows. If only one vacancy-type defect is present, the positron lifetime spectra can be analyzed by fitting a sum of exponential decay components, $(I_1/\tau_V)\exp(-t/\tau_V) + (I_2/\tau_b)\exp(-t/\tau_b)$ convoluted with the Gaussian resolution function. The intensities, I_1 and I_2, and the lifetime components, τ_b and τ_V, are used to calculate the average positron lifetime as $\tau_{av} = I_1\tau_b + I_2\tau_V$. The positron lifetime in the defect-free GaN lattice ($\tau_b = 160 \pm 1$ ps) and the lifetime of positrons trapped at negatively charged V_{Ga} or V_{Ga}-containing complexes ($\tau_V = 235 \pm 5$ ps) have been determined using GaN samples with negligible and high concentrations of the V_{Ga}-containing defects, respectively. Then, the concentration of the V_{Ga}-containing defects, c_V, can be found as Oila et al. (2001), Reurings and Tuomisto (2007), and Saarinen et al., (1997, 1998)

$$c_V = \frac{N_{at}}{\mu_V \tau_b} \frac{\tau - \tau_b}{\tau_V - \tau} = \frac{N_{at}}{\mu_V \tau_b} \frac{S - S_b}{S_V - S} = \frac{N_{at}}{\mu_V \tau_b} \frac{W - W_b}{W_V - W}, \tag{14}$$

where N_{at} is the atomic density of GaN (8.775 × 10^{22} cm^{-3}) and μ_V is the specific trapping coefficient for the V_{Ga}-containing defect. This coefficient was assumed to be 1 × 10^{15} s^{-1} in early works (Oila et al., 2001; Saarinen et al., 1997) and adjusted to 3 × 10^{15} s^{-1} in later works (Reurings and Tuomisto, 2007; Saarinen et al., 2001a). The detection limit for the V_{Ga}-containing defects in GaN can be as low as 1 × 10^{15} cm^{-3} (Oila et al., 2003).

The PAS studies indicate that no Ga vacancies are present in Mg-doped GaN, but V_{Ga} is commonly present in n-type GaN, where its concentration can reach 10^{19} cm^{-3} (Oila et al., 2001; Saarinen et al., 1997). The concentration of the V_{Ga}-containing defects increases with O-doping but decreases with Si-doping (Oila et al., 2001). The annealing experiments clearly indicate that the isolated V_{Ga} becomes mobile at temperatures exceeding 600 K, and the stable form of the V_{Ga}-containing defect in GaN is the $V_{Ga}O_N$ complex (Oila et al., 2003; Saarinen et al., 2001a,b; Tuomisto, 2007). Thus, in as-grown n-type GaN, the $V_{Ga}O_N$ complexes are the dominant vacancy-related defects, while the isolated V_{Ga} can be created by irradiation. The $V_{Ga}O_N$ complex is stable up to 1500 K and dissociates at higher temperatures (Tuomisto, 2007; Tuomisto et al., 2006). As the thickness of GaN grown by HVPE increases from 1 to 300 μm, the concentration of the $V_{Ga}O_N$ complexes decreases from 2×10^{19} to 2×10^{15} cm^{-3} (Oila et al., 2003). A comparison with Hall-effect data shows that the total concentration of acceptors roughly matches the concentration of the $V_{Ga}O_N$ acceptor; i.e., this complex is the dominant acceptor in HVPE-grown n-type GaN where the dominant shallow donor is oxygen (Oila et al., 2003). The typical concentration of the $V_{Ga}O_N$ defect in n-type GaN grown by different methods is 10^{17} cm^{-3} (HNP), 10^{15} cm^{-3} (HVPE and MOCVD), 10^{16} cm^{-3} (MBE), and 10^{18} cm^{-3} (AT) (Tuomisto et al., 2012). PAS studies also show that the V_NMg_{Ga} complexes are formed in GaN doped with Mg, and that these complexes dissociate at annealing temperatures in the range of 500–800 °C (Hautakangas et al., 2003).

A relation between the YL band and the V_{Ga} or $V_{Ga}O_N$ defects deserves closer attention. Saarinen et al. (1997) observed a correlation between the YL band intensity and the concentration of the V_{Ga}-related defects in GaN grown by the HNP method (Fig. 15). However, the YL band in this work was measured from the GaN/sapphire interface where the concentration of various defects is very high and may not be reproducible. Reurings and Tuomisto (2007) observed an opposite relation between the YL intensity and the concentration of the V_{Ga}-containing defects in C-doped GaN samples (Fig. 15). Moreover, Xu et al. (2010) noticed that, in GaN samples with an undetectable amount of Ga vacancies, the YL intensity was significantly *higher* than that in GaN samples containing V_{Ga} with a concentration of $\sim 10^{17}$ cm^{-3}. Suihkonen et al. (2013) also noted that the YL band in GaN is not related to Ga vacancies, because no increase of the YL intensity with increasing concentration of V_{Ga} was observed.

Another argument in favor of the $V_{Ga}O_N$ attribution of the YL band was the good agreement between the concentrations of all acceptors from the Hall-effect data (3×10^{15} cm^{-3}), YL/GL-related defects from PL (2×10^{15} cm^{-3}), and the $V_{Ga}O_N$ complexes from PAS (2×10^{15} cm^{-3}) in high-purity freestanding GaN grown by HVPE at the Samsung Institute of Advanced Technology (Reshchikov and Morkoç, 2005). However,

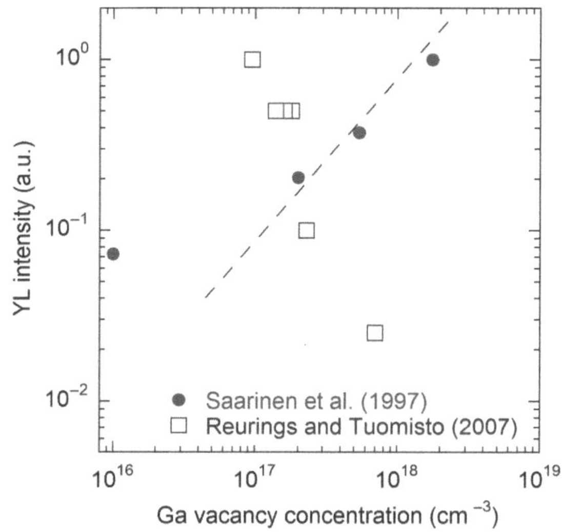

Figure 15 Dependence of the YL intensity (normalized for a sample with the strongest YL band) as a function of the concentration of the V_{Ga}-containing defects as determined from PAS. The dashed line is a linear dependence with a zero intercept. *Reproduced with permission from Saarinen et al. (1997), Copyright 1997 by The American Physical Society and Reurings and Tuomisto (2007), Copyright 2007 Society of Photo Optical Instrumentation Engineers.*

the "effective thickness" model, which was used to find the defect concentration from PL, can overestimate the defect concentration by a factor of 2–3 as compared to a more advanced model which accounts for the exponential decrease of the incident light intensity in semiconductor depth (Reshchikov, 2012a). The accuracy of the acceptors concentration from the Hall effect and the V_{Ga}-related defects from PAS is also not high. In particular, the parameter μ_V in Eq. (14) is not well known for the V_{Ga}-containing defects in GaN.

In summary, it appears that neither the isolated V_{Ga} nor the $V_{Ga}O_N$ complex is related to the YL band, but the idea that the YL band is caused by a V_{Ga}-containing defect at least in some GaN samples is still widespread (Lee et al., 2014; Uedono et al., 2014).

7. POINT DEFECTS REVEALED BY MAGNETIC RESONANCE TECHNIQUES

The $V_{Ga}O_N$ complex was detected by EPR in freestanding HVPE GaN (Son et al., 2009). After irradiating the samples with 2-MeV electrons at room temperature, the EPR experiments with below-bandgap illumination were performed at 77 K. A strong D2 signal with a characteristic fine structure was identified as the negatively charged $V_{Ga}O_N$ complex. The D2 signal started increasing at photon energy of 1.24 eV and reached a plateau at 1.8 eV. The signal persisted for at least 1 day after switching off the light at 77 K. The authors suggest that the D2 signal at photon energies between 1.24 and 2.1 eV is caused by the excitation of electrons from the 2−/− level of $V_{Ga}O_N$ to the higher lying electron traps, while the signal at higher photon energies was attributed to electron transitions from the same 2−/− level to the conduction band with the following capture of the electrons by electron traps. From this, the location of the 2−/− level of $V_{Ga}O_N$ was estimated to be at 1.2–1.4 eV above the valence band. We propose that the threshold at 1.24 eV may correspond to the beginning of transitions from the 2−/− level *to the conduction band*, because the intercenter transitions are much less likely. Then, the 2−/− level should be placed at 2.25 eV above the valence band. According to the hybrid functional calculations, the 2−/− level of the $V_{Ga}O_N$ complex is predicted to be at 1.75 eV above the valence band maximum (Section 2), which is between our estimate and the estimate of Son et al. (2009).

Glaser et al. (1995) conducted ODMR spectral studies of the YL band in undoped GaN. A sharp resonance with $g_{||}=1.9515$ and $g_{\perp}=1.9485$ was attributed to the effective-mass shallow donors, while a broad resonance with $g_{||}=1.989$ and $g_{\perp}=1.992$ was attributed to a deep donor. These resonances increased the YL intensity, and their spectral dependences matched the one of the YL band. Glaser et al. (1995) proposed that the YL band is caused by transitions from a deep donor to a shallow acceptor, whereas the spin-dependent transition from a shallow donor to the deep donor was assumed to be nonradiative. Hofmann et al. (1995) obtained similar results; however, they assigned the YL band to transitions from the shallow donor to the deep donor. An argument in favor of the deep donor (not a deep acceptor) was the value of the *g*-factor: the observed resonance has a lower *g* value than the free-electron value, while for acceptors a larger value is expected. Later, Glaser et al. (2002) agreed with the shallow donor-deep center model but concluded that the donor or acceptor character of the deep center remained an open question from a magnetic resonance viewpoint.

For the broad band peaking at 3.0 eV in high-resistivity GaN, in addition to the effective-mass signal, Glaser et al. (2002) observed two resonances: an isotropic signal with $g=1.978$ and a weaker feature with $g_{||}=2.022$ and $g_{\perp}=2.008$. This PL band could be the same as the BL2 band observed only in high-resistivity GaN (Section 4.3.6). For the RL band in freestanding GaN grown by HVPE, Glaser et al. (2001) observed an isotropic signal with $g = 2.019$. Finally, for the GL band in the freestanding GaN (see Section 4.3.4), Glaser et al. (2001) observed an effective-mass signal and a resonance labeled A_1 with $g_{||} = 1.975$ and $g_{\perp} = 1.969$.

Important results on native defects in GaN have been obtained from EPR studies. By using electron irradiation *in situ* at 4 K, Chow et al. (2004) detected the Ga interstitial defect in its 2+ paramagnetic state. The related signals, labeled L5 and L6, disappeared after several hours at room temperature, indicating that Ga_i is mobile at this temperature. A PL band with a maximum at 0.95 eV appears after the electron irradiation. The L5 signal competes with the PL, while the L6 signal is directly or indirectly involved in the production of the 0.95 eV band. The 0.95 eV band has been attributed to a radiative transition of a hole from the valence band to the triply negatively charged Ga vacancy located in the vicinity of the Ga_i defect. Such an assignment agreed with the expected creation of Ga_i–V_{Ga} Frenkel pairs under electron irradiation and with the energy levels of V_{Ga} calculated by Neugebauer and Van de Walle (1996).

8. SUMMARY

A significant progress is achieved in recent years in understanding of point defects in GaN. Theoretical predictions continue to improve, but to verify their accuracy, reliable experimental data, and unambiguous identification of defects are needed. In undoped GaN grown by different techniques, the concentrations of uncontrolled defects are in the 10^{15}–10^{16} cm^{-3} range or even higher. From detailed analysis of several point defects in undoped, Si-, and C-doped GaN, using luminescence and other experimental techniques and theoretical calculations, the following conclusions can be drawn.

The YL band with a maximum at 2.2 eV is related to carbon. In the majority of GaN samples, it is the C_NO_N complex with a transition level at 0.85 eV above the valence band. This defect is also the dominant trap H1 with the hole-capture cross-section of $\sigma_p = 7 \times 10^{-14}$ cm^2 detected in optical DLTS studies. In high-purity GaN (usually thick HVPE GaN), a similar YL band is apparently caused by the C_N defect. The capture of two holes by this defect in n-type GaN leads to the GL band with a maximum at 2.4 eV.

In high-resistivity GaN, either the RL2 band (maximum at 1.8 eV) and GL2 band (2.35 eV), or the BL2 band (3.0 eV) can be observed. The GL2 band is identified as the V_N defect. The BL2 band is a complex involving hydrogen and possibly carbon. The identity of the RL2 band is unknown. These bands are not observed in high-quality, conductive n-type GaN.

The BL band with a maximum at 2.9 eV is caused by transitions of electrons from the conduction band (or shallow donors) to the Zn_{Ga} acceptor (transition level at 0.4 eV above the valence band). It can be detected in undoped GaN grown by MOCVD or HVPE due to contamination from the sources or the reactor. The UVL band with a maximum at 3.27 eV is caused by the unidentified shallow acceptor. The Mg_{Ga} defects might be the source, while carbon should be excluded from the candidates.

The $V_{Ga}O_N$ complexes are present in undoped n-type GaN and can be detected in positron annihilation experiments. The hole trap H5 (with a level at 1.2 eV above the valence band), which is the dominant hole trap in thick HVPE GaN according to optical DLTS studies, is very likely the $V_{Ga}O_N$ complex. It appears that this defect is nonradiative, or its emission is in the infrared region.

ACKNOWLEDGMENTS

The author thanks Joy McNamara for assistance in the preparation of this chapter and Denis Demchenko for fruitful discussions. This work was supported by the National Science Foundation (DMR-1410125) and the Thomas F. and Kate Miller Jeffress Memorial Trust.

REFERENCES

Abakumov, V.N., Perel, V.I., Yassievich, I.N., 1991. Nonradiative Recombination in Semiconductors. Elsevier, Amsterdam.

Abernathy, C.R., MacKenzie, J.D., Pearton, S.J., Hobson, W.S., 1995. CCl_4 doping of GaN grown by metalorganic molecular beam epitaxy. Appl. Phys. Lett. 66, 1969–1971.

Alkauskas, A., Lyons, J.L., Steiauf, D., Van de Walle, C.G., 2012. First-principles calculations of luminescence spectrum line shapes for defects in semiconductors: the example of Ga and ZnO. Phys. Rev. Lett. 109, 267401.

Alkauskas, A., Yan, Q., Van de Walle, C.G., 2014. First-principles theory of nonradiative carrier capture via multiphonon emission. Phys. Rev. B 90, 075202.

Ambacher, O., Angerer, H., Dimitrov, R., Rieger, W., Stutzmann, M., Dollinger, G., Bergmaier, A., 1997. Hydrogen in gallium nitride grown by MOCVD. Phys. Status Solidi A 159, 105–119.

Arehart, A.R., Corrion, A., Poblenz, C., Speck, J.S., Mishra, U.K., Ringel, S.A., 2008. Deep level optical and thermal spectroscopy of traps in n-GaN grown by ammonia molecular beam epitaxy. Appl. Phys. Lett. 93, 112101.

Armitage, R., Hong, W., Yang, Q., Feick, H., Gebauer, J., Weber, E.R., Hautakangas, S., Saarinen, K., 2003. Contributions from gallium vacancies and carbon-related defects to the "yellow luminescence" in GaN. Appl. Phys. Lett. 82, 3457–3459.

Armitage, R., Yang, Q., Weber, E.R., 2005. Analysis of the carbon-related "blue" luminescence in GaN. J. Appl. Phys. 97, 073524.
Armstrong, A., Arehart, A.R., Moran, B., DenBaars, S.P., Mishra, U.K., Speck, J.S., Ringel, S.A., 2004. Impact of carbon on trap states in *n*-type GaN grown by metalorganic vapor deposition. Appl. Phys. Lett. 84, 374–376.
Armstrong, A., Arehart, A.R., Ringel, S.A., 2005a. A method to determine deep level profiles in highly compensated, wide band gap semiconductors. J. Appl. Phys. 97, 083529.
Armstrong, A., Arehart, A.R., Green, D., Mishra, U.K., Speck, J.S., Ringel, S.A., 2005b. Impact of deep levels on the electrical conductivity and luminescence of gallium nitride codoped with carbon and silicon. J. Appl. Phys. 98, 053704.
Asghar, M., Munet, P., Hussain, I., Beaumont, B., Gibart, P., Shahid, M., 2006. Luminescence properties of hole traps in homojunction gallium nitride diodes grown by metal organic vapour phase epitaxy. Mater. Sci. Eng. B 130, 173–176.
Auret, F.D., Meyer, W.E., Wu, L., Hayes, M., Legodi, M.J., Beaumont, B., Gibart, P., 2004. Electrical characterization of hole traps in n-type GaN. Phys. Status Solidi A 201, 2271–2276.
Bartram, R.H., Stoneham, A.M., 1975. On the luminescence and absence of luminescence of F centers. Solid State Commun. 17, 1593–1598.
Bell, A., Harrison, I., Cheng, T.S., Korakakis, D., Foxton, C.T., Novikov, S., Ber, B.Y., Kudriavtsev, Y.A., 2000. An investigation into the origin of the 3.424 eV peak in the low-temperature photoluminescence of GaN grown by molecular beam epitaxy. Semicond. Sci. Technol. 15, 789–793.
Bliss, D., Wang, B., Suscavage, M., Lancto, R., Swider, S., Eikenberry, W., Lynch, C., 2010. Ammonothermal GaN: morphology and properties. J. Cryst. Growth 312, 1069–1073.
Bogusławski, P., Bernholc, J., 1997. Doping properties of C, Si, and Ge impurities in GaN and AlN. Phys. Rev. B 56, 9496–9505.
Boles, T., Varmazis, C., Carlson, D., Palacios, T., Turner, G.W., Molnar, R., 2013. >1200 V GaN-on-silicon Schottky diode. Phys. Status Solidi C 10, 835–839.
Brunner, F., Bahat-Treidel, E., Cho, M., Netzel, C., Hilt, O., Würfl, J., Weyers, M., 2011. Comparative study of buffer designs for high breakdown voltage AlGaN/GaN HFETs. Phys. Status Solidi C 8, 2427–2429.
Buyanova, I.A., Wagner, Mt, Chen, W.M., Monemar, B., Lindström, J.L., Amano, H., Akasaki, I., 1998. Photoluminescence of GaN: effect of electron irradiation. Appl. Phys. Lett. 73, 2968–2970.
Calleja, E., Sánchez, F.J., Basak, D., Sánchez-García, M.A., Munoz, E., Izpura, I., Calle, F., Tijero, J.M.G., Sánchez-Rojas, J.L., Beaumont, B., Lorenzini, P., Gibart, P., 1997. Yellow luminescence and related deep states in undoped GaN. Phys. Rev. B 55, 4689.
Chantre, A., Vincent, G., Bois, D., 1981. Deep-level optical spectroscopy in GaAs. Phys. Rev. B 23, 5335–5359.
Chen, J.-T., Forsberg, U., Janzén, E., 2013. Impact of residual carbon on two-dimensional electron gas properties in $Al_xGa_{1-x}N$/GaN heterostructure. Appl. Phys. Lett. 102, 193503.
Chow, K.H., Vlasenko, L.S., Johannesen, P., Bozdog, C., Watkins, G.D., Usui, A., Sunakawa, H., Sasoka, C., Mizuta, M., 2004. Intrinsic defects in GaN. I. Ga sublattice defects observed by optical detection of electron paramagnetic resonance. Phys. Rev. B 69, 045207.
Demchenko, D.O., Reshchikov, M.A., 2013. Blue luminescence and Zn acceptor in GaN. Phys. Rev. B 88, 115204.
Demchenko, D.O., Diallo, I.C., Reshchikov, M.A., 2013. Carbon defect complex as a source of yellow luminescence in GaN. Phys. Rev. Lett. 110, 087404.

Dexter, D.L., Klick, C.C., Russel, G.A., 1955. Criterion for the occurrence of luminescence. Phys. Rev. 100, 603–605.

Dyba, P., Placzek-Popko, E., Zielony, E., Gumienny, Z., Grzanka, S., Czernecki, R., Suski, T., 2011. Deep levels in GaN p-n junctions studied by deep level transient spectroscopy and Laplace transform deep-level spectroscopy. Acta Phys. Pol. A 119, 669–671.

Eberlein, T.A.G., Jones, R., Öberg, S., Briddon, P.R., 2007. Shallow acceptors in GaN. Appl. Phys. Lett. 91, 132105.

Fischer, S., Wetzel, C., Haller, E.E., Meyer, B.K., 1995. On *p*-type doping in GaN—acceptor binding energies. Appl. Phys. Lett. 67, 1298–1300.

Freitas Jr., J.A., Mastro, M.A., Glaser, E.R., Garces, N.Y., Lee, S.K., Chang, J.H., Oh, D.K., Shim, K.B., 2012. Structural and optical studies of thick freestanding GaN films grown by hydride vapor phase epitaxy. J. Cryst. Growth 350, 27–32.

Freysoldt, C., Grabowski, B., Hickel, T., Neugebauer, J., Kresse, G., Janotti, A., Van de Walle, C.G., 2014. First-principles calculations for point defects in solids. Rev. Mod. Phys. 86, 253–305.

Fujito, K., Kubo, S., Nagaoka, H., Mochizuki, T., Namita, H., Nagao, S., 2009. Bulk GaN crystals grown by HVPE. J. Cryst. Growth 311, 3011–3014.

Garces, N.Y., Feigelson, B.N., Freitas Jr., J.A., Kim, J., Myers-Ward, R., Glaser, E.R., 2010. Characterization of bulk GaN crystals grown from solution at near atmospheric pressure. J. Cryst. Growth 312, 2558–2563.

Gillen, R., Robertson, J., 2013. A hybrid density functional view of native vacancies in gallium nitride. J. Phys. Condens. Matter 25, 405501.

Glaser, E.R., Kennedy, T.A., Doverspike, K., Rowland, L.B., Gaskill, D.K., Freitas Jr., J.A., Asif Khan, M., Olson, D.T., Kuznia, J.N., Wickenden, D.K., 1995. Optically detected magnetic resonance of GaN films grown by organometallic chemical vapor deposition. Phys. Rev. B 51, 13326–13336.

Glaser, E.R., Freitas Jr., J.A., Braga, G.C.B., Carlos, W.E., Twigg, M.E., Wickenden, D.K., Koleske, D.D., Henry, R.L., Leszczynski, M., Grzegory, I., Suski, T., Porowski, S., Park, S.S., Lee, K.Y., Molnar, R.J., 2001. Magnetic resonance studies of defects in GaN with reduced dislocation densities. Physica B 308–310, 51–57.

Glaser, E.R., Carlos, W.E., Braga, G.C.B., Freitas Jr., J.A., Moore, W.J., Shanabrook, B.V., Wickenden, D.K., Koleske, D.D., Henry, R.L., Bayerl, M.W., Brandt, M.S., Obloh, H., Kozodoy, P., DenBaars, S.P., Mishra, U.K., Nakamura, S., Haus, E., Speck, J.S., Van Nostrand, J.E., Sanchez, M.A., Calleja, E., Ptak, A.J., Myers, T.H., Molnar, R.J., 2002. Characterization of nitrides by electron paramagnetic resonance (EPR) and optically detected magnetic resonance (ODMR). Mater. Sci. Eng. B93, 39–48.

Hall, R.N., 1951. Germanium rectifier characteristics. Phys. Rev. 83, 228.

Hashimoto, T., Letts, E., Hoff, S., 2013. Current status and future prospects of ammonothermal bulk GaN growth. Sensor Mater. 25, 155–164.

Hautakangas, S., Oila, J., Alatalo, M., Saarinen, K., Liszkay, L., Seghier, D., Gislason, H.P., 2003. Vacancy defects as compensating centers in Mg-doped GaN. Phys. Rev. Lett. 90, 137402.

Heyd, J., Scuseria, G.E., Ernzerhof, M., 2003. Hybrid functionals based on a screened Coulomb potential. J. Chem. Phys. 118, 8207–8215.

Hikosaka, T., Koide, N., Honda, Y., Yamaguchi, M., Sawaki, N., 2006. P-type conduction in C-doped (1–101)GaN grown on a 7-degree-off oriented (001)Si substrate by selective MOVPE. Phys. Status Solidi C 3, 1425–1428.

Hofmann, D.M., Kovalev, D., Steude, G., Meyer, B.K., Hoffmann, A., Eckey, L., Heitz, R., Detchprom, T., Amano, H., Akasaki, I., 1995. Properties of the yellow luminescence in undoped GaN epitaxial layers. Phys. Rev. B 52, 16702–16706.

Honda, U., Yamada, Y., Tokuda, Y., Shiojima, K., 2012. Deep levels in n-GaN doped with carbon studied by deep level and minority carrier transient spectroscopies. Jpn. J. Appl. Phys. 51, 04DF04.

Huang, K., Rhys, A., 1950. Theory of light absorption and non-radiative transitions in F-centers. Proc. R. Soc. 204, 406–423.

Jenkins, D.W., Dow, J.D., 1989. Electronic structures and doping of InN, $In_xGa_{1-x}N$, and $In_xAl_{1-x}N$. Phys. Rev. B 39, 3317–3329.

Kamiura, Y., Kaneshiro, M., Tamura, J., Ishiyama, T., Yamashita, Y., Mitani, T., Mukai, T., 2005. Enhancement of blue emission from Mg-doped GaN using remote plasma containing atomic hydrogen. Jpn. J. Appl. Phys. 44, L926–L928.

Kamyczek, P., Placzek-Popko, E., Kolkovsky, V., Grzanka, S., Czernecki, R., 2012. A deep acceptor defect responsible for the yellow luminescence in GaN and AlGaN. J. Appl. Phys. 111, 113105.

Kaufmann, U., Kunzer, M., Obloh, H., Maier, M., Manz, Ch., Ramakrishnan, A., Santic, B., 1999. Origin of defect-related photoluminescence bands in doped and nominally undoped GaN. Phys. Rev. B 59, 5561–5567.

Koleske, D.D., Wickenden, A.E., Henry, R.L., Twigg, M.E., 2002. Influence of MOVPE growth conditions on carbon and silicon concentrations in GaN. J. Cryst. Growth 242, 55–69.

Kucheyev, S.O., Toth, M., Phillips, M.R., Williams, J.S., Jagadish, C., Li, G., 2002. Chemical origin of the yellow luminescence in GaN. J. Appl. Phys. 91, 5867–5874.

Laaksonen, K., Ganchenkova, M.G., Nieminen, R.M., 2009. Vacancies in wurtzite GaN and AlN. J. Phys. Condens. Matter 21, 015803.

Lang, D.V., 1974. Deep-level transient spectroscopy: a new method to characterize traps in semiconductors. J. Appl. Phys. 45, 3023–3032.

Lee, I.-H., Polyakov, A.Y., Smirnov, N.B., Govorkov, A.V., Kozhukhova, E.A., Kolin, N.G., Boiko, V.M., Korulin, A.V., Pearton, S.J., 2011a. Deep electron and hole traps in neutron transmutation doped GaN. J. Vac. Sci. Technol. 29, 041201.

Lee, I.-H., Polyakov, A.Y., Smirnov, N.B., Govorkov, A.V., Kozhukhova, E.A., Kolin, N.G., Boiko, V.M., Korulin, A.V., Pearton, S.J., 2011b. Carrier removal rates and deep traps in neutron irradiated n-GaN films. J. Electrochem. Soc. 158, H866–H871.

Lee, I.-H., Polyakov, A.Y., Smirnov, N.B., Govorkov, A.V., Kozhukhova, E.A., Zaletin, V.M., Gazizov, I.M., Kolin, N.G., Pearton, S.J., 2012. Electrical properties and radiation detector performance of free-standing bulk n-GaN. J. Vac. Sci. Technol. 30, 021205.

Lee, I.-H., Polyakov, A.Y., Smirnov, N.B., Govorkov, A.V., Usikov, A.S., Helava, H., Makarov, Yu.N., Pearton, S.J., 2014. Deep hole traps in undoped n-GaN films grown by hydride vapor phase epitaxy. J. Appl. Phys. 115, 223702.

Leszczynski, M., Grzegory, I., Teisseyre, H., Suski, T., Bockowski, M., Jun, J., Baranowski, J.M., Porowski, S., Domagala, J., 1996. The microstructure of gallium nitride monocrystals grown at high pressure. J. Cryst. Growth 169, 235–242.

Limpijumnong, S., Van de Walle, C.G., 2004. Diffusivity of native defects in GaN. Phys. Rev. B 69, 035207.

Look, D.C., Stutz, C.E., Molnar, R.J., Saarinen, K., Liliental-Weber, Z., 2001. Dislocation-independent mobility in lattice-mismatched epitaxy: application to GaN. Solid State Commun. 117, 571–575.

Lucovsky, G., 1965. On the photoionization of deep impurity centers in semiconductors. Solid State Commun. 3, 299–302.

Lyons, J.L., Janotti, A., Van de Walle, C.G., 2010. Carbon impurities and the yellow luminescence in GaN. Appl. Phys. Lett. 97, 152108.

Lyons, J.L., Janotti, A., Van de Walle, C.G., 2014. Effects of carbon on the electrical and optical properties of InN, GaN, and AlN. Phys. Rev. B 89, 035204.

Mattila, T., Nieminen, R.M., 1997. Point-defect complexes and broadband luminescence in GaN and AlN. Phys. Rev. B 55, 9571–9576.

McNamara, J.D., Foussekis, M.A., Baski, A.A., Li, X., Avrutin, V., Morkoç, H., Leach, J.H., Paskova, T., Udwary, K., Preble, E., Reshchikov, M.A., 2013. Electrical and optical

properties of bulk GaN substrates studied by Kelvin probe and photoluminescence. Phys. Status Solidi C 10, 536–539.
Mohajerani, M.S., Reshchikov, M.A., Behrends, A., Bakin, A., Waag, A., 2013. Determination of zinc concentration in GaN:Zn, Si from photoluminescence. Phys. Status Solidi C 10, 523–526.
Monemar, B., Paskov, P.P., Paskova, T., Bergman, J.P., Pozina, G., Chen, W.M., Hai, P.N., Buyanova, I.A., Amano, H., Akasaki, I., 2002. Optical characterization of III-nitrides. Mater. Sci. Eng. B93, 112–122.
Morkoç, H., 2008. Handbook of Nitride Semiconductors and Devices: Materials Properties, Physics and Growth, , vol. 1. Wiley, New York.
Muret, P., Philippe, A., Monroy, E., Munoz, E., Beaumont, B., Omnès, F., Gibart, P., 2002. Properties of a hole trap in n-type hexagonal GaN. Appl. Phys. Lett. 91, 2998–3001.
Murthy, M., Freitas Jr., J.A., Kim, J., Glaser, E.R., Storm, D., 2007. Residual impurities in GaN substrates and epitaxial layers grown by various techniques. J. Cryst. Growth 305, 393–398.
Muth, J.F., Lee, J.H., Shmagin, I.K., Kolbas, R.M., Casey, H.C., Keller, B.P., Mishra, U.K., DenBaars, S.P., 1997. Absorption coefficient, energy gap, exciton binding energy, and recombination lifetime of GaN obtained from transmission measurements. Appl. Phys. Lett. 71, 2572–2574.
Neugebauer, J., Van de Walle, C.G., 1996. Gallium vacancies and the yellow luminescence in GaN. Appl. Phys. Lett. 69, 503–505.
Ogino, T., Aoki, M., 1980. Mechanism of yellow luminescence in GaN. Jpn. J. Appl. Phys. 19, 2395–2405.
Oila, J., Ranki, V., Kivioja, J., Saarinen, K., Hautojärvi, P., Likonen, J., Baranowski, J.M., Pakula, K., Suski, T., Leszczynski, M., Grzegory, I., 2001. Influence of dopants and substrate material on the formation of Ga vacancies in epitaxial GaN layers. Phys. Rev. B 63, 045205.
Oila, J., Kivioja, J., Ranki, V., Saarinen, K., Look, D.C., Molnar, R.J., Park, S.S., Lee, S.K., Han, J.Y., 2003. Ga vacancies as dominant intrinsic acceptors in GaN grown by hydride vapor phase epitaxy. Appl. Phys. Lett. 82, 3433–3435.
Polenta, L., Fang, Z.-Q., Look, D.C., 2000. On the main irradiation-induced defect in GaN. Appl. Phys. Lett. 76, 2086–2088.
Polyakov, A.Y., Shin, M., Freitas Jr., J.A., Skowronski, M., Greve, D.W., Wilson, R.G., 1996. On the origin of electrically active defects in AlGaN alloys grown by organometallic vapor phase epitaxy. J. Appl. Phys. 80, 6349–6354.
Polyakov, A.Y., Smirnov, N.B., Usikov, A.S., Govorkov, A.V., Pushniy, V., 1998a. Studies of the origin of the yellow luminescence band, the nature of nonradiative recombination and the origin of persistent photoconductivity in n-GaN films. Solid State Electron. 42, 1959–1967.
Polyakov, A.Y., Smirnov, N.B., Govorkov, A.V., Shin, M., Skowronski, M., Greve, D.W., 1998b. Deep centers and their spatial distribution in undoped GaN films grown by organometallic vapor phase epitaxy. J. Appl. Phys. 84, 870–876.
Polyakov, A.Y., Smirnov, N.B., Govorkov, A.V., Fang, Z.-Q., Look, D.C., Park, S.S., Han, J.H., 2002. Deep electron and hole traps in freestanding n-GaN grown by hydride vapor phase epitaxy. J. Appl. Phys. 92, 5241–5247.
Polyakov, A.Y., Lee, I.-H., Smirnov, N.B., Kozhukhova, E.A., Pearton, S.J., 2011. Comparison of hole traps in n-GaN grown by hydride vapor phase epitaxy, metal organic chemical vapor deposition, and epitaxial lateral overgrowth. J. Appl. Phys. 109, 123701.
Polyakov, A.Y., Smirnov, N.B., Govorkov, A.V., Yugova, T.G., Cox, H., Usikov, A.S., Helava, H., Makarov, Yu.N., 2014. Hydride vapor phase GaN films with reduced density of residual electrons and deep traps. J. Appl. Phys. 115, 183706.

Rebane, K.K., 1970. Impurity Spectra of Solids. Plenum Press, New York.
Reshchikov, M.A., 2006. Determination of acceptor concentration in GaN from photoluminescence. Appl. Phys. Lett. 88, 202104.
Reshchikov, M.A., 2012a. Internal quantum efficiency of photoluminescence in wide-bandgap semiconductors. In: Case, M.A., Stout, B.C. (Eds.), Photoluminescence: Applications, Types and Efficacy. Nova Science Publishers, Inc, New York. ISBN: 978-1-61942-426-5, pp. 53–120.
Reshchikov, M.A., 2012b. Two-step thermal quenching of photoluminescence in Zn-doped GaN. Phys. Rev. B 85, 245203.
Reshchikov, M.A., 2014a. Time-resolved photoluminescence from defects in GaN. J. Appl. Phys. 115, 103503.
Reshchikov, M.A., 2014b. Temperature dependence of defect-related photoluminescence in III–V and II–VI semiconductors. J. Appl. Phys. 115, 012010.
Reshchikov, M.A., 2014c. Carrier-capture characteristics of point defects in GaN and ZnO. AIP Conf. Proc. 1583, 127–131. http://dx.doi.org/10.1063/1.4865619.
Reshchikov, M.A., 2014d. Temperature dependence of photoconductivity in Zn-doped GaN. AIP Conf. Proc. 1583, 292–296. http://dx.doi.org/10.1063/1.4865655.
Reshchikov, M.A., Korotkov, R.Y., 2001. Analysis of the temperature and excitation intensity dependencies of photoluminescence in undoped GaN films. Phys. Rev. B 64, 115205.
Reshchikov, M.A., Morkoç, H., 2005. Luminescence properties of defects in GaN. J. Appl. Phys. 97, 061301.
Reshchikov, M.A., Morkoç, H., 2006. Luminescence from defects in GaN. Physica B 376–377, 428–431.
Reshchikov, M.A., Yi, G.-C., Wessels, B.W., 1999. Behavior of 2.8- and 3.2-eV photoluminescence bands in Mg-doped GaN at different temperatures and excitation densities. Phys. Rev. B 59, 13176–13183.
Reshchikov, M.A., Shahedipour, F., Korotkov, R.Y., Ulmer, M.P., Wessels, B.W., 2000. Photoluminescence band near 2.9 eV in undoped GaN epitaxial layers. J. Appl. Phys. 87, 3351–3354.
Reshchikov, M.A., Morkoç, H., Park, S.S., Lee, K.Y., 2002a. Two charge states of dominant acceptor in unintentionally doped GaN: evidence from photoluminescence study. Appl. Phys. Lett. 81, 4970–4972.
Reshchikov, M.A., Morkoç, H., Park, S.S., Lee, K.Y., 2002b. Photoluminescence study of deep-level defects in undoped GaN. Mater. Res. Soc. Symp. Proc. 693, I6.19.
Reshchikov, M.A., Zafar Iqbal, M., Park, S.S., Lee, K.Y., Tsvetkov, D., Dmitriev, V., Morkoç, H., 2003. Persistent photoluminescence in high-purity GaN. Physica B 340–342, 444–447.
Reshchikov, M.A., Patillo, R.H., Travis, K.C., 2006a. Photoluminescence in wurtzite GaN containing carbon. Mater. Res. Soc. Symp. Proc. 892, FF23.12.
Reshchikov, M.A., Moon, Y.T., Gu, X., Nemeth, B., Nause, J., Morkoç, H., 2006b. Unstable luminescence in GaN and ZnO. Physica B 376–377, 715–718.
Reshchikov, M.A., Kvasov, A., McMullen, T., Bishop, M.F., Usikov, A., Soukhoveev, V., Dmitriev, V.A., 2011. Tunable and abrupt thermal quenching of photoluminescence in high-resistivity Zn-doped GaN. Phys. Rev. B 84, 075212.
Reshchikov, M.A., Foussekis, M.A., McNamara, J.D., Behrends, A., Bakin, A., Waag, A., 2012. Determination of the absolute internal quantum efficiency of photoluminescence in GaN co-doped with Zn and Si. J. Appl. Phys. 111, 073106.
Reshchikov, M.A., Demchenko, D.O., McNamara, J.D., Fernández-Garrido, S., Calarco, R., 2014a. Green luminescence in Mg-doped GaN. Phys. Rev. B 90, 035207.
Reshchikov, M.A., Usikov, A., Helava, H., Makarov, Y., 2014b. Fine structure of the red luminescence band in GaN. Appl. Phys. Lett. 104, 032103.

Reshchikov, M.A., Demchenko, D.O., Usikov, A., Helava, H., Makarov, Y., 2014. Carbon defects as sources of the yellow and green luminescence bands in GaN grown by HVPE. Phys. Rev. B 90, 235203.
Reurings, F., Tuomisto, F., 2007. Interplay of Ga vacancies, C impurities and yellow luminescence in GaN. Proc. SPIE 6473, 64730M.
Saarinen, K., Laine, T., Kuisma, S., Nissilä, J., Hautojärvi, P., Dobrzynski, L., Baranowski, J.M., Pakula, K., Stepniewski, R., Wojdak, M., Wysmolek, A., Suski, T., Leszszynski, M., Grzegory, I., Porowski, S., 1997. Observation of native Ga vacancies in GaN by positron annihilation. Phys. Rev. Lett. 79, 3030–3033.
Saarinen, K., Hautojärvi, P., Corbel, C., 1998. Positron annihilation spectroscopy of defects in semiconductors. In: Stavola, M. (Ed.), Identification of Defects in Semiconductors. Academic Press, New York, pp. 209–285.
Saarinen, K., Suski, T., Grzegory, I., Look, D.C., 2001a. Thermal stability of isolated and complexed Ga vacancies in GaN bulk crystals. Phys. Rev. B 64, 233201.
Saarinen, K., Suski, T., Grzegory, I., Look, D.C., 2001b. Ga vacancies in electron irradiated GaN: introduction, stability and temperature dependence of positron trapping. Physica B 308–310, 77–80.
Saitoh, Y., Sumiyoshi, K., Okada, M., Horii, T., Miyazaki, T., Shiomi, H., Ueno, M., Katayama, K., Kiyama, M., Nakamura, T., 2010. Extremely low on-resistance and high breakdown voltage observed in vertical GaN Schottky barrier diodes with high-mobility drift layers on low-dislocation-density GaN substrates. Appl. Phys. Express 3, 081001.
Seager, C.H., Wright, A.F., Yu, J., Götz, W., 2002. Role of carbon in GaN. J. Appl. Phys. 92, 6553–6560.
Seager, C.H., Tallant, D.R., Yu, J., Götz, W., 2004. Luminescence in GaN co-doped with carbon and silicon. J. Lumin. 106, 115–124.
Shockley, W., Read, J.W.T., 1952. Statistics of the recombinations of holes and electrons. Phys. Rev. 87, 835–842.
Son, N.T., Hemminingsson, C.G., Paskova, T., Evans, K.R., Usui, A., Morishita, N., Ohshima, T., Isoya, J., Monemar, B., Janzén, E., 2009. Identification of the gallium vacancy-oxygen pair defect in GaN. Phys. Rev. B 80, 153202.
Stoneham, A.M., 1975. Theory of Defects in Solids. Clarendon Press, Oxford.
Suihkonen, S., Nykänen, H., Tanikawa, T., Yamaguchi, M., Honda, Y., Amano, H., 2013. Effects of low energy e-beam irradiation on cathodoluminescence from GaN. Phys. Status Solidi A 210, 383–385.
Tansley, T.L., Egan, R.J., 1992. Point-defect energies in the nitrides of aluminum, gallium, and indium. Phys. Rev. B 45, 10942–10950.
Tokuda, Y., Yamada, Y., Shibata, T., Yamaguchi, S., Ueda, H., Uesugi, T., Kachi, T., 2011. Hole traps in n-GaN detected by minority carrier transient spectroscopy. Phys. Status Solidi C 8, 2239–2241.
Tuomisto, F., 2007. Defect studies of HVPE GaN by positron annihilation spectroscopy. Proc. SPIE 6473, 647312.
Tuomisto, F., Saarinen, K., Lucznik, B., Grzegory, I., Teisseyre, H., Suski, T., Porowski, S., Hageman, P.R., Likonen, J., 2005. Effect of growth polarity on vacancy defect and impurity incorporation in dislocation-free GaN. Appl. Phys. Lett. 86, 031915.
Tuomisto, F., Saarinen, K., Paskova, T., Monemar, B., Bockowski, M., Suski, T., 2006. Thermal stability of in-grown vacancy defects in GaN grown by hydride vapor phase epitaxy. J. Appl. Phys. 99, 066105.
Tuomisto, F., Mäki, J.-M., Rauch, C., Makkonen, I., 2012. On the formation of vacancy defects in III-nitride semiconductors. J. Cryst. Growth 350, 93–97.
Uedono, A., Fujishima, T., Cao, Y., Zhang, Y., Yoshihara, N., Ishibashi, S., Sumiya, M., Laboutin, O., Johnson, W., Palacios, T., 2014. Optically active vacancies in GaN grown

on Si substrates probed using a monoenergetic positron beam. Appl. Phys. Lett. 104, 082110.

Urbach, F., 1953. The long-wavelength edge of photographic sensitivity and of the electronic absorption of solids. Phys. Rev. 92, 1324.

Usikov, A., Soukhoveev, V., Kovalenkov, O., Syrkin, A., 2013. Accumulation of background impurities in hydride vapor phase epitaxy grown GaN layers. Jpn. J. Appl. Phys. 52, 08JB22.

Wang, L., Richter, E., Weyers, M., 2007. Red luminescence from freestanding GaN grown on $LiAlO_2$ substrate by hydride vapor phase epitaxy. Phys. Status Solidi A 204, 846–849.

Wright, A.F., 2002. Substitutional and interstitial carbon in wurtzite GaN. J. Appl. Phys. 92, 2575–2585.

Xu, F.J., Shen, B., Lu, L., Miao, Z.L., Song, J., Yang, Z.J., Zhang, G.Y., Hao, X.P., Wang, B.Y., Shen, X.Q., Okumura, H., 2010. Different origins of the yellow luminescence in as-grown high-resistance GaN and unintentional-doped GaN films. J. Appl. Phys. 107, 023528.

Zhang, R., Kuech, T.F., 1998. Photoluminescence of carbon in situ doped GaN grown by halide vapor phase epitaxy. Appl. Phys. Lett. 72, 1611–1613.

Zhang, Z., Hurni, C.A., Arehart, A.R., Yang, J., Myers, R.C., Speck, J.S., Ringel, S.A., 2012. Deep traps in nonpolar m-plane GaN grown by ammonia-based molecular beam epitaxy. Appl. Phys. Lett. 100, 052114.

CHAPTER TEN

Point Defects in Silicon Carbide

Naoya Iwamoto[1], Bengt G. Svensson[1]
University of Oslo, Physics Department, Center for Materials Science and Nanotechnology, Oslo, Norway
[1]Corresponding authors: e-mail address: naoya.iwamoto@smn.uio.no; b.g.svensson@fys.uio.no

Contents

1. INTRODUCTION

Silicon carbide (SiC) is a binary compound with exceptional thermal, mechanical, and electrical properties, making it the material of choice for a wide range of different applications. With the exception of a few meteorites, SiC does not exist naturally on the earth, but it has been synthesized artificially since the 1890s (Acheson, 1892). Today, single-crystalline SiC wafers with diameter size up to 150 mm (6 in.) are commercially available, and

Semiconductors and Semimetals, Volume 91
ISSN 0080-8784
http://dx.doi.org/10.1016/bs.semsem.2015.02.001

especially during the past ~25 years a tremendous development has taken place in the area of crystal growth, primarily driven by the potential use of SiC for power electronic devices.

General hallmarks of SiC are an extremely high thermal stability and a high thermal conductivity, and in crystalline form, it becomes a wide energy bandgap semiconductor with attractive inherent properties like an exceptionally high electric breakdown field, high saturation velocity of charge carriers, and high hardness with respect to transient ionizing radiation. Already in 1907 (Round, 1907), a light emitting diode (LED) was demonstrated using SiC, and for several decades, short-wavelength LEDs were considered as a major (opto)electronic use of SiC. However, the efficiency of these LEDs was low because of the indirect bandgap of SiC, and it was realized that the inherent properties of SiC are more advantageous for devices operating at high powers, high frequencies and/or high temperatures, and in harsh environment. Especially, rectifiers and switches employed in systems for transmission, distribution, and regulation of high power electricity benefit greatly from the high electric breakdown field and the high thermal conductivity, making bulky and costly cooling installations redundant as well as reducing the switching losses by orders of magnitude relative to those in other semiconductors, like Si. Hence, the potential for energy saving by utilizing SiC-based device technology in power electronics is huge; for instance, in the late 1990s it was forecasted that these saving can amount to ~10% of the electricity production worldwide (Breitholtz, 1998). In a more recent forecast, the SiC power device market is projected to reach a size of 800 M$ per year in 2020 and capture a ~10% share over Si power devices (mainly, insulated-gate bipolar transistors) (Yole Développement, 2013). In fact, the replacement has already begun, and especially in areas like "inverters for photovoltaics" and "(hybrid) electric vehicles ((H)EVs)," the volume growth of SiC devices is exceptionally high. In addition to power electronics, also "RF and microwave communication," "chemical gas sensors in harsh environment," and "detectors for ionizing particles" are examples of technology fields relevant to SiC.

SiC is the only stable compound in the Si–C phase diagram, but it can (in principle) crystallize in an infinite number of different atomic structures, so-called polytypes; experimentally, more than 200 different polytypes have been reported each with its own specific properties (Harris, 1995), such as lattice constants, energy bandgap, charge carrier mobility, and density of states in valence and conduction bands. The basic building block of all the polytypes is the tetrahedral arrangement between one Si atom and four neighboring C atoms, and vice versa, as depicted in Fig. 1a and b. The atomic bonds are of covalent character but with a distinct ionicity where the electron density is strongly located around the C atoms (Park et al., 1994; Pensl and Helbig, 1990), and any SiC crystal can be described as stacked layers of Si–C atom pairs forming close-packed hexagonal planes. There are three equivalent positions in the stacking of the hexagonal Si–C bilayers, commonly labeled as A, B, and C (see Fig. 1c and d), and the different polytypes arise from different stacking sequences of these layers. Accordingly, all SiC crystals can be described by a hexagonal unit cell but it may not be the smallest convenient one, and in addition to hexagonal (H) (wurtzite structure), also one cubic (3C, zinc blende structure) and several rhombohedral (R) variants exist. Usually, the polytypes are labeled following the Ramsdell notation giving the unit cell and the number of planes in the periodicity of the stacking sequence. For example, some of the most studied polytypes are 4H, 6H, and 3C, and they have the stacking sequences of ABAC, ABCACB, and ABC, respectively. 3C–SiC is often also referred to as β-SiC while the hexagonal and rhombohedral polytypes are sometimes called α-SiC. In 4H, two inequivalent lattice sites occur, hexagonal (*h*) and quasi-cubic (*k*), and in 6H, three inequivalent sites exist, one *h* and two quasi-cubic (k_1 and k_2).

The remarkable development in growth of high-quality (device-worthy) single-crystalline SiC wafers holds primarily for the 4H polytype which is technologically favored for electronic devices because of its wide bandgap,

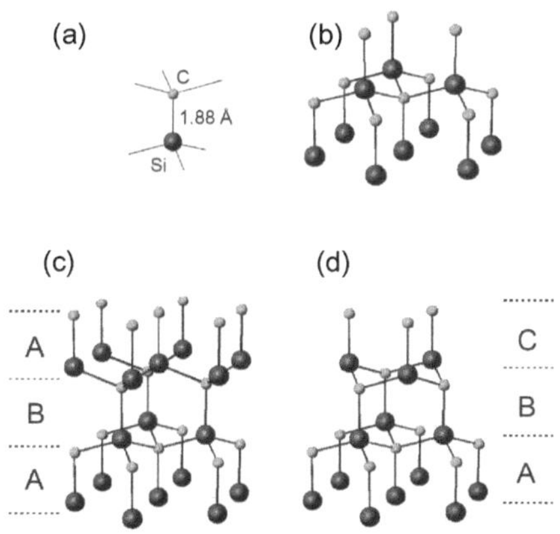

Figure 1 Illustration of the SiC crystal structure by formation of stacked, hexagonal layers of Si–C pairs (a), the positions of the first two bilayers are defined as A and B (b), and the third layer can take either position A (c) or C (d).

high and isotropic electron mobility, and high thermal stability. A comparison of some critical properties for the most frequently studied polytypes is given in Table 1. The SiC wafers consist typically of a bulk substrate with so-called physical vapor transport as the prevailing (but not exclusive) principle growth method (Tairov and Tsvetkov, 1978) and then a high-purity epitaxial layer commonly realized via variants of chemical vapor deposition (CVD), such as hot-wall CVD (Kordina et al., 1994). Historically, the wafer quality has been hampered by micropipes and inclusions of different polytypes as major obstacles (Berkman et al., 2009; Gupta et al., 2012) but because of a gradual and persistent improvement of growth reactor design, source material morphology, temperature control, pressure and gas flow control, etc., 150-mm diameter 4H–SiC epitaxial wafers which are essentially free of micropipes and polytype inclusions can now be purchased, albeit at high cost.

With the advancement in materials growth and increasing level of sophistication, point defects, dopants, impurities, as well as extended structural defects have evolved as crucial issues within the SiC community. Especially, realization of large-area/high-voltage/high-power devices is demanding, requiring thick epitaxial layers ($\geq$50 μm) with long minority charge carrier lifetime ($\geq$10 μs) and well-controlled dopant distributions in the concentration range of $\sim10^{13}$–10^{15} cm^{-3}. Further, several of these issues link not only to the materials growth processes but also to the processes for device fabrication, e.g., oxidation, selective area doping by ion implantation, surface passivation, and contact formation.

In the following sections, we will critically review recent progress in the understanding and control of (i) the silicon vacancy (V_{Si}) and the carbon

Table 1 Some Critical Properties of 4H–SiC, 6H–SiC, 3C–SiC, and Si

	4H–SiC		6H–SiC		3C–SiC	Si
Melting point (°C)	2830		2830		1825	1415
Bandgap (eV)	3.26		3.02		2.39	1.11
Electron mobility (cm^2/V/s)	880 // *c*	800 ⊥ *c*	97 // *c*	360 ⊥ *c*	750	1100
Hole mobility (cm^2/V/s)	115		90		40	420
Breakdown field (10^6 V/cm)	~3		~3		~4	~0.25

Refer to room temperature (RT) and doping concentration of 10^{16} cm^{-3} (when applicable). *c* is direction perpendicular to the basal plane of the H polytypes.
Data are mainly taken from Harris (1995).

vacancy (V_C), (ii) the carbon interstitial (C_I) and the silicon interstitial (Si_I), (iii) antisite defects (C_{Si} and Si_C), and (iv) hydrogen and transition metal impurities. Emphasis is put on the electronic, paramagnetic, and optical properties, but also the thermal stability and mechanisms of formation, diffusion, and annihilation will be discussed. At the end, some final remarks and suggestion for future research tasks are made.

2. THE SILICON VACANCY (V_{Si}) AND THE CARBON VACANCY (V_C)

In spite of substantial efforts, rather few point defects in SiC are unambiguously identified both experimentally and theoretically with respect to atomic structure, optical, paramagnetic, and electronic properties. However, V_{Si} and V_C are exceptions and most of their characteristics are now established with a high degree of confidence; Figs. 2A and C illustrate there unrelaxed geometries in a cubic SiC structure. Theoretically, the formation energies of V_{Si} and V_C have been estimated by several authors (Bockstedte et al., 2003; Hornos et al., 2011; Torpo et al., 2001) and irrespective of the calculation method used, it was generally found that V_C is more abundant than V_{Si} by several orders of magnitude. This reflects the different atomic bond structure involved in the two cases and holds for both Si-rich and C-rich stoichiometry as well as *k* and *h* lattice sites. Both V_{Si} and V_C are electrically active, and results from recent bandgap error-free density-functional theory (DFT) calculations employing the HSE06 functional show charge states of 0, −1, and −2 for V_{Si} and +2, +1, 0, −1, and −2 for V_C in the bandgap of the 4H polytype (Hornos et al., 2011) (see Fig. 3).

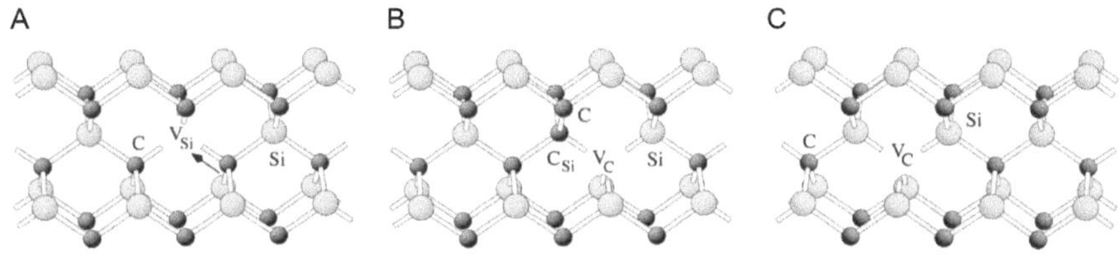

Figure 2 Vacancy defects in cubic SiC structure. (A) V_{Si}, (B) the carbon antisite–vacancy pair (C_{Si}–V_C), and (C) V_C. The arrow in (A) illustrates a jump of a C neighbor that transforms V_{Si} to C_{Si}–V_C in (B). *Reprinted with permission from Bockstedte et al. (2003). Copyright 2003 by the American Physical Society.*

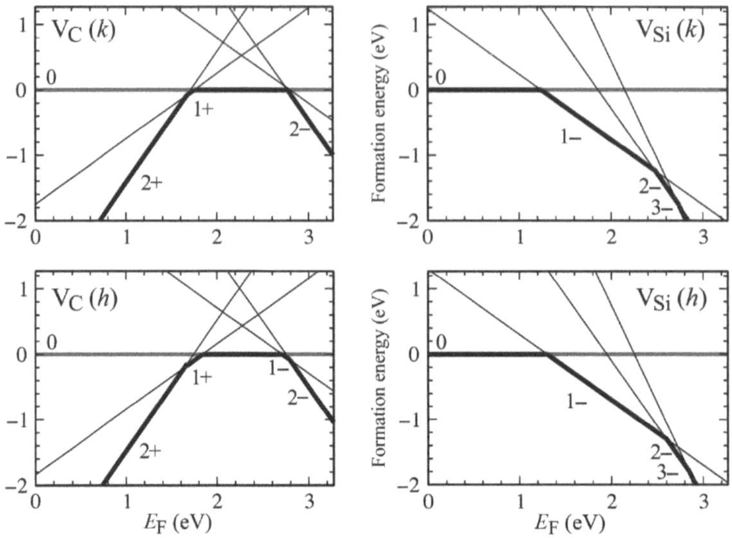

Figure 3 Relative formation energies of V_C (left column) and V_{Si} (right column). Charge state transitions are at the intersection of the lines. The Fermi energy position (E_F) is relative to the valence band edge (E_V). *Reprinted with permission from Hornos et al. (2011).*

2.1 V_{Si}

V_{Si} exhibits formation of nondegenerate high-spin states promoted by the strongly localized dangling bonds of the surrounding C atoms. This applies to both $V_{Si}(k)$ and $V_{Si}(h)$ and the point group symmetry is C_{3v} with the positions of the four nearest-neighbor C atoms slightly distorted from a regular tetrahedron (cf. Fig. 2A), as unveiled by electron paramagnetic resonance (EPR) studies of the ^{13}C hyperfine interaction in $V_{Si}^{-}(k)$ and $V_{Si}^{-}(h)$ (Mizuochi et al., 2003) and high-precision X-ray structure analysis (Bauer et al., 1998). For a vacancy system with no Jahn–Teller distortion, exchange interactions can result in high-spin ground states as is the case for the monovacancy in diamond (Isoya et al., 1992). Considering that V_{Si} and the monovacancy in diamond have similar local atomic environment, it can be anticipated that also V_{Si} has a high-spin ground state (effective spin $S=3/2$), as indeed confirmed for V_{Si}^{-} by early EPR and electron nuclear double resonance measurements of irradiated 3C, 4H, and 6H samples (Itoh et al., 1987, 1997; Wimbauer et al., 1997). In 4H samples, there are two main (and deep) no-phonon photoluminescence (PL) lines originating from V_{Si}, commonly labeled V1 and V2 with positions at 1352 and 1458 meV, respectively (Sörman et al., 2000). In 6H samples, three corresponding PL lines occur: V1 (1366 meV), V2 (1398 meV), and V3 (1433 meV). All these lines are also found in absorption measurements and they do not split in magnetic field. After some discussion in the literature on the interpretation of the V-lines, Janzén et al. (2009) presented strong evidence for a model where they originate from internal transitions between ground and excited states (multiplets) of $V_{Si}^{-}(k)$ (V1 line) and $V_{Si}^{-}(h)$ (V2 line) in 4H samples and $V_{Si}^{-}(k_1)$ (V1 line), $V_{Si}^{-}(h)$ (V2 line), and $V_{Si}^{-}(k_2)$ (V3 line) in 6H samples. In the model derivation by Janzén et al. (2009), optical selection rules for transitions between initial and final states having the same spin and optically detected magnetic resonance (ODMR) signals via each PL line were important factors.

As summarized above, the structural, paramagnetic, and optical properties of V_{Si} are known to a large extent, and because of its high formation energy (low abundance), V_{Si} is exclusively studied in irradiated samples. In contrast, the positions in the bandgap of the electrical charge state transitions of V_{Si} are not well established experimentally, and no confident assignment of deep levels to V_{Si} has been made. For the 4H polytype, a very tentative suggestion was put forward by David et al. (2004) ascribing the so-called S_1 and S_2 levels, with positions at 0.40 and 0.71 eV below the conduction band minimum (E_C), to the singly and doubly negative charge state of V_{Si}, respectively. However, this assignment needs to be corroborated by more extensive investigations and correlated with results from other techniques, ideally using the same samples. Furthermore, experimental data on the formation energy of V_{Si} are scarce in the literature, presumably because of the high temperatures required to obtain a detectable concentration. Also, the annealing mechanisms and kinetics for V_{Si} are not well determined and the amount of quantitative data is limited, especially for the 4H and 6H polytypes. Itoh et al. (1987, 1997) studied the disappearance of V_{Si}^{-} in electron-irradiated n-type 3C samples and observed two main annealing stages, as displayed in Fig. 4. The final one (at ~750 °C) was reported to obey first-order kinetics with an activation energy of ~2.2 eV, but no value was given for the pre-exponential factor of the rate constant. A possible annealing mechanism may be migration of V_{Si} with a barrier of 2.2 eV and subsequent trapping by a sink of substantially higher concentration than V_{Si}. For the low-temperature stage (at ~150 °C), a process invoking annihilation of V_{Si} with migrating self-interstitials was discussed.

For 6H and 3C samples of n-type, attempts were made by Kawasuso et al. (1996, 1998) to resolve the V_{Si} annealing kinetics using positron annihilation spectroscopy (PAS), and the results indicated that V_{Si} (or V_{Si}-impurity complexes) were present to much higher temperatures in the 6H samples

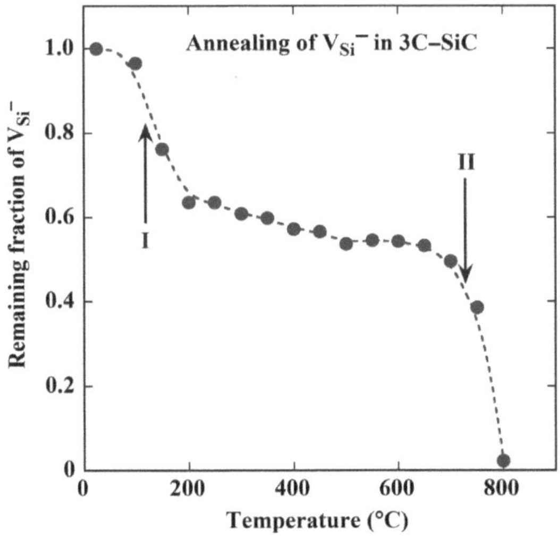

Figure 4 Remaining fraction of V_{Si}^- during isochronal (5 min) annealing of electron-irradiated 3C–SiC samples. *Reprinted with permission from Itoh et al. (1987). Copyright 1987, AIP Publishing LLC.*

(~1400 °C) than in the 3C samples (~800 °C). Kawasuso et al. (1996, 1998) proposed that V_{Si} formed complexes with nitrogen (used as dopant) in the 6H samples which remained stable up to 1400 °C and could not be readily resolved from the pure V_{Si} signal in their PAS lifetime spectra. Obviously, there is an urge for detailed investigations of the prevailing V_{Si} annealing mechanisms. In particular, isothermal studies using high-purity epitaxial layers of both n- and p-type are encouraged to elucidate the migration behavior of V_{Si} and the influence of the Fermi level position.

Finally, most recently and quite fascinating, single V_{Si}'s and ensembles thereof have been promoted as defects with long spin coherence times, enabling long-lived qubits even at room temperature (RT) (Widmann et al., 2015). Moreover, an ultra-bright single-photon source operating at RT has been demonstrated in high-purity semi-insulating 4H–SiC samples (Castelleto et al., 2014) and attributed to the C_{Si}–V_C antisite defect formed via transformation of V_{Si} (cf. Fig. 2B and also Section 3.3). Accordingly, SiC and its device processing technology are emerging as a potential platform for integrating spintronics, photonics, and electronics in a single quantum system.

2.2 V_C

Experimentally, V_C in 4H–SiC was first identified by EPR measurements using p-type material irradiated with MeV electrons to enhance the defect concentration (Son et al., 2001a; Umeda et al., 2004a,b). Two EPR signals labeled as EI5 and EI6 occur and they are today unambiguously ascribed to the singly positive charge state of $V_C(k)$ and $V_C(h)$, respectively (Bockstedte et al., 2003; Umeda et al., 2004a,b). Theoretically and in direct contrast to V_{Si}, V_C was shown by Zywietz et al. (1999) and Torpo et al. (2001) to exhibit a considerable Jahn–Teller distortion driven by the overlap of the Si dangling bonds, which have an extended character, causing a lowering of the defect symmetry. These relaxations resemble those of the Si vacancy in single-crystalline Si, but they were initially found to be remarkably different for the EI5 ($V_C(k)$) and EI6 ($V_C(h)$) spectra; for EI6, the four surrounding Si atoms rearranged to a C_{3v} symmetry but with a strongly increasing distortion at low temperatures leading to the low C_{1h} symmetry, while for EI5 no large distortion and no strong temperature dependence appeared. However, it was subsequently realized that EI5 arises from a motion-averaged state caused by thermally activated electronic bond switching (reorientation) between the surrounding Si atoms yielding an effective (high) symmetry of C_{3v}. At temperatures well below 50 K, the rate of reorientation decreases eventually and the observed symmetry of EI5 becomes lowered to C_{1h}, as expected from *ab initio* calculations (Bockstedte et al., 2003). The energy barrier for the transition from EI5 (C_{1h}) to EI5(C_{3v}) was deduced to be only 14 meV (Umeda et al., 2004b).

Later, also the EPR spectrum of $V_C^-(h)$ was reported by Umeda et al. (2005) using n-type samples irradiated with MeV electrons at elevated temperature and characterized under photon excitation (photo-EPR). The spectrum was labeled as HEI1 and arises from an unpaired electron in a Si–Si antibonding state of $V_C^-(h)$. Similar to that of $V_C^+(k)$, but not of $V_C^+(h)$, $V_C^-(h)$ exhibits thermal reorientation due to jumping of the unpaired electron between three equivalent Si–Si bonds (barrier height ~22 meV). The symmetry is C_{3v} (also during thermal averaging) but the unpaired electron always stays on one of the four Si atoms while jumping between the three other ones. As a result, the hyperfine structure of the HEI1 spectrum does not become fully "developed" and this complicates the detection of $V_C^-(k)$, which was not found in the pioneering work by Umeda et al. (2005).

Very recently, Son et al. (2012) and Trinh et al. (2013) were able to detect the "missing" $V_C^-(k)$ signal using relatively highly n-doped 4H–SiC epilayers irradiated with 250 keV electrons and combining EPR and photo-EPR measurements. The electron energy of 250 keV is close to the threshold for displacement of Si atoms and primarily V_C's and C_I's were generated, selectively promoting formation of carbon-related defects only.

Accordingly, in the absence of other EPR signals, which are dominant after high MeV electron irradiation, both $V_C^-(h)$ and $V_C^-(k)$ were resolved including their hyperfine structure as well as C_{1h} symmetry at low temperatures. The latter result facilitated comparison with DFT-based calculations of the V_C atomic and electronic structure, and the interpretation of the EPR spectra was firmly corroborated by theory (Trinh et al., 2013).

In their recent work, Son et al. (2012) also showed for the first time that V_C is a negative-U center, as perhaps might be expected because of the large Jahn–Teller distortion involved. However, this has been a standing controversy in the literature; results from early calculations by Torpo et al. (2001) indicated no negative-U behavior of V_C after charge correction, while without correction Zywietz et al. (1999) found evidence for negative-U of both the negative and positive charge states of V_C. Later, Bockstedte et al. (2010) reported negative-U for only the negative charge state of V_C after charge correction. More recent calculations by Hornos et al. (2011) using the HSE06 functional revealed negative-U properties of $V_C(k)$ only (and not $V_C(h)$) after charge correction with the double-negative charge state being about ~0.1 eV deeper than the single negative one (see Fig. 3). Compared with the EPR data of V_C, it is obvious that the schemes used for charge correction by Torpo et al. (2001) and Hornos et al. (2011) give rise to an overcorrection, and as assessed in some detail by Trinh et al. (2013), the appropriate charge correction technique is still under intensive research. For instance, if only potential alignment corrections are applied to the results by Hornos et al. (2011) and not so-called Freyholdt ones, also $V_C(h)$ displays a negative-U character (Trinh et al., 2013) albeit weaker than for $V_C(k)$, which is in full accordance with the EPR results.

Furthermore, the EPR data by Son et al. (2012) were also correlated with results from deep-level transient spectroscopy (DLTS) and capacitance–voltage (CV) measurements, partly using the same physical samples, and an unambiguous assignment of the so-called and very prominent $Z_{1/2}$ and EH_7 DLTS signals as different charge state transitions of V_C was demonstrated. In Fig. 5, a typical DLTS spectrum of an as-grown n-type 4H–SiC epitaxial layer is depicted and compared with that of a similar sample implanted with MeV Si ions to a fluence of $\sim 3 \times 10^8$ cm^{-2} and then annealed at 1150 °C. Both spectra are dominated by the $Z_{1/2}$ and $EH_{(6)/7}$ peaks while other defects are of minor importance having more than one order of magnitude lower intensity. $Z_{1/2}$ and $EH_{(6)/7}$ are located at ~0.7 and ~1.5 eV below the E_C, where $EH_{(6)/7}$ is somewhat broad with two contributions, EH_6 and EH_7. High-energy resolution DLTS studies have

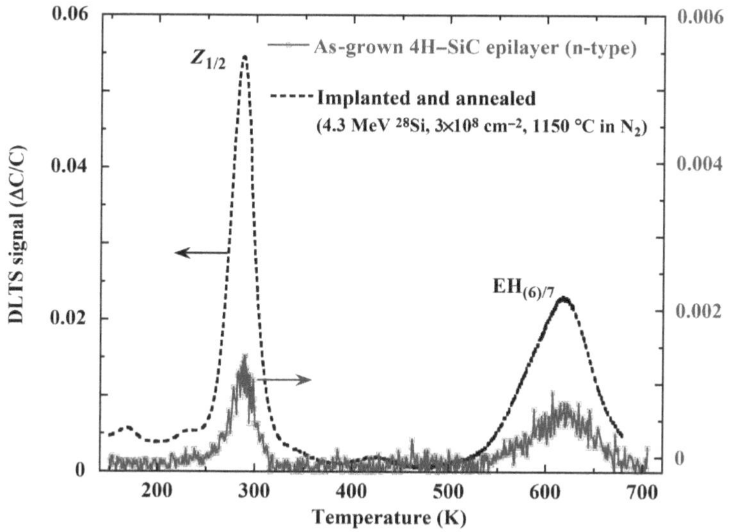

Figure 5 DLTS spectra of an as-grown n-type 4H–SiC epitaxial sample and one implanted with 4.3 MeV Si ions to a fluence of 3×10^8 cm^{-2} and then annealed at 1150 °C in flowing nitrogen for 3.5 h. *Reprinted with permission from Ayedh et al. (2014). Copyright 2014, AIP Publishing LLC.*

unveiled that EH_7 is the dominant contribution and prevails over EH_6 by a factor of typically 3–4 (Alfieri and Kimoto, 2013). In the implanted and annealed sample, $Z_{1/2}$ and EH_7 are enhanced by almost a factor of 40 relative to the as-grown sample which has a uniform concentration of $Z_{1/2}$ and EH_7 in the mid 10^{12} cm^{-3} range. Already in the late 1990s, Hemmingsson et al. (1998) discovered that the $Z_{1/2}$ peak is due a negative-U defect and has two components, Z_1 and Z_2. The negative-U character was revealed by illuminating the samples with light of ~470 nm wavelength before each DLTS filling pulse, and as a result, the $Z_{1/2}$ peak at $\sim E_C - 0.7$ eV disappeared completely while two new and more shallow levels emerged at ~0.52 and ~0.45 eV below E_C, respectively. A close one-to-two proportionality was unveiled between the sum of the two shallow peak amplitudes and the amplitude of the $E_C - 0.7$ eV peak recorded for the same sample with and without illumination, respectively. Hemmingsson et al. (1998) attributed the two shallow levels to single donor states (0/+) and the $E_C - 0.7$ eV peak to two overlapping single acceptor states (−/0); because of the inverted level ordering, emission from the acceptor states is immediately followed by emission from the donor states and the apparent $E_C - 0.7$ eV peak concentration becomes twice that of the actual defect concentration. The failure to detect the shallow levels in conventional DLTS measurements (cf. Fig. 5) is due to the fast capture of two electrons during each filling pulse leading to increased electron binding and "disappearance" of the shallow levels in the "altered" defect configuration. In the work by Son et al. (2012), it was shown that the $Z_{1/2}$ center does not exhibit any shallow donor levels and the $E_C-0.52$ and $E_C-0.45$ eV levels were conclusively reassigned as single acceptor states $(-/0)$ and the $E_C-0.7$ eV level as double acceptor state $(2-/-)$.

Several authors have reported a two-to-one relation between the strengths of the $Z_{1/2}$ and $EH_{(6)/7}$ peaks in a wide variety of different as-grown, irradiated, and annealed samples, as illustrated in Fig. 6 which shows the comprehensive compilation by Danno and Kimoto (2006). The relation holds over three orders of magnitude in concentration and provides almost unambiguous evidence that the two peaks arise from different charge states of the same defect. Further, high-temperature CV data of ion-implanted samples acquired by Son et al. (2012) showed that EH_7 is a deep single donor state.

Combining all these findings from EPR, photo-EPR, DLTS, and CV measurements, Son et al. (2012) were able to present the energy level scheme depicted in Fig. 7, correlating V_C with the $Z_{1/2}$ and EH_7 levels. The shallow acceptor states at $\sim E_C-0.52$ eV (Z_1) and $\sim E_C-0.45$ eV (Z_2) are identified as $V_C^-(h)$ and $V_C^-(k)$, respectively, while the $E_C-0.7$ eV level ($Z_{1/2}$) is due to the double acceptor states $V_C^{2-}(h)$ and $V_C^{2-}(k)$ and the $E_C-1.5$ eV level (EH_7) is due to the single donor states $V_C^+(h,k)$. In conjunction with Fig. 7, it should be pointed out that the

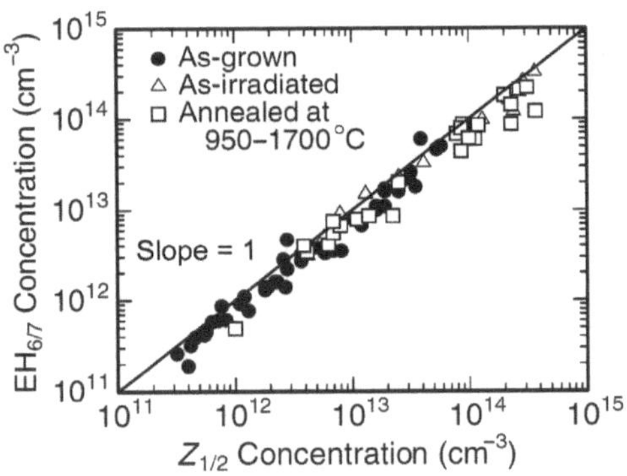

Figure 6 $EH_{(6)/7}$ concentration versus $Z_{1/2}$ concentration in as-grown epitaxial 4H–SiC samples, as-electron-irradiated samples (100–400 keV), and irradiated samples annealed at 950–1700 °C. *Reprinted with permission from Kimoto et al. (2008). Copyright 2008 Wiley-VCH Verlag GmbH & Co. KGaA, Weinheim.*

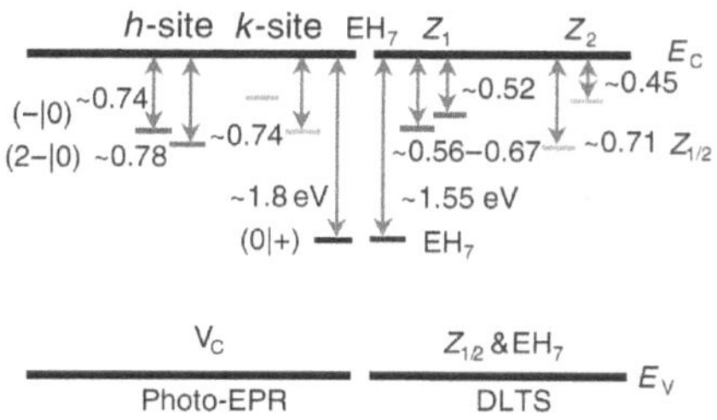

Figure 7 Energy level scheme of V_C determined by photo-EPR compared with the $Z_{1/2}$ and EH_7 levels determined by DLTS. The optical transitions obtained by photo-EPR involve a Franck–Condon shift in the range of 0.03–0.3 eV. For clarity, the location of the states in the scheme does not follow the relative scale. *Reprinted with permission from Son et al. (2012). Copyright 2012 by the American Physical Society.*

optical transitions in the photo-EPR measurements involve a Frank–Condon shift in the range of ~0.03–0.3 eV, as also supported by calculated values (Bockstedte et al., 2010). Some ambiguity has existed in the literature regarding the optical transition associated with the threshold of ~1.8–1.9 eV found in photo-EPR studies for detection of $V_C{}^{+}(h,k)$; Bockstedte et al. (2010) assigned the transition to electron excitation from E_V to the (+/2+) level of V_C rather than an excitation from the single donor level (0/+) to E_C, as originally proposed by Son et al., (2002) and Zvanut and Konovalov (2002) and depicted in Fig. 7. However, this issue has now been resolved since the same threshold occurs also in n-type material with the Fermi level located relatively close to E_C, and hence, the original assignment was correct.

The identification of the $Z_{1/2}$ and EH_7 peaks as different charge state transitions of V_C enables studies of V_C down to concentrations of $\sim 10^{10}$ to 10^{11} cm^{-3}, i.e., several orders of magnitude lower than the detection limit of EPR (and photo-EPR). Very recently, Ayedh et al. (2014) exploited this possibility and investigated the V_C concentration in high-purity epitaxial layers after heat treatment at different temperatures in the range of 1600–2000 °C. Both as-grown and as-oxidized layers were studied, and as illustrated in Fig. 8, $[V_C]$ increases by two orders of magnitude in the as-oxidized samples when the temperature increases from 1600 to 1950 °C (brackets denote concentration value). Also, the as-grown samples exhibit a similar increase in $[V_C]$ but as will be discussed in some detail later (Section 3.2), the oxidized samples have a lower initial content of V_C, and thus, the relative increase in $[V_C]$ (but not the absolute one) becomes larger in the oxidized samples. These results show that V_C is thermally very

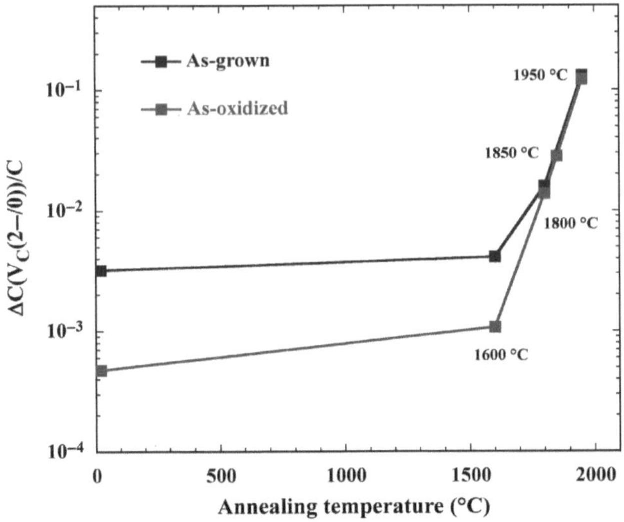

Figure 8 Amplitude of the $V_C(2-/0)$ DLTS peak (or $Z_{1/2}$) as a function of annealing temperature for as-grown and as-oxidized 4H epitaxial n-type samples. *Reprinted with permission from Ayedh et al. (2014). Copyright 2014, AIP Publishing LLC.*

stable, consistent with isochronal annealing data for electron-irradiated samples where $[Z_{1/2}]$ gradually decreased at temperatures above ~1000 °C but did not disappear fully and persisted even at 2000 °C (Alfieri et al., 2005).

The results in Fig. 8 were obtained from samples heat-treated isochronally (10 min) in an inductively heated furnace under C-rich conditions where the samples were protected by a pyrolyzed resist film (C-cap) and hosted in a graphite box with glassy carbon-coated surfaces under high-purity Ar ambient. Steady-state conditions (thermal equilibrium) were expected to apply above 1600 °C as evidenced by (i) the very similar values of $[V_C]$ in the as-grown and as-oxidized samples despite quite different initial conditions, and (ii) rather uniform concentration versus depth profiles of V_C (Ayedh et al., 2014). In fact, the depth profiles after the 1950 °C treatment exhibited a small decrease (~30%) toward the surface favoring a bulk formation process of V_C rather than Schottky formation at the surface with subsequent in-diffusion of V_C. On the basis of the data in Fig. 8, the following relation can be deduced

$$\lfloor V_C \rfloor \cong 3 \times 10^{25}\, e^{-4.95\,(\mathrm{eV})/kT}\, \mathrm{cm}^{-3} \tag{1}$$

where k is Boltzmann's constant and T is the absolute temperature. The value of ~4.95 eV represents the formation energy of V_C, in excellent agreement with recent theoretical predictions for C-rich conditions

(Hornos et al., 2011). The exponential prefactor value exceeds the number of available C-sites in the SiC crystal by almost three orders of magnitude and indicates a considerable increase of the crystal entropy, ΔS, during the V_C formation. Ayedh et al. (2014) extracted a value of $\Delta S \approx (5\text{–}6)k$ which is rather substantial and implies that atomic rearrangement occurs around V_C, in full accordance with the pronounced Jahn–Teller distortion of V_C. A similar dependence of $[V_C]$ (or $[Z_{1/2}]$) on temperature as given by Eq. (1) has also been reported by Zippelius et al. (2012) but with a slightly higher formation energy (5.7–6.6 eV) using different experimental conditions.

Finally, because of the firm assignment of the $Z_{1/2}$ peak, it can be concluded that V_C is a defect of decisive technological impact, controlling the minority charge carrier lifetime in n-type 4H–SiC wafers. Several authors (Danno et al., 2007; Kimoto et al., 2008, 2010; Klein et al., 2006) have identified $Z_{1/2}$ as a carrier lifetime killer in n-type 4H epitaxial layers and Fig. 9 shows the inverse of the carrier lifetime determined by microwave photoconductance decay measurements versus $[Z_{1/2}]$. For $[Z_{1/2}]$ above $\sim 10^{13}$ cm^{-3}, a linear correlation holds implying that Shockley–Read–Hall recombination via V_C is the dominant/controlling process, while below 10^{13} cm^{-3} other processes start to dominate. The data in Fig. 9 demonstrate that the presence of V_C has to be limited to concentrations in the low 10^{12} cm^{-3} range in order to reach carrier lifetimes of $\sim$10 μs, enabling realization of low-loss bipolar devices with blocking voltages in the range of several tens of kV.

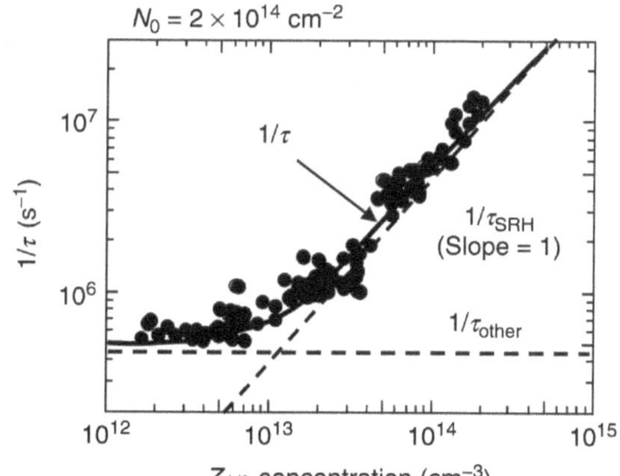

Figure 9 Relation between the inverse of the minority carrier lifetime and $[Z_{1/2}]$ in 4H samples. The injection level used was 2×10^{14} photons/cm^2. *Reprinted with permission from Kimoto et al. (2008). Copyright 2008 Wiley-VCH Verlag GmbH & Co. KGaA, Weinheim.*

3. THE SILICON INTERSTITIAL (Si_I), THE CARBON INTERSTITIAL (C_I), AND ANTISITE DEFECTS (C_{Si}, Si_C, C_{Si}–V_C)

Our understanding of the elementary self-interstitials and antisite defects stems primarily from theory and modeling while experiments are lacking behind. The so-called P-U and DII PL lines have been interpreted as C_I's or interstitial clusters because of their C-related local vibrational modes appearing at frequencies above those of the SiC bulk spectrum (Evans et al., 2002; Patrick and Choyke, 1973). However, no definite conclusion on the origin of these lines was possible, and in general, the limited amount of experimental data for self-interstitials can be largely associated with the following two issues:

(i) self-interstitials seem to have high formation energies typically exceeding those of their vacancy counterparts by up to several eV, with a possible exception for Si_I in p-type samples (Bockstedte et al., 2003), and this holds irrespective of stoichiometry conditions. Thus, the abundance of interstitials is low under thermal equilibrium and irradiation is required to obtain detectable concentrations.

(ii) self-interstitials are by far more mobile than the vacancies and are estimated to have a migration barrier of only ~1 eV in their low energy configurations (Bockstedte et al., 2003). Accordingly, they tend to form high-order complexes with other defects (or impurities) and become challenging to characterize in their isolated forms even in irradiated samples.

3.1 Si_I

Various configurations of Si_I exist in SiC, some of them differing only by C-like or Si-like local environment. Relevant Si_I configurations are displayed in Fig. 10, taken from the comprehensive *ab initio* DFT-based modeling study on intrinsic defects in 3C–SiC by Bockstedte et al. (2003). The hexagonal sites in Fig. 10 are found to be unstable and the same is true for bond-centered sites as well as split interstitials on the C sublattice, which have been omitted in the figure. The abundance of the different Si_I's depends strongly on the Fermi level position; in p-type material, the tetrahedrally carbon-coordinated configuration (Si_{TC}) in 4+ charge state dominates, while in intrinsic and n-type material the neutral <110>-oriented split one ($Si_{sp\langle 110\rangle}$) prevails. This has rather drastic implications on the mobility of Si_I which becomes much slower in p-type than in n-type material since

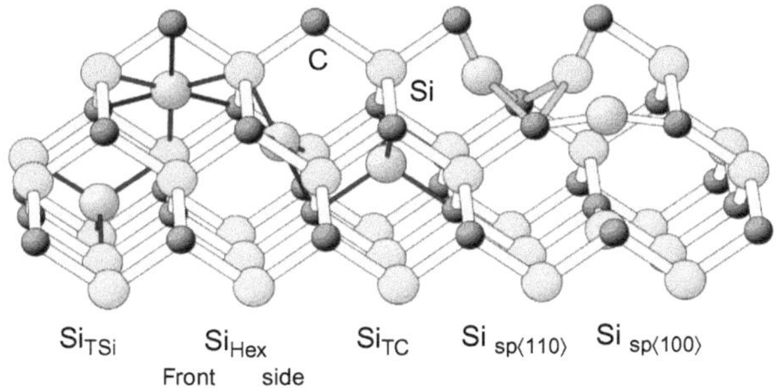

Figure 10 Configurations of the silicon interstitial in 3C–SiC. The hexagonal site is not stable. *Reprinted with permission from Bockstedte et al. (2003). Copyright 2003 by the American Physical Society.*

the migration barrier of Si_{TC} is calculated to be ~2.0 eV higher than that of $Si_{sp\langle 110\rangle}$ (~3.4 eV vs. ~1.4 eV). These results are anticipated to be of rather general validity and should hold also for other polytypes than 3C, such as 4H (Bockstedte et al., 2003).

3.2 C_I

Relative to Si_I, a different hierarchy of configurations occurs and the stable ones predicted by Bockstedte et al. (2003) are shown in Fig. 11. Irrespective of the Fermi level position, the split interstitials with $\langle 100\rangle$-orientation, i.e., $C_{sp\langle 100\rangle}$ on the C sublattice and $C_{spSi\langle 100\rangle}$ on the Si sublattice, have the lowest formation energy and they can appear in four or five different charge states ranging from 2+ to 2−. In the neutral and negative charge states of $C_{sp\langle 100\rangle}$, the axis of the C-pair is not fully oriented along the $\langle 100\rangle$-direction (as assumed in Fig. 11) but becomes tilted slightly in the $\{011\}$ plane and one of the C atoms in the pair bonds to the three Si atoms. This rebonding yields an energy gain of ~0.6 eV with a low energy barrier (<0.1 eV) separating the fully oriented configuration from the tilted one. Generally, the hierarchy of configurations reflects the prevalence of C to form short bonds in the $C_{sp\langle 100\rangle}$ and $C_{spSi\langle 100\rangle}$ split interstitials.

The estimated energy barriers for migration of $C_{sp\langle 100\rangle}$, which is energetically the most favorable configuration, are only about 1 eV and even well below that for the neutral and negative charge states. This implies a high mobility of C_I already at RT and suggests the employment of particle irradiation at cryogenic temperatures with online characterization as a viable future experimental concept to unveil spectroscopic signals of isolated C_I's.

In 2009, Hiyoshi and Kimoto (2009a,b) reported stunning results on how the concentration of the $Z_{1/2}$ and $EH_{(6)/7}$ deep levels in epitaxial

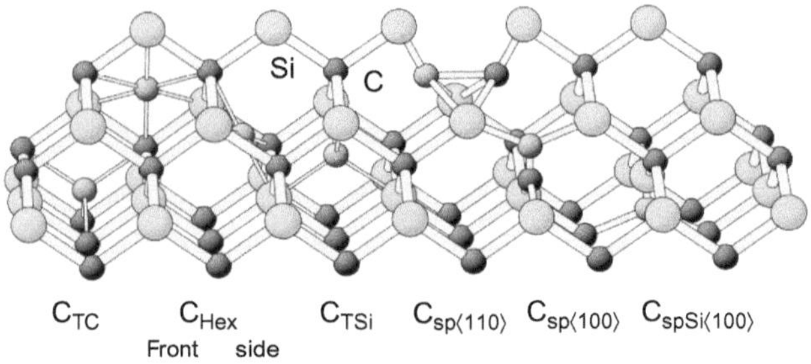

Figure 11 Configurations of the carbon interstitial in 3C–SiC. *Reprinted with permission from Bockstedte et al. (2003). Copyright 2003 by the American Physical Society.*

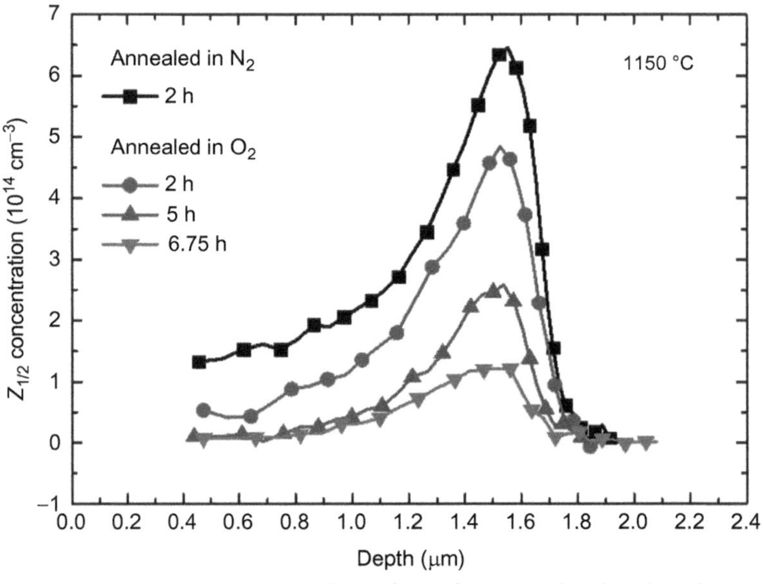

Figure 12 Concentration versus depth profiles of the $Z_{1/2}$ level in low-dose Si ion-implanted 4H samples, after annealing in O_2 and N_2 atmospheres at 1150 °C. *Reprinted with permission from Løvlie and Svensson (2012). Copyright 2012 by the American Physical Society.*

n-type 4H layers was drastically reduced below the DLTS detection limit after thermal oxidation of the Si-face of the samples. These results have later been confirmed by several authors and also in ion-implanted layers where the concentration versus depth distribution of $Z_{1/2}$ was followed as a function of oxidation time and temperature (Løvlie and Svensson, 2011, 2012) (see Fig. 12). It was conclusively shown that the reduction is due to injection of mobile species from the oxidizing surface which eliminate the $Z_{1/2}$ and $EH_{(6)/7}$ levels, and already in their first studies, Hiyoshi and Kimoto

(2009a,b) ascribed these species to C_I. This assignment was, indeed, later corroborated by the firm identification of the $Z_{1/2}$ and $EH_{(6)/7}$ levels as different charge state transitions of V_C (Son et al., 2012), and it is now general consensus that C_I's are injected from the oxidizing surface and annihilate V_C's in the bulk of the layers. Using low-fluence ion-implanted samples, Løvlie and Svensson (2012) undertook a detailed study of the reaction kinetics involved in the annihilation process of V_C, and at a given depth, the decay in $[V_C]$ was found to be of first order. Combining numerical simulations of the reaction kinetics with measured depth-resolved data for $[V_C]$, quantitative estimates for the (steady-state) concentration profile of injected C_I's and their migration energy were obtained. The value extracted for the latter was ~3.0 eV, i.e., about 2 eV higher than that predicted by Bockstedte et al. (2003) for the migration of $C_{sp\langle 100\rangle}$. This discrepancy of ~2 eV needs to be resolved but may indicate that an additional barrier exists at the SiO_2/SiC interface and has to be surmounted before the C_I's are injected into the SiC epitaxial layer. On the other hand, also self-diffusion experiments of ^{12}C and ^{13}C in intrinsic 4H samples (Linnarsson et al., 2004) yielded an activation energy for the diffusion of C about 1.5–2.0 eV higher than calculated for the interstitial mechanism (Bockstedte et al., 2003).

3.3 C_{Si}, Si_C, and the Carbon Antisite–Vacancy Pair (C_{Si}–V_C)

According to Torpo et al. (2001), both C_{Si} and Si_C are quite favorable defects with formation energies between 3 and 5 eV in the 4H polytype. C_{Si} is electrically inactive and exhibits inward relaxation of the nearest neighbors with a shortening of the C–C bonds by ~10% relative to the ideal Si–C bond length. No clear experimental evidence of isolated C_{Si} is reported, and because of its "inert" character, any properties enabling direct observation may be challenging to identify. In contrast, Si_C is expected to be electrically active with a shallow single donor state close to E_V for both the *h*-site and *k*-site in 4H but solid proof of its existence by spectroscopic techniques is still scarce in the literature. In a relatively early EPR study, Son et al. (2001b) tentatively attributed the EI6 spectrum to Si_C^+, but only a few years later EI6 was conclusively shown to arise from $V_C^+(h)$, as discussed in Section 2.2.

Perhaps the most unambiguous experimental support for the presence of antisite defects is the so-called AB PL lines appearing in the wavelength range of 648.7–676.5 nm in irradiated 4H and 6H samples and being enhanced after annealing up moderate temperatures (~400 °C) (Steeds, 2009). Using first-principles density-functional calculations, Eberlein et al. (2002) argued that the AB lines may be due to pairs of Si_C and C_{Si} in close proximity. However, in a comprehensive study compiling results from a wide variety of different 4H samples and a smaller amount of 6H samples, Steeds (2009) concluded that the AB lines arise from the neutral on-axis and off-axis C_{Si}–V_C pair (see Fig. 2B for the on-axis configuration). The conclusion was drawn on the basis of the AB lines spectral details, temperature dependence, emission energies, introduction, and annealing characteristics. This identification is also corroborated by theoretical results showing that V_{Si} is metastable under p-type and intrinsic conditions and transforms to the more energetically favorable C_{Si}–V_C pair via a barrier of ~1.8–2.4 eV (Bockstedte et al., 2004; Rauls et al., 2000). Moreover, in the recent work by Castelleto et al. (2014) exploiting C_{Si}–V_C as a RT single-photon source (see Section 2.1), it was found that the AB lines grow on the expense of the V1 line (arising from V_{Si}) during annealing up to 800 °C. Thus, the assignment of the AB lines to the C_{Si}–V_C pair can be considered as firm, especially since recent results from *ab initio* super cell calculations of the C_{Si}–V_C structure in conjunction with group theory are able to describe the spectral features of the AB lines (Castelleto et al., 2014).

4. HYDROGEN

Hydrogen (H) affects strongly the electronic properties of semiconductors because of its ability to passivate electrically active defects by forming neutral complexes. The effect of hydrogen has been studied in various semiconductors since H is one of the most common contaminants. Indeed, a significant amount of H is anticipated to be incorporated in SiC epitaxial layers since the standard CVD growth technique employs silane (SiH_4) and propane (C_3H_8) as precursors and H_2 as carrier gas. Device fabrication processes such as wet oxidation, forming gas annealing, and dry etching can also introduce H. Therefore, it is technologically important to understand the effects of hydrogen in SiC. Further, since hydrogen is the simplest atom of all, studying the properties of hydrogen will aid a better fundamental understanding of impurities in SiC.

4.1 Incorporation

A clear indication of hydrogen incorporation in boron (B)-doped p-type 6H–SiC samples was first reported by Larkin et al. (1995). Figure 13A shows depth profiles of B and H in p-type 6H–SiC epilayers measured by secondary ion mass spectrometry (SIMS). Boron was doped by adding diborane

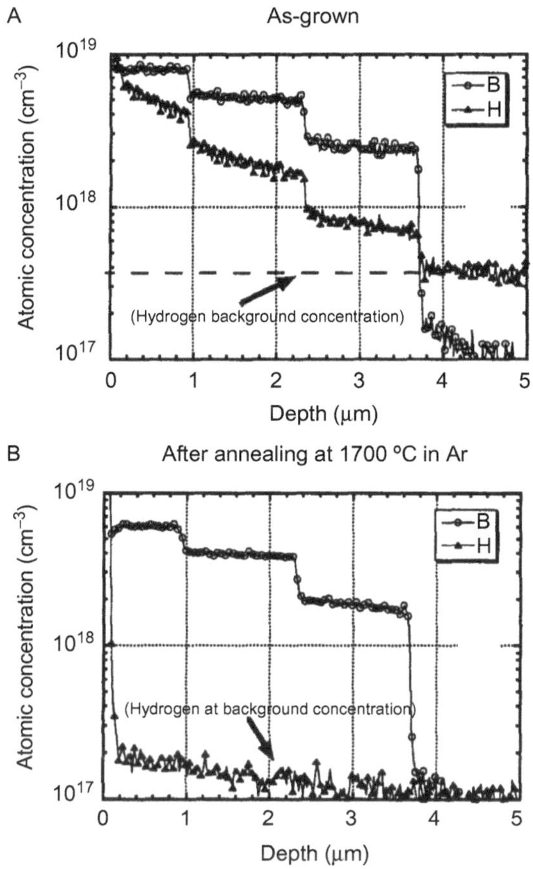

Figure 13 Depth profiles of B and H in (A) as-grown p-type 6H–SiC epilayer and (B) after annealing at 1700 °C in Ar. The step-like B concentration profile was formed by the site-competition effect. The H incorporated in the as-grown epilayer also forms a step-like profile corresponding to the B profile. The incorporated H is removed by annealing at 1700 °C in Ar. *Reprinted from Larkin et al. (1995). With kind permission from Springer Science and Business Media.*

(B_2H_6) in the standard precursor and carrier gases, and its step-like doping profile was accomplished via site competition during the CVD growth (i.e., by changing the Si/C ratio in the precursors). As can be seen in Fig. 13A, also H forms a step-like depth profile corresponding to the B profile although the amount of H in the CVD system was not changed. This implies that H is incorporated under the presence of B. The incorporated H can be removed by high-temperature annealing in the absence of H, as shown in Fig. 13B. The incorporation and removal of H have also been confirmed by low-temperature PL measurement (Larkin et al., 1995; Schöner et al., 1997). With the removal of the incorporated H, the net hole concentration increases indicating that H passivates the B acceptor by forming a neutral B–H complex. The incorporation of H was also observed in aluminum (Al)-doped p-type 4H and 6H samples, and the H concentration profile correlated with that of Al (Schöner et al., 1997). Since also Al is one of the shallow acceptor dopants in SiC, formation of Al–H complexes can be anticipated. In direct contrast to p-type material, no clear evidence of donor–H complexes in n-type material has been reported so far, possibly reflecting Coulomb repulsion between donors and H.

Hydrogen is considered to occur in either positive or negative charge state depending on the position of Fermi level (Aradi et al., 2001). The positively charged H is experimentally confirmed in p-type SiC (Janson et al., 2004; Samiji et al., 2001). Figure 14 shows the net hole concentration profiles of a deuterium (^{2}H)-passivated Al-doped p-type SiC Schottky barrier diode (SBD) recorded as a function of time after the deuterium treatment (Janson et al., 2004). In between the measurements, the SBD was kept at 260 °C and reverse biased at 5 V. The net hole concentration increases with time in the surface vicinity (<1.5 μm), while it decreases in the deeper region (>1.5 μm). This indicates that Al–H complexes at the surface vicinity dissociate with time, and released H^+ ions migrate to deeper regions assisted by the electric field. The migrated H^+ ions form Al–H complexes again at

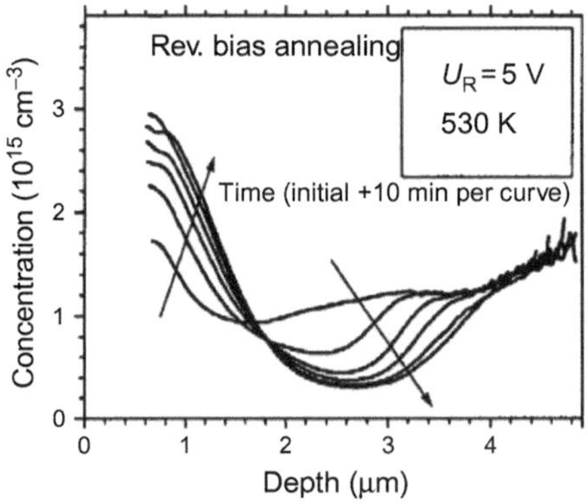

Figure 14 Net hole concentration profiles of a deuterium (^{2}H)-passivated Al-doped p-type SiC SBD measured as a function of time. In between the measurements, the SBD was kept at 260 °C and reverse biased at 5 V. The change of the net hole concentration profile proves the presence of positively charged H ions in p-type SiC. *Reprinted from Janson et al. (2004). Copyright The Royal Swedish Academy of Sciences. Reproduced by permission of IOP Publishing. All rights reserved.*

the edge of depletion region. It is readily understood that H^+ prefers to form a complex with a negatively charged acceptor in p-type SiC, and the resulting complex becomes neutral. Contrary to H^+, H^- has not been experimentally detected in SiC so far.

An interesting application utilizing the effect of acceptor passivation in SiC by hydrogen is H_2 gas sensors which can be operated at elevated temperature. As mentioned above, the net hole concentration decreases as H atoms/ions form neutral complexes with acceptors. This phenomenon can be detected by monitoring the increase in the series resistance of an SBD. Figure 15 shows the forward current–voltage characteristics of an SBD fabricated on a p-type 4H–SiC epilayer (Janson et al., 2004). Platinum (Pt) was used as the Schottky metal since its catalytic property leads H_2 molecules to dissociate and the resulting H atoms can diffuse into the epilayer. The SBD was annealed at 600 °C repeatedly in N_2 ambient with the presence of either 2% of O_2 or H_2. After the annealing in H_2/N_2 ambient, an increase in the series resistance is clearly observed (the forward current

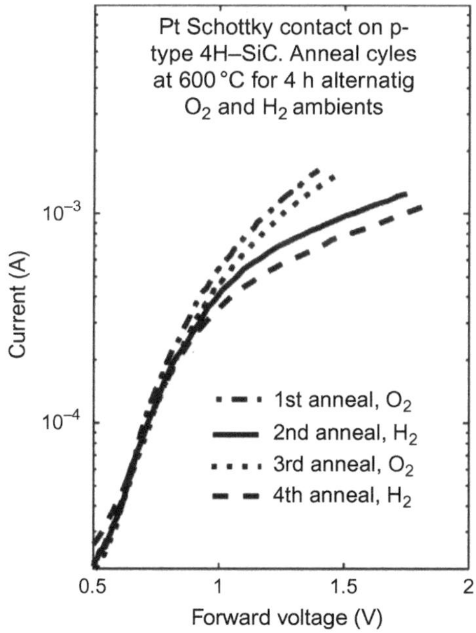

Figure 15 Forward current–voltage characteristics of an SBD fabricated on a p-type 4H–SiC epilayer. Forward current is decreased by annealing in H_2, and it is restored by subsequent annealing in O_2. *Reprinted from Janson et al. (2004). Copyright The Royal Swedish Academy of Sciences. Reproduced by permission of IOP Publishing. All rights reserved.*

decreases). This can be explained by that a significant amount of H atoms diffuse into the p-type epilayer and the shallow acceptors are partially passivated. However, after the subsequent annealing in O_2/N_2 ambient, the forward current is restored due to out diffusion of H. Thus, this is a reversible reaction and the SBD can be used as a sustainable H_2 gas sensor.

4.2 Acceptor–Hydrogen Complexes

Physical properties of B–H and Al–H complexes in 4H–SiC were investigated theoretically by using *ab initio* supercell calculations (Aradi et al., 2001; Deák et al., 2001). Here, both the B and Al acceptors were assumed to substitute Si lattice site (B_{Si} and Al_{Si}) (Larkin, 1997). Among the investigated possible B–H configurations, the complex with H in a bond center position (H_{BC}) between B_{Si} and one of its neighboring C atoms was found to be the most stable one. For the H–Al complex, on the other hand, a configuration with H at an antibonding position (H_{ABc}) to one of the first neighbor C atoms of Al_{Si} was the most stable one. The B_{Si}–H_{BC} and Al_{Si}–H_{ABc} geometries are shown in Fig. 16A and B, respectively. The difference in H positions between the B_{Si}–H_{BC} and Al_{Si}–H_{ABc} complexes is attributed to the size of the acceptor atom. The H atom would prefer to occupy a bond center position to have a normal bond with C also in the Al–H complex, similar to the B_{Si}–H_{BC} complex, but since the Al atom is big, this configuration involves too large lattice relaxation. Consequently, it is not energetically favorable. The binding energies (E_b) of B_{Si}–H_{BC} and Al_{Si}–H_{ABc} were calculated to be 1.6 and 0.7 eV, respectively. The difference in E_b (0.9 eV) is due to their different configurations. Hydrogen in B_{Si}–H_{BC} bonds to the

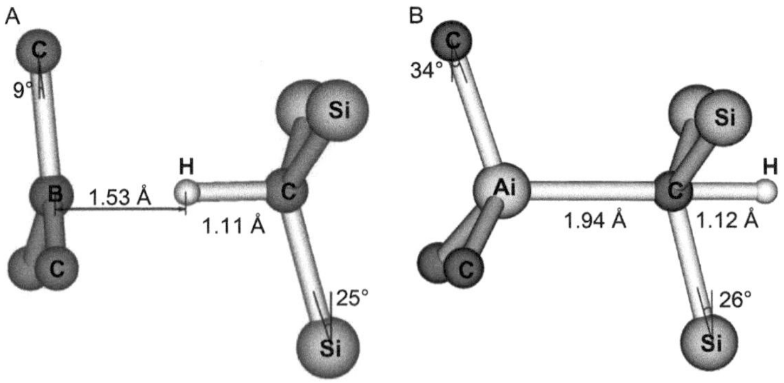

Figure 16 Geometries of (A) B_{Si}–H_{BC} and (B) Al_{Si}–H_{ABc} determined by *ab initio* calculations. These are the most stable configurations for B–H and Al–H complexes in 4H–SiC. *Reprinted from Aradi et al. (2001). Copyright 2001, with permission from Elsevier.*

C atom as one of four neighbors of the tetrahedral coordinated C atom, giving rise to a strong H–C bond, while H in Al_{Si}–H_{ABc} bonds weakly to the C atom from an antibonding position. Both the B_{Si}–H_{BC} and Al_{Si}–H_{ABc} complexes were found to be electrically inactive, consistent with the acceptor passivation effect observed experimentally.

The difference in E_b between B_{Si}–H_{BC} and Al_{Si}–H_{ABc} can be revealed by their dissociation energies (E_d) which have been determined experimentally. In general, E_d of an acceptor–H complex is given by the sum of the binding energy (E_b) and the migration energy (E_m) of a mobile H, and thus $E_d = E_b + E_m$. E_m is expected to be more or less the same for B- and Al-doped p-type SiC, and hence, any difference in E_d can be attributed to a difference in E_b. E_d for B_{Si}–H_{BC} and Al_{Si}–H_{ABc} were deduced by measuring the effective diffusion constant (D_{eff}) of H in B- and Al-doped 4H–SiC, respectively (Janson et al., 2001). Under the condition in which local equilibrium is established, H should diffuse according to Fick's law with D_{eff}. By fitting a Gaussian function to the redistributed H profile measured after annealing at a certain temperature and time, D_{eff} can be deduced and subsequently also the dissociation rate (ν) of the acceptor–H complex. From an Arrhenius analysis of ν, E_d can then be determined. A detailed discussion of this method can be found in Janson et al. (2001, 2004). Janson et al. (2001, 2004) studied the redistribution of H profiles in B- and Al-doped 4H samples annealed at a large set of temperatures and times, and they obtained $E_d \approx 2.5$ eV for the B–H complex and $E_d \approx 1.6$ eV for the Al–H complex. Thus, this difference of 0.9 eV in E_d between the two complexes can be attributed to their different E_b's and the 0.9 eV value is identical with that obtained theoretically (Aradi et al., 2001; Deák et al., 2001). From the experiments by Janson et al. (2001), it was not possible to determine the exact atomic structures of the B–H and Al–H complexes; however, the large difference in E_d suggests that their configurations are significantly different in accordance with the results from *ab initio* calculations (see Fig. 16) (Aradi et al., 2001; Deák et al., 2001).

4.3 Carrier Lifetime Improvement by Hydrogen Annealing

As a beneficial effect of hydrogen, improvement of the minority carrier lifetime in p-type SiC by H_2 annealing has recently been demonstrated (Okuda et al., 2013, 2014). Figure 17 shows the photoconductivity decay curves of a p-type 4H epilayer sample measured before and after Ar and H_2 annealing (Okuda et al., 2013). The epilayer was doped with Al, and the net hole

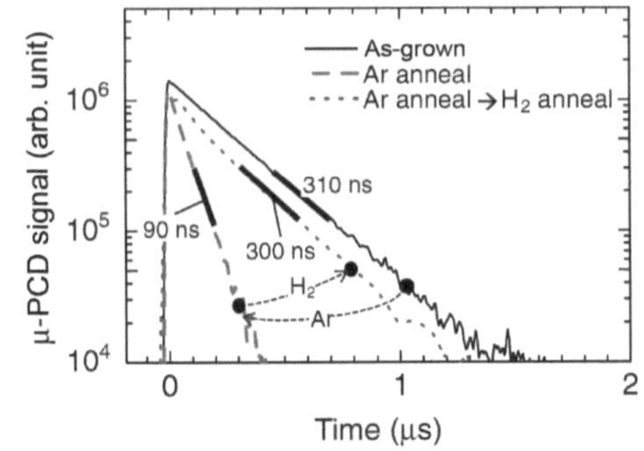

Figure 17 Photoconductivity decay curves of a p-type 4H–SiC epilayer measured before and after Ar and H_2 annealing. The carrier lifetimes deduced from the photoconductivity curves are shown in the figure. *Reprinted with permission from Okuda et al. (2013). Copyright 2013 The Japan Society of Applied Physics.*

concentration, as determined by CV measurements, was 1×10^{18} cm^{-3}. The carrier lifetime in the as-grown epilayer extracted from the photoconductivity decay curve was 310 ns, which is actually quite long at such a high doping concentration. After annealing at 1000 °C in Ar ambient, the carrier lifetime decreased to 90 ns. The carrier lifetime was then restored to the initial level by annealing in H_2 at 1000 °C, and the behavior was confirmed to be reversible. The exact physics processes behind the phenomenon have not been identified yet, but they might be qualitatively explained as follows: There are some unknown carrier lifetime killing defects in p-type 4H–SiC which can be passivated by H. In the as-grown epilayer, those defects are already passivated because H is incorporated during the CVD growth and the subsequent cooling process. At high temperatures and in the absence of H_2 ambient, H breaks up from the defects and diffuses out. However, the defects can be passivated by H again in H_2 ambient at high temperatures. The incorporation of H in the epilayer by H_2 annealing was confirmed by using deuterium ($^2H_2$) annealing and subsequent SIMS analysis. The SIMS analysis identified that 2H #Deutrium with a concentration of (2–3) $\times 10^{15}$ cm^{-3} was incorporated in the epilayer during the $^2H_2$ annealing, while the ^2H concentration was below the detection limit (5×10^{14} cm^{-3}) after the Ar annealing. The incorporated H concentration is significantly lower than the acceptor doping concentration (1×10^{18} cm^{-3}), and the net hole concentration did not change after the H_2 annealing. In this context, it can be mentioned that the Al–H complex starts to dissociate at much lower temperature (~300 °C) (Janson et al., 2001; Samiji et al., 2001) than 1000 °C, and acceptor passivation by

H cannot be a reason for the carrier lifetime improvement. Hence, there seem to be deep-level defects which act as carrier lifetime killers in p-type layers, similar to that in n-type layers. Further, the H_2 annealing was recently reported to be efficient for carrier lifetime improvement also in low-doped p-type 4H–SiC layers (Okuda et al., 2014).

A frequent use of highly doped p-type SiC epilayers is as base layer in bipolar junction transistors (BJT), and a carrier lifetime of 300 ns is presently considered to be sufficient for such transistors. However, since various treatments at high temperatures are required for the SiC BJT fabrication, it is crucial that the H_2 annealing for carrier lifetime improvement is performed at the appropriate order.

5. TRANSITION METAL IMPURITIES

Transition metals in semiconductors are known to strongly influence the physical properties even if their concentrations are low. Because of incompletely occupied d-orbits, transition metals can capture electrons and thus energy states (defect levels) can be formed in the bandgap. These defect levels act as carrier traps and/or recombination centers, and they are, therefore, generally harmful for device operation (Weber, 1983). Transition metals, such as titanium (Ti), vanadium (V), chromium (Cr), and iron (Fe), are common residual impurities in SiC crystals (Hobgood et al., 1995; Schulz et al., 1999). In particular, the SiC source material (powder) and the graphite parts used for crystal growth contain certain amounts of transition metals, and these impurities are readily incorporated during high-temperature growth. In fact, effects of transition metals in SiC have been investigated since the 1990s, and up to date, some of the defect levels related to transition metals have been firmly identified, as will be discussed below.

5.1 Identification and Characterization of Transition Metal-Related Defect Levels

Defect levels related to Ti, V, and Cr in the bandgap of 4H–SiC were identified by DLTS, utilizing implantation of radioactive isotopes (Achtziger and Witthuhn, 1997, 1998). A radioactive isotope (parent isotope) decays to its daughter isotope with time following an exponential function of the characteristic nuclear transmutation half-life. By repeating DLTS measurements to monitor the defect concentrations which decrease and increase with time, the parent and the daughter isotope-related defect levels can be identified, respectively. If the defect level involves either a single parent or single daughter

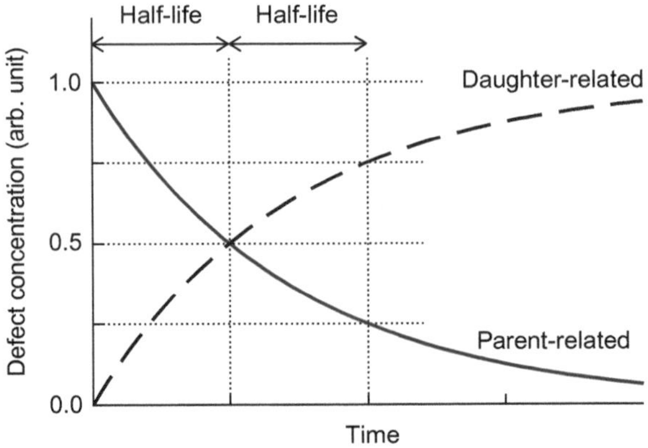

Figure 18 The concentration of a single parent- and single daughter-related defect increases and decreases, respectively, with time following an exponential function of the characteristic nuclear transmutation half-life.

isotope, the change in the defect concentration follows exactly the characteristic half-life as illustrated in Fig. 18. If two isotopes are involved in one defect level, the defect concentration changes faster (Achtziger et al., 1999).

In order to identify Ti-, V-, and Cr-related defect levels, Achtziger and Witthuhn (1997, 1998) implanted parent isotopes of ^{51}Cr and ^{48}V into n-type 4H–SiC epilayers and repeated DLTS measurements with several delay times. ^{51}Cr decays to ^{51}V with the half-life of 27.7 days, while ^{48}V decays to ^{48}Ti with the half-life of 16 days. These parent isotopes were produced by recoil implantations using appropriate target foils and primary ion beams. Figures 19A and B show DLTS spectra of ^{51}Cr- and ^{48}V-implanted n-type 4H–SiC epilayers, respectively. Note that the samples were annealed at 1600 °C after the implantation in order to reduce implantation-induced intrinsic defects.

As can be seen in Fig. 19A, the peaks labeled #1, #2, and #3 decrease with time while the peak #4 increases. Since the changes of these peak heights follow the half-life of the ^{51}Cr decay, the peaks #1, #2, and #3 are unambiguously identified as single Cr-related levels while the peak #4 is a single V-related level. The energy positions and electron capture cross sections of the levels were deduced from their Arrhenius plots, and the results are summarized in Table 2. The levels #1 and #2 are considered as donor type since their emission rates depend on the electric field in the range from 15 to 90 kV/cm (Poole–Frenkel effect), while #3 and #4 are likely to be acceptor levels. The peak #4 was assigned as a V-related level and it occurred already from the beginning; this is because some amount

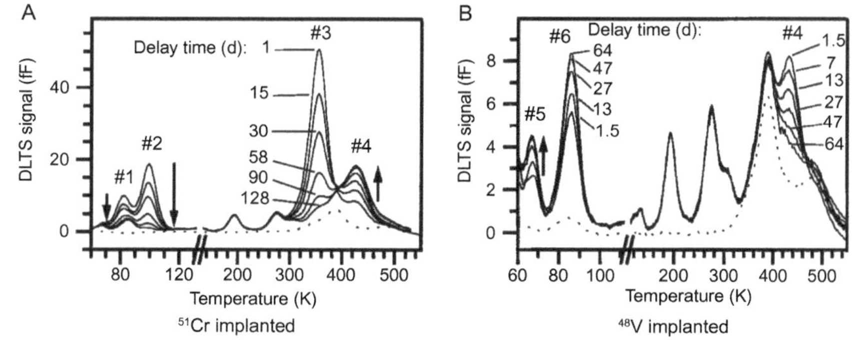

Figure 19 DLTS spectra of (A) ^{51}Cr-implanted and (B) ^{48}V-implanted n-type 4H–SiC epilayers measured after several delay times as indicated in the figures. The dashed line in each figure is a reference spectrum for an unimplanted sample. The half-life of ^{51}Cr and ^{48}V are 27.7 and 16 days, respectively. The DLTS signal unit is 10^{-15F}. *Reprinted with permission from Achtziger and Witthuhn (1997). Copyright 1997, AIP Publishing LLC.*

Table 2 Energy Positions (Activation Enthalpies) and Capture Cross Sections of Deep Levels Related to Cr, V, Ti, and Fe, Deduced from Arrhenius Plots of the Emission Rates

Label	Energy Position (eV)	Capture Cross Section (cm^2)	Element	Reference
#1	$E_C - 0.15$	2×10^{-15}	Cr	Achtziger and Witthuhn (1997)
#2	$E_C - 0.18$	8×10^{-16}	Cr	
#3	$E_C - 0.74$	2×10^{-15}	Cr	
#4	$E_C - 0.97$	6×10^{-15}	V	
#5	$E_C - 0.13$	7×10^{-15}	Ti	
#6	$E_C - 0.17$	1×10^{-15}	Ti	
Fe1	$E_C - 0.39$	2×10^{-15}	Fe	Beyer et al. (2011)
Fe2	$E_V + 1.46$	3×10^{-14}	Fe	
Fe3	$E_V + 0.97$	7×10^{-16}	Fe	

The capture cross sections of Fe2 and Fe3 are for holes while the others are for electrons.

of the stable ^{51}V isotope was also present together with ^{51}Cr during the recoil implantation. Other peaks seen at 180 and 280 K do not change with time and are considered as intrinsic defect levels (cf. Section 2.2).

In the DLTS spectra of ^{48}V-implanted samples (Fig. 19B), the V-related peak #4 decreases with time while the peaks labeled #5 and #6 increase following the half-life of ^{48}V decay. Consequently, peaks #5 and #6 are identified as single Ti-related defect levels (Achtziger and Witthuhn,

1997). The energy positions and electron capture cross sections are $E_C - 0.13$ eV and 7×10^{-15} cm^2 for peak #5, and $E_C - 0.17$ eV and 1×10^{-15} cm^2 for peak #6 (as listed in Table 2). A Poole–Frenkel effect was not found for the peaks #5 and #6 under an electric field ranging from 10 to 140 kV/cm, and thus, they are likely to be of acceptor type (Dalibor et al., 1997). Ti in 4H–SiC is known to exhibit PL emission with zero-phonon lines (ZPLs) at 2.85 and 2.79 eV, and this has been attributed to a radiative recombination of an exciton bound to the Ti level (Patrick and Choyke, 1974). In a detailed ODMR study, the ZPLs at 2.85 and 2.79 eV were assigned to Ti acceptors substituting hexagonal and cubic Si sites ($Ti_{Si}(h)$ and $Ti_{Si}(k)$), respectively (Lee et al., 1985). Taking these PL and ODMR results into account, the Ti-related levels #5 and #6 were assigned to $Ti_{Si}(h)$ and $Ti_{Si}(k)$, respectively, where the shallower level corresponds to the larger photon energy, and vice versa (Dalibor et al., 1997).

Fe-related deep levels were studied by DLTS measurements on Fe-doped n- and p-type 4H–SiC epilayers (Beyer et al., 2011; Trapaidze et al., 2011). Figure 20A compares DLTS spectra of n-type epilayers grown with and without Fe doping. In the Fe-doped sample, a peak labeled as Fe1 is detected in addition to the well-known V_C levels $Z_{1/2}$ and $EH_{(6)/7}$ (cf. Section 2.2). The energy position and electron capture cross section deduced for Fe1 were $E_C - 0.39$ eV and 2×10^{-15} cm^2, respectively. The concentration of Fe1 obtained from the DLTS peak height was 7.6×10^{12} cm^{-3}, while the chemical Fe concentration in the epilayer determined by SIMS was 6.5×10^{14} cm^{-3}. Therefore, only a small fraction of Fe

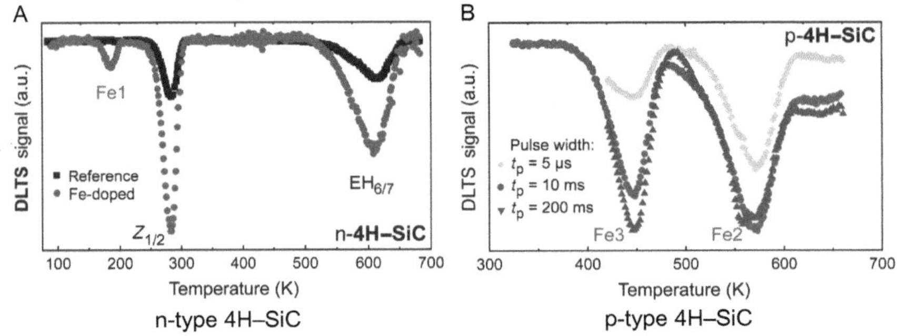

Figure 20 (A) DLTS spectra of n-type 4H–SiC epilayers with and without Fe doping during the epitaxial growth. Fe1 is detected only in the Fe-doped sample. (B) DLTS spectra of p-type 4H–SiC epilayers doped with Fe and measured with different DLTS pulse width. *Reprinted with permission from Beyer et al. (2011). Copyright 2011, AIP Publishing LLC.*

in the epilayer contributes to the Fe1 level. Figure 20B shows DLTS spectra of p-type 4H–SiC epilayers doped with Fe during growth. Two peaks labeled as Fe2 and Fe3 were detected with energy positions and hole capture cross sections of $E_V + 1.46$ eV and 3×10^{-14} cm^2 for Fe2, and $E_V + 0.97$ eV and 7×10^{-16} cm^2 for Fe3 (as also listed in Table 2). Earlier DLTS studies on p-type 4H–SiC without Fe doping revealed deep levels with similar energy positions as either Fe2 or Fe3, e.g., HK4 ($E_V + 1.49$ eV) (Danno and Kimoto, 2007) and h2 ($E_V + 0.96$ eV) (Han and Lee, 2004). However, the actual emission rates of Fe2 and Fe3 are very different from those of HK4 and h2, and hence, they must be of different origin. Moreover, Trapaidze et al. (2011) detected two levels identical to Fe2 and Fe3 in Fe-implanted p-type 4H–SiC epilayers but not in Ar-implanted ones. In addition, the peak heights of these two levels were almost proportional to the Fe implantation dose, and thus, these results by Trapaidze et al. (2011) corroborate the assignment of Fe2 and Fe3 to iron. Finally, from the DLTS results depicted in Figure 20B, a total concentration of Fe2 and Fe3 of 3.9×10^{14} cm^{-3} was determined and found to be comparable to the chemical Fe concentration (6.2×10^{14} cm^{-3}) determined by SIMS. Hence, Fe2 and Fe3 account for a major fraction of the introduced Fe concentration.

5.2 Semi-insulating SiC by V-Doping

Vanadium is historically one of the most studied transition metals in SiC because of its use to realize semi-insulating (high-resistivity) substrates (or layers) by intentional doping (Hobgood et al., 1995; Jenny et al., 1995; Kimoto et al., 1996). Semi-insulating substrates are essential for fabricating high-power microwave devices, leading to a minimum of dielectric losses caused by parasitic capacitance. In general, semi-insulating semiconductor materials are produced by forming defect levels near the middle of the bandgap such that residual shallow dopant impurities are efficiently compensated. Consequently, the Fermi level should be pinned at the mid-gap level(s).

EPR studies by Schneider et al. (1990) unveiled that V appears in semi-insulating SiC with one of three possible charge states: positive V^{5+} ($3d^0$), neutral V^{4+} ($3d^1$), and negative V^{3+} ($3d^2$). Therefore, V occurs predominantly as either an acceptor V^{4+}/V^{3+} ($-/0$) or a donor V^{5+}/V^{4+} ($0/+$), depending on the Fermi level position. As shown in Figs. 19A and B, the DLTS peak #4 ($E_C - 0.97$ eV) was assigned to V, and since its emission rate did not depend on the electric field strength (i.e., a Poole–Frenkel effect was not observed), the level was considered to be acceptor-like (Achtziger and Witthuhn, 1997). From the temperature dependence of the resistivity of 4H–SiC samples co-doped with V and N, the thermal activation energy was determined to be 0.8 eV, which is in fair agreement with the position of the assigned V acceptor level ($E_C - 0.97$ eV) (Jenny et al., 1996) and indicates that the Fermi level was pinned at the V acceptor state. Contrary to the V acceptor level, no V donor level has been detected by DLTS so far. This is presumably because the anticipated V donor level position is close to midgap and only possible to detect by very high-temperature DLTS measurements. However, for semi-insulating 4H–SiC substrates, an activation energy of ~1.6 eV has often been found employing temperature-dependent resistivity measurements. In addition, EPR data of such substrates showed no existence of the V^{3+} signal, corroborating that the Fermi level is pinned to a deeper V-related electron state (Mitchel et al., 1999, 2007).

Dominant residual shallow dopants in semi-insulating SiC are N (donor) and B (acceptor). In order to obtain semi-insulation, [V] must be higher than the net concentration of N and B. In the case of [N] < [B], the Fermi level is pinned at the V donor level while for [N] > [B], it is pinned at the V acceptor level (Mitchel et al., 2007; Son et al., 2007a,b). In addition, EPR studies by Son et al. (2007a,b) suggested that the semi-insulating behavior of 4H–SiC is not only due to V but that also intrinsic defects, such as divacancies ($V_C V_{Si}$), can play a role if [V] is below $\sim 10^{17}$ cm^{-3}.

Finally, the use of V for commercial production of semi-insulating SiC wafers is decreasing. Electrons trapped by V in the semi-insulating substrate can be thermally emitted to E_C and cause a shift in the threshold voltage of field-effect devices at relatively modest operation powers (the so-called back-gate effect). Instead, the use of intrinsic defects with very deep levels is gradually becoming more common (Jenny et al., 2002, 2003; Son et al., 2007c) and so-called high-purity semi-insulating (HPSI) wafers are commercially available today.

5.3 Diffusion of Transition Metals

SiC crystal growth and device fabrication require very high temperatures (1500–2000 °C), and it is decisive to understand and control impurity diffusion. Most of the common dopant atoms exhibit a low diffusivity even at high temperatures (except for B). However, some of the transition metals seem to be relatively mobile, but up to date there are only a few reports

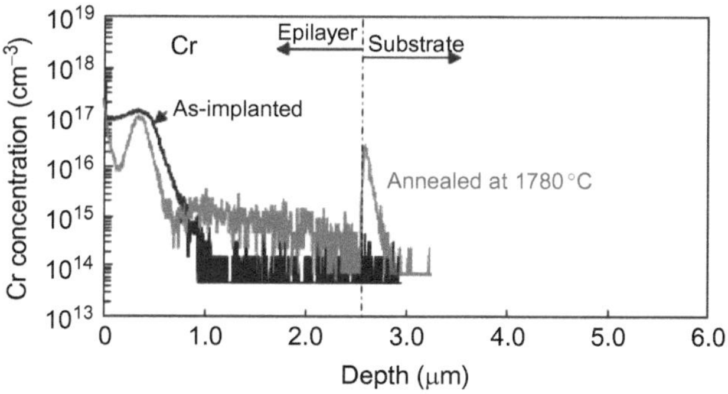

Figure 21 Concentration versus depth profiles of Cr implanted in n-type 4H–SiC epilayers before and after annealing at 1780 °C for 30 min. Diffusion of Cr into the epilayer is clearly seen after the annealing. *Reprinted with permission from Danno et al. (2012). Copyright 2012 The Japan Society of Applied Physics.*

available in the literature. Systematic experimental results for Ti, Cr, Fe, and Ni have recently been presented by Danno et al. (2012), and Figure 21 shows depth profiles of Cr implanted in n-type 4H–SiC epilayers before and after annealing at 1780 °C (30 min). Diffusion of Cr into the epilayer is evidenced by an increase of [Cr] at depths exceeding ~1.0 μm and a decrease in the implanted region. Also, diffusion of Fe and Ni was observed under the same annealing conditions. Although quantitative diffusion coefficients have not been reported yet, the diffusivity seems to be in the order of Ti < Cr < Fe < Ni. In fact, this order coincides with the order of atomic numbers (Ti-22, Cr-24, Fe-26, Ni-28), and this tendency is the same as for transition metals in Si (Weber, 1983). Figure 21 unveils also a strong accumulation of Cr at the epilayer/substrate interface, suggesting trapping at the interface and/or low mobility of Cr in the substrate. Since the defect concentration is high in the substrate (and at the interface), it appears likely that Cr forms stable Cr-defect complexes with low mobility. Further, this hints a possibility of controlling transition metal distributions by intentionally introducing defects in localized regions, i.e., gettering. Indeed, Danno et al. (2012) have shown that the diffusion of Ti, Cr, and Fe can be successfully suppressed by He implantation-induced defects.

6. FINAL REMARKS

"Point defects in SiC" is a vivid field of research driven by both scientific curiosity and rapidly growing technological impact, especially within high power electronics. As illustrated in the five preceding sections of this chapter, substantial advancement has been made during the past 15 years regarding the understanding and control of fundamental intrinsic point defects, especially V_C and V_{Si}. Both V_C and V_{Si} are well characterized with the former exhibiting large Jahn–Teller distortion and negative-U character while the latter is a high-spin ground-state defect ($S=3/2$) with $\sim$3 eV higher formation energy than V_C and much lower abundance under thermal equilibrium. However, in contrast to V_C (with the exception of its double donor state), the positions of the electronic (deep) levels of V_{Si} are not well established and need further experimental attention. In particular, this may become highly relevant for future and fascinating applications where V_{Si} (and ensembles thereof) are utilized as defects with long spin coherence times enabling long-lived qubits at RT. Further, the metastability of V_{Si} with respect to the carbon antisite–vacancy pair (C_{Si}–V_C) is interesting where C_{Si}–V_C shows promise as an ultra-bright single-photon source (operating at RT). In general, SiC and its device processing technology may emerge as a platform for integrating spintronics, photonics, and electronics in a single quantum system.

Isolated self-interstitials (Si_I and C_I) and antisite defects are less investigated experimentally than the vacancies, and perhaps with some exception, no clear spectroscopic signals have been identified. One reason may be that the interstitials are rather mobile with a migration energy of only $\sim$1 eV, as predicted by theory, and online experiments of irradiated samples at cryogenic temperatures are encouraged.

Charge carrier lifetimes in excess of $\sim$10 μs are required for bipolar devices with blocking voltages above $\sim$15–20 kV, and the concept of C_I injection from an oxidizing surface (or an implanted surface layer) to annihilate V_C, which is a lifetime-limiting defect, has been proven quite successful. However, further advances are needed and the influence of residual impurities, e.g., transition metals, and higher-order intrinsic defects, e.g., divacancies, should be addressed at the concentration range of 10^{13} cm^{-3} and below. These impurities/defects may also play a decisive role for HPSI wafers.

Last but not least, *ab initio* defect modeling based on DFT and utilizing high-performance computers (enabling large-sized super cells) has undergone a stunning development and is today an indispensable tool for both interpretation of measured data and prediction of defect properties aiding experiments in SiC.

ACKNOWLEDGMENTS

Numerous discussions and lasting collaboration with G. Alfieri, H.M. Ayedh, V. Bobal, A. Gali, J. Grillenberger, U. Grossner, A. Hallén, C. Jagadish, M.S. Janson, E. Janzén, T. Kimoto, M.K. Linnarsson, L.S. Løvlie, A-M.M. Moe, E.V. Monakhov, R. Nipoti, I. Pintilie, N.T. Son, L. Vines, and J. Wong-Leung are gratefully acknowledged. Financial support was partly and kindly provided by the Norwegian Research Council through the programs FRINATEK (WEDD project) and BIA (ICE 2016 project).

REFERENCES

Acheson, A.G., 1892. Production of artificial crystalline carbonaceous materials, carborundum. Engl. Pat. 17911.

Achtziger, N., Witthuhn, W., 1997. Band gap states of Ti, V, and Cr in 4H-silicon carbide. Appl. Phys. Lett. 71, 110–112.

Achtziger, N., Witthuhn, W., 1998. Band-gap states of Ti, V, and Cr in 4H-SiC: identification and characterization by elemental transmutation of radioactive isotopes. Phys. Rev. B 57, 12181–12196.

Achtziger, N., Grillenberger, J., Witthuhn, W., 1999. Radiotracer spectroscopy of deep levels in the semiconductor bandgap. Hyperfine Interact. 120 (121), 69–79.

Alfieri, G., Kimoto, T., 2013. Laplace transform deep level transient spectroscopy study of the $EH_{6/7}$ center. Mater. Sci. Forum 740–742, 645–648.

Alfieri, G., Monakhov, E.V., Svensson, B.G., Linnarsson, M.K., 2005. Annealing behavior between room temperature and 2000 °C of deep level defects in electron-irradiated *n*-type 4H silicon carbide. J. Appl. Phys. 98, 043518-1-5.

Aradi, B., Gali, A., Deák, P., Son, N.T., Janzén, E., 2001. Passivation of p-type dopants in 4H-SiC by hydrogen. Physica B 308–310, 722–725.

Ayedh, H.M., Bobal, V., Nipoti, R., Hallen, A., Svensson, B.G., 2014. Formation of carbon vacancy in 4H silicon carbide during high-temperature processing. J. Appl. Phys. 115, 012005-1-6.

Bauer, A., Kräußlich, J., Dressler, L., Kuschnerus, P., Wolf, J., Goetz, K., Käckell, P., Furthmüller, J., Bechstedt, F., 1998. High-precision determination of atomic positions in crystals: the case of 6H- and 4H-SiC. Phys. Rev. B 57, 2647–2650.

Berkman, E., Leonard, R.T., Paisley, M.J., Khlebnikov, Y., O'Loughlin, M.J., Burk, A.A., Powell, A.R., Malta, D.P., Deyneka, E., Brady, M.F., Khlebnikov, I., Tsvetkov, V.F., Hobgood, H.M., Sumakeris, J., Basceri, C., Balakrishna, V., Carter Jr., C.H., Balkas, C., 2009. Defect status in SiC manufacturing. Mater. Sci. Forum 615–617, 3–6.

Beyer, F.C., Hemmingsson, C.G., Leone, S., Lin, Y.-C., Gällström, A., Henry, A., Janzén, E., 2011. Deep levels in iron doped n- and p-type 4H-SiC. J. Appl. Phys. 110, 123701-1-5.

Bockstedte, M., Mauttausch, A., Pankrov, O., 2003. Ab initio study of the migration of intrinsic defects in 3C-SiC. Phys. Rev. B 68 205201-1-17.

Bockstedte, M., Mattausch, A., Pankrov, O., 2004. Ab initio study of the annealing of vacancies and interstitials in cubic SiC: vacancy-interstitial recombination and aggregation of carbon interstitials. Phys. Rev. B 69, 235202-1-13.

Bockstedte, M., Marini, A., Pankrov, O., Rubio, A., 2010. Many-body effects in the excitation spectrum of a defect in SiC. Phys. Rev. Lett. 105, 026401-1-4.

Breitholtz, B., 1998. Private communication, ABB, Sweden.

Castelleto, S., Johnson, B.C., Ivady, V., Stavrias, N., Uneda, T., Gali, A., Ohshima, T., 2014. A silicon carbide room-temperature single-photon source. Nat. Mater. 13, 151–156.

Dalibor, T., Pensl, G., Nordell, N., Schöner, A., 1997. Electrical properties of the titanium acceptor in silicon carbide. Phys. Rev. B 55, 13618–13624.

Danno, K., Kimoto, T., 2006. Investigation of deep levels in n-type 4H-SiC epilayers irradiated with low-energy electrons. J. Appl. Phys. 100, 113728-1-6.

Danno, K., Kimoto, T., 2007. Deep level transient spectroscopy on as-grown and electron-irradiated p-type 4H-SiC epilayers. J. Appl. Phys. 101, 103704-1-5.

Danno, K., Nakamura, D., Kimoto, T., 2007. Investigation of carrier lifetime in 4H-SiC epilayers and lifetime control by electron irradiation. Appl. Phys. Lett. 90, 202109-1-3.

Danno, K., Saitoh, H., Seki, A., Shirai, T., Suzuki, H., Bessho, T., Kawai, Y., Kimoto, T., 2012. Diffusion of transition metals in 4H-SiC and trials of impurity gettering. Appl. Phys. Express 5, 031301-1-3.

David, M.L., Alfieri, G., Monakhov, E.M., Hallén, A., Blanchard, C., Svensson, B.G., Barbot, J.F., 2004. Electrically active defects in irradiated 4H-SiC. J. Appl. Phys. 95, 4728–4732.

Deák, P., Aradi, B., Gali, A., 2001. Boron and aluminium doping in SiC and its passivation by hydrogen. J. Phys. Condens. Matter 13, 9019–9026.

Eberlein, T.A.G., Fall, C.J., Jones, R., Briddon, P.R., Öberg, S., 2002. Alphabet luminescence lines in 4H-SiC. Phys. Rev. B 65, 184108-1-4.

Evans, G.A., Steeds, J.W., Ley, L., Hundhausen, M., Schulze, N., Pensl, G., 2002. Identification of carbon interstitials in electron-irradiated 6H-SiC by use of a ^{13}C enriched specimen. Phys. Rev. B 66, 035204-1-9.

Gupta, A., Wu, P., Rengarajan, V., Xu, X., Yoganathan, M., Martin, C., Emorhokpor, E., Souzis, A., Zwieback, I., Anderson, T., 2012. Status of large diameter SiC single crystals. Mater. Sci. Forum 717–720, 3–8.

Han, S.Y., Lee, J.-L., 2004. Interpretation of Fermi level pinning on 4H-SiC using synchrotron photoemission spectroscopy. Appl. Phys. Lett. 84, 538–540.

Harris, G.I. (Ed.), 1995. Properties of Silicon Carbide. In: EMIS Data Review No. 13, INSPEC, IEE, London.

Hemmingsson, C.G., Son, N.T., Ellison, A., Zhang, J., Janzén, E., 1998. Negative-U centers in 4H silicon carbide. Phys. Rev. B 58, R10119–R10122.

Hiyoshi, T., Kimoto, T., 2009a. Reduction of deep levels and improvement of carrier lifetime in n-type 4H-SiC by thermal oxidation. Appl. Phys. Express 2, 041101-1-3.

Hiyoshi, T., Kimoto, T., 2009b. Elimination of the major deep levels in n- and p-type 4H-SiC by two-step thermal treatment. Appl. Phys. Express 2, 091101-1-3.

Hobgood, H.M., Glass, R.C., Augustine, G., Hopkins, R.H., Jenny, J., Skowronski, M., Mitchel, W.C., Roth, M., 1995. Semi-insulating 6H-SiC grown by physical vapor transport. Appl. Phys. Lett. 66, 1364–1366.

Hornos, T., Gali, A., Svensson, B.G., 2011. Large scale electronic structure calculations in 4H-SiC using the Heyd-Scuseria-Ernzerhof screened hybrid density functional. Mater. Sci. Forum 679–680, 261–264.

Isoya, J., Kanda, H., Uchida, Y., Lawson, S.C., Yamasaki, S., Itoh, H., Morita, Y., 1992. EPR identification of the negatively charged vacancy in diamond. Phys. Rev. B 45, 1436–1439.

Itoh, H., Hayakawa, N., Nashiyama, I., Sakuma, E., 1987. Electron spin resonance in electron-irradiated 3C-SiC. J. Appl. Phys. 66, 4529–4531.

Itoh, H., Kawasuso, A., Ohshima, T., Yoshikawa, M., Nashiyama, I., Tanigawa, S., Misawa, S., Okimura, H., Yoshida, S., 1997. Intrinsic defects in cubic silicon carbide. Phys. Status Solidi A 162, 173–198.

Janson, M.S., Hallén, A., Linnarsson, M.K., Svensson, B.G., 2001. Hydrogen diffusion, complex formation, and dissociation in acceptor-doped silicon carbide. Phys. Rev. B 64, 195202-1-12.

Janson, M.S., Linnarsson, M.K., Hallén, A., Svensson, B.G., Achtziger, N., Unéus, L., Lloyd Spetz, A., Forsberg, U., 2004. Hydrogen in the wide bandgap semiconductor silicon carbide. Phys. Scr. T108, 99–112.

Janzén, E., Gali, A., Carlsson, P., Gällström, A., Magnusson, B., Son, N.T., 2009. The silicon vacancy in SiC. Physica B 404, 4354–4358.

Jenny, J.R., Skowronski, M., Mitchel, W.C., Hobgood, H.M., Glass, R.C., Augustine, G., Hopkins, R.H., 1995. On the compensation mechanism in high-resistivity 6H-SiC doped with vanadium. J. Appl. Phys. 78, 3839–3842.

Jenny, J.R., Skowronski, J., Mitchel, W.C., Hobgood, H.M., Glass, R.C., Augustine, G., Hopkins, R.H., 1996. Deep level transient spectroscopic and Hall effect investigation of the position of the vanadium acceptor level in 4H and 6H SiC. Appl. Phys. Lett. 68, 1963–1965.

Jenny, J.R., Müller, S.G., Powell, A., Tsvetkov, V.F., Hobgood, H.M., Glass, R.C., Carter Jr., C.H., 2002. High-purity semi-insulating 4H-SiC grown by the seeded-sublimation method. J. Electron. Mater. 31, 366–369.

Jenny, J.R., Malta, D.P., Müller, S.G., Powell, A.R., Tsvetkov, V.F., Hobgood, H.M., Glass, R.C., Carter Jr., C.H., 2003. High-purity semi-insulating 4H-SiC for microwave device applications. J. Electron. Mater. 32, 432–436.

Kawasuso, A., Itoh, H., Okada, S., Okumura, H., 1996. Annealing processes of vacancy-type defects in electron-irradiated and as-grown 6H-SiC studied by positron annihilation spectroscopy. J. Appl. Phys. 80, 5639–5645.

Kawasuso, A., Itoh, H., Morishita, N., Yoshikawa, M., Oshima, T., Nashiyama, I., Okada, S., Okumura, H., Yoshida, S., 1998. Silicon vacancies in 3C-SiC observed by positron lifetime and electron spin resonance. Appl. Phys. A 67, 209–212.

Kimoto, T., Nakajima, T., Matsunami, H., Nakata, T., Inoue, M., 1996. Formation of semi-insulating 6H-SiC layers by vanadium ion implantations. Appl. Phys. Lett. 69, 1113–1115.

Kimoto, T., Danno, K., Suda, J., 2008. Lifetime-killing defects in 4H-SiC epilayers and lifetime control by low-energy electron irradiation. Phys. Status Solidi B 245, 1327–1336.

Kimoto, T., Feng, G., Hiyoshi, T., Kawahara, K., Noborio, M., Suda, J., 2010. Defect control in growth and processing of 4H-SiC for power device applications. Mater. Sci. Forum 645–648, 645–650.

Klein, P.B., Shanabrook, B.V., Huh, S.W., Polyakov, A.Y., Skowronski, M., Sumakeris, J.J., O'Loughlin, M.J., 2006. Lifetime-limiting defects in n^- 4H-SiC epilayers. Appl. Phys. Lett. 88, 052110-1-3.

Kordina, O., Hallin, C., Glass, R.C., Janzén, E., 1994. A novel hot-wall CVD reactor for SiC epitaxy. Inst. Phys. Conf. Ser. 137, 41.

Larkin, D.J., 1997. SiC dopant incorporation control using site-competition CVD. Phys. Status Solidi B 202, 305–320.

Larkin, D.J., Sridhara, S.G., Devaty, R.P., Choyke, W.J., 1995. Hydrogen incorporation in boron-doped 6H-SiC CVD epilayers produced using site-competition epitaxy. J. Electron. Mater. 24, 289–294.

Lee, K.M., Dang, L.S., Watkins, G.D., Choyke, W.J., 1985. Optically detected magnetic resonance study of SiC:Ti. Phys. Rev. B 32, 2273–2284.

Linnarsson, M.K., Janson, M.S., Zhang, J., Janzén, E., Svensson, B.G., 2004. Self-diffusion of ^{12}C and ^{13}C in intrinsic 4H-SiC. J. Appl. Phys. 95, 8469–8471.

Løvlie, L.S., Svensson, B.G., 2011. Enhanced annealing of implantation-induced defects in 4H-SiC by thermal oxidation. Appl. Phys. Lett 98, 052108-1-3.

Løvlie, L.S., Svensson, B.G., 2012. Oxidation-enhanced annealing of implantation-induced Z1/2 centers in 4H-SiC: reaction kinetics and modeling. Phys. Rev. B 86, 075205-1-8.

Mitchel, W.C., Perrin, R., Goldstein, J., Saxler, A., Roth, M., Smith, S.R., Solomon, J.S., Evwaraye, A.O., 1999. Fermi level control and deep levels in semi-insulating 4H–SiC. J. Appl. Phys. 86, 5040–5044.

Mitchel, W.C., Mitchell, W.D., Landis, G., Smith, H.E., Lee, W., Zvanut, M.E., 2007. Vanadium donor and acceptor levels in semi-insulating 4H- and 6H-SiC. J. Appl. Phys. 101, 013707-1-7.

Mizuochi, N., Yamasaki, S., Takizawa, H., Morishita, N., Ohshima, T., Itoh, H., Isoya, J., 2003. EPR studies of isolated negatively charged silicon vacancies in 4H- and 6H-SiC: identification of C_{3v} symmetry and silicon sites. Phys. Rev. B 68, 165206-1-11.

Okuda, T., Kimoto, T., Suda, J., 2013. Improvement of carrier lifetimes in highly Al-doped p-type 4H-SiC epitaxial layers by hydrogen passivation. Appl. Phys. Express 6, 121301-1-3.

Okuda, T., Miyazawa, T., Tsuchida, H., Kimoto, T., Suda, J., 2014. Enhancement of carrier lifetime in lightly Al-doped p-type 4H-SiC epitaxial layers by combination of thermal oxidation and hydrogen annealing. Appl. Phys. Express 7, 85501-1-3.

Park, C.H., Cheong, B.H., Lee, K.H., Chang, K.J., 1994. Structural and electronic properties of cubic, 2H, 4H, and 6H SiC. Phys. Rev. B 49, 4485–4493.

Patrick, L., Choyke, W.J., 1973. Localized vibrational modes of a persistent defect in ion-implanted SiC. J. Phys. Chem. Solids 34, 565–567.

Patrick, L., Choyke, W.J., 1974. Photoluminescence of Ti in four SiC polytypes. Phys. Rev. B 10, 5091–5094.

Pensl, G., Helbig, R., 1990. Silicon carbide (SiC)—recent results in physics and in technology. Festkörperbleme 30, 133.

Rauls, E., Lingner, T., Hajnal, Z., Greulich-Weber, S., Frauenheim, T., Spaeth, J.M., 2000. Metastability of the neutral silicon vacancy in 4H-SiC. Phys. Status Solidi B 217, R1–R3.

Round, H.J., 1907. A note on carborundum. Electr. World 19, 309.

Samiji, M.E., Venter, A., Wagener, M.C., Leitch, A.W.R., 2001. Hydrogen passivation and reactivation of the Al-acceptors in p-type 6H-SiC. J. Phys. Condens. Matter 13, 9011–9017.

Schneider, J., Müller, H.D., Maier, K., Wilkening, W., Fuchs, F., Dörnen, A., Leibenzeder, S., Stein, R., 1990. Infrared spectra and electron spin resonance of vanadium deep level impurities in silicon carbide. Appl. Phys. Lett. 56, 1184–1186.

Schöner, A., Rottner, K., Nordell, N., Linnarsson, M., Peppermüller, C., Helbig, R., 1997. Hydrogen incorporation in epitaxial layers of 4H-and 6H-silicon carbide grown by vapor phase epitaxy. Diam. Relat. Mater. 6, 1293–1296.

Schulz, D., Wagner, G., Dolle, J., Irmscher, K., Müller, T., Rost, H.-J., Siche, D., Wollweber, J., 1999. Impurity incorporation during sublimation growth of 6H bulk SiC. J. Cryst. Growth 198 (199), 1024–1027.

Son, N.T., Hai, P.N., Janzén, E., 2001a. Carbon vacancy-related defect in 4H and 6H SiC. Phys. Rev. B 63, 201201(R)-1–4.

Son, N.T., Hai, P.N., Janzén, E., 2001b. Silicon antisite in 4H-SiC. Phys. Rev. Lett. 87, 045502-1-4.

Son, N.T., Magnusson, B., Janzén, E., 2002. Photoexcitation-electron-paramagnetic-resonance studies of the carbon vacancy in 4H-SiC. Appl. Phys. Lett. 81, 3945–3947.

Son, N.T., Carlsson, P., Gällström, A., Magnusson, B., Janzén, E., 2007a. Deep levels and carrier compensation in V-doped semi-insulating 4H-SiC. Appl. Phys. Lett 91, 202111-1-3.

Son, N.T., Carlsson, P., Gällström, A., Magnusson, B., Janzén, E., 2007b. Prominent defects in semi-insulating SiC substrates. Physica B 401–402, 67–72.

Son, N.T., Carlsson, P., ul Hassan, J., Magnusson, B., Janzén, E., 2007c. Defects and carrier compensation in semi-insulating 4H-SiC substrates. Phys. Rev. B 75, 155204-1-8.

Son, N.T., Trinh, X.T., Løvlie, L.S., Svensson, B.G., Kawahara, K., Suda, J., Kimoto, T., Umeda, T., Isoya, J., Makino, T., Oshima, T., Janzén, E., 2012. Negative-U system of carbon vacancy in 4H-SiC. Phys. Rev. Lett 109, 187603-1-5.

Sörman, E., Son, N.T., Chen, M.W., Kordina, O., Hallin, C., Janzén, E., 2000. Silicon vacancy related defect in 4H and 6H SiC. Phys. Rev. B 61, 2613–2620.

Steeds, J.W., 2009. Photoluminescence study of the carbon antisite-vacancy pair in 4H-SiC and 6H-SiC. Phys. Rev. B 80, 245202-1-14.

Tairov, T.M., Tsvetkov, V.F., 1978. Investigation of growth processes of ingots of silicon carbide single crystals. J. Cryst. Growth 43, 209–212.

Torpo, L., Marlo, M., Staab, T.E.M., Nieminen, R.M., 2001. Comprehensive ab initio study of properties of monovacancies and antisites in 4H-SiC. J. Phys. Condens. Matter 13, 6203–6231.

Trapaidze, L., Hollweck, R., Beljakowa, S., Zippelius, B., Waber, H.B., Pensl, G., Krieger, M., 2011. Iron-related defect centers in 4H-SiC detected by deep level transient spectroscopy. Mater. Sci. Forum 679–680, 257–260.

Trinh, X.T., Szasz, K., Hornos, T., Kawahara, K., Suda, J., Kimoto, T., Gali, A., Janzén, E., Son, N.T., 2013. Negative-U carbon vacancy in 4H-SiC: assessment of charge correction schemes and identification of the negative carbon vacancy at the quasicubic site. Phys. Rev. B 88, 235209-1-13.

Umeda, T., Isoya, J., Morishita, N., Oshima, T., Kamiya, T., 2004a. EPR identification of two types of carbon vacancies in 4H-SiC. Phys. Rev. B 69, 121201(R)-1-4.

Umeda, T., Isoya, J., Morishita, N., Oshima, T., Kamiya, T., Gali, A., Deák, P., Son, N.T., Janzén, E., 2004b. EPR and theoretical studies of positively charged carbon vacancy in 4H-SiC. Phys. Rev. B 70, 235212-1-6.

Umeda, T., Ishitsuka, Y., Isoya, J., Son, N.T., Janzén, E., Morishita, N., Oshima, T., Itoh, H., Gali, A., 2005. EPR and theoretical studies of negatively charged carbon vacancy in 4H-SiC. Phys. Rev. B 71, 193202-1-4.

Weber, E.R., 1983. Transition metals in silicon. Appl. Phys. A 30, 1–22.

Widmann, M., Lee, S.Y., Rendler, T., Son, N.T., Fedder, H., Paik, S., Yang, L.P., Zhao, N., Yang, S., Booker, I., Denisenko, A., Jamali, M., Momenzadeh, S.A., Gerhardt, I., Ohshima, T., Gali, A., Janzén, E., Wrachtrup, J., 2015. Coherent control of single spins in silicon carbide at room temperature. Nat. Mater. 14, 164–168.

Wimbauer, T., Meyer, B.K., Hofstaetter, A., Scharmann, A., Overhof, H., 1997. Negatively charged Si vacancy in 4H-SiC: a comparison between theory and experiment. Phys. Rev. B 56, 7384–7388.

Yole Développement, 2013. SiC market 2013. http://www.yole.fr.

Zippelius, B., Suda, J., Kimoto, T., 2012. On the formation of intrinsic defects in 4H-SiC by high temperature annealing steps. Mater. Sci. Forum 717–720, 247–250.

Zvanut, M.E., Konovalov, V.V., 2002. The level position of a deep intrinsic defect in 4H-SiC studied by photoinduced electron paramagnetic resonance. Appl. Phys. Lett. 80, 410–412.

Zywietz, A., Furthmüller, J., Bechstedt, F., 1999. Vacancies in SiC: influence of Jahn-Teller distortions, spin effects, and crystal structure. Phys. Rev. B 59, 15166–15180.

INDEX

Note: Page numbers followed by "*f*" indicate figures and "*t*" indicate tables.

E

F

G

H

I

J

K

L

M

T

U

V

W

Y

Z

CONTENTS OF VOLUMES IN THIS SERIES

Volume 1 **Physics of III–V Compounds**

Volume 2 **Physics of III–V Compounds**

Volume 3 **Optical Properties of III–V Compounds**

Volume 4 **Physics of III–V Compounds**

Volume 5 **Infrared Detectors**

Volume 6 **Injection Phenomena**

Volume 7 **Application and Devices**

Part A

Part B

Volume 8 **Transport and Optical Phenomena**

Volume 9 **Modulation Techniques**

Volume 10 **Transport Phenomena**

Volume 11 **Solar Cells**

H. J. Hovel, Introduction; Carrier Collection, Spectral Response, and Photocurrent; Solar Cell Electrical Characteristics; Efficiency; Thickness; Other Solar Cell Devices; Radiation Effects; Temperature and Intensity; Solar Cell Technology

Volume 12 **Infrared Detectors (II)**

W. L. Eiseman, J. D. Merriam, and R. F. Potter, Operational Characteristics of Infrared Photodetectors
P. R. Bratt, Impurity Germanium and Silicon Infrared Detectors
E. H. Putley, InSb Submillimeter Photoconductive Detectors
G. E. Stillman, C. M. Wolfe, and J. O. Dimmock, Far-Infrared Photoconductivity in High Purity GaAs
G. E. Stillman and C. M. Wolfe, Avalanche Photodiodes
P. L. Richards, The Josephson Junction as a Detector of Microwave and Far-Infrared Radiation
E. H. Putley, The Pyroelectric Detector – An Update

Volume 13 **Cadmium Telluride**

K. Zanio, Materials Preparations; Physics; Defects; Applications

Volume 14 **Lasers, Junctions, Transport**

N. Holonyak, Jr., and M. H. Lee, Photopumped III–V Semiconductor Lasers
H. Kressel and J. K. Butler, Heterojunction Laser Diodes
A. Van der Ziel, Space-Charge-Limited Solid-State Diodes
P. J. Price, Monte Carlo Calculation of Electron Transport in Solids

Volume 15 **Contacts, Junctions, Emitters**

B. L. Sharma, Ohmic Contacts to III–V Compounds Semiconductors
A. Nussbaum, The Theory of Semiconducting Junctions
J. S. Escher, NEA Semiconductor Photoemitters

Volume 16 **Defects, (HgCd)Se, (HgCd)Te**

H. Kressel, The Effect of Crystal Defects on Optoelectronic Devices
C. R. Whitsett, J. G. Broerman, and C. J. Summers, Crystal Growth and Properties of $Hg_{1-x}Cd_xSe$ Alloys
M. H. Weiler, Magnetooptical Properties of $Hg_{1-x}Cd_xTe$ Alloys
P. W. Kruse and J. G. Ready, Nonlinear Optical Effects in $Hg_{1-x}Cd_xTe$

Volume 17 **CW Processing of Silicon and Other Semiconductors**

J. F. Gibbons, Beam Processing of Silicon
A. Lietoila, R. B. Gold, J. F. Gibbons, and L. A. Christel, Temperature Distributions and Solid Phase Reaction Rates Produced by Scanning CW Beams

Volume 18 **Mercury Cadmium Telluride**

Volume 19 **Deep Levels, GaAs, Alloys, Photochemistry**

Volume 20 **Semi-Insulating GaAs**

Volume 21 **Hydrogenated Amorphous Silicon**

Part A

Part E

Volume 23 **Pulsed Laser Processing of Semiconductors**

Volume 24 **Applications of Multiquantum Wells, Selective Doping, and Superlattices**

Volume 25 **Diluted Magnetic Semiconductors**

W. M. Becker, Band Structure and Optical Properties of Wide-Gap $A^{II}_{1-x}Mn_xB_{IV}$ Alloys at Zero Magnetic Field
S. Oseroff and P. H. Keesom, Magnetic Properties: Macroscopic Studies
T. Giebultowicz and T. M. Holden, Neutron Scattering Studies of the Magnetic Structure and Dynamics of Diluted Magnetic Semiconductors
J. Kossut, Band Structure and Quantum Transport Phenomena in Narrow-Gap Diluted Magnetic Semiconductors
C. Riquaux, Magnetooptical Properties of Large-Gap Diluted Magnetic Semiconductors
J. A. Gaj, Magnetooptical Properties of Large-Gap Diluted Magnetic Semiconductors
J. Mycielski, Shallow Acceptors in Diluted Magnetic Semiconductors: Splitting, Boil-off, Giant Negative Magnetoresistance
A. K. Ramadas and R. Rodriquez, Raman Scattering in Diluted Magnetic Semiconductors
P. A. Wolff, Theory of Bound Magnetic Polarons in Semimagnetic Semiconductors

Volume 26 **III–V Compound Semiconductors and Semiconductor Properties of Superionic Materials**

Z. Yuanxi, III–V Compounds
H. V. Winston, A. T. Hunter, H. Kimura, and R. E. Lee, InAs-Alloyed GaAs Substrates for Direct Implantation
P. K. Bhattacharya and S. Dhar, Deep Levels in III–V Compound Semiconductors Grown by MBE
Y. Ya. Gurevich and A. K. Ivanov-Shits, Semiconductor Properties of Supersonic Materials

Volume 27 **High Conducting Quasi-One-Dimensional Organic Crystals**

E. M. Conwell, Introduction to Highly Conducting Quasi-One-Dimensional Organic Crystals
I. A. Howard, A Reference Guide to the Conducting Quasi-One-Dimensional Organic Molecular Crystals
J. P. Pouqnet, Structural Instabilities
E. M. Conwell, Transport Properties
C. S. Jacobsen, Optical Properties
J. C. Scolt, Magnetic Properties
L. Zuppiroli, Irradiation Effects: Perfect Crystals and Real Crystals

Volume 28 **Measurement of High-Speed Signals in Solid State Devices**

J. Frey and D. Ioannou, Materials and Devices for High-Speed and Optoelectronic Applications
H. Schumacher and E. Strid, Electronic Wafer Probing Techniques
D. H. Auston, Picosecond Photoconductivity: High-Speed Measurements of Devices and Materials
J. A. Valdmanis, Electro-Optic Measurement Techniques for Picosecond Materials, Devices and Integrated Circuits
J. M. Wiesenfeld and R. K. Jain, Direct Optical Probing of Integrated Circuits and High-Speed Devices
G. Plows, Electron-Beam Probing
A. M. Weiner and R. B. Marcus, Photoemissive Probing

Volume 29 **Very High Speed Integrated Circuits: Gallium Arsenide LSI**

M. Kuzuhara and T. Nazaki, Active Layer Formation by Ion Implantation
H. Hasimoto, Focused Ion Beam Implantation Technology
T. Nozaki and A. Higashisaka, Device Fabrication Process Technology
M. Ino and T. Takada, GaAs LSI Circuit Design
M. Hirayama, M. Ohmori, and K. Yamasaki, GaAs LSI Fabrication and Performance

Volume 30 **Very High Speed Integrated Circuits: Heterostructure**

H. Watanabe, T. Mizutani, and A. Usui, Fundamentals of Epitaxial Growth and Atomic Layer Epitaxy
S. Hiyamizu, Characteristics of Two-Dimensional Electron Gas in III–V Compound Heterostructures Grown by MBE
T. Nakanisi, Metalorganic Vapor Phase Epitaxy for High-Quality Active Layers
T. Nimura, High Electron Mobility Transistor and LSI Applications
T. Sugeta and T. Ishibashi, Hetero-Bipolar Transistor and LSI Application
H. Matsuedo, T. Tanaka, and M. Nakamura, Optoelectronic Integrated Circuits

Volume 31 **Indium Phosphide: Crystal Growth and Characterization**

J. P. Farges, Growth of Discoloration-Free InP
M. J. McCollum and G. E. Stillman, High Purity InP Grown by Hydride Vapor Phase Epitaxy
I. Inada and T. Fukuda, Direct Synthesis and Growth of Indium Phosphide by the Liquid Phosphorous Encapsulated Czochralski Method
O. Oda, K. Katagiri, K. Shinohara, S. Katsura, Y. Takahashi, K. Kainosho, K. Kohiro, and R. Hirano, InP Crystal Growth, Substrate Preparation and Evaluation
K. Tada, M. Tatsumi, M. Morioka, T. Araki, and T. Kawase, InP Substrates: Production and Quality Control
M. Razeghi, LP-MOCVD Growth, Characterization, and Application of InP Material
T. A. Kennedy and P. J. Lin-Chung, Stoichiometric Defects in InP

Volume 32 **Strained-Layer Superlattices: Physics**

T. P. Pearsall, Strained-Layer Superlattices
F. H. Pollack, Effects of Homogeneous Strain on the Electronic and Vibrational Levels in Semiconductors
J. Y. Marzin, J. M. Gerárd, P. Voisin, and J. A. Brum, Optical Studies of Strained III–V Heterolayers
R. People and S. A. Jackson, Structurally Induced States from Strain and Confinement
M. Jaros, Microscopic Phenomena in Ordered Superlattices

Volume 33 **Strained-Layer Superlattices: Material Science and Technology**

R. Hull and J. C. Bean, Principles and Concepts of Strained-Layer Epitaxy

Volume 34 **Hydrogen in Semiconductors**

Volume 35 **Nanostructured Systems**

Volume 36 **The Spectroscopy of Semiconductors**

Volume 37 **The Mechanical Properties of Semiconductors**

Volume 38 **Imperfections in III/V Materials**

Volume 39 **Minority Carriers in III–V Semiconductors: Physics and Applications**

Volume 40 **Epitaxial Microstructures**

Volume 41 **High Speed Heterostructure Devices**

Volume 42 **Oxygen in Silicon**

Volume 43 **Semiconductors for Room Temperature Nuclear Detector Applications**

Volume 44 **II–IV Blue/Green Light Emitters: Device Physics and Epitaxial Growth**

Volume 45 **Effect of Disorder and Defects in Ion-Implanted Semiconductors: Electrical and Physiochemical Characterization**

Volume 46 **Effect of Disorder and Defects in Ion-Implanted Semiconductors: Optical and Photothermal Characterization**

Volume 47 **Uncooled Infrared Imaging Arrays and Systems**

Volume 48 **High Brightness Light Emitting Diodes**

Volume 49 **Light Emission in Silicon: from Physics to Devices**

Volume 50 **Gallium Nitride (GaN)**

Volume 51A **Identification of Defects in Semiconductors**

Volume 51B **Identification Defects in Semiconductors**

Volume 52 **SiC Materials and Devices**

Volume 53 **Cumulative Subjects and Author Index Including Tables of Contents for Volumes 1–50**

Volume 54 **High Pressure in Semiconductor Physics I**

Volume 55 **High Pressure in Semiconductor Physics II**

Volume 56 **Germanium Silicon: Physics and Materials**

Volume 57 **Gallium Nitride (GaN) II**

Volume 58 **Nonlinear Optics in Semiconductors I**

Volume 59 **Nonlinear Optics in Semiconductors II**

J. B. Khurgin, Second Order Nonlinearities and Optical Rectification
K. L. Hall, E. R. Thoen, and E. P. Ippen, Nonlinearities in Active Media
E. Hanamura, Optical Responses of Quantum Wires/Dots and Microcavities
U. Keller, Semiconductor Nonlinearities for Solid-State Laser Modelocking and Q-Switching
A. Miller, Transient Grating Studies of Carrier Diffusion and Mobility in Semiconductors

Volume 60 **Self-Assembled InGaAs/GaAs Quantum Dots**

Mitsuru Sugawara, Theoretical Bases of the Optical Properties of Semiconductor Quantum Nano-Structures
Yoshiaki Nakata, Yoshihiro Sugiyama, and Mitsuru Sugawara, Molecular Beam Epitaxial Growth of Self-Assembled InAs/GaAs Quantum Dots
Kohki Mukai, Mitsuru Sugawara, Mitsuru Egawa, and Nobuyuki Ohtsuka, Metalorganic Vapor Phase Epitaxial Growth of Self-Assembled InGaAs/GaAs Quantum Dots Emitting at 1.3 μm
Kohki Mukai and Mitsuru Sugawara, Optical Characterization of Quantum Dots
Kohki Mukai and Milsuru Sugawara, The Photon Bottleneck Effect in Quantum Dots
Hajime Shoji, Self-Assembled Quantum Dot Lasers
Hiroshi Ishikawa, Applications of Quantum Dot to Optical Devices
Mitsuru Sugawara, Kohki Mukai, Hiroshi Ishikawa, Koji Otsubo, and Yoshiaki Nakata, The Latest News

Volume 61 **Hydrogen in Semiconductors II**

Norbert H. Nickel, Introduction to Hydrogen in Semiconductors II
Noble M. Johnson and Chris G. Van de Walle, Isolated Monatomic Hydrogen in Silicon
Yurij V. Gorelkinskii, Electron Paramagnetic Resonance Studies of Hydrogen and Hydrogen-Related Defects in Crystalline Silicon
Norbert H. Nickel, Hydrogen in Polycrystalline Silicon
Wolfhard Beyer, Hydrogen Phenomena in Hydrogenated Amorphous Silicon
Chris G. Van de Walle, Hydrogen Interactions with Polycrystalline and Amorphous Silicon–Theory
Karen M. McManus Rutledge, Hydrogen in Polycrystalline CVD Diamond
Roger L. Lichti, Dynamics of Muonium Diffusion, Site Changes and Charge-State Transitions
Matthew D. McCluskey and Eugene E. Haller, Hydrogen in III–V and II–VI Semiconductors
S. J. Pearton and J. W. Lee, The Properties of Hydrogen in GaN and Related Alloys
Jörg Neugebauer and Chris G. Van de Walle, Theory of Hydrogen in GaN

Volume 62 **Intersubband Transitions in Quantum Wells: Physics and Device Applications I**

Manfred Helm, The Basic Physics of Intersubband Transitions
Jerome Faist, Carlo Sirtori, Federico Capasso, Loren N. Pfeiffer, Ken W. West, Deborah L. Sivco, and Alfred Y. Cho, Quantum Interference Effects in Intersubband Transitions
H. C. Liu, Quantum Well Infrared Photodetector Physics and Novel Devices
S. D. Gunapala and S. V. Bandara, Quantum Well Infrared Photodetector (QWIP) Focal Plane Arrays

Volume 68 **Isotope Effects in Solid State Physics**

Volume 69 **Recent Trends in Thermoelectric Materials Research I**

Volume 70 **Recent Trends in Thermoelectric Materials Research II**

Volume 71 **Recent Trends in Thermoelectric Materials Research III**

Volume 72 **Silicon Epitaxy**

Volume 73 **Processing and Properties of Compound Semiconductors**

Volume 79 **Semiconducting Chalcogenide Glass II**

Volume 80 **Semiconducting Chalcogenide Glass III**

Volume 81 **Conducting Organic Materials and Devices**

Volume 82 **Semiconductors and Semimetals**

Volume 83 **Semiconductors and Semimetals**

Volume 84 **Semiconductors and Semimetals**

Volume 85 **Semiconductors and Semimetals**

Volume 86 **Advances in Semiconductor Lasers**

Volume 87 **Advances in Photovoltaics: Volume 1**

Volume 88 **Oxide Semiconductors**

Volume 89 **Advances in Photovoltaics: Part 2**

Volume 90 **Advances in Photovoltaics: Part 3**

Kolibri 360 GmbH
Ferdinand-Jühlke-Straße 7
99095 Erfurt, Deutschland
produktsicherheit@kolibri360.de